Space Science Series of ISSI

For other titles published in this series, go to
www.springer.com/series/6592

L. Bengtsson · R.-M. Bonnet · M. Calisto ·
G. Destouni · R. Gurney · J. Johannessen ·
Y. Kerr · W.A. Lahoz · M. Rast
Editors

The Earth's Hydrological Cycle

Previously published in *Surveys in Geophysics*, Volume 35, Issue 3, 2014

Springer

Editors
L. Bengtsson
International Space Science Institute,
Hallerstrasse 6, 3012 Bern, Switzerland
lennart.bengtsson@mpimet.mpg.de

R.-M. Bonnet
International Space Science Institute,
Hallerstrasse 6, 3012 Bern, Switzerland
rmbonnet@issibern.ch

M. Calisto
International Space Science Institute,
Hallerstrasse 6, 3012 Bern, Switzerland
MCalisto@issibern.ch

G. Destouni
Stockholm University, Svante Arrhenius
väg 8, 10691 Stockholm, Sweden
georgia.destouni@natgeo.su.se

R. Gurney
University of Reading, Earley Gate 3,
RG6 6AL Reading, United Kingdom
r.j.gurney@reading.ac.uk

J. Johannessen
Nansen Environmental and Remote
Sensing Center, Thormøhlens gate 47,
5006 Bergen, Norway
johnny.johannessen@nersc.no

Y. Kerr
Center for the Study of the Biosphere
from Space, Avenue Edouard Belin 18,
31401 Toulouse, France
yann.kerr@cesbio.cnes.fr

W.A. Lahoz
Norwegian Institute for Air Research,
Instituttveien 18,
2027 Kjeller, Norway
wal@nilu.no

M. Rast
ESA-ESRIN, Via Galileo Galilei,
00044 Frascati, Italy
Michael.Rast@esa.int

ISSN: 1385-7525
ISBN 978-94-017-8788-8
ISBN 978-94-017-8789-5 (eBook)
DOI 10.1007/978-94-017-8789-5
Springer Dordrecht Heidelberg New York London

Library of Congress Control Number: 2014936873

Cover illustration: Selected image from "The dramatic retreat of the Aral Sea", released 10/07/2009 10:34 am. See http://www.esa.int/spaceinimages/Images/2009/07/The_dramatic_retreat_of_the_Aral_Sea. Reproduced with the kind permission of ESA © ESA

Printed on acid-free paper

Springer is part of Springer Science+Business Media (www.springer.com)

Contents

DOI 10.1007/978-94-017-8789-5_1
Reprinted from *Surveys in Geophysics* Journal, DOI 10.1007/s10712-013-9265-8

Foreword: International Space Science Institute (ISSI) Workshop on the Earth's Hydrological Cycle

Lennart Bengtsson

Received: 6 November 2013 / Accepted: 7 November 2013 / Published online: 13 December 2013

Water is a central component in the Earth's system. It is indispensable for life on Earth in its present form and influences virtually every aspect of our planet's life support system. On relatively short time scales, atmospheric water vapor interacts with the atmospheric circulation and is crucial in forming the Earth's climate zones that determine where habitable areas can exist. On the longest time scales of hundreds of millions of years, water contributes to the lubrication of the movements of the tectonic plates, creating a pattern of change that has shaped and is continuing to shape the Earth.

In the atmosphere, water vapor plays a key role in the Earth's energy balance and regulates the Earth's climate in a significant way. Water vapor is the most powerful of the greenhouse gases and serves to enhance the tropospheric temperature because water vapor is physically and dynamically controlled by atmospheric temperature and atmospheric circulation. The total amount of available water on the Earth amounts to some 1.5×10^9 km^3. The dominant part of this, 1.4×10^9 km^3, resides in the oceans. About 29×10^6 km^3 are locked up in the land ice on Greenland and Antarctica, and some 15×10^6 km^3 are estimated to exist as groundwater. If all the ice over the land and all the glaciers were to melt, as has happened several times in the Earth's history, the sea level would rise by some 80 m. In comparison, the total amount of water vapor in the atmosphere is small; it amounts to $\sim$25 kg/m^2, or the equivalent of 25 mm water for each column of air. Yet atmospheric water vapor is crucial for the Earth's energy balance.

The annual mean global values of evaporation and precipitation are $\sim$1,000 mm of water/m^2. However, these values vary enormously in space and time from areas that are almost completely dry to areas where the annual precipitation is more than an order of magnitude larger than the global mean value. An evaporation of 1,000 mm of water/year corresponds to 80 W/m^2 in energy loss for the surface and a corresponding gain for the

This Foreword is a slightly amended version of DOI 10.1007/s10712-013-9265-8

L. Bengtsson (✉)
Environmental Systems Science Centre, University of Reading, Reading, West Berkshire RG6 6UR, UK
e-mail: lennart.bengtsson@zmaw.de; lennart.bengtsson@mpimet.mpg.de

atmosphere when condensation takes place. This is the single largest component for heating the atmosphere; it is even larger than the direct solar energy absorbed by the atmosphere. This statement highlights the importance of the hydrological cycle for the energy balance the atmosphere.

As water vapor is also an effective absorber of terrestrial radiation, it contributes significantly to the regulation of the temperature of the lower atmosphere. The greenhouse effect of water vapor is estimated to be $\sim$24 °C. However, water vapor has a residence time of 7–8 days in the atmosphere and responds effectively to temperature through the Clausius–Clapeyron (CC) relation. In the present situation, it is the increase of the long-lasting greenhouse gases, namely CO_2, CH_4, N_2O and the CFCs, that are the drivers of climate change while water vapor generally acts as a positive feedback factor (Lacis et al. 2013). This is a fundamental factor in climate change. Model simulations suggest that water vapor feedback can more than double the initial effect of the long-lasting greenhouse gases.

With a population that has increased more the fourfold over the last 100 years and with an infrastructure that has grown by more than a factor of ten, society at large has become more exposed, in particular, to extremes in precipitation with associated flooding damages. Society has also over time significantly increased the amount of water that is needed, primarily for agriculture as well as for different kinds of industrial usage. This has contributed to a severe lack of water in exposed regions, even affecting major water bodies such as the Aral Sea and Lake Chad that have almost dried up completely during the last decades because of excessive extraction of water.

Other potential problems are disruptions related to climate change. The most severe prospects are systematic changes in weather zones such as a tendency for a poleward shift of the extra-tropical storm tracks that is indicated in climate simulation studies; others are the likelihood of more intense precipitation that will increase severe flooding. The poleward shift of weather systems is expected to create regional water problems with increased precipitation in some areas and decreased precipitation in other regions. Most severe here are the increasing risks of persistent periods of droughts, preferentially in the subtropics of both hemispheres (IPCC 2013).

In summary, society will have to cope with a multitude of disruptive events related to the water cycle due both to natural and anthropogenic effects such as (1) extreme events of heavy and persistent precipitation as well as extended periods of drought, which are all possible within the present climate, (2) anthropogenic actions unrelated to weather and climate, such as large-scale environmental changes caused by changing practices in large-scale agriculture and forestry, and (3) changes in the water cycle as a consequence of climate change. Presently the first two are dominant but gradually, as the climate system is getting warmer, the third factor is expected to be of increasing importance.

A scientific assessment of the Earth's hydrological cycle is a complex task which covers a multitude of areas and applications. The scientific papers in the present volume address a broad area of research related to the Earth's Hydrological Cycle. They represent the outcome of the third workshop within the ISSI Earth's Science Programme. The workshop took place from 6 to 10 February, 2012, in Bern, Switzerland, with the objective of providing an in-depth overview of the Earth's hydrological cycle. The participants in the workshop were experts in a wide range of disciplines; they included geophysicists, meteorologists, hydrologists, oceanographers and climate modelers.

The increase in the world's population and the increasing need for food, energy and natural resources have put increasing stress on the water requirements.

Perhaps the most extreme effect of such a water stress is the almost complete destruction of the world's previously fourth largest lake, the Aral Sea. This is one of the planet's worst environmental disasters. Over a period of some 40 years, the lake has lost more than 90 % of its area. The most likely reason for this is the massive agricultural developments that have used more and more of the water in the catchment areas of the Amu Darya and Syr Darya (Asokan and Destouni 2014).

As the extreme opposite effect, excessive flooding in rivers and in coastal areas is causing significant economic costs and hardship to their populations. Two kinds of events stand out. The first is mostly coupled to excessive precipitation in river catchments. This occurs over all continents during the rainy season, often with devastating consequences. The main cause is not necessarily higher precipitation but enhanced exposure to heavy precipitation mainly due to increased population and increased occupation in exposed areas. The second most important cause is related to coastal flooding that occurs in relation to intense weather systems, in particular tropical cyclones. The coastal flooding in New Orleans in 2005 and in New York in 2012 are prime examples of such events. Fortunately due to the advances made in weather prediction in recent decades, the public was warned several days in advance and evacuated the areas most at risk.

There is overwhelming evidence from theory (Held and Soden 2006, Stevens and Bony 2013) and from empirical and model studies (e.g., Allan 2014) that the atmospheric water vapor content increases with increasing temperature since it varies according to the CC relation. It also follows from theory that the horizontal transport of water vapor also scales with the CC. This is a serious consequence, as it will affect regional precipitation; in areas of convergence (such as the tropical convergence zone and at high latitudes), there will be increased precipitation. Alternatively, it will also affect areas of divergence (such as the subtropical regions), with the consequence of reduced precipitation. This response of the water cycle to climate warming is probably one of the most severe consequences of climate change, as it will amplify the extremes of the hydrological cycle. So far, because of the large variability of precipitation in time and space, we are not yet able to show this from observations but model results demonstrate this response clearly (e.g., Bengtsson 2011).

Even ocean circulation is affected as heavy precipitation will reduce salinity and thus diminish ocean convection. And, in an analogous way, reduced precipitation over ocean areas will increase salinity and increase ocean convection. Observations from ESA's Soil Moisture and Ocean Salinity satellite (SMOS) are providing new and exciting knowledge (Reul et al. 2014) in this area. There is a wide range of methodological studies, mainly related to the handling of the water cycle in climate models due to the fine structure of precipitation processes in the atmosphere that present climate models cannot handle very well (Foufola-Giorgiou 2014). Both satellite and ground-based radar information are of special importance here and highlight the need for higher resolution in weather and climate models.

References

Allan R (2014) Physically consistent responses of the global atmospheric cycle in models and observations. Surv Geophys (this issue)

Asokan S, Destouni G (2014) Irrigation effects on hydro-climatic change: basin-wide water balance-constrained quantification and cross-regional comparison. Surv Geophys (this issue)

Bengtsson L (2011) The global atmospheric water cycle, Environ Res Lett, 5:025001 (8 pp) doi:10.1088/1748-9326/5/2/025001, 2010

Foufola-Giorgiou E (2014) Downscaling satellite precipitation with emphasis on extremes, Surv Geophys (this issue)

Held I, Soden B (2006) Robust response of the hydrological cycle to global warming. J Climate 19:5686–5699

Intergovernmental Panel on Climate Change, IPCC (2013), Climate Change 2013: the physical science basis, Working Group I Contribution to the IPCC Fifth Assessment Report

Lacis A, Hansen J, Jonas J, Oinas V, Russell G (2013) The role of long-lived greenhouse gases as principal LW control knob that governs the global surface temperature for past and future climate change. Tellus (accepted)

Reul N, Fournier S, Boutin J, Hernandes O, Maes C, Chapron B, Alory G, Quilfen Y, Tenerelli J, Morisset S, Kerr Y, Mecklenburg S, Delwart S (2014) Sea surface salinity observations from space with SMOS satellite: new tool to better monitor the marine branch of the water cycle. Surv Geophys (this issue)

Stevens B, Bony S (2013) What are climate models missing? Science 340:1053–1054

DOI 10.1007/978-94-017-8789-5_2
Reprinted from *Surveys in Geophysics* Journal, DOI 10.1007/s10712-014-9279-x

Review of Understanding of Earth's Hydrological Cycle: Observations, Theory and Modelling

Michael Rast · Johnny Johannessen · Wolfram Mauser

Received: 20 January 2014 / Accepted: 22 January 2014 / Published online: 11 March 2014

Abstract Water is our most precious and arguably most undervalued natural resource. It is essential for life on our planet, for food production and economic development. Moreover, water plays a fundamental role in shaping weather and climate. However, with the growing global population, the planet's water resources are constantly under threat from overuse and pollution. In addition, the effects of a changing climate are thought to be leading to an increased frequency of extreme weather causing floods, landslides and drought. The need to understand and monitor our environment and its resources, including advancing our knowledge of the hydrological cycle, has never been more important and apparent. The best approach to do so on a global scale is from space. This paper provides an overview of the major components of the hydrological cycle, the status of their observations from space and related data products and models for hydrological variable retrievals. It also lists the current and planned satellite missions contributing to advancing our understanding of the hydrological cycle on a global scale. Further details of the hydrological cycle are substantiated in several of the other papers in this Special Issue.

Keywords Earth observation · Satellite remote sensing · Water cycle · Hydrological cycle

1 Introduction

The water cycle of the Earth system and its variability at global, regional and local scales are influenced by a range of processes and mutual interactions, feedback mechanisms and

M. Rast (✉)
ESA/ESRIN, Frascati, Italy
e-mail: Michael.Rast@esa.int

J. Johannessen
Nansen Environmental & Remote Sensing Center (NERSC), Bergen, Norway

W. Mauser
Ludwig Maximilian University of Munich, Munich, Germany

Fig. 1 Schematic illustration of the hydrological cycle (courtesy of European Space Agency, Earth Observation Graphic Bureau)

as well as affected by anthropogenic processes. The scales at which processes interact both spatially and temporally vary across the atmosphere, the hydrosphere, the cryosphere and the biosphere in a complex manner. The hydrological cycle is composed of different components, which include evaporation from water surfaces and bare soil, evapotranspiration from vegetated land, transport of water vapour in the atmosphere, cloud droplet formation and cloud dynamics, the mechanisms leading to liquid and solid precipitation, the movement of water and change in soil moisture in the unsaturated soil, including root dynamics, surface and river run-off, and groundwater flow as schematically illustrated in Fig. 1. Oki and Kanae (2006) give an overview of current knowledge on the world water resources.

The hydrological cycle is the basic purification mechanism for water on Earth as any water constituents are left behind during the phase change from liquid water to water vapour. For billions of years, this process supplied the land surface with freshwater, which was and is the basis for life. The precipitated water dissolved minerals on its way through the hydrological compartments to the oceans and thereby gradually increased ocean salinity. Reliable and adequate supply of clean freshwater is essential to the survival of humankind as well as to the maintenance of terrestrial biotic systems worldwide. However, the rapidly growing human population increasingly stresses available water resources (Ferguson and Maxwell 2012). Human water demand consists of (1) drinking water, (2) sanitation and (3) industrial water and agricultural water to produce biomass for food, fibre, energy and industrial materials. On the global average, the ratio between these three sources of human water demand is approximately 3:4:92 (Turner et al. 2004). Human-induced stresses on available water resources therefore are mainly connected to biomass production through agriculture. Human biomass demand as well as the closely linked agricultural water demand is expected to double by 2050 in order to supply a growing and more prosperous global population with food, fibre, energy and industrial materials (Alexandratos and Bruinsma 2012; Mauser 2009). At this point, it becomes clear that agricultural activities modify the closely coupled global hydrological and carbon cycle through land-use changes, optimization of rainwater use by plants, irrigation, emission of greenhouse gases, changing seasonal albedo and changing plant physiology caused by fertilizers.

Importantly, the expected increase in human water demand and the related change in land use seem to occur faster than any anticipated effect of a changing climate on the availability of freshwater (Piao et al. 2007). To comprehend the combined effect of (and interactions between) increasing human stresses on water supplies, land use and climate change, the understanding of the hydrological cycle needs to be advanced on all interconnected scales from local to global.

Many regional and local studies already exist on how climate change is expected to impact the different compartments of the hydrological cycle and how societies may best adapt to these changes by changing land use, water storage or even water use itself (Ludwig et al. 2008; Christensen et al. 2004; Fowler et al. 2007; Ludwig et al. 2008; Prasch et al. 2013). The majority of these hydrological cycle studies, with few exceptions (e.g., Strengers et al. 2010; Zabel and Mauser 2013), treat this issue as a unidirectional cause–effect chain, where a changing climate influences the regional and local hydrological cycle and thereby affects water availability. Possible effects of dynamic, human-induced land surface changes on the carbon cycle and regional and global water cycle, which in turn may result in a changing climate, are, however, neglected here.

All components of the hydrological cycle are involved in different ways in climate change, either by causing it or by reacting to it, sometimes amplifying each other's action, sometimes giving rise to negative feedbacks such as atmospheric cooling through larger Sun shielding cloud formations from increased evaporation. Variations in the hydrological cycle often take place at regional or even local scale (such as orographic variations in precipitation, small-scale variations of soil physical properties, ecosystem composition or run-off processes), but can still trigger modifications that have an upscale effect possibly leading to global changes in the hydrological cycle. The mechanisms and magnitudes of the feedbacks of local interferences with the hydrological cycle with the regional and global hydrological cycle are complex, largely unknown and cannot satisfactorily be explored with current Earth system models. The feedbacks, e.g., between precipitation, land use, soil moisture and the resulting evapotranspiration, are especially challenging and exhibit strong intermittency and interactions at all scales, which makes them often hard to model with contemporary Earth system models. The strong, small-scale and nonlinear dependency of the participating processes on topography, soil physical properties and plant physiology makes it also hard to estimate their magnitudes from sparse rain gauge, sparser soil moisture and even sparser evapotranspiration measurement networks.

Nevertheless, there are important research issues waiting to be addressed, which are related to the hydrological cycle and its relation both to the carbon cycle and the global climate system. They centre on the feedbacks between the accumulated effects of increasing local human interferences with the hydrological cycle through more intense agricultural use and the global climate system. Most relevant and probably most difficult to understand and model is a possible change in precipitation patterns, which results from land use and evapotranspiration changes in one region and which affects the local and regional hydrological cycle as well as the water availability elsewhere through teleconnections in the global circulation. The described mechanism may be of extreme relevance to the affected region in cases when rainfall is reduced below a critical value, where agriculture becomes impossible. Besides the global climate, a substantial part of the global population may thereby indirectly be affected by collective land-use activities in another part of the globe. Current understanding of the hydrological cycle, its components and the feedbacks with the carbon cycle and the climate system is not sufficient to approach this and similar research questions related to the multi-scale hydrological cycle. Currently available Earth system models both lack spatial resolution to be able to accumulate the nonlinear effects of the local human interferences with the hydrological cycle through land-use changes and do not contain a detailed and realistic representation especially of the terrestrial hydrological cycle and the strongly nonlinear response mainly of its components soil and vegetation during times of water shortage. These important research issues can hardly be addressed by most present hydrological models, which are often conceptual in nature, calibrated to measured data to various degrees and often data poor in design. This is the underlying reason why several authors recently suggest developing new hyper-resolution hydrological land surface models, which can be fully coupled to regional and global climate models to combine global with regional hydrology. These models should be based on first-order principles, close the terrestrial energy, water and carbon balance at all scales, should be able to overcome these scale issues by coupling the very small with the very large processes and should be able to tap the rich global data archives available through remote sensing (Hibbard et al. 2010; Wood et al. 2011; Mauser and Bach 2009; Su et al. 2010).

However, research on these multi-scale issues is currently also severely hampered by the lack of observations of key variables with adequate resolution related to hydrological cycle change in space and time. The measurement of precipitation, for instance, either by ground stations, radars or satellites, is still a challenging task at all spatial and temporal scales that could hamper efforts devoted to understanding and modelling the hydrological cycle and its variability. Surface soil moisture is one of the least observed variables and only recently, with the advent of ESA's Soil Moisture and Ocean Salinity (SMOS) mission, became accessible to global spatio-temporal measurements with a very coarse spatial but almost adequate temporal resolution.

Vegetation response to water stress and its effect on water release to the atmosphere has to be inferred from models, which simulate evapotranspiration and vegetation surface temperature, which can then be compared to remote sensing measurements under cloud-free conditions. Moreover, the terrestrial snow and ice masses are important components of the global climate system. Snow and ice influence the radiation and surface energy budget, the moisture balance, gas and particle fluxes, precipitation, hydrology, and atmospheric and oceanic circulation. These processes are coupled with the global climate system through complex feedbacks that are not yet well understood. Improved observational data are therefore needed for a better understanding and accurate quantification of the main cryospheric processes and the corresponding representation of the cryosphere in climate models (Lemke et al. 2007).

Remote sensing offers the possibility of delivering the kind of data that allows observing with adequate resolution how key variables related to hydrological cycle change in space and time. Most remote sensing measurements indirectly observe the hydrological cycle and can only be utilized to their full potential when assimilated into appropriate hydrological models. Some recent hydrological model developments have taken up this task (Mauser and Bach 2009) and have shown how to assimilate remote sensing data streams into hydrological models (Bach 2003). Nevertheless, hydrological land surface models face the challenge of a future data-rich environment. They have to learn how to assimilate on all scales globally available, high spatial resolution, frequent coverage remote sensing data, e.g., from the operational Sentinel satellites of the European Commission - European Space Agency (EC-ESA) Copernicus program. This is expected to generate the necessary knowledge to successfully approach issues of the local to global hydrological cycle as outlined above. Critical elements to accomplish this are related to (1) the quality of the observing system, jointly combining the ground-based network and the Earth observation (EO); (2) removal of major knowledge gaps; and (3) development and implementation of high-quality hydrological models with the ability to assimilate EO data.

Table 1 GCOS essential climate variables, variables directly related to the hydrological cycle are marked in italics

Atmosphere	
Surface	Air temperature; *precipitation* Air pressure; Surface radiation budget; Wind speed and direction; *water vapour*
Upper air	Earth radiation budget; Upper air temperature; *Water vapour;* *Cloud properties*; Wind speed and direction
Composition	Carbon dioxide, methane and other greenhouse gases (GHGs); Ozone; Aerosol properties
Ocean	
Surface	Sea-surface temperature, *sea-surface salinity*, *sea level*, sea state, *sea ice*, ocean colour; Carbon dioxide partial pressure
Sub-surface	Temperature, salinity; current; nutrients; carbon; ocean tracer; phytoplankton
Terrestrial	*Lake level; snow cover; glacier and ice caps; albedo*; land-cover fraction of absorbed photo-synthetically active radiation; *Leaf Area Index (LAI) biomass, fire disturbance; soil moisture* *Water use, ground water, river discharge* *Permafrost and seasonally frozen ground*

The usefulness of remote sensing for assessing products beneficial for hydrological monitoring dates back to the 1970s, when the potential of the infrared geostationary data for rainfall and vegetation monitoring was first demonstrated as an important technology for complementing and enhancing information from in situ observational systems. Over the last three decades, the amount of relevant hydrological products derived from satellites has increased, and they have been implemented in derivation of the majority of the Global Climate Observing System's (GCOS) and the Global Terrestrial Observing System's (GTOS) terrestrial and atmospheric Essential Climate Variables (ECVs) (see Table 1). These include, in accordance with GCOS recommendations, precipitation, water vapour, cloud properties, soil moisture, leaf area index, water level and estimates of ground water, evapotranspiration, river discharge, ocean salinity, snow cover, albedo, glaciers and ice caps, ice sheets, permafrost extent and seasonally frozen ground (GCOS 2003).

Each of these variables shown in italics can currently be estimated with the use of at least one Earth observation system although possibly not in all cases with the requisite spatial and temporal resolution for advancing the understanding of local to global feedbacks in the hydrological cycle. However, adequate validation of the satellite data is challenging and often limited implying retrievals lacking satisfactory uncertainty estimates. Commonly, several systems have capabilities to derive identical parameters, and implementation techniques for their merging and blending were suggested. As a result, new integrated, multi-year data sets are being generated, taking advantage of the opportunity presented by the simultaneous operation of key satellites by Europe, Japan and the USA. For instance, the ESA Climate Change Initiative (CCI) aims to demonstrate the full potential of the long-term global Earth Observation archives as a significant and timely contribution to the GCOS Essential Climate Variables (ECVs) databases required by UNFCCC (United Nations Framework for Combating Climate Change). In its first phase, the initiative is targeting the following hydrology-related variables: sea level, sea ice, glaciers and ice caps, ice sheets and clouds. The ongoing discussion on possible operations of convoys and constellations is also expected to strengthen the value and use of the satellite data for understanding and monitoring of the hydrological cycle.

The state-of-the-art overview of currently available EO products for hydrological monitoring and modelling is provided in the following. Importantly, it is not aimed for listing methodologies on how to implement such products in models, as this have been extensively summarized elsewhere (Van Dijk and Renzullo 2011), but it clearly emphasizes the multidisciplinary aspects and complexity that jointly with the wide range of spatial and temporal scales establish a significant demand on the observing system. The individual papers presented in this issue provide further quantitative evidence of the capability and limitation of the observing system.

2 Clouds and Precipitation

Cloud and precipitation systems tend to be somewhat random in character, and they are usually small scale and also evolve very rapidly, especially during the summer in convection regimes. These factors make clouds and precipitation difficult to quantify. Reliable ground-based precipitation measurements are difficult to obtain over regional and global scales because more than 70 % of the Earth's surface is covered by ocean and lakes, and additionally many countries are not equipped with precise rain measuring sensors (i.e., rain gauges and/or radars). In such regions, regional and global scale precipitation measurements from Earth Observation satellite systems are extremely valuable.

Over its lifetime of more than 15 years, the Tropical Rain Monitoring Mission (TRMM) satellite has provided a wealth of information on tropical cyclones and short-duration climate shifts such as El Niño (Curtis et al. 2007) and has proved to be an essential tool for the measurement of precipitation. Current operational and research platforms form a constellation that can be used for the routine generation of precipitation with nominal 3-h temporal and 0.25-degree spatial resolution. Estimates derived at full resolution (4 km) are available up to instantaneously, albeit more sporadically. However, TRMM misses low-rate precipitation, e.g., drizzle, which is expected to contribute significantly to the total precipitation in regions with low rainfall amounts. The upcoming Global Precipitation Measurement (GPM) mission [e.g., http://pmm.nasa.gov/GPM] is a network of satellites and will provide solid and liquid precipitation, including light precipitation. Full vertical profiles of clouds and light and solid precipitation are being observed by CloudSat [http://cloudsat.atmos.colostate.edu/] and will be further improved by EarthCARE (see, e.g., http://www.esa.int/Our_Activities/Observing_the_Earth/The_Living_Planet_Programme/Earth_Explorers/EarthCARE/ESA_s_cloud_aerosol_and_radiation_mission). EarthCARE will provide extended vertical cloud profiles with significantly higher sensitivity compared to CloudSat and added Doppler capability for observation of vertical motion within clouds. Solid precipitation and light precipitation will be measured together with full cloud profiles, and heavy precipitation will be detected. In contrast to TRMM and GPM radar satellites, CloudSat and EarthCARE which travel on polar orbits provide full global coverage.

The usefulness of radar systems capable of measuring precipitation (e.g., TRMM Precipitation Radar and CloudSat Cloud Profiling Radar) has been demonstrated (Lonfat et al. 2004). They provide a unique and crucial addition to our observational capabilities for precipitation. The retrieval of precipitation at higher latitudes remains an open challenge. Problems include contamination by the surface background and low-level, frozen precipitation.

In the tropical oceans, large vertical salinity gradients can develop in the upper few metres of the ocean after heavy rainfall as evidenced by Soloviev and Lukas (1996); Schlüssel et al. (1997); and Wijesekera et al. (1999). Signatures of these intense precipitation regimes can be detected in the SMOS Sea-Surface Salinity (SSS) data in the form of freshwater patches, as clearly shown by Reul et al. (2013).

3 Soil Moisture

Numerous soil moisture products are available from active experiments (ERS, ASCAT) (Naeimi et al. 2009; Wagner et al. 1999; Loew et al. 2006), or from passive sensors (AMSR-E, SMOS) (Kerr et al. 2010; Njoku et al. 2003; Loew et al. 2013) and will be available as well from combined passive/active microwave remote sensors (e.g., the planned Soil Moisture Active Passive (SMAP) mission). These operate at coarser (>25 km) resolutions and span altogether more than three decades of data. Recently, a multi-decadal blended dataset was developed that is expected to further enhance the understanding of the water balance in hydrological models (Dorigo et al. 2012; Liu et al. 2011).

The application of coarse resolution soil moisture data in hydrological models is controversial. There seems to be no obvious approach in river run-off studies that would explain under which conditions an improvement could be achieved. Recently, however, several studies demonstrated positive impact when Soil Water Index (SWI) was assimilated (Brocca et al. 2010a, b; Matgen et al. 2011; Meier et al. 2011; Wagner et al. 1999). SWI represents the profile of soil moisture in the root zone which is the hydrological most important zone in terms of run-off generation (Parajka et al. 2006).

Semi-operational products are also available at medium-resolution scale (>1 km) (Pathe et al. 2009). Nevertheless, assimilation of such data into models was restricted by poor radiometric resolution and revisit period. A soil moisture product from Sentinel-1 has been foreseen with coverage every 6 days globally, nearly daily over Europe and Canada (depending on latitude) (Hornacek et al. 2012). With its remarkably improved radiometric accuracy, it has the potential to be of great benefit for data assimilation, anomaly and threshold detection as well as direct input into models operating at medium-resolution scales (Doubková et al. 2012).

4 Evapotranspiration

Neither evapotranspiration (ET) nor any of its components can be directly sensed from satellites, as heat fluxes do not absorb nor emit electromagnetic signals directly. Nonetheless, the last three decades have seen substantial progress in the combined field of evaporation and remote sensing. Current methodologies concentrate on the derivation of ET by combining some of the satellite-observable physical variables that are linked to the evaporation process. Some of the existing algorithms differ in their purpose of application, which to a certain extent defines the type of remote sensing data used and the amount of required ancillary data. The majority use some form of thermal and visible data, with only a few applying microwave observations. Some of these methodologies are fully empirical, and others are based on more physically based calculations of ET via formulations like the ones of Penman (1948), Monteith (1965), and Priestley and Taylor (1972), or focus on solving the surface energy balance targeting the accurate determination of the sensible heat flux (H). Most of the early methods were designed for local-scale studies and agricultural and water management practices, while more recent methodologies have started to pursue the coverage of the entire globe. A general review of these methodologies can be found in Courault et al. (2005); Kalma et al. (2008); Wang and Dickinson (2012) and Su et al. (2010). Notice also that other methodologies based on relatively complex land surface models are also producing global ET estimates for climatological applications (Strengers et al. 2010; Zabel and Mauser 2013). Currently, they use remote sensing-derived parameter fields in a rudimentary way, usually as temporally irregular medium spatial resolution land use, the fraction of Absorbed Photosynthetically Active Radiation (fAPAR) or LAI fields from the MODIS or MERIS instruments. With the advent of higher spatial resolution imaging spectrometers like the German Environmental Mapping and Analysis Program (EnMAP), the assimilation of higher accuracy and more physiological parameters like, e.g., chlorophyll content from remote sensing sources will potentially improve vegetation parameterization (Rodriguez et al. 2011).

In the framework of the Global Energy and Water Cycle Experiment (GEWEX) Data Assimilation Panel (GDAP) LandFlux-Eval initiative, the first satellite-based ET products (reported as latent heat fluxes) and these other estimates have been inter-compared (Jiménez et al. 2011; Mueller et al. 2013). As a contribution to LandFlux, the ESA WACMOS-ET (WAter Cycle Multi-mission Observation Strategy-EvapoTranspiration) project aims at advancing the improvement and characterization of ET estimates from satellite observations, both at continental and regional scales. To this end, a cross-comparison, error assessment and validation exercise of a selection of state-of-art algorithms will be undertaken at different spatial domains and resolutions (Mueller et al. 2013).

Over the ocean, changes in the evaporation can be indirectly inferred from observed changes in sea-surface salinity derived from SMOS and Aquarius in areas of strong precipitation. Further details on this are found in Reul et al. (2013).

5 Ground Water Observations

An emerging application area is the use of the GRACE satellite mission, and its gravimetric measurements of mass changes, which are being used to quantify changes in groundwater storages (Rodell and Famiglietti 2002) as well as melting of ice sheets and glaciers. Plans are currently being formulated for GRACE Follow On and GRACE-II missions, and it is expected that this area of research will continue to expand. The multiple applications of GRACE data were summarized by Cazenave and Chen (2010).

6 Water Extent and Levels

Optical (MERIS, MODIS and AVHRR), active (ERS and Envisat) as well as passive (SSM/I) microwave data were employed to estimate the extent of water bodies, floods and volumes (Prigent et al. 2007), and the first global estimate of wetland extent and dynamics over almost a decade was presented. Importantly, it was the combination of several observation techniques and capitalizing on the strength of each of them that allowed extracting the most from inundation characteristics.

Several semi-operational products exist at local to regional scales ($\sim$100 m–5 km) mostly derived from synthetic aperture radar (SAR) sensors (ESA 2013). SAR sensors demonstrated a great potential for the monitoring of open water bodies at medium-resolution scale (Bartsch et al. 2007). Nevertheless, there is no operational product providing water body extent. In addition, the potential of monitoring inland water levels by using radar altimeters mainly from ERS and Envisat became apparent to the extent that river and lake heights have been produced on a global scale (ESA 2013).

Given the high revisit period, the Sentinel-1 sensor, which is planned to be launched towards the end of 2014, holds a great potential for high-resolution water extent mapping (http://www.esa.int/Our_Activities/Observing_the_Earth/GMES/Sentinel-1).

As for the oceans, areas of interest including the use of optical wavelengths to assess ocean colour, i.e., phyto-plankton and other water borne materials (e.g., MERIS, MODIS, SeaWIFS), and the exploration of radar altimetry to measure water levels in lakes and rivers are aimed to be continued into the future. Regarding the latter, our knowledge of the global dynamics of terrestrial surface waters and their interactions with coastal oceans in estuaries is expected to significantly advance with the planned launch of the joint NASA–CNES–CSA Surface Water Topography Mission (SWOT) in 2020 (http://swot.jpl.nasa.gov/). By measuring water storage changes in all wetlands, lakes and reservoirs and making it possible to estimate discharge in rivers more accurately, SWOT will contribute to a fundamental understanding of the terrestrial branch of the global water cycle. SWOT will also map wetlands and non-channelized flow.

7 Vegetation Stage

Optical vegetation indices and land-cover classifications, as well as passive and active microwave derived estimates of vegetation water content, biomass and vegetation structure can be used to initialize hydrological models. There seems to be a good understanding and variety of independent algorithms that estimate vegetation stage by using data acquired in optical, near-infrared and thermal-infrared spectrum or derived products such as fPAR or LAI. Also, a variety of land-cover classification approaches have been employed in land surface models that implement Normalized Difference Vegetation Index (NDVI) data from AVHRR or SPOT/Vegetation. (DeFries 2008) gives an excellent review of the current status and role of remote sensing on observing the terrestrial vegetation.

Synthetic data experiments undertaken with simulated Sentinel-2 data showed a reduction in the uncertainty in Leaf Area Index (LAI) (Richter et al. 2012; Bach et al. 2012). Severe improvements are expected also in the land-cover classification in the future.

A variety of products is also derived from passive and active microwave observations that include estimates of vegetation water content, biomass, or vegetation height and structure. The latter can be used to estimate variables such as emissivity, canopy conductance and vegetation roughness, which affect the partitioning of radiation into ET and other terms (Van Dijk and Renzullo 2011). Further potentials for greater use of satellite microwave observations include parameterization of biomass, height or aerodynamic roughness. The possibility to observe forest biomass has been proposed by the new Earth Explorer Mission Biomass that uses P-band synthetic aperture polarimetric radar (ESA 2012).

Lastly, to gain a detailed knowledge about the observed medium and to improve understanding of upcoming high-resolution Sentinel and potential Biomass mission, a combination of airborne and terrestrial LiDAR observations is investigated (ESA 2012).

8 Water Vapour

A large variety of space-borne sensors are used to retrieve atmospheric profiles of humidity or the water vapour column amount (microwave, infrared, optical, UV). SSM/I total column water vapour over ocean is mature for climate analysis. In addition, MERIS observations have shown over time a significant potential for high spatial resolution total column water vapour over land during daytime in clear sky, e.g., ESA DUE (Data User Element) GlobVapour.

UV/VIS instruments provide independent means for total column water vapour retrievals, but these measurements are biased towards clear sky. Upper tropospheric humidity (UTH) data sets provide a data source with high value for climate research.

Microwave sounding data sets from AMSU-B, MHS and SSM/T2 hold a great potential to improve our knowledge on UTH—also allowing estimates of absolute humidity. The availability of atmospheric temperature and humidity profiles for more than 30 years has been identified as a critical issue (see, e.g., WMO, 2012 [http://www.wmo.int/pages/prog/www/OSY/Meetings/Wshop-Impact-NWP-5/index.html).

9 Snow

Snow plays an important role in the regional and global hydrological system, since it acts as temporary water storage. As a cause of large land surface albedo changes and because it is highly variable both in time and space, dedicated measurements/monitoring systems are needed. Satellite products have advantage over point-based measurements of the snow-related parameters, especially due to their spatial coverage. The mass of seasonal snow (the snow water equivalent, SWE) accumulated on land surfaces and the extent of the snow-covered area are the principal variables in hydrology and for water resources applications (Rott 2013). They are also essential for determining and modelling surface/atmosphere exchange of mass and energy, and therefore of great importance for numerical weather prediction. Satellite-based snow sensing techniques use visible/infrared (optical), active microwave (SAR) and passive microwave sensing techniques (Frei et al. 2012; Botteron et al. 2013).

Operational snow cover products are commonly related to the fractional snow cover (percentages of the coverage) and snow albedo. They are derived from operational geostationary and numerous polar-orbiting satellite sensors, e.g., AVHRR, MODIS and VIIRS. The use of active microwave (SAR) on ENVISAT and RADARSAT enables the detection of wet snow, indicating melting processes. This information can be integrated in regional hydrological monitoring activities (Bach et al. 2010). Imaging microwave radiometry (SSM/I, AMSR-E) allows the global snow mass to be mapped every day or two at a spatial resolution of about 25 km. SWE retrieval exploits the scattering losses in the (dry) snow pack of the microwave radiation emitted by the soil below snow. The accuracy of SWE retrieval is impaired by uncertainty in snow morphology (grain size, stratification). Satellite scatterometer measurements at Ku-band are also sensitive to morphology of the snow volume (Nghiem and Tsai 2001), suggesting that the active microwave measurements support the characterization of snow scattering properties and thus improve the retrieval of SWE by means of microwave radiometry. The Ku-band scatterometer aboard flying with the second generation MetOp satellite is also used to measure ice sheet snow accumulation, for the measurement of land snow mass at medium/low spatial resolutions and for the characterization of the soil freeze/thaw cycle.

In order to account for improve remote sensing capabilities to measure snow parameters, the validation with ground-based measurements and regional optimized algorithms to account for variable landscape and physical properties effects are still a scientific challenge (IGOS 2007). It is expected that the Sentinels will be beneficial due to their increased temporal coverage, accessibility and ability to monitor snow cover and snow melting process.

10 Permafrost

In the Northern Hemisphere, permafrost regions extend over about 23 million km^2 (Zhang et al. 2001).

Permafrost is currently monitored mainly by means of ground-based point measurements. Remote sensing systems are used as complementary tools, to map surface features of permafrost terrain and monitor their changes driven by climate warming. Surface indicators of permafrost terrains that can be identified by remote sensing images include pingos, thaw lakes and basins, retrogressive thaw slumps, thermo-erosional valleys, thermokarst mounds, ice wedge polygons, beaded drainage, palsa fields, slope failures, and rock glaciers (IGOS 2007). Precise topographical data are required for accurate geocoding of the remote sensing imagery (optical and SAR) so that changes in permafrost features can be tracked accurately. High resolution Digital Elevation Models (DEMs) are also required for modelling hydrology, permafrost distribution, erosion and matter fluxes resulting from permafrost degradation (McNamara et al. 1999). DEMs derived from current satellite systems (e.g., ASTER DEM) are lacking the accuracy needed for these tasks.

11 Glaciers and Ice Caps

Precise data on surface topography of mountain glaciers, ice caps and outlet glaciers of ice sheets are needed as basic information for ice dynamic models, mass balance models, and regional hydrological and climate models. Vertical accuracy on the order of 5 m is acceptable for most of these applications, except for some special ice dynamic models. The requirements in vertical accuracy are more stringent for measuring changes in surface topography to infer glacier mass balance through annual (goal) or multi-annual volume changes. The typical requirement in vertical accuracy for this application is ≤ 1 m (elevation change). There is still high uncertainty in the mass balance of the world's glaciers and ice caps (Lemke et al. 2007). This is due to the fact that accurate mass balance measurements are made only on few glaciers worldwide (Dyurgerov et al. 2005). The representativeness of this small sample is rather questionable, as there is a strong bias towards small glaciers that are easily accessible. Extrapolating from these glaciers to global numbers causes large uncertainty. To overcome this deficit requires spatially detailed, precise repeat measurements of temporal changes in glacier surface topography for a large sample of the glaciers worldwide. For calving glaciers, these measurements need to be complemented by estimates of the calving flux to obtain the mass balance. In this context, radar altimetry and SAR interferometry are providing highly important observations. The sustainability of these observations is also promising in view of the Sentinel-1 and Sentinel-3 missions. The limiting factor for use of SAR interferometry is the temporal variability of the radar signals due to snow fall, drift and melting (Rott and Siegel 1997) as well as due to signal decorrelation in zones of strong sea ice deformation such as along glacier margins.

12 Ice Sheet

Radar altimetry (ERS-RA, Envisat-RA), IceSat-1 and CryoSat 2 have been the main sensors for precise measurements of surface topography on the ice sheets for estimating volume changes. Because the accumulation rate on the main accumulation zones of the ice sheets is rather small, the requirements in vertical accuracy of the repeat measurements are rather high ($\leq$10 cm/year minimum, $\leq$5 cm/year goal). Uncertainty of radar signal penetration in firn and temporal trends in firn properties are reasons for differences in surface elevation measured by radar and lidar (Brenner et al. 2007). Further uncertainties in computing mass changes of the ice sheets result from variability of firn layer thickness, caused by regional variability of accumulation and temperature (Helsen et al. 2008). These are reasons for the rather large error bars in the mass balance estimates for Antarctica and Greenland in the IPCC report (Lemke et al. 2007). Significant reduction in the uncertainty can be expected by applying different altimetric systems in synergy.

A combination of SAR interferometry and satellite altimetry strengthens the retrieval method and reduces the uncertainties that in turn allows for more detailed studies of topographical and mass changes as well as surface velocity and associated deformation as reported by Rignot et al. (1998) and Shepherd and Wingham (2002) showing that dynamically related thinning is penetrating deep into the interior of the West Antarctic, Pine Island and Thwaites drainage basins. The temporal variability and corresponding signal decorrelation time are again a limiting factor regarding appropriate use of interferometry.

CryoSat (Wingman et al. 2006) is a separate mission developed mainly for measuring ice sheet elevation and sea ice thickness and their changes. By accurately measuring thickness change in both types of ice, CryoSat-2 will provide information to complete the picture and lead to a better understanding of the role ice plays in the Earth system.

The planned Sentinel-1 mission is expected to offer unique operational and scientific capacity due to its increase revisit period. Also, by a synergistic use of Sentinel-1 and the other space-borne SAR missions, often operating at different wavelengths and modes, certain ice types can be easily identified.

13 Sea Ice

Microwave satellite observations are routinely providing essential data on large-scale ice concentration, area, type and large-scale motion. Moreover, measurement of the vertical dimension of sea ice (ridges, freeboard, thickness, snow thickness) and thermodynamic properties (temperature, heat flux) is possible by use of altimeters (e.g., IceSAT, CryoSat 2 for sea ice thickness >0.5 m) and infrared/microwave radiometers (e.g., SMOS for sea ice thickness <0.5 m), although at varying degree of maturity with respect to retrieval accuracies. Many small-scale processes and phenomena related to sea ice deformation and marginal ice zone thermodynamics can also be observed by high-resolution SAR (coupled with optical/infrared images under cloud-free conditions), but there are no systematic and long-term observations because the data coverage is insufficient. Snow depth and snow water equivalent are also important variables in the presence of sea ice that need to be retrieved more reliably from satellites. Data on snow depth can be obtained from satellite sensors such as by combined use of IceSAT and CryoSat, and from optical (snow cover) and passive microwave data (snow depth), or higher frequency (Ka band) SAR data, but the methods are not adequately validated and need to be further carefully examined and improved.

14 Sea Level

Sea levels are rising in several places around the world potentially impacting human populations (e.g., those living in coastal regions and on islands) and the natural environment (e.g., marine ecosystems). Global average sea level rose at an average rate of around

1.7 mm per year in the twentieth century and at a satellite-measured average rate of about 3.2 mm per year from 1993 to 2009 (Meyssignac and Cazenave 2012), but no acceleration has been noted in the period 1993–2013. It is unclear whether the increased rate reflects an increase in the underlying long-term trend. Two main factors contribute to the observed sea-level rise, notably thermal expansion from general warming of the ocean and enhanced freshwater run-off from land-based ice due to increased melting. As such, long-term observations of the global mean and spatial sea-level change jointly with mass changes of the ice sheets and glaciers become of paramount importance for monitoring of the hydrological cycles and constraining of the water budget between the glaciers and the ocean.

15 Sea-surface Salinity

A total of 86 % of the total global evaporation and 78 % of the precipitation occur over the ocean (Schmitt 1995). As such, the ocean surface salinities have proven to be a much more reliable indicator of the water cycle than many of the land-based measurements. Salinity, moreover, is a fundamental ocean state variable and a major determinant, along with Sea-Surface Temperature (SST), of the density of seawater; hence it is a crucial factor in ocean circulation, which in turn has a major impact on climate (Schmitt 2008). Salinity variability at the sea surface also modulates or is modulated by heat, momentum and CO_2 exchange between the ocean and atmosphere. Salinity is an important constraint in ocean models and an indicator of freshwater capping. Sea-surface salinity (SSS) is correlated with differences between precipitation and evaporation (P–E), and improved knowledge of P–E would provide a better estimation of latent heat flux and improve characterization of stratification of the near-surface ocean layer. Besides, SSS variability is also related to the freezing and melting of sea ice and to freshwater river run-off. A better understanding of all these phenomena will be fostered by the recent availability of synoptic measurements of SSS thanks to the ESA SMOS (Font 2010) and NASA/CONAE Aquarius (Le Vine et al. 2010) satellites. Concurrently with the continuous improvements in the accuracy and reliability of these data, a routine monitoring of Sea Surface Salinity (SSS) is becoming possible for the first time thus allowing a quantitative characterization of the above-mentioned processes and their mutual relationships in the context of the hydrological cycle.

16 Freshwater Discharge for Large Mid-Latitude and Tropical Rivers

Occurrences of low salinity surface patches in tropical regions are closely related to the presence of the estuaries of the world's largest rivers in terms of fresh-water discharge (Amazon, Congo, Orinoco, Niger) and the subsequent spreading of freshwater by the surface oceanic circulation. The largest tropical river discharge regions have been studied using satellite altimetry, SST and ocean colour, but each technique has limits in these fresh pool regions, since salinity is the main controller of surface density in those areas. Now the satellite sea surface salinity (SSS) missions bring the unique capability to directly detect and track freshwater spatial gradients and lateral advection across the tropical oceans. The spatial extent of the buoyant plumes of freshwater that form in the tropical seas due to discharges from these world largest rivers can thus be temporally traced by SMOS/Aquarius imagery with an unprecedented resolution (Reul et al. 2013) In particular, river-influenced low salinity waters behave as excellent tracers of the local oceanic circulation and can be a very interesting proxy of subsurface properties (stratification, ocean heat content).

17 Conclusions and Observational Needs for the Future

Since the first experiments performed on assimilating data products derived from EO data into hydrological modelling for practical purposes of water resources management in the early 1980s (Ramamoorthi 1983), there seemed to be a prolonged pause. It is only now, more than 30 years later, that there are signs that application into operational hydrological models progresses rapidly (Van Dijk and Renzullo 2011). This is due to the fact that remote sensing data availability for the complex and interdisciplinary task of hydrology has long remained subcritical to capture the key multidisciplinary variables and their mutual interactions and feedback. EO data availability is now, especially with the prospect of the high-frequency global coverage with high-resolution satellite data, such as the ones from the Sentinels, approaching a point where monitoring of the hydrological variability in the context of a balanced Earth system approach is gradually becoming feasible (see Table 2). Also, hydrological models of operational character and spanning longer time intervals are increasingly becoming available. These models are optimized more and more for the use and assimilation of EO products.

Nevertheless, still today, the main shortcomings in many cases are that derived hydrological indices from so-called black-box models or conceptual models are not based on physical principles and relationships. These black-box models have to be calibrated to unknown and unconsidered circumstances in local and regional watersheds. Because of calibration with historical data, they more or less lack predictive power. This makes their use for any kind of long-range forecast difficult and for the necessary formulation of a global hydrology impossible. In addition, the climate system is governed by the global water and energy cycle, which is constituted of many interdependent and complex processes, interactions and mutual feedbacks in the atmosphere, hydrosphere, cryosphere and biosphere. Only a sound understanding of these processes will allow a quantitative and accurate determination of hydrological variables from satellites. The new wealth of data relevant for hydrological land surface processes studies, which originates from remote sensing sources in turn allows us to abandon conceptual model approaches and further develop hydrological models, which are based on first-order principles in the representation of hydrological land surface processes both in the physical and physiological domain.

The understanding of the complex hydrology of the Earth and its linkage with the carbon cycle and the atmosphere will drive our ability to realistically model the local, regional and global water cycle with high predictive skill and to thereby further reduce uncertainties in climate and Earth system models. Besides the general need for better and longer-term global data coverage at higher temporal and spatial resolutions to constrain model projections, observational needs and areas of continuing difficulties to obtain consistent observations and measurements include

- Improved observations of precipitation as the basic driver both for numerical weather prediction and hydrological land surface models to quantify global and regional trends,
- Increased and continuous precipitation observations over the oceans,
- Improved satellite-based global measurements of land surface parameters and their assimilation into dedicated high spatial resolution hydrological land surface models to better quantify stream flow, soil moisture and evapotranspiration and the carbon cycle,

Table 2 Availability of in situ and remotely sensed data products for hydrological variable retrievals and their status

Variable	In situ	Remote sensing	Product	Status
Soil moisture	Mesonets, climate reference networks, regional soil moisture networks	SMMR, AMSR, HYDROS, SMOS, *ASCAT* METOP-A, *ERS-Scat*	Surface soil moisture (SSM) Root-zone soil moisture Global Soil Wetness Project (GSWP) and Land Data Assimiliation Schemes (LDAS) Products	Products available from a) active (scatterometers and radars), b) passive (radiometers) microwave remote sensors and c) blended soil moisture product Global operational products available at coarse resolution scale (>25 km) Semi-operational, regional, products available at medium-resolution scale (>1 km) 30+ year historical soil moisture product available for long-term analyses (WACMOS) Potential future mission to deliver operational soil moisture SAR product from Sentinel-1
Sea-surface salinity	ARGO profilers, mooring networks, drifters, gliders, cruises, voluntary observing ships (VOS)	SMOS, Aquarius	Sea-Surface Salinity (SSS)	First ever synoptic monitoring for large-scale oceanography Large-scale river plumes and their extension into the open ocean tracking Freshwater barrier layers in tropical regions of intense rain detection Large-scale ocean currents and their salinity fronts characterization Surface processes monitoring in strong evaporation areas
Evapotranspiration	Flux Towers, Flux measurement aircraft, gradient observations, pan evaporation networks	Derived from models by inputting or assimilating vegetation indices from environmental satellites (MODIS, MERIS) and polar meteorological satellites (SUOMI NPP, EPS) as well as land surface temperature (AMSR), soil moisture (AMSR) and precipitation (NOAA's CMORPH technique)	No direct estimate of evapotranspiration	High spatial and temporal resolution product available (WACMOS) Global product available from MODIS and GLEAM

Table 2 continued

Variable	In situ	Remote sensing	Product	Status
Water level	Coastal gauge stations	Radar Altimeters	Satellite altimetry measures water level in large water bodies Optical, active and passive microwave RS data measure the extent of water bodies, floods and volumes	No operational product existing for water bodies extent Semi-operational products exist at local to regional scales ($\sim$ 100 m to 5 km)
Estimates of groundwater	Monitoring bores	A residual product of GRACE total water storage (TWS) after removal of other factors (i.e., SSM, ice) Interpretative use of optical and microwave data (e.g., thermal anomalies)	No direct measurement of groundwater	It is essential to have a good estimate of TWS errors to derive a good groundwater estimates GRACE provides an estimate of all water storages; GOCE complements GRACE to derive improved model of geoid
Vegetation stage	Field surveys, Aircraft surveys	Environmental satellites (MODIS, MERIS, MISR, SPOT, Landsat, Sentinels)	Optical vegetation indices (fPAR and LAI) passive or active microwave derived estimates of vegetation water content, biomass, vegetation structure Land-cover classifications	Open opportunities in use of radar and microwave data to estimate biomass and canopy height A good understanding and numerous algorithms for optical vegetation greenness indices
Precipitation	Surface Gages (manual and automatic)	SSM/I, TRMM, AMSR, AMSR-E, Geostationary environmental satellites, polar-orbiting meteorological satellites, GPM, NPOESS, Doppler radar	A routine generation of precipitation products—hourly 0.25-degree resolution via operational and research platforms Estimates from full-resolution observations available up to instantaneous, 4 km resolution, albeit more sporadically. Corrected numerical weather prediction (NWP) derived fields for structure and distribution	The Radar systems (e.g., TRMM PR and CloudSat CPR) provide a unique and crucial addition to our observational capabilities for precipitation. A retrieval at higher latitudes remains an open challenge due to contamination by the surface background as well as by the detection of light intensity, low-level, frozen precipitation.

Table 2 continued

Variable	In situ	Remote sensing	Product	Status
Clouds	Radiosondes, meteorological surface networks	Meteorological geostationary satellites, e.g., Meteosat Second Generation (MSG) Spinning Enhanced Visible and InfraRed Imager (SEVIRI), complemented by lower orbiting satellites (CloudSat, Calipso)	Cloud mask, cloud classification, cloud optical depth, liquid and ice water path, cloud top temperature and infrared emissivity	High spatiotemporal retrievals from SEVIRI provide powerful means to analyse model performance, in particular, the diurnal cycle New observations from CloudSat and Calipso provide powerful tools to validate other instruments series and also improve model parameterizations However, major efforts are needed to make cloud Climate Data Records (CDRs) fit for climate trend analysis Upcoming sensors include EarthCARE at 0.5 km resolution
Water vapour	Radiosondes, meteorological surface networks	A variety of space-borne sensors are used to retrieve atmospheric profiles of humidity or the column amount (microwave, infrared, optical, UV)	Atmospheric profiles of humidity or the column amount	SSM/I total column water vapour over ocean is mature for climate analysis MERIS observations have a significant potential for high spatial resolution total column water vapour over land during daytime in clear sky, e.g., ESA DUE GlobVapour. UV/VIS instruments provide independent means for total column water vapour retrievals but these measurements are biased towards clear sky Upper Tropospheric Humidity (UTH) data sets provide a data source with high value for climate research Microwave sounding data sets from AMSU-B, MHS and SSM/T2 hold a great potential to improve our knowledge on UTH—also allowing estimates of absolute humidity. The availability of atmospheric temperature and humidity profiles for 30 and more years is a critical issue The GEWEX Radiation Panel plans to carry out an intercomparison exercise

Table 2 continued

Variable	In situ	Remote sensing	Product	Status
Streamflow	Streamflow gages, field observation, global runoff data centre	Laser, radar altimeter, InSAR systems	River runoff	Experimental
Water quality	In stream sampling Buoys	MERIS, MODIS, SeaWIFS, Landsat	Ocean Colour, water quality indicators	Semi-operational coastal water quality assessments
Snow	Snow pillow networks, snow surveys	Geostationary environmental satellites, polar-orbiting meteorological satellites ATSR-2/ATSR, MERIS, MODIS, SSM/I, AMSR, NPOESS	Snow cover and snow water equivalent water storage from optical and passive microwave observations	Experimental
Permafrost	Temperature- and moisture probes	Only indirect, complementarity through SAR interferogrammes together with high precision digital elevation models (DEMs) and Thermal InfraRed (TIR) observations	Differential Interfergrammes, TIR maps	Experimental, terrain movement observations over time

Sources CEOS handbook (2013); Doubková et al. (2011); Van Dijk and Renzullo (2011)

- Enhanced ground water monitoring from satellite gravity observations,
- Enhanced monitoring of water quality, not only in coastal zones, but also over inland water bodies and large rivers,
- Improved inputs from higher resolution space data for land snow- and ice inventories as important water storage and frozen soil/permafrost monitoring,
- Inventories of data needed to do broad assessments of socio-economic trends of water use (e.g., agricultural water demands),
- Improved assessment of the insights offered by the recent satellite monitoring of sea-surface salinity regarding the oceanic branch of the hydrological cycle,
- Improved quantitative observation of river discharge,
- Enhanced monitoring of the surface albedo (from changes in snow, cover, sea ice extent) and its influence on evaporation, cloud formation and precipitation.

The new generation of operational satellites is expected through their increased coverage and temporal repeat observation capability to augment data availability and as such further our understanding of the hydrological cycle, also helping to move towards a quantitative closure of the water budget. Here, increased international collaboration and the use of observations from many satellites and/or satellite constellations will constitute important assets.

Acknowledgments The authors gratefully acknowledge the contributions by Florian Appel, Heike Bach, Marcela Doubkova, Diego Fernández-Prieto, Paul Ingmann, Roberto Sabia and Tobias Wehr.

References

Alexandratos N, Bruinsma J (2012) World agriculture towards 2030/2050, ESA Working paper No. 12–03, Agricultural Development Economics Division, Food and Agricultural Organisation of the United Nations, June 2012

Bach H, Mauser W (2003) Methods and examples for remote sensing data assimilation in land surface process modelling. IEEE Trans Geosci Remote Sens (TGARSS) 41(7):1629–1637

Bach H, Appel F, Rust F, Mauser W (2010) Polar View Snow Service—operational snow cover mapping for downstream Runoff modeling and hydropower predictions. Proceedings of the ESA Living Planet Symposium. ESA Special Publication SP-686, CD-Rom

Bach H, Migdall S, Spannraft K, Hank T, Mauser W (2012) Potential and challenges of using sentinel-2 for smart farming; sentinel-2 preparatory symposium, ESA–ESRIN, Frascati, Italy, 23–27

Bartsch A, Kidd RA, Pathe C, Scipal K (2007) Satellite radar imagery for monitoring inland wetlands in boreal and sub-arctic environments. Methods 317(3):305–317. doi:10.1002/aqc.836

Botteron C, Dawes N, Leclère J, Skaloud J, Weijs SV, Farine P-A (2013) Soil moisture and snow properties determination with GNSS in alpine environments: challenges, status, and perspectives. Remote Sens 5:3516–3543

Brenner AC, DiMarzio JP, Zwally HJ (2007) Precision and accuracy of satellite radar and laser altimeter data over the continental ice sheets. IEEE Trans Geosci Remote Sens 45(2). doi:10.1109/TGRS.2006.887172

Brocca L, Melone F, Moramarco T, Morbidelli R (2010a) Spatial-temporal variability of soil moisture and its estimation across scales. Water Resour Res, 46(2), art. no. W02516

Brocca L, Melone F, Moramarco T, Wagner W, Naeimi V, Bartalis Z, Hasenauer S (2010b) Improving runoff prediction through the assimilation of the ASCAT soil moisture product. Hydrol Earth Syst Sci Dis 14(10):1881–1893. doi:10.5194/hess-14-1881-2010

Cazenave A, Chen J (2010) Time-variable gravity from space and present-day mass redistribution in the Earth system. Earth Plan Sci Lett 298(3–4):263–274. doi:10.1016/j.epsl.2010.07.035

CEOS (2013) The Earth observation handbook—special edition for rio + 20 Updated for 2014. http://www.eohandbook.com

Christensen NS, Wood AW, Voisin N, Lettenmaier DP, Palmer RN (2004) The effects of climate change on the hydrology and water resources of the colorado river basin. Clim Change 62:337–363

Courault D, Seguin B, Olioso A (2005) Review on estimation of evapotranspiration from remote sensing data: from empirical to numerical modelling approaches. Irrigat Drain Syst 19(3):223–249

Curtis S, Salahuddin A, Adler RF, Huffman GJ, Gu G, Hong Y (2007) Precipitation extremes estimated by GPCP and TRMM: ENSO relationships. J Hydrometeorol. doi:10.1175/JHM601.1

DeFries R (2008) Terrestrial vegetation in the coupled human-Earth system: contributions of remote sensing. Ann Rev Environ Resour 2008(33):369–390. doi:10.1146/annurev.environ.33.020107.113339

Dorigo WA, De Jeu R, Chung D, Parinussa R, Liu Y, Wagner W, Fernández-Prieto D (2012) Evaluating global trends (1988–2010) in harmonized multi-satellite surface soil moisture. Geophys Res Lett 39:L18405. doi:10.1029/2012GL052988

Doubková M, Wagner W, De Jeu RAM (2011) Water from space: soil moisture and landscape dynamics. WIRADA Science Symposium, Melbourne

Doubková M, Van Dijk AIJM, Sabel D, Wagner W, Blöschl G (2012) Evaluation of the predicted error of the soil moisture retrieval from C-band SAR by comparison against modelled soil moisture estimates over Australia. Remote Sens Environ 120:188–196

Dyurgerov MB, Mark FM (2005) Glaciers and the changing Earth system: a 2004 Snapshot. Institute of Arctic and Alpine Research. Occasional paper 58

ESA (2012) The biomass mission. Report for mission selection—ESA SP-1324/1. http://esamultimedia.esa.int/docs/EarthObservation/SP1324-1_BIOMASSr.pdf

ESA (2013) ERS missions, 20 years of observing Earth. ESA SP-1326, ESA Communications, Fletcher K (ed), ISBN 978-92-9221-424-1, Leiden July 2013

Ferguson IM, Maxwell RM (2012) Human impacts on terrestrial hydrology: climate change versus pumping and irrigation. Environ Res Lett 7(2012) 044022 (8pp). doi:10.1088/1748-9326/7/4/044022

Font J, Camps A, Borges A, Martín-Neira M, Boutin J, Reul N, Kerr YH, Hahne A, Mecklenburg S (2010) SMOS: the challenging sea surface salinity measurement from space. Proc IEEE 98:649–665

Fowler HJ, Blenkinsop S, Tebaldi C (2007) Linking climate change modelling to impacts studies: recent advances in downscaling techniques for hydrological modelling—review. Int J Climatol 27:1547–1578. doi:10.1002/joc.1556

Frei A, Tedesco M, Lee S, Foster J, Hall DK, Kelly R, Robinson DA (2012) A review of global satellite-derived snow products. Adv Space Res 50:1007–1029

GCOS (2003) The second report on the adequacy of the global observing systems for climate in support of the UNFCCC. GCOS- 82 (WMO/TD No. 1143) April 2003. (http://www.wmo.int/pages/prog/gcos/Publications/gcos-82_2AR.pdf)

Helsen MM, van den Broeke MR, van de Wal RSW, van de Berg WJ, van Meijgaard E, Davis CH, Li Y, Goodwin I (2008) Elevation changes in Antarctica mainly determined by accumulation variability. Science 320:1626–1629

Hibbard K, Janetos A, van Vuuren DP, Pongratz J, Rose SK, Betts R, Herold M, Feddema JJ (2010) Research priorities in land use and land-cover change for the Earth system and integrated assessment modelling. Int J Climatol 30: 2118–2128. doi:10.1002/joc.2150

Hornacek M, Wagner W, Sabel D, Truong H, Snoeij P, Hahmann T, Diedrich E et al (2012) Potential for high resolution systematic global surface soil moisture retrieval via change detection using sentinel-1. IEEE J Sel Top Appl Earth Observation Remote Sens 5(4):1303–1311

IGOS (2007) Integrated Global Observing Strategy Cryosphere Theme Report, August 2007, WMO/TD No 1405

Jiménez C et al (2011) Global intercomparison of 12 land surface heat flux estimates. J Geophys Res 116:D02102. doi:10.1029/2010JD014545

Kalma JT, McVicar M, McCabe M (2008) Estimating land surface evaporation: a review of methods using remotely sensed surface temperature data. Surveys Geophys 29(4):421–469

Kerr YH, Waldteufel P, Wigneron JP, Delwart S, Cabot F, Boutin J, Escorihuela MJ et al (2010) The SMOS L: new tool for monitoring key elements of the global water cycle. Proc IEEE 98(5):666–687

Le Vine DM, Lagerloef GSE, Torrusio SE (2010) Aquarius and remote sensing of sea surface salinity from Space. Proceedings of the IEEE 98(5) 688–703. doi:10.1109/JPROC.2010.2040550

Lemke P, Ren J, Alley RB, Allison I, Carrasco J, Flato G, Fujii Y, Kaser G, Mote P, Thomas RH, Zhang T (2007) Observations: changes in snow, ice and frozen ground. Climate Change: the physical science basis; summary for policymakers, technical summary and frequently asked questions. Part of the Working Group I contribution to the Fourth Assessment Report of the Intergovernmental Panel on Climate Change, ISBN: 92-9169-121-6. hdl:10013/epic.30870

Liu YY, Parinussa RM, Dorigo WA, De Jeu RAM, Wagner WM, Van Dijk AIJ, McCabe MF et al (2011) Developing an improved soil moisture dataset by blending passive and active microwave satellite-based retrievals. Hydrol Earth Syst Sci 15(2):425–436

Loew A, Ludwig R, Mauser W (2006) Derivation of surface soil moisture from ENVISAT ASAR Wide-Swath and image mode data in agricultural areas. IEEE Trans Geosci Remote Sens 44(4):889–899

Loew A, Stacke T, Dorigo W, de Jeu R, Hagemann S (2013) Potential and limitations of multidecadal satellite moisture observations for climate model evaluation studies. Hydrol Earth Syst Sci 17:3523–3542. doi:10.5194/hess-17-3523-2013

Lonfat M, Marks FD, Chen SS (2004) Precipitation distribution in tropical cyclones using the tropical rainfall measuring mission (TRMM) microwave imager: a global perspective. Mon Weather Rev. doi:10.1175/1520-0493(2004)132<1645:PDITCU>2.0.CO;2

Ludwig F, Kabat P, van Schaik H, van der Valk M (eds) (2008) Climate change adaptation in the water sector, Earthscan, London

Matgen P, Hasenauer S, Hissler C, Brocca L, Hoffmann L, Wagner W, Savenije HHG (2011) On the potential of MetOp ASCAT-derived soil wetness data hydrological monitoring and due to limited prediction: SAR evaluation over Luxembourg lack of coverage. Hydrol Process. doi:10.1002/hyp

Mauser W (2009) Water resources: efficient, sustainable and equitable use, the sustainability project. House Publishing, London

Mauser W, Bach H (2009) PROMET—large scale distributed hydrological modelling to study the impact of climate change on the water flows of mountain watersheds. J Hydrol 376(2009):362–377. doi:10.1016/j.jhydrol.2009.07.046

McNamara JP, Kane DL, Hinzman LD (1999) An analysis of an arctic channel network using digital elevation model. Geomorphology 29:339–353

Meier P, Frömelt A, Kinzelbach W (2011) Hydrological real-time modelling in the Zambezi river basin using satellite-based soil moisture and rainfall data. Hydrol Earth Syst Sci 15(3):999–1008. doi:10.5194/hess-15-999-2011

Meyssignac B, Cazenave A (2012) Sea level: a review of present-day and recent-past changes and variability. J Geodyn 58:96–109

Monteith JL (1965) Evaporation and Environment. In: The state and movement of water in living organism. 19th Symp Soc Exptl Biol 205–234

Mueller B, Hirschi M, Jimenez C, Ciais P, Dirmeyer PA, Dolman AJ, Fisher JB, Jung M, Ludwig F, Maignan F, Miralles D, McCabe MF, Reichstein M, Sheffield J, Wang KC, Wood EF, Zhang Y, Seneviratne SI (2013) Benchmark products for land evapotranspiration: landFlux-EVAL multi-dataset synthesis. Hydrol Earth Syst Sci 17:3707–3720

Naeimi V, Bartalis Z, Wagner W (2009) ASCAT soil moisture: an assessment of the data quality and consistency with the ERS scatterometer heritage. J Hydrometeorol 10(2):555–563

Nghiem S, Tsai W-Y (2001) Global snow cover monitoring with space-borne Ku-band scatterometer. IEEE Trans Geosci Remote Sens 39(10):2118–2134

Njoku EG, Jackson TJ, Lakshmi V, Chan TK, Nghiem SV (2003) Soil moisture retrieval from AMSR-E. IEEE Trans Geosci Remote Sens 41(2):215–229

Oki T, Kanae S (2006) Global hydrological cycles and world water resources. Science 313:1068–1072. doi:10.1126/science.1128845

Parajka J, Naeimi V, Blöschl G, Wagner W, Merz R, Scipal K (2006) Assimilating scatterometer soil moisture data into conceptual hydrologic models at the regional scale. Hydrol Earth Syst Sci 10(3):353–368. doi:10.5194/hess-10-353-2006

Pathe C, Wagner W, Sabel D, Doubkova M, Basara J (2009) Using ENVISAT ASAR global mode data for surface soil moisture retrieval over Oklahoma, USA. IEEE Trans Geosci Remote Sens 47(2):468–480

Penman HL (1948) Natural evaporation from open water, bare soil and grass. P Roy Soc Lon Ser- A 193(1032):120–145

Piao S, Friedlingstein P, Ciais P, de Noblet-Ducoudre N, Labat D, Zaehle S (2007) Changes in climate and land use have a larger direct impact than rising CO_2 on global river runoff trends. PNAS 104(39): 15242–15247. doi:10.1073/pnas.0707213104

Prasch M, Mauser W, Weber M (2013) Quantifying present and future glacier melt-water contribution to runoff in a central Himalayan river basin. *The Cryosphere* 7:889–904. doi:10.5194/tc-7-889-2013

Priestley CHB, Taylor RJ (1972) On the assessment of surface heat fluxes and evaporation using large scale parameters. Mon Weather Rev 100:81–92

Prigent C, Papa F, Aires F, Rossow WB, Matthews E (2007) Global inundation dynamics inferred from multiple satellite observations, 1993–2000. J Geophys Res 112(D12):1993–2000. doi:10.1029/2006JD007847

Ramamoorthi AS (1983) Snow-melt run-off studies using remote sensing data. Proceedings of the Indian Academy of Sciences, Section C: Engineering Sciences, Sept 1983, 6(3): 279–286. doi:10.1007/BF02842888

Reul N, Fournier S, Boutin J, Hernandez O, Maes C, Chapron B, Alory G, Quilfen Y, Tenerelli J, Morisset S, Kerr Y, Mecklenburg S, Delwart S (2013) Sea surface salinity observations from space with the SMOS satellite: a new means to monitor the marine branch of the water cycle. Surv Geophys. doi:10.1007/s10712-013-9244-0

Richter K, Hank T, Voulo F, Mauser W, D'Urso G (2012) Optimal exploitation of the sentinel-2 spectral capabilities for crop leaf area index mapping. Remote Sens 4(3):561–582

Rignot E, MacAyeal DR (1998) Ice-shelf dynamics near the front of the Filchner-Ronne Ice Shelf, Antarctica, revealed by SAR interferometry. J Glaciol 44(147):405–418

Rodell M, Famiglietti JS (2002) The potential for satellite-based monitoring of groundwater storage changes using GRACE: the High Plains aquifer, Central US. J Hydrol 263(1–4):245–256. doi:10.1016/S0022-1694(02)00060-4

Rodriguez JM, Ustin SL, Riano D (2011) Contributions of imaging spectroscopy to improve estimates of evapotranspiration. Hydrol Process 25:4069–4081. doi:10.1002/hyp.8368

Rott (2013) Contribution to EO sentinel convoy—ocean and ice theme final report, produced under ESA contract, ESA, Astrium Limited Gunnels Wood Road, Stevenage, Hertfordshire, SG1 2AS, UK

Rott H, Siegel A (1997) Glaciological studies in the alps and in Antarctica using ERS interferometric SAR. ESA SP-406: Conference proceedings of Fringe'96 (Zürich, Switzerland 30 Sept–2 Oct). European Space Agency

Schlüssel P, Soloviev AV, Emery WJ (1997) Cool and freshwater skin of the ocean during rainfall. Boundary-Layer Meteorol 82(3):439–474

Schmitt RW (1995) The ocean component of the global water cycle: US National Report to International Union of Geodesy and Geophysics, 1991–1994. Rev Geophys 33(Supplement):1395–1409

Schmitt RW (2008) Salinity and the global water cycle. Oceanography 21(1):12

Shepherd A, Wingham D (2002) Recent Sea-level contributions of the Antarctic and Greenland Ice Sheets. Science 315:1529–1533. doi:10.1126/science.1136776

Shepherd A, Ivins ER, Valentina GA, Barletta R, Bentley MJ, Bettadpur S, Briggs KH, Bromwich DH, Forsberg R, Galin NH, Jacobs S, Joughin I, King Lenaerts, JTM, Li J, Ligtenberg SRM, Luckman A, Luthcke SB, McMillan M, Meister R, Milne G, Mouginot J, Muir A, Nicolas JP, Paden J, Payne AJ, Pritchard H, Rignot E, Rott H, Sørensen LS, Scambos TA, Scheuchl B, Schrama EJO, Smith B, Sundal AV, van Angelen JH, van de Berg WJ, van den Broeke MR, Vaughan DG, Velicogna I, Wahr J, Whitehouse PL, Wingham DJ, Yi D, Young D, Zwally HJ, (2012) A reconciled estimate of ice-sheet mass balance. Science 338:1183. doi:10.1126/science.1228102

Soloviev A, Lukas R (1996) Observation of spatial variability of diurnal thermocline and rain-formed halocline in the western Pacific Warm Pool. J Phys Oceanogr. doi:10.1175/1520-0485(1996

Strengers BJ, Müller C, Schaeffer M, Haarsma RJ, Severijns C, Gerten D, Schaphoff S, van den Houdt R, Oostenrijk R (2010) Assessing 20th century climate–vegetation feedbacks of land-use change and natural vegetation dynamics in a fully coupled vegetation–climate model. Int J Climatol 30: 2055–2065. doi:10.1002/joc.2132

Su Z, Wen J, Wagner W (2010) Advances in land surface hydrological processes—field observations, modeling and data assimilation. Hydrol Earth Syst Sci 14:365–367. www.hydrol-earth-syst-sci.net/14/365/2010/

Turner K, Georgiou S, Clark R, Brouwer R, Burke K (2004) Economic valuation of water resources in agriculture, FAO Water Reports 27. Food and Agriculture Organization of the United Nations, Rome

Van Dijk AIJM, Renzullo LJ (2011) Water resource monitoring systems and the role of satellite observations. Hydrol Earth Syst Sci 15(1):39–55. doi:10.5194/hess-15-39-201

Wagner W, Lemoine G, Rott H (1999) A method for estimating soil moisture from ERS scatterometer and soil data. Remote Sens Environ 70(2):191–207

Wang K, Dickinson RE (2012) A review of global terrestrial evapotranspiration. Rev Geophys 50(2): RG2005. doi:10.1029/2011RG000373

Wijesekera HW, Paulson CA, Huyer A (1999) The effect of rainfall on the surface layer during a Westerly Wind Burst in the Western Equatorial Pacific. J Phys Oceanogr. doi:10.1175/1520-0485

Wingham DJ, Francis CR, Baker S, Bouzinac C, Brockley D, Cullen R, de Chateau-Thierry P, Laxon SW, Mallow U, Mavrocordatos B, Phalippou L, Ratier G, Rey L, Rostan F, Viau P, Wallis DW (2006) CryoSat: a mission to determine the fluctuations in Earth's land and marine ice fields. Adv Space Res 37(4):841–871

Wood EF, Roundy JK, Troy TJ, van Beek R, Bierkens M, Blyth E, de Roo A, Doell P, Ek M, Famiglietti J, Gochis D, van de Giesen N, Houser P, Jaffe P, Kollet S, Lehner B, Lettenmaier DP, Peters-Lidard C, Sivapalan M, Sheffield J, Wade A, Whitehead P (2011) Hyper-resolution global land surface modeling: meeting a grand challenge for monitoring Earth's terrestrial water. Water Resour Res 47:W05301. doi:10.1029/2010WR010090

Zabel F, Mauser W (2013) 2-way coupling the hydrological land surface model PROMET with the regional climate model MM5. Hydrol Earth Syst Sci 17:1705–1714. doi:10.5194/hess-17-1705-2013

Zhang T, Barry RG, Haeberli W (2001) Numerical simulation of the influence of the seasonal snow cover on the occurrence of permafrost at high latitudes. Nor Geogr Tidsskr 55(2001):261–266

DOI 10.1007/978-94-017-8789-5_3
Reprinted from *Surveys in Geophysics* Journal, DOI 10.1007/s10712-012-9214-y

Challenges and Opportunities in Water Cycle Research: WCRP Contributions

Kevin E. Trenberth · Ghassem R. Asrar

Received: 1 May 2012 / Accepted: 10 November 2012 / Published online: 4 December 2012

Abstract The state of knowledge and outstanding challenges and opportunities in global water cycle observations, research and modeling are briefly reviewed to set the stage for the reasons behind the new thrusts promoted by the World Climate Research Programme (WCRP) as Grand Challenges to be addressed on a 5- to 10-year time frame. Those focused on water are led by the GEWEX (Global Energy and Water Exchanges) project. A number of GEWEX science questions are being brought forward within GEWEX and the WCRP under guidance of the Joint Scientific Committee. Here, we describe what are some imperatives and opportunities for major advancements in observations, understanding, modeling and product development for water resources and climate that will enable a wide range of climate services and inform decisions on water resources management and practices.

Keywords Global water cycle · Hydrological Cycle · WCRP · Precipitation · GEWEX · Water resources · Climate Change · Climate extremes

1 Introduction

Driven mainly by solar heating, water is evaporated from ocean and land surfaces, transported by winds, and condensed to form clouds and precipitation which falls to land and oceans. Precipitation over land may be stored temporarily as snow or soil moisture, while excess rainfall runs off and either forms streams and rivers, which discharge the freshwater into the oceans, or infiltrates into the soil and percolates to depths to re-charge the underground aquifers thereby completing the global water cycle (Trenberth et al. 2007a; Fig. 1). Associated with this water cycle, energy, salt within the oceans, and nutrients and minerals over land are all transported and redistributed within the Earth

K. E. Trenberth (✉)
National Center for Atmospheric Research, Boulder, CO 80307, USA
e-mail: trenbert@ucar.edu

G. R. Asrar
World Climate Research Programme, Geneva, Switzerland

climate system. Moreover, water is vital for human existence and is irreplaceable. It is more than a natural resource that we exploit and often take for granted. Water plays a crucial role in Earth's climate, functioning of ecosystems and environment.

Many studies on the global water cycle deal with only specific aspects (see the review by Trenberth et al. 2007a and other chapters in this monograph). Reliable data on the surface water budget are often available only over certain regions. Relatively few studies (e.g., Trenberth et al. 2007a, 2011; Oki and Kanae 2006) have attempted to provide a synthesized, quantitative view of the global water cycle, and our quantitative knowledge of the various components, and their variability of the global water cycle is still fairly limited because of a lack of reliable data for surface evapotranspiration, oceanic precipitation, terrestrial runoff and several other fields. Regional closure of the water cycle over many large river basins has been attempted by Vinukollu et al. (2011) and Sahoo et al. (2011) using satellite data but, unless adjusted, they do not adequately close the water budget, and the imbalances highlight the outstanding observational and modeling limitations.

Satellite-based observations provide global coverage but may lack continuous coverage in time and generally require some kind of algorithm to produce a geophysical product that inevitably has limitations, so the result must be verified against other independent measurements such as in situ observations. However, as the number of analyzed fields grows with ever increasing satellite data products, it is vital for these to be properly evaluated and documented for their strengths and weaknesses along with quantifying their uncertainties. Some of the satellite-based observations limitations may be overcome with in situ observations because they measure directly the quantity desired, but it is likely a spot

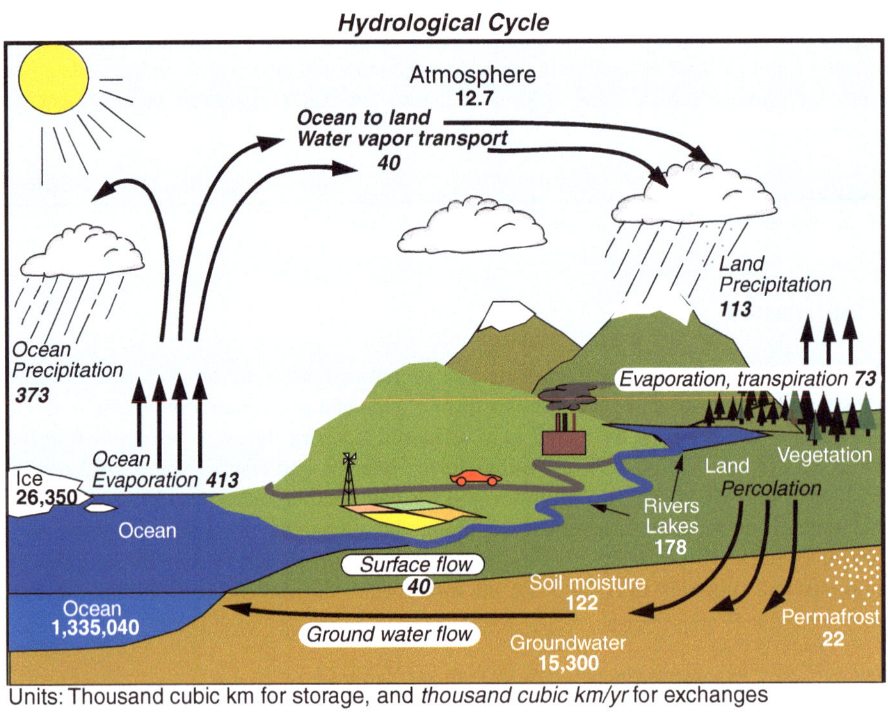

Fig. 1 The global annual mean Earth's water cycle for the 1990s. The arrows indicate the schematic flow of water substance in various forms. From Trenberth et al. (2007a, b)

measurement and its representativeness and calibration may not be sufficient to capture the spatial characteristics, unless a sufficiently large number of such measurements are obtained. Building and maintaining such large measurement networks have been challenging, especially over the developing regions of the world where such measurements are needed most. Blended or hybrid satellite and in situ products are also growing in number and attempt to capitalize on the strengths of each. Some are produced in a model framework and may involve data assimilation. Nevertheless, with multiple products synthesized in the framework of the overall water cycle, it is possible to make use of physical constraints inherent in a closed water budget, and physical models to help refine all components that are not well observed, by taking their uncertainty into account.

2 The WCRP

The World Climate Research Programme (WCRP) mission is to facilitate analysis and prediction of Earth system variability and change for use in an increasing range of practical applications of direct relevance, benefit and value to society. The two overarching objectives of the WCRP are

(1) to determine the predictability of climate and
(2) to determine the effect of human activities on climate.

Progress in understanding climate system variability and change makes it possible to address its predictability and to use this predictive knowledge in developing adaptation and mitigation strategies. Such strategies assist the global communities in responding to the impacts of climate variability and change on major social and economic sectors including food security, energy and transport, environment, health and water resources (Asrar et al. 2012a). The main foci of WCRP research are

- Observing changes in the components of the Earth system (atmosphere, oceans, land and cryosphere) and in the interfaces among these components;
- Improving our knowledge and understanding of global and regional climate variability and change and of the mechanisms responsible for this change;
- Assessing and attributing significant trends in global and regional climates;
- Developing and improving numerical models that are capable of simulating and assessing the climate system for a wide range of space and time scales;
- Investigating the sensitivity of the climate system to natural and human-induced forcing and estimating the changes resulting from specific disturbing influences.

The WCRP is sponsored by the World Meteorological Organization (WMO), the International Council for Science (ICSU) and the Intergovernmental Oceanographic Commission (IOC) of the United Nations Educational, Scientific and Cultural Organization (UNESCO).

The World Climate Research Programme (WCRP) is organized as a network of core and co-sponsored projects, working groups and cross-cutting initiatives. The current core projects of WCRP are

- *Climate and Cryosphere* (*CliC*): The principal goal of CliC is to assess and quantify the impacts of climatic variability and change on components of the cryosphere and their consequences for the climate system and to determine the stability of the global cryosphere.

- *Climate Variability and Predictability* (*CLIVAR*): CLIVAR's mission is to observe, simulate and predict the Earth's climate system with a focus on ocean–atmosphere interactions in order to better understand climate variability, predictability and change.
- *Global Energy and Water EXchanges* (*GEWEX*): GEWEX was previously known as the Global Energy and Water cycle Experiment but has recently been renamed although with the same acronym. It focuses on the atmospheric, terrestrial, radiative, hydrological and coupled processes and interactions that determine the global and regional hydrological cycle, radiation and energy transitions and their involvement in global changes such as increases in greenhouse gases.
- *Stratospheric Processes And their Role in Climate* (*SPARC*): SPARC has as its principal focus research on the significant role played by stratospheric processes in the Earth's climate, with a particular emphasis on the interaction between chemistry and climate.

There are also several working groups or councils on modeling and data that coordinate climate observations, modeling and prediction activities across the entire WCRP. The coordination of research among the physical, biogeochemical, socio-economic dimension of global change research has been achieved through Earth System Science Partnership (ESSP) which is being succeeded by a new initiative entitled "Future Earth: research for global sustainability" http://www.icsu.org/future-earth.

The Joint Scientific Committee of the WCRP is considering several scientific Grand Challenges that emerged from the consultation with the global scientific community at a recent WCRP Open Science Conference to be the major foci for the WCRP activities during the next decade (Asrar et al. 2012b). They include:

- Provision of skillful future climate information on regional scales
- Regional Sea-Level Rise
- Cryosphere response to climate change
- Improved understanding of the interactions of clouds, aerosols, precipitation, and radiation and their contributions to climate sensitivity
- Past and future changes in water availability
- Science underpinning the prediction and attribution of extreme events.

Although global water cycle is affected by and affects all of these, we focus only on the last two challenges that involve water and the hydrological cycle for this monograph.

3 The Global Water Budget and Hydrological Cycle

As the climate changes partly from human activities, the water cycle is also changing (Trenberth 2011). Moreover, demand for water continues to increase owing to growing population, enhanced agricultural and industrial development, and other human activities such as transformation of landscape and construction of dams and reservoirs, so that very little of the land surface remains in a natural state. This affects the disposition of water when it hits the ground: how much runs off, and how much finds its way to rivers or infiltrates into the soil and percolates to depths to replenish the underground water reservoirs.

The adverse impact of such activities is not confined to quantity and distribution of water, but also increasingly affects water quality. Water is used in various ways: such as through irrigation or by consumption in other human activities; reservoirs and artificial lakes are used to store water, while dams and other structures are used to control water flows in rivers. Water is heated and cooled and, as a strong solvent, it is polluted in many areas.

Many physical scientists have tended to ignore the latter aspects and deal mainly with the climate system either in its "natural" state or as changed by human activities by mainly accounting for increased greenhouse gases and changing atmospheric particulates (IPCC 2007). Even in this somewhat simplified framework, it has been challenging to simulate the hydrological cycle. For example, global reanalyses of most existing observations have substantial shortcomings in representing the hydrological cycle (Trenberth et al. 2011). Such shortcomings arise because, while observations are assimilated to ensure a realistic representation of atmosphere and some Earth surface processes, the analysis increment ensures that water is not conserved and sources of moisture for precipitation may come from the increment and not evapotranspiration. Models generally have a lifetime of water in the atmosphere that is too short, and this affects their ability to transport water vapor onto land while they tend to recycle moisture locally more than observed.

The main impacts of a warmer climate on global water cycle include the following:

- With warming, higher atmospheric temperatures increase the water holding capacity of the atmosphere by about 7 % per degree Celsius (e.g., Trenberth et al. 2003).
- Over the ocean where there is ample water supply, the relative humidity remains about the same and hence the observed moisture goes up at about this rate: an increase in total column water vapor of about 4 % since the 1970s (Trenberth et al. 2007b).
- Over land the response depends on the moisture supply.
- With more heat in the Earth system, the evaporation is enhanced resulting in more precipitation. The rate of increase is estimated to be about 2 % per degree Celsius warming (Trenberth 2011).
- Locally this means increased potential evapotranspiration, and in dry areas this means drying and more intense and longer lasting droughts.
- Larger warming over land versus the ocean further changes monsoons.
- Precipitation occurs mainly from convergence of atmospheric moisture into the weather system producing the precipitation, and hence increased water vapor leads to more intense rains and snow, and potentially to more intense storms.
- More precipitation occurs as rain rather than snow.
- However, higher temperatures in winter over continents favor higher snowfalls.
- Snow pack melts quicker and sooner, leading to less snow pack in the spring.
- These conditions lead to earlier runoff and changes in peak streamflow. Hence, there is a risk of more extremes, such as floods and droughts.

The pattern of observed changes, so far, indicates wetter conditions in higher latitudes across Eurasia, east of the Rockies in North America, and in Argentina, but drier conditions across much of the tropics and subtropics (IPCC 2007; Dai et al. 2009, 2011; Trenberth 2011), and this pattern is referred to as "The rich get rich and the poor get poorer" syndrome (the wet areas get wetter while the arid areas get drier). This pattern is projected to continue into the future (IPCC 2007), including an increase in probability of the water-related extremes (IPCC 2012).

Over land there is a strong negative correlation between precipitation and temperature throughout the tropics and over continents in summer, but a positive correlation in the extratropics in winter (Trenberth and Shea 2005). The latter arises from the baroclinic storms that advect warm moist air ahead of and into the storm, combined with the ability of warmer air to hold much more moisture. The former arises from the nature of the atmospheric circulation interactions with land. In cyclonic conditions, increased cloud and rain provide more soil moisture and thus partitions the decreased surface energy more into latent energy (higher evaporation) instead of sensible heat (lower temperatures). Anticyclonic conditions favor sunshine (more available energy), less rain and soil moisture, and the larger surface energy raises temperatures instead of evaporating moisture. The result is more likely either hot and dry or cool and wet conditions, but not the other options.

On global land, there is large variability in precipitation from year to year and decade to decade associated especially with the El Niño-Southern Oscillation (ENSO) but there has been an increase overall in land precipitation (Fig. 2). The two wettest years are 2010 and 2011. In particular, major flooding in Pakistan, Australia, and Colombia was associated with record high sea surface temperatures (SSTs) in the second half of 2010 into 2011 (Trenberth 2012) and led to a dramatic drop in sea level of about 5 mm (Fig. 3). The prospects for more intense precipitation but longer dry spells lead to the increased risk of flooding and drought, which pose major challenges for the society at large and those who have to manage water resources for food, fiber and energy production, and human consumption and leisure. We therefore view observing, understanding, modeling and predicting the global water cycle as a grand science challenge.

4 Grand Challenges

A Grand Challenge should inspire the community to want to be involved; it needs to be specific and focused while identifying barriers and ways to advance the science, and it must capture the imaginations of funding agencies, science program managers and the public. It should also provide a vehicle to encourage the different WCRP panels to interact in pursuing a common goal. It must provide a way forward that is tractable, perhaps via new observations (e.g., from satellites), computer and model advancements, and ideas. It must matter, as shown by answers to questions on possible benefits to society by providing

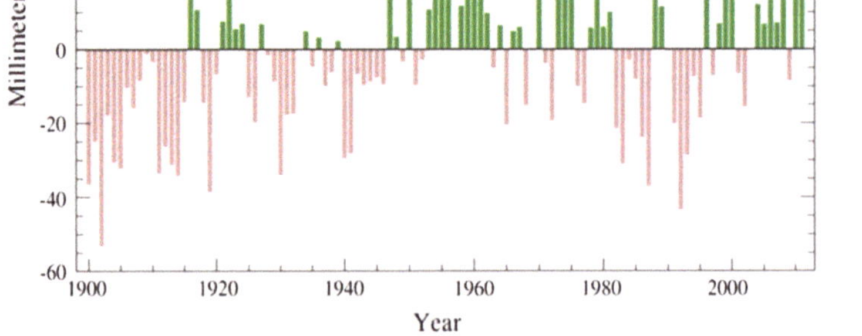

Fig. 2 Annual mean anomalies in global land precipitation from 1900 to 2011 in mm; from NOAA. http://www.ncdc.noaa.gov/sotc/global/2011/13

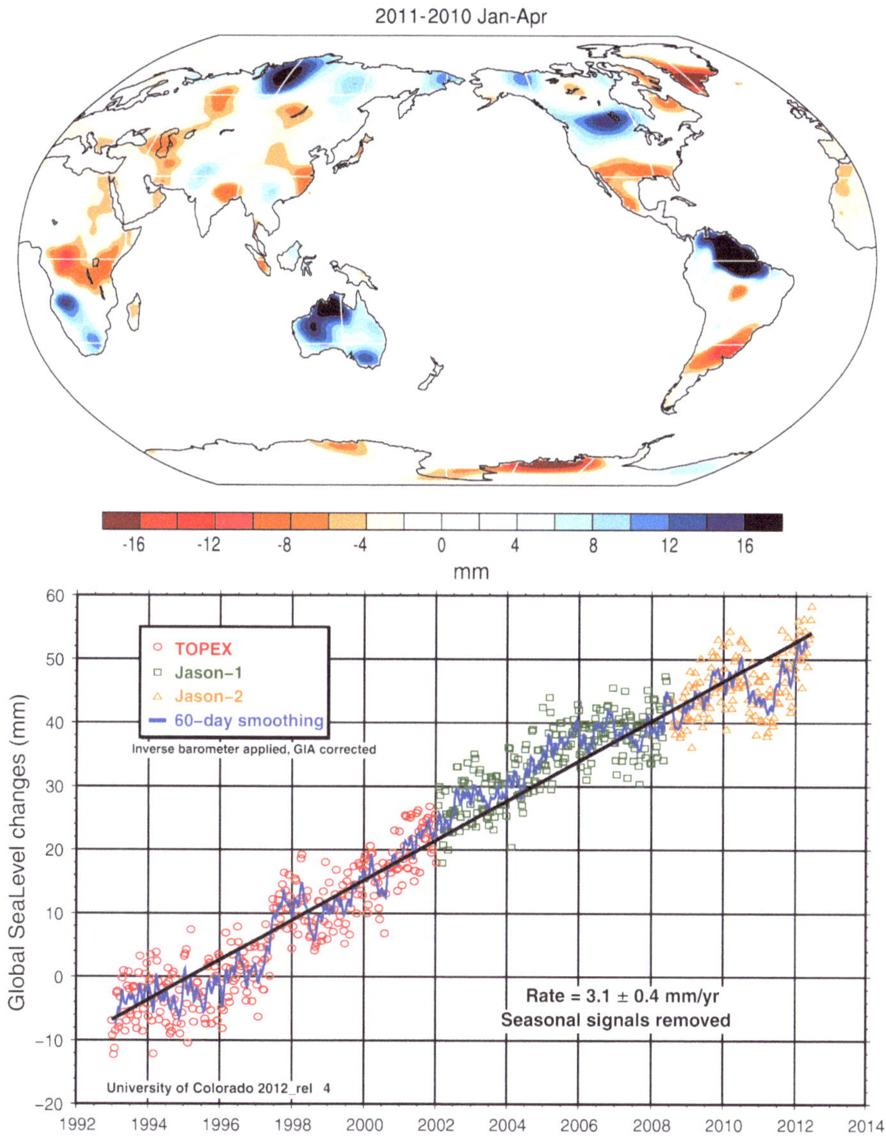

Fig. 3 (*top*) Differences between GRACE microgravity estimates of changes in mass on land in mm of land water equivalent from January to April 2011 versus 2010. (*bottom*) Estimates of global sea-level anomalies from University of Colorado based upon satellite altimetry. The inverse barometer and postglacial rebound adjustments have been applied. The *dots* show individual values, and the *dark blue line* shows the 2-month smoothed changes (Nerem et al. 2010)

the science-based information to address impacts of climate variability and change to food, water, health, energy, biodiversity and so on.

The GEWEX Science Steering Group (SSG) has identified four GEWEX Science Questions (GSQs). These emerged from in depth discussions and subsequent circulation to all GEWEX Panel members for commentary. They were then posted on the GEWEX web site for open commentary. They have also been presented to the WCRP Joint Scientific Committee for comment, and the outcome is what we present here. Three of these GSQs deal with water and two of them are combined into a more general water resource Grand Challenge for WCRP that also encompass scientific activities coordinated by the CliC, CLIVAR and SPARC projects. This is outlined in Sect. 4.1 along with a number of more specific questions. The third GSQ is part of a WCRP-wide theme of extremes, and those extremes related to water are discussed in Sect. 4.2. The other GSQ relates to energy and processes.

4.1 Grand Challenge on Water Resources

How can we better understand and predict precipitation variability and changes, and how do changes in land surface and hydrology influence past and future changes in water availability and security?

These questions focus on the exploitation of improved data sets of precipitation, soil moisture, evapotranspiration, and related variables such as water storage and sea surface salinity expected in the next 5 to 10 years. These will allow us to help close the water budget over land and provide improved information for products related to water availability and quality for decision makers and for initializing climate predictions from seasons to years in advance. The improvements will come from ongoing and planned satellite missions (see below) as well as greater use of in situ observations; their evaluation and analysis to document means, variability, patterns, extremes and probability density functions; their use to confront models in new ways and to improve our understanding of atmospheric and land surface processes that in turn improve simulations of precipitation; and new techniques of data assimilation and forecasts that can lead to improved predictions of the hydrological cycle across scales, from catchments to regions to the entire globe, including hydrogeological aspects of ground water recharge. In particular, attention is needed on the use of realistic land surface complexity with all anthropogenic effects taken into account, instead of a fictitious natural environment. This encompasses all aspects of global change, including water management, land-use and land-cover change, and urbanization. The ecosystem response to climate variability and responsive vegetation to such changes must be included, as must cryospheric changes such as dynamics of permafrost, thawing and changes in mountain glaciers. The focus on these scientific questions should lead to improved understanding and prediction of precipitation and water variability, enhance the evaluation of the vulnerability of water systems, especially to extremes, which are vital for considerations of water security and can be used to increase resilience through good management and governance.

The 21st century poses formidable challenges for the sustainable management of water resources at all levels, from the local, regional to the global scale. Water is a basic requirement for life, and effective water management is needed to provide some of society's most basic needs. However, demand for water resources is increasing, due to population growth and economic development, while water resources are under pressure globally from over-abstraction and pollution. This is increasingly leading to competition for water, at local, regional and international levels. Environmental change is adding additional pressures. Consequently, there are growing issues of vulnerability and accessibility to water, both of which are highly relevant for society. Anthropogenic influences are changing land and water systems, redefining the state of drainage basins and the rivers and groundwater aquifers that supply the bulk of renewable freshwater to society. Widespread land-use changes, associated with population increases, urbanization, agricultural intensification and industrialization are changing hydrological systems in complex ways, and on many of the world's major rivers, water management is changing flows, often with severe effects on downstream users, aquatic ecosystems and freshwater discharges to the world's seas and oceans. Superposed on these pressures, expected climate change and climate variability can combine to create extreme and perhaps unprecedented conditions which have high impact consequences for human populations, economic assets and critical physical infrastructure. This unique combination of pressures has exposed weaknesses in current water governance and management. It has increased the awareness of uncertainties, the complexity of the systems to be managed, and the need for profound changes in policy and management paradigms, as well as governance systems.

The World Climate Research Programme (WCRP) has a unique role to play in developing the new scientific understanding and modeling and prediction tools needed for a new era of global water management. WCRP mainly through GEWEX, and based on significant contributions from CLIVAR and CliC projects, is well poised to motivate a new generation of land surface and global hydrological models, building on recent developments in Earth observations, that represent the dynamics of major managed water systems. The modeling activities have an equally important role in motivating a new generation of weather-resolving climate models that are capable of simulating and potentially predicting the basic modes of variability, whether arising from sea surface temperature and ocean, land surface moisture, sea ice or other sources that are known to drive global precipitation variability and extremes on seasonal to decadal time scales. Such prediction systems are increasingly necessary to address regional impacts of climate change.

The vast majority of water comes from precipitation—either directly or indirectly through runoff from distant locations. From a climate perspective, it is therefore an imperative to understand the natural variability of precipitation in the system, as well as its susceptibility to change from external forcings. Within GEWEX, the Global Precipitation Climatology Project (GPCP) (Huffman et al. 2009) has been a focus of improving estimates of precipitation. Because of its inherently intermittent nature, it is a major challenge to determine precipitation amounts reliably with a few instantaneous observations of rates such as from available satellites. Improved observations and analysis products related to precipitation and the entire hydrological cycle and their use in evaluating and improving weather, climate and hydrological models are important and tractable over the next 5 to 10 years.

The specific questions that will be addressed over the next 5–10 years include:

- *How well can precipitation be described by various observing systems, and what basic measurement deficiencies and model assumptions determine the uncertainty estimates at various space and time scales?* Despite the significant improvements in many observing systems during the past two decades, the uncertainty in precipitation estimates lies not only in the measurement error itself, but in the space/time interpolation of a naturally discontinuous and intermittent field and/or in the assumptions needed to convert a physical measurement from remote sensing into a precipitation amount. Critical water source regions often reside in complex terrain where sampling issues, remote sensing artifacts and limitations are compounded. The errors are not static but instead depend on the nature of the precipitation itself. Focusing on the large-scale environment responsible for the precipitation therefore holds hope to build not only better rainfall products, but characterizing the uncertainties in a verifiable manner as well. Regional hydroclimate projects provide detailed understanding that translate the large-scale information into usable information for decision makers.

- *How do changes in climate affect the characteristics —(distribution, amount, intensity, frequency, duration, type) of precipitation with particular emphasis on extremes of droughts and floods?* Increased temperatures and associated increases in lower tropospheric water vapor, by making more water vapor available to storms, will very likely increase the intensity of rains and snows, increasing risk of severe floods. Changes in seasonality, shifts in monsoons, changes in snow-melt and runoff, and so on are also part of this question which is elaborated on in the "extremes" science question.
- *How do models become better and how much confidence do we have in global and regional climate predictions of precipitation?* A challenge to the Earth System science community is to develop improved global models. Scientists are beginning to run global climate models at sub-10-km resolution, resolving meso-scale weather including the most extreme tropical storms. These need to be coupled to the ocean and land and will require a new generation of parameterizations that better reflect what processes are and are not resolved in such models. These models can potentially revolutionize our ability to correct long-standing model biases, minimize the need for downscaling and provide predictions of regional impacts and changes in extremes from months to decades ahead. There is great need to quantify the uncertainty in precipitation projections and predictions, especially at regional scales. Starting with improved uncertainties in the climate observations of precipitation, new and improved diagnostics must be developed to test the robustness of model predictions in different regimes. Knowing the uncertainties is critical if predictions of the mean precipitation and its distribution are to be used in local planning efforts.
- *How do changes in the land surface and hydrology influence past and future changes in water availability and security?* While the land surface has small heat capacity and heat moves slowly via conduction, the water flow and storage vary enormously. Land has a wide variety of features, topography, vegetation cover and soil types and consists of a mixture of natural and managed systems. Land plays a vital role in carbon and water cycles, and ecosystems functions and services. Of particular need of attention is use of realistic land surface complexity in hydrological models with all anthropogenic effects included instead of a fictitious natural environment. This includes all aspects of global change including water management, land-use and land-cover change and urbanization, and their feedbacks to the climate system. There is a need to address terrestrial water storage changes and close the water budget over land through exploitation of new datasets, data assimilation, improved physical and biogeochemical understanding and modeling skill across scales, from catchments to regional to global with links to the entire hydrological cycle.
- *How do changes in climate affect terrestrial ecosystems, hydrological processes, water resources and water quality, especially water temperature?* The ecosystem response to climate variability and responsive vegetation must be included but is mostly neglected in today's climate models. Cryospheric changes such as permafrost thawing, changes in the extent, duration and depth of seasonal snowpacks, and changes in mountain glaciers must also be included. How changes in vegetation affect the hydrological cycle and climate in turn are vital. Feedbacks, tipping points and extremes are of particular concern to all economic sectors and regions, globally. The scientific knowledge of water cycle should enhance the evaluation of the vulnerability of water systems, especially to extremes, which is vital for considerations of water and food security and can be used to increase their resilience through good management practices and governance.

- *How can new observations lead to improvements in water management?* Over the last few decades, in situ observations of land surface hydrological variables, such as streamflow, rainfall and snow have generally been in decline. Regional estimation of evapotranspiration remains a significant challenge. At the same time, new observation methods, such as weather radars, flux towers and satellite sensors have led to different types of measurements, and challenges for their incorporation in the hydrological models used for hydrological prediction and water management. One example is soil moisture, which in most models essentially acts as a buffer between the land forcings (mostly precipitation and evapotranspiration) and runoff, and whose characteristics are defined by the internal model parameterizations that control runoff production. Sustained measurements of soil moisture are critically important to understanding, modeling and prediction of the water cycle.
- *How can better climate models contribute to improvements in water management?* Regional precipitation predictions remain a challenge at all timescales from seasonal forecasting out to centennial climate change. However, there are limited regions with forecast skill on seasonal timescales, associated mainly with ENSO, and broad scale, zonally averaged precipitation changes associated with climate change appear to be detectable. The challenge now is to maximize the skill and reliability of predictions of regional rainfall changes on all timescales, for all regions around the world. This requires better understanding and model simulation of the tele-connections and drivers of regional climate such as changes in the oceans and cryosphere that are relevant to regional precipitation. Subsequent improved climate prediction systems and better dissemination of climate prediction information must be developed to deliver the envisioned information and their ultimate benefit to society.

Prospects for advancements are excellent on this Grand Challenge because of new observations already underway and those planned for the ensuing decades and the growing interest in climate predictions on all timescales. Key areas of development include

1. A new Global Precipitation Mission as detailed at http://pmm.nasa.gov/GPM. *"Through improved measurements of precipitation globally, the GPM mission will help to advance our understanding of Earth's water and energy cycle, improve forecasting of extreme events that cause natural hazards and disasters, and extend current capabilities in using accurate and timely information of precipitation to directly benefit society."* The joint US National Aeronautics and Space Administration (NASA)/Japan Aerospace Exploration Agency (JAXA) mission's Core Observatory is scheduled for launch in 2014. Most of the world's major space agencies will participate in this mission through the contribution of constellation satellites to obtain the desired revisit times to roughly 3 h, over the entire Earth.
2. Closely related missions such CloudSat (a NASA mission with components from the Canadian Space Agency to measure clouds and light precipitation) and EarthCARE, a European Space Agency (ESA) mission (http://www.esa.int/esaLP/SEM75KTWLUG_LPearthcare_0.html) to advance our understanding of the role that clouds and aerosols play in the climate system), due for launch late 2015, that will make important contributions to the global precipitation estimates.
3. New satellite sensors such as soil moisture and ocean salinity (SMOS) (an ESA mission to map soil moisture and sea surface salinity), Aquarius (a NASA/Space Agency of Argentina mission to improve sea surface salinity) and future soil moisture active passive (SMAP) data (a NASA mission dedicated to measuring soil

moisture and the freeze/thaw cycle), produce or will produce estimates of near-surface soil moisture that can be used to diagnose or update model estimates, and Gravity Recovery and Climate Experiment (GRACE) (a joint NASA/German Aerospace Center (DLR) mission to map gravity anomalies and thus detect changes in water storage), now provides a nearly decade-long record of total water storage, albeit at coarse spatial resolutions. The GRACE follow on mission is intended to enhance the spatial resolution of such measurements and provide continuity of measurements over the future decade. The planned surface water and ocean topography (SWOT) mission will provide observations of lake and reservoir surface area and levels, from which changes in storage of over 7,000 km^3 of the estimated 8,000 km^3 of reservoir storage globally will be available at 1–2-week intervals. In addition, in situ observations from buoys to Argo floats will help close the water and energy budgets over the oceans.

4. A dedicated snow hydrology mission such as ESA's Cold Regions Hydrology High-Resolution Observatory (CoReH2O) will enable better understanding of the role snow hydrology plays in the regional/global water cycle, especially in mountainous regions of the globe that depend mainly on snow as a source of fresh water for human consumption, food production and industrial activities (e.g., California, Tibetan Plateau, La Plata Basin, etc.).
5. Improvements in communication and data exchange policies help create higher resolution global surface maps of precipitation and soil moisture based upon both local very dense networks of high-resolution measurements as well as surface radar networks where these are available. Significant gains are expected from high-resolution gridded products being developed by GEWEX and other projects based on in situ data as well as inventories of long-term in situ precipitation time series focused on engagement of these data into validation, error estimation and intercomparison efforts. The use of improved error statistics to develop new blending algorithms and fusion techniques capable of bringing together precipitation measurements with distinct error characteristics (e.g., gauges, radar, satellites and models) into a consistent physical framework. Advances in data assimilation techniques allow more precipitation information to be incorporated into Numerical Weather Prediction models.
6. Estimates of fluxes of moisture from surface are improving through the use of flux tower and other observations over land, feeding into improved estimates of evapotranspiration as part of the GEWEX Landflux and ocean flux projects.
7. The production of an Integrated Water and Energy product by the GEWEX data and assessments panel (GDAP) can be used to explore linkages between hydrology and energy variables in the Earth System which in turn provides a much improved basis for evaluating models on all aspects of the water cycle. Advanced diagnostic methods that use the observed variables and their co-variability are used to diagnose not only problems in the model output, but also assess model processes and potential improvements to these processes in order to better represent the observed climate behavior.
8. Incorporate more realistic land surface hydrology into land surface models, including water management, land management and land-use and land-cover change, as well as improved process representation (including cryospheric processes). The envisioned new information is expected to be revolutionary in terms of the management of trans-boundary rivers, but current climate models have no mechanisms for use of this information, since most do not represent the effects of water management.

9. New methods must be developed to address water system vulnerability, particularly to extremes. Quantification of the uncertainty in each of the elements of the global water-balance, including the managed aspects, in a consistent manner is required. Further, there is a need to communicate uncertainties, manage expectations, address the needs of water management under uncertainty (e.g., building resilience).
10. Several other developments in modeling are progressing and advances appear likely. These include development of improved precipitation downscaling methods, particularly for mountainous and arid regions; evaluation of the hydrological dynamics of land surface models with newly available data; prediction of stream temperature as a diagnostic tool in land surface models; improving freshwater fluxes to the world's seas and oceans; and including the known climate feedbacks in off-line land surface change assessments. Water demand models and assessments to land surface and hydrological models must be linked at the global scale.
11. Demonstration of the usefulness of GEWEX, and Global Climate Observing System (GCOS) and WCRP coordinated data products is required along with new tools and provision of derived information for water resources management. The new tools include cross-scale modeling, ensemble hydrological prediction, data assimilation, and data analysis and visualization.

There are multiple benefits, and the results are critically important for society. In addition to greatly improved knowledge about land water resources and ocean salinity, and the causes of their variations, much improved models will allow better predictions of the variability and change on all time scales from seasonal to centennial and from global to continental to basin scales. Predictions with quantified uncertainties provide invaluable information for water managers and users, including decision makers at many levels associated with food and water security. These developments would naturally serve to push WCRP research and development priorities, as users provide feedback on weaknesses and further needs for information.

The information provided also feeds into the development of a "Global Drought Information System". Such a system would provide a user anywhere in the world access to information on our current understanding of drought in that region (e.g., role of ENSO, Pacific Decadal Oscillation, global warming, etc.), the history of drought in that region (with access to various data, time series, indices, etc.), current conditions (monitoring results), the results of near real-time attribution (our understanding of the current conditions) and regularly updated forecasts from months to years in advance (with consistent estimates of uncertainties).

The system would naturally build on the various investments being made in observations (including reanalysis), drought research, and modeling and forecasting capabilities (e.g., the various national and international multi-model ensemble (MME) efforts such as the WMO lead center for long range forecasts: http://www.wmolc.org). The system would be built hand-in-hand with the user community and would have to be sustainable and refreshable as new datasets, better understanding and better modeling capabilities become available. It would naturally serve to push WCRP research and development priorities, as users provide feedback on weaknesses and further needs (analogous to how the weather community is continuously being pushed for better weather forecasts). These are the envisioned products and information to be provided by the network of organizations and centers through Global Framework for Climate Services (GFCS) and Future Earth (WMO 2011; Asrar et al. 2012a).

4.2 Grand Challenge on Water Extremes

How does a warming world affect climate extremes, especially droughts, floods and heat waves, and how do land area processes, in particular, contribute?

A warming world is expected to alter the occurrence and magnitude of extremes such as droughts, heavy rainfalls and floods, as well as the geographic distribution of rain and snow. Such changes are related to an acceleration of the hydrological cycle and circulation changes and include the direct impact of warmer conditions on atmospheric water vapor amounts, rainfall intensity and snow-to-rain occurrence. How well are models able to handle extremes and how can we improve their capability? New improved and updated data sets at high frequency (e.g., hourly) are needed to properly characterize many of these facets of Earth's climate and to allow for assessment against comparable model data sets. New research activities are needed to promote analyses quantifying which changes are consistent with our expectations and how we can best contribute to improving their prediction in a future climate. Confronting models with new observationally based products will lead to new metrics of performance and highlight shortcomings and developmental needs that will focus field experiments, process studies, numerical experimentation and model development. New applications should be developed for improved tracking and warning systems, and assessing changes in risk of drought, floods, river flow, storms, coastal sea-level surges and ocean waves.

There is major concern that the occurrence, character and intensity of extremes will change in the future as the climate changes due to human activities, and this will have enormous consequences for society and the environment. Yet addressing changing extremes satisfactorily is a daunting task, and it will be difficult to keep up with society's expectations. As noted above, huge improvements in near-global spatial and temporal coverage for precipitation, soil moisture and other hydrological variables provide opportunities for new datasets, products, improved models and model applications, making it an opportune time to fully address extremes.

The climate system does not neatly package such extremes. Extremes may be highly localized in time and in space. Drought in one region frequently means heavy precipitation not that far away. The worst extremes are generally compound events which often are consequences of a chain-of-events that may be related at the global scale despite their regional implications. Flooding may be accentuated due to saturated soils from previous storms and/or from snowmelt. Furthermore, coastal flooding may involve storm surge effects, local precipitation and remote snowmelt signals.

Because of its importance, there are many efforts focusing at least in part on extremes within WCRP. One focus is on drought, although there is certainly interest in other hydrometeorological extremes and related issues, such as statistical analyses. WCRP, mainly through CLIVAR, also addresses tropical and extratropical cyclones and associated marine storms as well as extreme sea-level variability and change that is connected to storm surges. GEWEX with its focus on the water cycle and on land surface processes with strong observational capabilities from global to local and with numerous links with society is a natural 'home' for addressing many types of extremes. The question is what is missing and what can be done within GEWEX to move ahead? The main GEWEX focal point is to increase efforts on hydrometeorological extremes including drought, heat waves, cold outbreaks, floods, storms and heavy precipitation events including hazardous winter snowfalls and hail.

The specific questions that will be addressed over the next 5–10 years include

- *What are the short-term, mid-term and strategic requirements for the existing observing systems and datasets, and which observations are needed to accurately quantify trends in the intensity and frequency of extremes on different space/time scales?* Despite a continuous improvement in most observing systems, high-frequency information (e.g., hourly precipitation) required to properly assess extremes is often not made available and shared. New satellite observations and the synthesis of all observations will help and may free up some data. Metrics for quantifying extremes need to be assessed, and new ones should be introduced to improve diagnostics of extremes and scale them to different areas. It is necessary to determine for which regions (national observing systems) the requirements are close to being satisfied and where they are not. There is an urgent need for research on design, development and maintaining optimum observing systems, the regular analysis of their adequacy/inadequacy for future investments in such systems.

- *How can models be improved in their simulation and predictions or projections of the magnitude and frequency of extremes?* Current models have difficulty in simulating the hydrological cycle, and they typically have problems handling the diurnal cycle. Model resolution is insufficient in most cases to simulate many of the extremes of interest, including floods with scales of a few kilometers and even drought whose worst-affected areas are typically in areas only of order a few 100 km or less. Model parameterizations addressing precipitation, convection and clouds are insufficient for accurate simulation and timing of many extreme events. Models need to be confronted with the new observational products in innovative analyses and with new diagnostics and metrics of performance. This includes numerical weather prediction and climate models. There are conceptual difficulties in validating model results against observations, first of all associated with (but not limited to) co-location in space and grid cell data versus point measurements. Many observational products are developed independent of models so that gridding projections and associated error characteristics are often different from model-derived data products thus making their direct inter-comparisons very difficult if not impossible. Focused investments by space agencies (e.g., ESA and NASA) to make the observational products consistent and inter-comparable is quite timely. Such efforts facilitate research on observations and make inter-comparisons with models much easier and enhance the use of observations by the modeling community.
- *How can the phenomena responsible for extremes be better simulated in models?* Many phenomena that are responsible for extremes are not well simulated in models; some because of resolution (such as tropical storms and highly localized precipitation events), but also others that are resolved (such as blocking anticyclones). As well as statistical analyses, studies should examine the phenomena responsible for extremes, whether and how well they are depicted in models, and how to overcome incompatible resolution requirements. Developmental needs should be used to focus field programs, process studies and numerical experimentation.
- *How can we promote development of applications for improved tracking and warning systems arising from extremes?* It is essential to develop ways to better assess changes in risk of drought, floods, river flow, storms, coastal sea-level surges and ocean waves. Such information has the greatest benefit to society for management of risks associated with these events to reduce their adverse impacts. In most cases, such applications will be done in conjunction with the CLIVAR and CliC projects and made available through networks sponsored by GFCS and other regional climate information systems.

Prospects for advancements are excellent on this question because of new observations, research, modeling and prediction activities already underway and planned. A number of specific, short and near-term activities are envisioned that will enable progress on this Grand Challenge. Key areas of development include

1. Utilization of the new global and regional datasets outlined above and from improved data assessment (within the GEWEX Data and Assessments Panel) to better characterize extremes on different spatial scales and, with the WCRP Modeling Council, promote evaluations of model results, potentially with one or more workshops in 2014–15.
2. Ensure strong involvement in the Global Drought Information System. This focuses on one particular type of extreme but the effort may also act as a prototype for dealing with all types of extremes in the future. In particular, GEWEX and CLIVAR will develop trackable actions on monitoring and quantification of the global distribution of droughts and their trends using observational information, model development, land area factors governing drought and societal interactions.
3. Facilitate a number of inter-comparison projects aimed at comparison of characteristics of extremes in different data sets (in situ, reanalyses and satellites), and revealed by different models.
4. Initiate a parallel activity centered on capabilities of statistical methodologies to deal with the complexity of extremes, including their clustering in space and time and with sparse and regionally unevenly distributed data.
5. Initiate multi-methods activities and encourage documentation and data inventory centered on a few mega-extreme events (for example, catastrophic flooding, droughts, unusual storm patterns) to enable further analysis with observations and models, ensure that all their aspects are comprehensively addressed, and with special attention on assessing their likelihood in the future. This activity may be facilitated by bringing teams together and should build in flexibility with adaptable approaches as one learns by doing. It has the advantage that the results are immediately relevant.
6. Examine cold season extremes such as snowstorms, rain-on-snow episodes, freezing precipitation and prolonged cold weather events with CliC and other international and national research programs/projects.

There are multiple benefits, and the results are important for society. Drought has devastating consequences whenever and wherever it occurs. Water resources can be strained, and adverse effects occur in agriculture. Heat waves are often but not always linked with drought. Health effects can be profound. Prolonged cold weather episodes are a critical feature of mid- and sub-polar latitudes in winter. They are disruptive and costly. Isolated extreme rainfalls as well as continuous periods of heavy and moderate precipitation occur everywhere with numerous impacts including flooding, devastation of ecosystems and havoc in urban regions. Storms in different parts of the world are the means by which precipitation, often linked with strong winds, occurs, and changes in their paths, intensity and frequency have enormous consequences, sometimes devastating. Warming conditions imply that regions accustomed to receiving snow should experience more rain, and changing times of runoff and peak stream flow, with large consequences for ecosystems, hydrological risks and water resources.

These examples highlight the importance of progress in the area of climate extremes, both in terms of their observations and analysis, and in terms of improved modeling and prediction. In summary, WCRP through GEWEX, CLIVAR and CliC and its seamless modeling framework across space and time scales (e.g., Working Groups on Coupled Modeling and Numerical Experimentation (WGCM and WGNE)) will focus great attention on extremes, including research on detection and attributions of causes and consequences of such events over the next 5–10 years. By doing so, it will be carrying out its very natural role of addressing the estimation, modeling, understanding and future projection of extremes with a particular focus over land.

5 Conclusions

The successful implementation of WCRP Grand Challenges and associated science questions described in this chapter depend significantly on GEWEX Imperatives: observations and data sets, their analysis, process studies, model development and exploitation, applications, technology transfer to operationalize results, and research capacity development and training of the next generation of scientists. They involve all of the GEWEX panels and will benefit greatly from strong interactions with other WCRP projects such as CLIVAR and CliC and other sister global change research programs such as the International Geosphere-Biosphere Programme (IGBP), International Human Dimensions (IHDP), etc.

Closure of the observed regional and global water budget over the past decade has progressed significantly, but remains a major challenge. Thus it continues to be a science imperative for the research community to better observe and understand all aspects of the water cycle in order to improve models that can predict reliably its future variability and change as a major source of information for decision makers for water resources, food production and management of risks associated with extreme events. Many potential products could be invaluable to water resource managers on several time horizons, extending well beyond the 1-week weather scale to seasonal, inter-annual and decadal predictions, and climate change projections.

Acknowledgments The research of Trenberth is partially sponsored by NASA under grant NNX09AH89G. We thank Howard Wheater and many other WCRP colleagues for discussions, and especially we thank GEWEX scientists who contributed to the GSQs. The National Center for Atmospheric Research is sponsored by the National Science Foundation.

References

Asrar GR, Ryabinin V, Detemmerman V (2012a) Climate science and services: providing climate information for adaptation, sustainable development and risk management. Current Opinion Environ Sust 41:1–13

Asrar GR, A Busalacchi A, Hurrell J (2012b) WCRP Open science conference. AGU Eos Trans 93(12)

Dai A (2011) Drought under global warming: a review. Wiley Interdisciplinary Rev: Climate Change 2:45–65

Dai A, Qian T, Trenberth KE, Milliman JD (2009) Changes in continental freshwater discharge from 1949–2004. J Clim 22:2773–2791

Huffman GJ, Adler RF, Bolvin DT, Gu G (2009) Improving the global precipitation record: GPCP version 2.1. Geophys Res Lett 36:L17808. doi:10.1029/2009GL040000

IPCC (Intergovernmental Panel on Climate Change) (2007) Climate change 2007. The physical science basis. In: Solomon S et al (Eds.) Cambridge University Press: New York, p 996

IPCC (2012) Managing the risks of extreme events and disasters to advance climate change adaptation. In: Field CB et al (eds) Cambridge University Press: New York, p 582

Nerem RS, Chambers D, Choe C, Mitchum GT (2010) Estimating mean sea level change from the TOPEX and Jason altimeter missions. Mar Geodesy 33(supp 1):435

Oki T, Kanae S (2006) Global hydrological cycles and world water resources. Science 313:1068–1072

Sahoo AK, Pan M, Troy TJ, Vinukollu RK, Sheffield J, Wood EF (2011) Reconciling the global terrestrial water budget using satellite remote sensing. Remote Sens Environ 115:1850–1865

Trenberth KE (2011) Changes in precipitation with climate change. Clim Res 47:123–138. doi: 10.3354/cr00953

Trenberth KE (2012) Framing the way to relate climate extremes to climate change. Clim Change 115:283–290. doi:10.1007/s10584-012-0441-5

Trenberth KE, Shea DJ (2005) Relationships between precipitation and surface temperature. Geophys Res Lett 32:L14703. doi:10.1029/2005GL022760

Trenberth KE, Dai A, Rasmussen RM, Parsons DB (2003) The changing character of precipitation. Bull Amer Meteor Soc 84:1205–1217

Trenberth KE, Smith L, Qian T, Dai A, Fasullo J (2007a) Estimates of the global water budget and its annual cycle using observational and model data. J Hydrometeor 8:758–769

Trenberth KE et al (2007b) Observations: surface and atmospheric climate change. In: Solomon S et al (eds) Climate change 2007. The physical science basis. New York, Cambridge University Press, pp 235–336

Trenberth KE, Fasullo JT, Mackaro J (2011) Atmospheric moisture transports from ocean to land and global energy flows in reanalyses. J Clim 24:4907–4924. doi:10.1175/2011JCLI4171.1

Vinukollu RK, Meynadier R, Sheffield J, Wood EF (2011) Multi-model, multi-sensor estimates of global evapotranspiration: climatology, uncertainties and trends. Hydrol Process 25:3993–4010

World Meteorological Organization (2011) Climate knowledge for action: a global framework for climate services—Empowering the most vulnerable, WMO/TD-No. 1065, p 240

DOI 10.1007/978-94-017-8789-5_4
Reprinted from *Surveys in Geophysics* Journal, DOI 10.1007/s10712-012-9213-z

Physically Consistent Responses of the Global Atmospheric Hydrological Cycle in Models and Observations

Richard P. Allan · Chunlei Liu · Matthias Zahn · David A. Lavers · Evgenios Koukouvagias · Alejandro Bodas-Salcedo

Received: 30 August 2012 / Accepted: 6 November 2012 / Published online: 10 January 2013

Abstract Robust and physically understandable responses of the global atmospheric water cycle to a warming climate are presented. By considering interannual responses to changes in surface temperature (*T*), observations and AMIP5 simulations agree on an increase in column integrated water vapor at the rate 7 %/K (in line with the Clausius–Clapeyron equation) and of precipitation at the rate 2–3 %/K (in line with energetic constraints). Using simple and complex climate models, we demonstrate that radiative forcing by greenhouse gases is currently suppressing global precipitation (*P*) at ~ -0.15 %/decade. Along with natural variability, this can explain why observed trends in global *P* over the period 1988–2008 are close to zero. Regional responses in the global water cycle are strongly constrained by changes in moisture fluxes. Model simulations show an increased moisture flux into the tropical wet region at 900 hPa and an enhanced outflow (of smaller magnitude) at around 600 hPa with warming. Moisture transport explains an increase in *P* in the wet tropical regions and small or negative changes in the dry regions of the subtropics in CMIP5 simulations of a warming climate. For AMIP5 simulations and satellite observations, the heaviest 5-day rainfall totals increase in intensity at ~ 15 %/K over the ocean with reductions at all percentiles over land. The climate change response in CMIP5 simulations shows consistent increases in *P* over ocean and land for the highest intensities, close to the Clausius–Clapeyron scaling of 7 %/K, while *P* declines for the lowest percentiles, indicating that interannual variability over land may not be a good proxy for climate change. The local changes in precipitation and its extremes are highly dependent upon small shifts in the large-scale atmospheric circulation and regional feedbacks.

Keywords Precipitation · Water · Climate models · Satellite data · Global change

R. P. Allan (✉) · C. Liu · M. Zahn · D. A. Lavers · E. Koukouvagias
Department of Meteorology, National Centre for Atmospheric Sciences,
University of Reading, Reading, UK
e-mail: r.p.allan@reading.ac.uk

A. Bodas-Salcedo
Met Office Hadley Centre, Exeter, UK

1 Introduction

Recent occurrence of extreme weather across the globe (including flooding and drought) relates to shifts in atmospheric and oceanic circulation patterns (e.g., the mid-latitude jet stream or El Niño Southern Oscillation, ENSO) yet heightens concern upon the potential for human influence on and vulnerability to such events (e.g., Peterson et al. 2012). The frequency of damaging flooding and drought are projected to increase in the future based upon detailed computer simulations of the climate system (Meehl et al. 2007) and backed up by basic physics (Held and Soden 2006; O'Gorman et al. 2012); it is also vital that projected responses of the water cycle are verified, where possible, by careful use of homogeneous, well-characterized observations (Trenberth 2011).

The magnitude and rate of change in the regional hydrological cycle determine the impacts suffered by infrastructure, agriculture and health. The distribution of rainfall (in time and space), in particular for the extremes, is crucial in determining damage from particular events. Time and space means of hydrological quantities may reflect aspects of these changes, in particular for the amount of water potentially available for a region. However, the relevance of changes in global mean quantities to local impacts is unclear. Yet, without appreciation for the driving mechanisms at the largest spatial and temporal scales, the robust nature of local projections is questionable at best. In this paper, our aim is to identify the most important, robust and physically understandable responses in the atmospheric hydrological cycle, exploiting climate model simulations and confronting these with the globally available observational record.

2 Constraints upon Global Mean Precipitation Responses

The most robust climatic response to increasing temperatures is a rise in mean water vapor near to Earth's surface at ~7 %/K, in line with the Clausius–Clapeyron equation (see Sect. 2.3). Although changes in moisture are an important constraint upon regional changes in precipitation and its extremes (discussed in Sect. 3), it has been known for some time that the total amount of precipitation (P) increases with warming at a slower rate than water vapor (~2–3 %/K), responding instead to a changing heat balance of the atmosphere (Manabe and Wetherald 1975; Mitchell et al. 1987; Allen and Ingram 2002). The primary physical basis for this is that a warming atmosphere radiates energy away more effectively, particularly to the surface (e.g., Allan 2006; Stephens and Ellis 2008; Prata 2008).

To maintain energy balance in the atmosphere, this additional atmospheric net radiative cooling to space and to the surface (ΔQ_{atm}) is primarily compensated for by extra latent heating via precipitation ($L\Delta P$) with changes in sensible heating of the atmosphere by the surface, ΔSH, playing a more minor role (but see Lu and Cai (2009)):

$$L\Delta P = \Delta Q_{\mathrm{atm}} - \Delta \mathrm{SH}, \quad (1)$$

Surface evaporation (E) is similarly constrained by energy balance; Richter and Xie (2008) find that climate models simulate small adjustments to boundary layer temperature, humidity and wind speed that cause E to increase with warming below the rate expected from the Clausius–Clapeyron equation (Lu and Cai 2009) and almost identically to the P changes (as expected from the trivial moisture holding capacity of the atmosphere relative to the moisture fluxes).

As discussed by O'Gorman et al (2012), radiative cooling above the lifting condensation level is in fact more directly related to precipitation since, when heating is applied within the boundary layer, adjustments through SH do become important (e.g., Ming et al. 2010). For simplicity, we here assume that the balance between atmospheric radiative cooling and latent heating dominates. However, this simple balance between ΔP and $\Delta Q_{\mathrm{atm}}(\Delta T)$ is complicated by a direct influence on atmospheric radiative cooling by the radiative forcings responsible for determining temperature response in the first place (Andrews et al. 2010; O'Gorman et al. 2012).

Global mean precipitation response, ΔP, is determined by a "slow" component set by the global mean surface temperature change, ΔT, (denoted slow since it takes a long time to heat up the oceans due to their dominating heat capacity) and a "fast" component ($f\Delta F$) in which the atmosphere (with its small heat capacity relative to the ocean) rapidly adjusts to changes in top of atmosphere radiative forcing, ΔF (for simplicity, here defined as the downward radiative heating into the top of the atmosphere), that is independent of ΔT (Allen and Ingram 2002; Bala et al. 2010; Andrews et al. 2010; O'Gorman et al. 2012):

$$L\Delta P \sim k\Delta T - f\Delta F. \tag{2}$$

In (2), $L = 2.5 \times 10^6$ J kg^{-1} is the latent heat of vaporization and $k \sim 2$ W m^{-2} K^{-1} is the response of atmospheric radiative cooling to surface temperature, $\partial Q_{\mathrm{atm}}/\partial T$ (e.g., Allan 2006; Lambert and Webb 2008; Andrews et al. 2010), set by the atmospheric temperature and humidity lapse rates (e.g., moist adiabatic lapse rate with near-constant mean relative humidity is a reasonable approximation).

The fast scaling parameter, $f = \Delta F_{\mathrm{atm}}/\Delta F$, is the instantaneous radiative forcing experienced by the atmosphere, $\Delta F_{\mathrm{atm}} = -\Delta Q_{\mathrm{atm}}(\Delta F)$, normalized by the top of atmosphere radiative forcing (ΔF) and is specific to the nature of each radiative forcing component (Andrews et al. 2010; Ming et al. 2010). For example, increases in atmospheric CO_2 concentrations produce an instantaneous increase in radiative forcing at the top of the atmosphere that is considerably larger than the increase in instantaneous (downward) radiative forcing at the surface, ΔF_{sfc} (e.g., Ramanathan 1981; Allan 2006), where $\Delta F = \Delta F_{\mathrm{atm}} + \Delta F_{\mathrm{sfc}}$. This causes a direct reduction in ΔP through the last term in (2), since $f_{CO_2} \sim 0.8$ (Andrews et al. 2010) and a slower increase in ΔP through the resulting rises in ΔT brought about by the positive radiative forcing. The timescale for $k\Delta T$ is increased for smaller ocean heat uptake and a more positive overall climate feedback (see also McInerney and Moyer 2012). The interaction between these two effects is fundamental in determining the transient response of ΔP to ΔF, or hydrological sensitivity (Ming et al. 2010).

2.1 Simple Model of Global Precipitation

To illustrate the global constraint upon ΔP, a simple zero-dimensional energy budget model is employed, based upon the approach of Hansen et al. (1981). A mixed-layer ocean temperature perturbation ΔT_{m} is computed as

$$\frac{\mathrm{d}\Delta T_{\mathrm{m}}}{\mathrm{d}t} = \frac{1}{C_{\mathrm{m}}}(\Delta F - Y\Delta T_{\mathrm{m}} - D), \tag{3}$$

where $C_{\mathrm{m}} = 4.218 \times 10^8$ J K^{-1} m^{-2} is the ocean mixed-layer heat capacity, $Y = 1.3$ W m^{-2} K^{-1} is the climate feedback parameter, and

$$D = c(\Delta T_{\mathrm{m}} - \Delta T_{\mathrm{D}})/d, \tag{4}$$

is the diffusion of energy into the deep ocean ($d = 500$ m, $c = 421.8$ W K^{-1} m^{-1}) where deep ocean temperature, ΔT_{D} is determined by $\mathrm{d}\Delta T_{\mathrm{D}}/\mathrm{d}t = D/C_{\mathrm{D}}$, where $C_{\mathrm{D}} = 3.7962 \times$

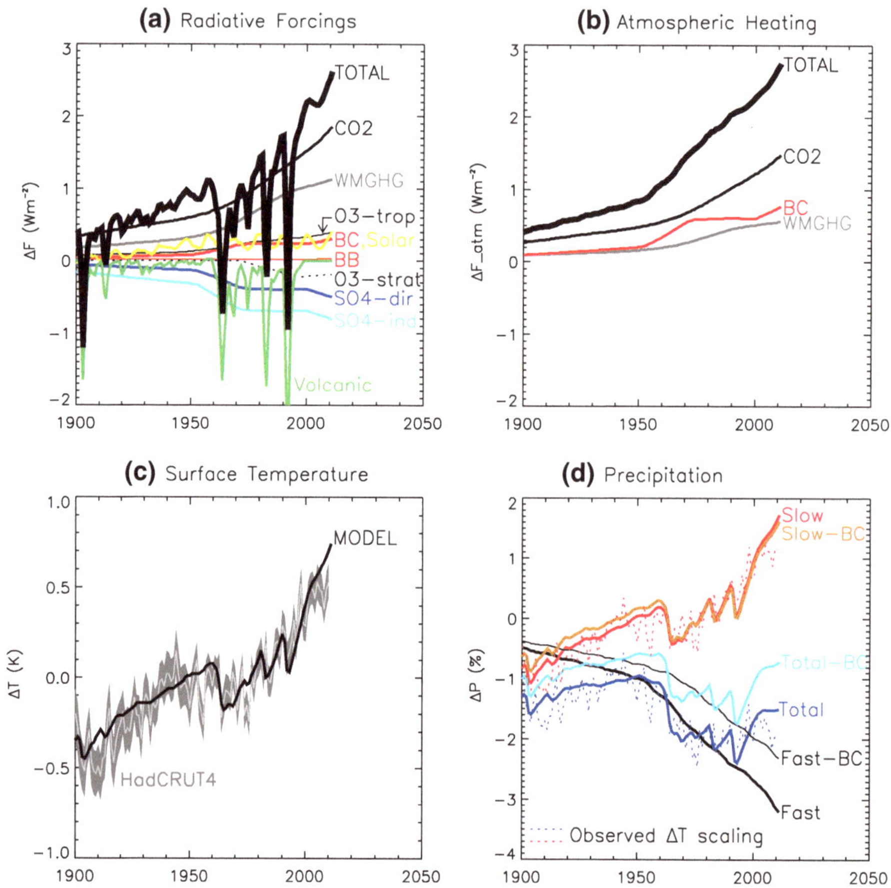

Fig. 1 Changes in **a** radiative forcings (ΔF), **b** atmospheric radiative heating (ΔF_{atm}) due to the top of atmosphere radiative forcings using scaling factors from Andrews et al. (2010), **c** simulated temperature response using a simple energy balance model compared with HadCRUT4 observations (95 % uncertainty range) and **d** precipitation response (ΔP) using the simple energy balance model decomposed into fast and slow components ("-BC" denotes simulations without the effects of Black Carbon aerosol). In **d**, observed slow response is also estimated using the HadCRUT4 ΔT; this is combined with the model fast responses to provide an estimated total scaling (*dotted lines*)

10^9 J $K^{-1}m^{-2}$ is the deep ocean heat capacity. Thus, ΔT_m is determined by the radiative forcing, the restoring flux relating to feedbacks, and diffusion of energy into the deep ocean.

Observed ΔF is prescribed over the period 1850–2010, based upon the Model for the Assessment of Greenhouse gas-Induced Climate Change (MAGICC) without carbon cycle for anthropogenic forcings,[1] and natural forcings are based upon the Goddard Institute for Space Studies[2] (e.g., Hansen et al. 2007) with simple extrapolation of some fields from 1990 onward (Fig. 1a). The model is initialized at 1850 as $\Delta T_m = \Delta T_D = 0$. Since the

[1] http://www.cgd.ucar.edu/cas/wigley/magicc/.

[2] http://data.giss.nasa.gov/modelforce/.

Table 1 Prescribed values of atmospheric forcing scaling parameter $f = \Delta F_{atm}/\Delta F$

Forcing	CO_2	Other WMGHG	O_3 trop.	O_3 strat.	SO_4 (all)	BB	BC	Solar
f	0.8	0.5	−0.3	0.0	0.0	−0.9	2.5	0.2

Well-Mixed Greenhouse Gases (WMGHG) includes CH_4, N_2O and CFCs; SO_4 includes all sulfate aerosol forcings (direct, indirect and volcanic). *BB* biomass burning aerosol, *BC* black carbon aerosol

purpose of this analysis is illustrative only, inaccuracies in the prescribed radiative forcings and other model parameters are not considered a serious impediment.

To compute ΔP using our simple model, we combine (2) with (3–4) where f is prescribed as in Table 1 and assuming that $\Delta T \equiv \Delta T_m$. Thus, ΔF_{atm} is computed from ΔF at each 1 year time step (Fig. 1b); forcing agents that do not interact with the atmosphere (e.g., purely scattering SO_4 aerosol) are assumed not to influence the fast ΔP responses. The model simulated ΔT is depicted in Fig. 1c and compares favorably with observed estimates from the HadCRUT4 observations (Morice et al. 2012). The fast and slow (T-dependent) components of ΔP are computed from (2) using the model simulated ΔT and ΔF_{atm} in Fig. 1d. Also shown are estimates neglecting the fast response from Black Carbon aerosol (-BC) and prescribing the HadCRUT4 ΔT for the slow component (dotted lines).

The slow ΔP component rises with ΔT at the rate 2 W m^{-2} K^{-1}, as prescribed by the parameter k, and is relatively insensitive to the inclusion of BC aerosol or error in simulated ΔT (although the recent hiatus in global temperature rises since ~2000 are not captured). The fast component leads to reduced ΔP, which is enhanced by the inclusion of BC (additional atmospheric absorption of radiative energy leading to atmospheric stabilization and declining global ΔP). Overall, the total ΔP response shows little global trend, consistent with recent observationally based estimates (Adler et al. 2008), although responses to volcanic forcings can introduce apparent trends over decadal timescales. Inaccuracies in the model parameters, including the scaling factors, k and f, will reduce the realism of the simple model estimates. It is, therefore, instructive to progress from the simple model on to the considerably more detailed depiction of global precipitation changes simulated by coupled climate model simulations.

2.2 Transient Response in Global Precipitation in CMIP5 Models

The simulated transient climate response is illustrated in Fig. 2 which shows ΔP simulated by fully coupled climate models (see details in Table 2) as part of the Coupled Model Intercomparison Project-phase 5 (CMIP5; Taylor et al. 2011). Fig. 2a shows increases in ΔP over the twenty-first century, rising by around 3−11 % over the period 2000−2100. The RCP8.5 (Representative Concentration Pathways) scenario (an emissions pathway leading to a radiative forcing of 8.5 W m^{-2} by 2100) simulates a larger response than the more mitigating RCP4.5 scenario (as illustrated by the thick ensemble mean lines) although there is considerable inter-ensemble spread. This is partly explained by the larger ΔT response simulated by RCP8.5 as illustrated in Fig. 2b. A simple linear fit between ΔP and ΔT produces a sensitivity of ~1 %/K; this is smaller than implied by the scaling parameter k in (2) since rising greenhouse gas concentrations are muting the overall ΔP response to warming by heating the atmosphere.

For each scenario, there is also a large inter-model spread, since models with higher climate sensitivities (and/or slower rates of ocean heat uptake) tend to warm more rapidly

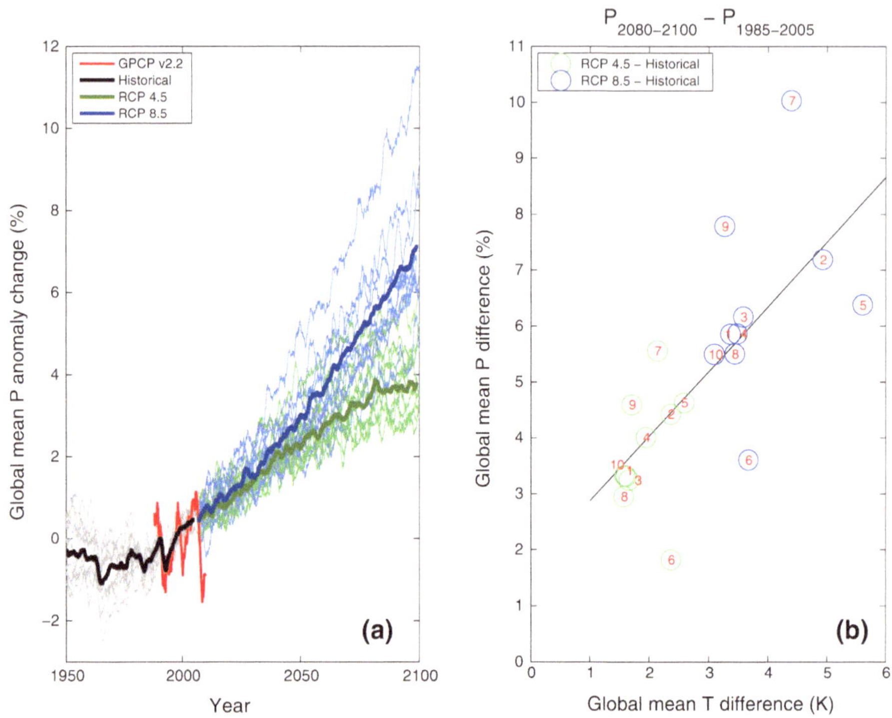

Fig. 2 **a** Changes in global mean precipitation (relative to 1988–2005) for CMIP5 climate simulations (RCP4.5 and RCP8.5 future projections relative to the parent simulation from the historical simulations, models 1–10 in Table 2). Observationally based GPCP precipitation anomalies are plotted for comparison. Twenty-four month running means are applied. **b** 2080–2100 minus 1985–2005 global mean precipitation change plotted relative to changes in global mean surface temperature for RCP4.5-historical and RCP8.5-historical simulations. Model numbers in Table 2 are displayed, and a line of best fit to these values is shown in *black*

and therefore increase P through the slow response term in (2). However, the relationship between ΔT and ΔP is not robust, particularly for the RCP8.5 scenario; this may be explained by the fact that the RCP8.5 scenario is far from equilibrium by 2100 (e.g., McInerney and Moyer 2012), while the model simulations in the RCP4.5 scenario are showing signs of a levelling off in ΔP (as $\Delta F \rightarrow 4.5\,\mathrm{W\,m^{-2}}$ after $\sim$2060). The INMCM4 model simulates a muted $\Delta P/\Delta T$ response relative to other models, while the IPSL-CM5A-LR and MRI-CGCM3 models simulate a slightly larger $\Delta P/\Delta T$ response relative to the other models (see Table 2 for definition of climate models). As noted, the actual ΔP is also related to the direct influence of the radiative forcings upon atmospheric radiative cooling (Andrews et al. 2010) while the timescale for change in ΔT is lengthened for models with weaker ocean heat uptake and more positive feedback (e.g., Hansen et al. 2011). The contribution of natural variability, model diversity and scenario uncertainty to precipitation projections is discussed further, in the context of the older CMIP3 model simulations, by Hawkins and Sutton (2011)

To investigate the direct influence of radiative forcing on transient climate change, a present-day AMIP5 simulation (observed sea surface temperature, SST, and sea ice distribution are prescribed over the period 1979–2008) of the atmospheric component of the

Table 2 Details of CMIP5 climate models simulations employed

Modelling centre	Model	References
(1) Beijing Climate Center	BCC-CSM1-1	Wu et al. (2012)
(2) Canadian Centre for Climate Modelling and Analysis	CanESM2 CanAM4	Arora et al. (2011)
(3) National Center for Atmospheric Research, USA	CCSM4	Gent et al. (2011)
(4) Centre National de Recherches Meteorologiques, France	CNRM-CM5	Voldoire et al. (2012)
(5) Met Office Hadley Centre, UK	HadGEM2-ES HadGEM2-A	Collins et al. (2011)
(6) Institute for Numerical Mathematics, Russia	INMCM4	Volodin et al. (2010)
(7) Institut Pierre Simon Laplace, France	IPSL-CM5A-LR	Hourdin et al. (2012)
(8) Max Planck Institute for Meteorology, Germany	MPI-ESM-LR	Raddatz et al. (2007)
(9) Meteorological Research Institute, Japan	MRI-CGCM3	Yukimoto et al. (2012)
(10) Norwegian Climate Centre	NorESM1-M	Zhang et al. (2012)
(11) Model for Interdisciplinary Research on Climate, Japan	MIROC5	Watanabe et al. (2010)

Hadley Centre Global Environment Model version 2 (HadGEM2) model (Collins et al. 2011) is conducted. A standard Atmospheric Models Intercomparison Project (AMIP5) control run is used (CTL), in which realistic radiative forcings are prescribed, and compared with an identical simulation but with greenhouse gases fixed at their 1978 levels (fGHG).

Figure 3a shows declining CTL minus fGHG global mean clear-sky outgoing longwave radiation, OLRc ($dOLRc/dt = -0.37$ W m^{-2}/dec, $r = -0.79$). This is explained by the rising greenhouse gas concentrations in CTL (e.g., Allan 2006; Chung and Soden 2010) which reduce the longwave radiative cooling of the atmosphere to space. A decreasing trend in total (all-sky) atmospheric net radiative cooling to space and to the surface (Fig. 3b) is less marked ($dQ_{atm}/dt = -0.13$ W m^{-2}/dec, $r = -0.32$) since the presence of high-altitude cloud masks the enhanced greenhouse effect of rising CO_2 concentrations in CTL. There are also substantial differences associated with differing internal variability between experiments; this also explains most of the differences in P between CTL and fGHG (Fig. 3c), but a declining trend is also discernible ($dP/dt = -0.15$ %/dec ~ -0.14 W m^{-2}/dec, $r = -0.25$). The clear correlation between ΔQ_{atm} and ΔP ($r = 0.66$) is physically reasonable and primarily explained by internal variability differences between simulations with only a small, yet detectable, fraction relating to the CO_2 trend.

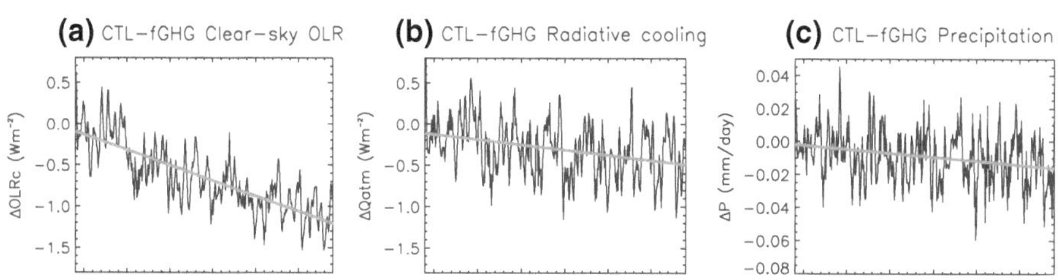

Fig. 3 Global mean differences in **a** clear-sky outgoing longwave radiation (OLR), **b** net all-sky atmospheric radiative cooling and **c** precipitation for HadGEM2-A AMIP5 simulations with observed greenhouse gases (CTL) and fixed greenhouse gases (fGHG) with least squares fit trend lines. A 3-month box-car average is applied to the time series

This simple experiment illustrates that the direct radiative effect of rising greenhouse gas concentrations is muting current trends in global P but that the influences may be difficult to isolate from natural variability in the present-day climate. We now examine in more detail how well simulations can capture current changes in the global hydrological cycle.

2.3 Current Changes in the Global Atmospheric Water Cycle

Observationally based estimates of global precipitation changes in Fig. 2a from the Global Precipitation Climatology Project (GPCP), which combines infrared and microwave satellite data over the ocean with infrared and rain-gauge observations over land (Huffman et al. 2009), show substantial variability but little discernible trend, consistent with the simple model results described in Fig. 1. Current changes in the global atmospheric water cycle are now examined in more detail.

Figure 4a–b shows that, on interannual timescales, global mean column integrated water vapor (W) is strongly coupled with T both in model simulations (here AMIP5 simulations with specified observed SST and sea ice) and in observations (combined microwave satellite data[3] over ocean between 50°S–50°N and ERA Interim reanalysis simulations[4] elsewhere) and also from the HadCRUH surface specific humidity (q) dataset (Willett et al. 2008). Table 3 shows that W rises at ~7 %/K when considering interannual variability over the period 1988–2008, similar to the Clausius–Clapeyron rate as previously demonstrated (Wentz and Schabel 2000; Willett et al. 2008). Correlation between SSM/I-ERA Interim W and independent HadCRUH q over the period 1989–2003 is remarkable ($r = 0.86$).

Reanalyses remain unable to adequately simulate global changes in the hydrological cycle (e.g., Trenberth et al. 2011; John et al. 2009) as illustrated by the changes in W and P in Fig. 4b–c. This primarily relates to changes in the observing system over the ocean (Dee et al. 2011). Nevertheless, changes in T globally and W over land are thought to be reasonable (Simmons et al. 2010) as are the simulated interannual changes in top of atmosphere net radiation (Loeb et al. 2012) and within the atmosphere as depicted in Fig. 4d. The satellite-based ISCCP (Zhang et al. 2004) and Surface Radiation Budget (SRB) (Stackhouse Jr. et al. 2011) estimates of Q_{atm} suffer from inaccurate changes in surface fluxes, for example clear-sky surface longwave fluxes over land.

While coupled model simulations in Fig. 2a appear to underestimate the observed GPCP variability, this is not the case for AMIP simulations in Fig. 4c. There is good agreement for the positive interannual dP/dT relationship between GPCP and AMIP5 over the period 1988–2008 (Table 3) although the robust nature of changes in precipitation from satellite data over the ocean remains questionable prior to 1995 (Liu et al. 2012). It is nevertheless encouraging that a global relationship between Q_{atm} from ERA Interim and P from GPCP emerges, broadly consistent with model simulations and close to unity (Table 3) as anticipated from physical grounds (O'Gorman et al. 2012).

[3] We use Scanning Multi-channel Microwave Radiometer (SMMR) and the Special Sensor Microwave Imager (SSM/I) data that provide ocean retrievals of W and also ocean retrievals of P from the SSM/I instruments on the F08/F11/F13 series of Defense Meteorological Satellite Program platforms (e.g., Wentz and Schabel 2000; John et al. 2009).

[4] Reanalyses combine weather forecast model simulations with available observations using data assimilation to provide a 3-dimensional depiction of atmospheric properties, usually every 6 hours. The European Centre for Medium-range Weather Forecasts (ECMWF) Interim reanalysis, ERA Interim, is a state of the art reanalysis system covering the period 1979 to present, which for example assimilates vertical temperature and humidity information from satellite and conventional observations. A detailed description is given by Dee et al. (2011).

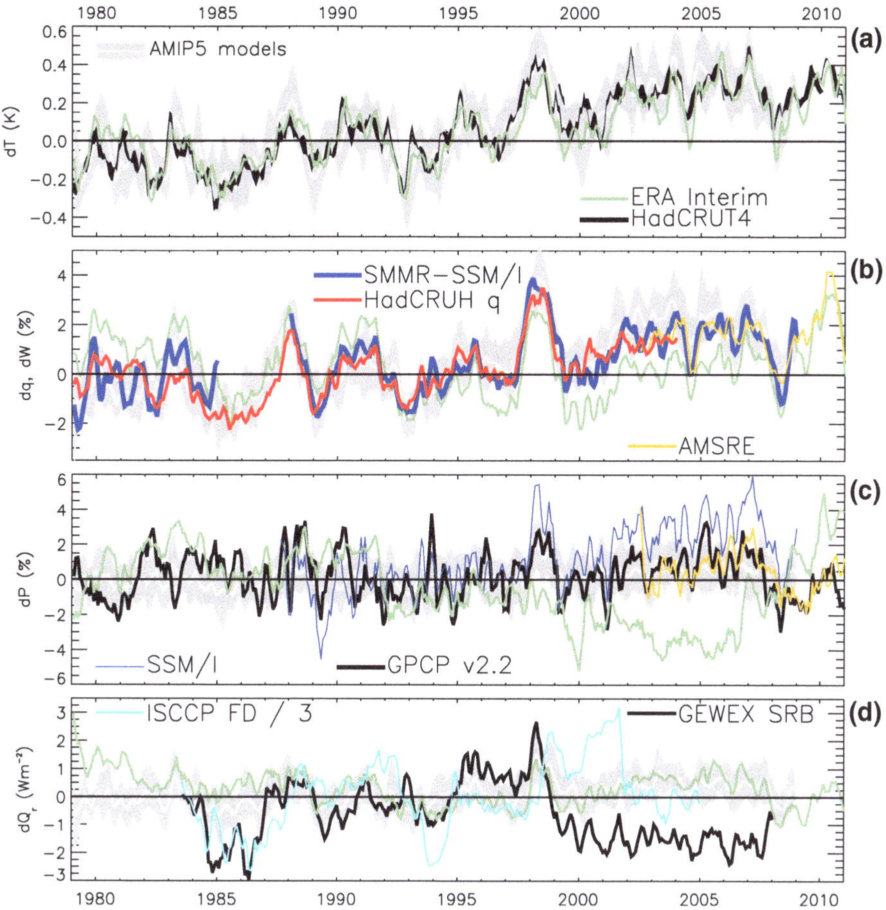

Fig. 4 Deseasonalized anomalies in **a** surface air temperature, **b** column integrated water vapor or specific humidity, **c** precipitation and **d** atmospheric net radiative cooling for AMIP5 simulations (±1 inter-model standard deviation, models 2 and 4–11 in Table 2) and observationally based estimates. All time series anomalies are relative to the 1988–1996 base period (models are adjusted relative to the ensemble mean) apart from SMMR (1983/1984 base period) and Advanced Microwave Scanning Radiometer-EOS (AMSR-E) (mean adjusted to agree with mean SSM/I *W* and GPCP *P* over the period 2003–2008); 3-month averages of anomalies are plotted. In **a**, the 95 % uncertainty range is shown for HadCRUT4 observations. In **b** and **c**, the SMMR and SSM/I record uses the satellite microwave record over the 50°S–50°N oceans and ERA Interim *W* or GPCP *P* data elsewhere. In **d**, the International Satellite Cloud Climatology Project - Flux Dataset (ISCCP-FD) atmospheric radiative cooling anomalies are reduced by a factor of 3

Trends documented in Table 3 cover too short a record (1988–2008) to be physically meaningful yet provide a useful comparison between AMIP5 simulations and observations. Rapid warming from 1993–2002 (~0.4 K), after the eruption of Mt. Pinatubo in 1991, is followed by static global mean *T*; it has been suggested that a shift in ocean circulation may have contributed to this variability (e.g., Merrifield 2011; Gu and Adler 2012). A rise in *W* over 1988–2008 of around 1 %/decade from SSM/I–ERA Interim and simulations is equivalent to ~5–6 %/K given the trend in *T*, slightly lower than anticipated from Clausius–Clapeyron. Trends in global *P* (Table 3) are not statistically significant over the period, consistent with expectations from the simple model in Sect. 2.1 However,

Table 3 Relationship between global water vapor, temperature, precipitation and radiative cooling and their trends based on linear regression over the 1988–2008 deseasonalized time series

Variables	*y*	*x*	dy/d*x*	*r*
d*W*/d*T*	SSM/I-ERAINT	HadCRUT4	6.42 ± 0.46 %/K	0.84*#
	AMIP5	AMIP5	7.14 ± 0.24 %/K	0.96*#
d*P*/d*T*	GPCP	HadCRUT4	2.83 ± 0.92 %/K	0.32*#
	AMIP5	AMIP5	2.52 ± 0.22 %/K	0.79*#
dQ_{atm}/dT	ERAINT	ERAINT	2.50 ± 0.29 W m^{-2}/K	0.68*#
	AMIP5	AMIP5	1.92 ± 0.16 W m^{-2}/K	0.79*#
dP/dQ_{atm}	GPCP	ERAINT	1.09 ± 0.17	0.57*#
	AMIP5	AMIP5	0.83 ± 0.03	0.95*#
d*T*/d*t*	HadCRUT4		0.18 ± 0.02 K/dec	0.70*#
	AMIP5		0.18 ± 0.02 K/dec	0.66#
d*W*/d*t*	SSM/I-ERAINT		0.84 ± 0.20 %/dec	0.42#
	AMIP5		1.17 ± 0.18 %/dec	0.59#
d*P*/d*t*	GPCP		0.18 ± 0.25 %/dec	0.08
	AMIP5		0.34 ± 0.09 %/dec	0.40

* Significant at the 95 % level allowing for autocorrelation; # correlations significant at the 95 % confidence level assuming 20 degrees of freedom

ERAINT is the ECMWF Interim reanalysis

normalizing by d*T*/d*t* gives a sensitivity of 1–2 %/K, slightly lower than the interannual relationship, d*P*/d*T* ~2–3 %/K, and consistent with the influence of greenhouse gases on supressing global rainfall via fast responses described in Sect. 2.2. The d*P*/d*t* response when using SSM/I data over the oceans is larger than the GPCP only product (Fig. 4c). Wentz et al. (2007) previously calculated a d*P*/d*T* sensitivity of 6 %/K over the period 1987–2006; the short length of this record, calibration uncertainty and lack of a mechanism for removing the extra latent heat associated with large increases in *P* would suggest that this is not a good proxy for the longer term response of *P* to warming.

3 Water Vapor and Regional Constraints on Precipitation

While global mean ΔP is found to rise in climate model projections, the changes are not uniformly distributed across the globe. As outlined by Meehl et al. (2007) in a warming climate, there is a tendency for

- the wet regions (inter-tropical convergence zones and higher latitudes) to become wetter and the dry subtropical regions drier,
- an increase in the intensity of precipitation over most continental regions,
- an increase in occurrence of consecutive dry days over many lower latitude regions.

This is broadly consistent with results from early climate model experiments (e.g., Manabe and Wetherald 1980; Mitchell et al. 1987) and implies more damaging flooding and drought in a warmer world (Trenberth 2011). Although there are large regions of the globe for which the sign of future projections of ΔP are ambiguous (Meehl et al. 2007), some of these regions are likely to experience only small ΔP relative to natural variability (Power et al. 2011). The physical basis for these projections is now discussed.

3.1 Moisture Flux into Tropical Wet Regions

Water vapor exerts a powerful physical constraint on the global water cycle. It provides a strongly positive feedback to climate change (Manabe and Wetherald 1967; Soden et al. 2005; Willett et al. 2008; Ingram 2010) but also controls the atmospheric temperature lapse rate and modulates radiative cooling (e.g., O'Gorman et al. 2012) and is central to the large-scale regional responses in precipitation. In the broadest sense, water is transported in the atmosphere by the tropical circulation from the dry, subtropical oceans (net moisture divergence) to the wet, moisture convergence zones, and also to the higher latitudes (e.g., Bengtsson et al. 2011) and the continents (e.g., Trenberth et al. 2011). Rising W with warming (Table 3) therefore indicates increased moisture flux (M_F). Assuming that the magnitude of changes in atmospheric moisture storage is negligible compared with the other terms, a simple moisture balance dictates that M_F balances precipitation minus evaporation ($P - E$):

$$P - E \sim M_F. \tag{5}$$

Assuming, for now, that wind flows are unchanged, M_F will simply vary with atmospheric moisture, at the rate $\alpha \sim 7$ %/K (Table 3). In fact, to reconcile the energetic constraints upon global precipitation with the thermodynamic contraints upon water vapor, climate models simulate a weakening of the tropical Walker circulation that is also evident in observations and simulations of the twentieth century (Vecchi et al. 2006), while decadal variability may cause temporary increases in the Walker circulation (Sohn et al. 2012). Nevertheless, as argued by Held and Soden (2006), the simple assumptions in (5) imply an enhancement of $P - E$ patterns in climate model simulations (at least over the ocean where moisture supply is unlimited) with warming

$$\delta(P - E) \sim -\nabla \cdot (\alpha \delta T M_F) \sim \alpha \delta T (P - E). \tag{6}$$

This effect is illustrated in Fig. 5 which shows the changes in M_F from the dry regions (column mean downward vertical motion) to the wet regions (column mean upward vertical motion) of the tropics defined by 6-hourly instantaneous fields (Zahn and Allan 2011). M_F is computed at each vertical level along the line dividing wet and dry regions.

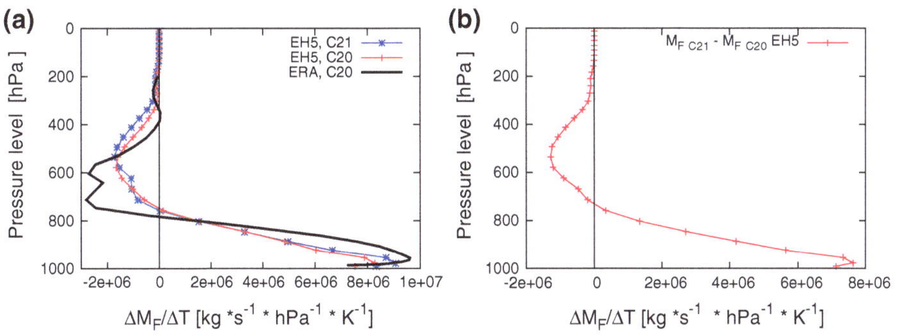

Fig. 5 Change in moisture flux from the dry to the wet regions of the tropics per K change in surface temperature between **a** warm and cold months in ERA Interim and ECHAM5 (twentieth and twenty-first centuries time slices) and for **b** the climate change response from twenty-first minus twentieth century simulated by ECHAM5 (EH5). In **a**, moisture fluxes were computed separately for months with above and below average tropical mean surface temperature; in forming the mean, each month of the year was ascribed equal weighting. For the climate change response, the mean moisture fluxes were calculated and differenced between (21C, 2069–2099) and (20C, 1959–1989)

Changes in M_F normalized by ΔT [kg/(s hPa K)] between cold and warm months are plotted for ERA Interim in Fig. 5a. This shows increased inflow at low levels (maximum at ~950 hPa), but a significant compensating increase in mid-level outflow at around 600–700 hPa as discussed by Zahn and Allan (2011). A similar response is evident for twentieth century (20C, 1959–1989) and twenty-first century (21C, 2069–2099) time slice simulations from a high resolution (0.5°) climate model, ECHAM5 (Roeckner et al. 2003), as described by Zahn and Allan (2012). However, the model simulations indicate a higher altitude for the maximum changes in moisture outflow and less vertical structure, which may be influenced by the higher vertical resolution below altitudes of 500 hPa in ERA Interim. Back and Bretherton (2006) suggest that meridional SST gradients determine the altitude of outflow regionally and the nature of the differences in Fig. 5a merits future analysis. Also shown in Fig. 5b is the 21C minus 20C change in M_F into the wet region of the tropics, normalized by $\Delta T = 3.0$K. The vertical structure of the climate change response of M_F is very similar to interannual variability shown in Fig. 5a where year to year changes in T are relatively small ($\Delta T \sim 0.6$ K).

3.2 Precipitation Response in the Wet and Dry Regions of the Tropics

It does not simply follow from (6) that P rises in the wet regions and declines in the dry regions. Nevertheless, since ΔE changes are expected to be more spatially uniform than ΔP changes, there is a strong expectation that the wet regions will become wetter and the dry regions drier in the tropics, borne out by recent analysis of climate models (e.g., Chou et al. 2009) and limited observational evidence (e.g., Allan et al. 2010).

Figure 6 demonstrates that climate models indeed simulate an increase in P in the wet regions (>70th percentile of P ordered by intensity, defined each month such that the precise location of the wet region varies with time) and static or declining P in other (dry) regions of the tropics. This definition of wet and dry regimes is based upon the analysis of Allan et al. (2010) but is somewhat arbitrary. Mean P is about 8 mm/day in the wet regions while the remaining regions are not completely dry ($P \sim 1$ mm/day). GPCP observations (displayed since 1988 due to inhomogeneities before this date when SSM/I data was not available) also show this contrasting wet/dry response as discussed by Liu et al. (2012). Since this response is contingent on warming (enhancing $P - E$), this explains the flattening off of P responses in the RCP4.5 simulations after around 2060 as T asymptotes. Nevertheless, as discussed previously, the precise transient ΔP response is strongly influenced by the fast forcing processes ($f\Delta F$) in addition to the slow response to warming, $k\Delta T$ (Andrews and Forster 2010; Wu et al. 2010).

3.3 Extremes of Precipitation

Within the wet regions, during heavy rainfall events, the atmosphere is typically precipitating more water over the course of a day than is contained in the atmospheric column at a particular location; this underlines the vital role of moisture convergence in determining intense rainfall rates (e.g., Trenberth 2011). While low-level moisture changes are indeed thought to provide a strong constraint upon P intensification in a warming world, changes in updraft velocity with warming explain the large range in simulated responses in the tropics (e.g., O'Gorman and Schneider 2009; Allan et al. 2010; Sugiyama et al. 2010). Observed relationships provide a powerful constraint upon future simulated responses in extreme precipitation (O'Gorman 2012). Figure 7 displays observed and simulated

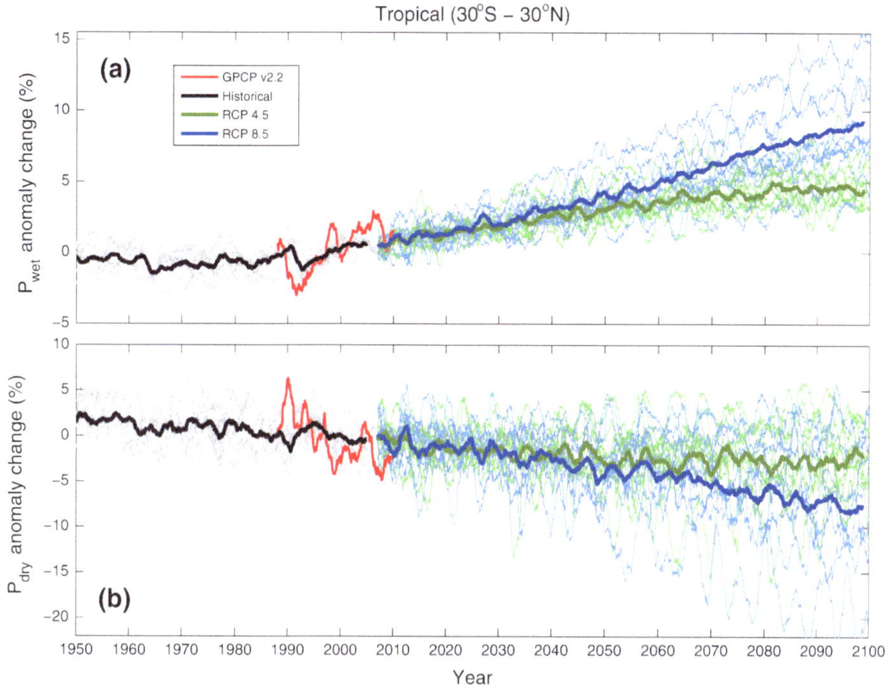

Fig. 6 Changes in tropical (30°S–30°N) mean precipitation (%) in **a** the wet regions (above the 70th percentile of monthly precipitation) and in **b** the dry regions (below the 70th percentile) relative to the 1988–2005 period for the historical, RCP4.5 and RCP8.5 scenario simulations (*thick lines* denote 10-model ensemble means, models 1–10 in Table 2) and the GPCP observations. A 24-month running mean is applied

precipitation responses, as a function of precipitation intensity (P_i), for the present day and also the simulated climate change responses.

The 5-day P averages are constructed, and a deseasonalized intensity (i) distribution constructed each month; 5-day averages are chosen to avoid structural inconsistencies between data sets that occur at higher time resolutions (Liu and Allan 2012). The linear regression of P_i is computed with respect to tropical mean T (deseasonalized anomalies) and normalized by mean P_i to give units of %/K. Figure 7a illustrates the intensification of P at the 99th percentile in both AMIP5 simulations and GPCP 1 Degree Daily (DD) V1.1 daily observations (Huffman et al. 2009) of around 15 %/K, larger than expected from Clausius–Clapeyron scaling. At the lowest percentiles, P_i decreases with tropic-wide warming, consistent with the monthly mean responses shown in Fig. 6.

The GPCP 1DD observations generally show a similar but more positive response than the AMIP5 simulations (Fig. 7a), explained primarily by ocean regions, for $P_{i<99\%}$ (Fig. 7b). The SSM/I observations display a more positive sensitivity still for these percentiles and do not capture the decline in $P_{i<60\%}$ with warming. This may be explained by the limited sampling of low rain rates by the instantaneous overpasses (e.g., Liu and Allan 2012). GPCP gauge-based observations and AMIP5 simulations display remarkable agreement over the land (Fig. 7c), consistent with monthly mean variability (Liu et al. 2012); for $P_{i<99\%}$ the response is negative and explained by ENSO variability: during warmer El Niño years, there is less rainfall on average over land (e.g., Gu et al. 2007).

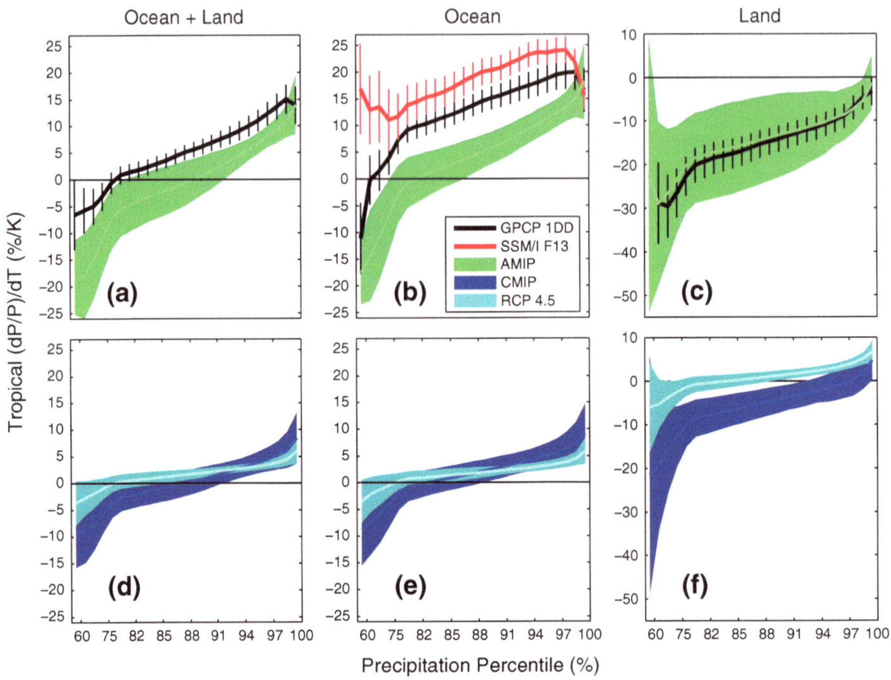

Fig. 7 Percentage responses of 5-day mean P to changes in tropical (30°S–30°N) mean T as a function of P intensity percentiles in the present day (1998–2008, linear regression) for **a** the tropics, **b** tropical ocean and **c** tropical land for AMIP5, GPCP 1DD and SSM/I and for the present and future in CMIP5 models (historical 1985–2005 using linear regression) and RCP4.5 minus historical (2080–2100 minus 1985–2005 using differences in mean intensity distribution) for the **d** tropics, **e** tropical ocean and **f** tropical land. In all cases, tropical (land and ocean) mean T is used. *Shaded areas* denotes ±1 standard deviation of inter-model spread; *vertical bars* denote ±1 standard error in the linear regression for GPCP and SSM/I. Models 1–10 in Table 2 are used

The climate change response of 5-day P extremes is compared with present-day simulations from the historical experiments in Fig. 7d–f. Consistent with the AMIP5 simulations and observations, there is a tendency for the wet percentiles to become wetter during warm months, while the dry percentiles become drier (Fig. 7d). Statistical uncertainty is large for $P_{i<60\,\%}$ over land (Fig. 7f), a likely result of the small and zero P totals in this bin. The tropic-wide percentile responses in Fig. 7d are dominated by the ocean regions (Fig. 7e). However, the climate change response (calculated as a percentile distribution difference between 2080 and 2099 RCP4.5 minus 1985–2005 historical simulations) displays a more modest response compared with the present-day linear regression applied to CMIP5 historical simulations (which are only slightly weaker than the AMIP5 simulations and observed relationships). The climate change response indicates a $P_{99\%}$ response of ~6 %/K, close to that expected from the Clausius–Clapeyron relation.

For land regions, the climate change response of P_i to warming is more positive than the present-day CMIP5 relationships (Fig. 7f) with $P_{99\%}$ responses out to 2080–2100 again close to that anticipated from Clausius–Clapeyron scaling. Thus, relationships derived from present-day variability are not good proxies for the climate change responses, in particular for the land regions. This is because ENSO variability enhances ocean P sensitivity to warming and cooling, due to changing rainfall patterns, while it introduces a

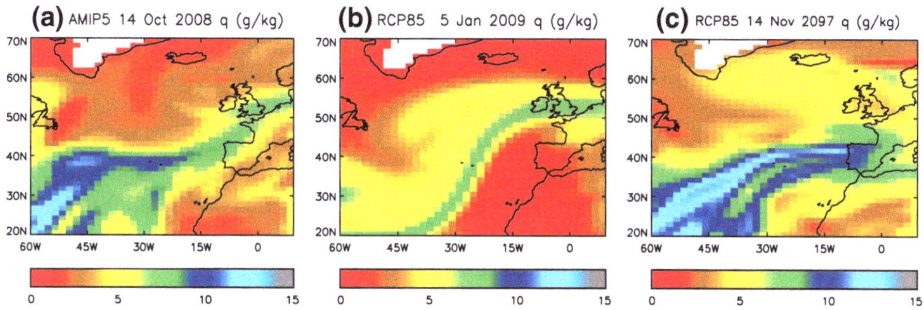

Fig. 8 Specific humidity (q) fields at 850 hPa associated with the highest 3-day precipitation total for *grid boxes* within the UK region (7°W–3°E, 50–60°N) during the winter-half year (October–March) for **a** HadGEM2-A AMIP simulations (1995–2008) and HadGEM2-ES RCP8.5 simulations over the periods **b** 2006–2019 and **c** 2085–2098

negative relationship between P and T over land (Gu et al. 2007; Liu et al. 2012). The climate change responses are consistent with an intensification of P at high percentiles, close to the water vapor responses of around 6 %/K, and reduced P for the dry regions (sub-tropical anticyclonic regimes) of net moisture export (e.g., Allan 2012), consistent with Fig. 6.

3.4 Influence of Dynamical Changes on Precipitation and Its Extremes

At finer time and space scales, it is less clear what the precise response of extreme precipitation to warming will be (Haerter et al. 2010). While the Clausius–Clapeyron relation provides a broad constraint, additional local latent heating may invigorate storms leading to a super-Clausius–Clapeyron scaling (Lenderink et al. 2011) while moisture limitation and atmospheric stabilization associated with latent heating at larger scales may result in responses below that anticipated from Clausius–Clapeyron (Haerter et al. 2010). Knowledge of the precise time and space scales associated with the most damaging flooding events is therefore crucial.

Changes in updraft velocity strongly influence the simulated $P_{99.9\,\%}$ response in the tropics (Turner and Slingo 2009; O'Gorman and Schneider 2009; Allan et al. 2010); the heaviest rainfall events are reliant on parametrized processes operating below the model grid scale. At higher latitudes, where large-scale processes become dominant in explaining intense precipitation, the simulated responses show greater consistency (O'Gorman and Schneider 2009) although there is evidence that they underestimate the observed responses (Min et al. 2011). In the UK, peak river flows during the winter-half of the year have been found to be associated with Atmospheric Rivers (Lavers et al. 2011). These are long ($\sim$1,000 km), narrow ($\sim$100 km) regions of strong water vapor transport in the warm sector of extra-tropical cyclones that can result in intense rainfall when the moisture is condensed, in particular following the uplift of the moisture-laden air-masses over mountains (e.g., Dettinger et al. 2011).

Atmospheric reanalyses, such as ERA Interim, and current climate models are typically able to resolve the large-scale processes associated with these flood-inducing events. Figure 8 illustrates the specific humidity (q) structure associated with 14-year maximum 3-day P totals for grid points around the UK for present-day simulations and future projections. The present-day simulations are indeed associated with Atmospheric River structures. The future projection in Fig. 8c simulates greater amounts of water vapor, consistent with warmer T, but the dynamical structure is unlike the present-day cases and in fact delivers a smaller 3-day P total (112 mm) than the present-day event simulated in Fig. 8b (152 mm). While this is just an illustration, it nevertheless highlights the potential importance of both dynamical and thermodynamic factors in influencing future changes in extreme rainfall. The thermodynamic climate change component of such events is likely to be robust with the Clausius–Clapeyron relation a reliable constraint. However, small changes in the jet stream and storm track regions (and the overall dynamical character of intense rainfall-producing events) may dominate the local response and therefore regional projections of the occurrence of damaging flooding and drought remain a substantial challenge.

4 Conclusions

Global precipitation is projected to rise in the future primarily to maintain balance with enhanced atmospheric radiative cooling as temperatures increase (Manabe and Wetherald 1975). This slow, well-constrained response of around ~2–3 %/K is modulated by more rapid adjustments to the radiative forcings, that are themselves responsible for current and future warming, but which directly influence atmospheric radiative cooling (Andrews et al. 2010). An additional influence appears to relate to how far from equilibrium the climate system is (McInerney and Moyer 2012). This is governed by the differing timescales of adjustment by the land and ocean and the associated changes in energy and moisture fluxes between them (Cao et al. 2012).

Regional changes in the hydrological cycle, of more importance for climate impacts, are more strongly tied to (i) changes in moisture transports which are well constrained by the Clausius–Clapeyron equation linking saturation vapor pressure with temperature and (ii) small spatial movements in large-scale circulation systems which are highly uncertain. This also applies to the local, extreme precipitation events which are strongly linked to the rises in low-level moisture of around 7 %/K but are also influenced by changes in the nature and spatial distribution of intense rainfall events.

The combination of the global changes in precipitation of around ~2–3 %/K and increased low-level moisture of ~7 %/K leads to a general trend toward the wet regions (Inter Tropical Convergence Zone and higher latitudes) becoming wetter and the dry subtropics getting drier. In the margins, although projected responses appear ambiguous, there are physical grounds for anticipating relatively small changes in P, signifying a greater consensus on regional projections than previously thought (Power et al. 2011). Nevertheless, local changes are highly dependent upon small shifts in position of large-scale atmospheric circulation patterns and in the physical nature of rainfall regimes, in particular for extremes. Hence, determining accurate responses in the hydrological cycle at the scales required for impact models may be beyond the predictive capacity of climate modelling. Thus, determining robust, large-scale, robust responses in the hydrological cycle (e.g., Martin 2012) remain a crucial tool in understanding and planning for a changing water cycle in the future.

Acknowledgments The comments and suggestions of Paul O'Gorman and two anonymous reviewers helped to improve the paper. This work was undertaken as part of the PAGODA and PREPARE projects funded by the UK Natural Environmental Research Council under grants NE/I006672/1 and NE/G015708/1 and was supported by the National Centre for Earth Observations and the National Centre for Atmospheric Science. A. Bodas-Salcedo was supported by the Joint DECC/Defra Met Office Hadley Centre Climate Programme (GA01101). GPCP v2.2 data were extracted from http://precip.gsfc.nasa.gov/gpcp_v2.2_data.html, and CMIP5 and AMIP5 data sets from the British Atmospheric Data Centre (http://badc.nerc.ac.uk/

home) and the Program for Climate Model Diagnosis and Intercomparison: pcmdi3.llnl.gov/esgcet. We acknowledge the World Climate Research Programme's Working Group on Coupled Modelling, which is responsible for CMIP, and we thank the climate modelling groups (models listed in Table 2) for producing and making available their model outputs; for CMIP, the U.S. Department of Energy's PCMDI provided coordinating support and led development of software infrastructure in partnership with the Global Organization for Earth System Science Portals. The scientists involved in the generation of these data sets are sincerely acknowledged. We thank also Prof Keith Shine for developing the zero-dimensional climate model.

References

Adler RF, Gu G, Wang JJ, Huffman GJ, Curtis S, Bolvin D (2008) Relationships between global precipitation and surface temperature on interannual and longer timescales (1979–2006). J Geophys Res 113:D22104. doi:10.1029/2008JD010536

Allan RP (2006) Variability in clear-sky longwave radiative cooling of the atmosphere. J Geophys Res 111:D22105. doi:10.1029/2006JD007304

Allan RP (2012) Regime dependent changes in global precipitation. Clim Dyn 39:827–840. doi:10.1007/s00382-011-1134-x

Allan RP, Soden BJ, John VO, Ingram WI, Good P (2010) Current changes in tropical precipitation. Environ Res Lett 5:025205. doi:10.1088/1748-9326/5/2/025205

Allen MR, Ingram WJ (2002) Constraints on future changes in climate and the hydrologic cycle. Nature 419:224–232

Andrews T, Forster PM (2010) The transient response of global-mean precipitation to increasing carbon dioxide levels. Environ Res Lett 5:025212. doi:10.1088/1748-9326/5/2/025212

Andrews T, Forster PM, Boucher O, Bellouin N, Jones A (2010) Precipitation, radiative forcing and global temperature change. Geophys Res Lett 37:L14701. doi:10.1029/2010GL043991

Arora VK, Scinocca JF, Boer GJ, Christian JR, Denman KL, Flato GM, Kharin VV, Lee WG, Merryfield WJ (2011) Carbon emission limits required to satisfy future representative concentration pathways of greenhouse gases. Geophys Res Lett 38:L05805. doi:10.1029/2010GL046270

Back LE, Bretherton CS (2006) Geographic variability in the export of moist static energy and vertical motion profiles in the tropical Pacific. Geophys Res Lett 33:L17810. doi:10.1029/2006GL026672

Bala G, Caldeira K, Nemani R (2010) Fast versus slow response in climate change: implications for the global hydrological cycle. Clim Dyn 35:423–434. doi:10.1007/s00382-009-0583-y

Bengtsson L, Hodges KI, Koumoutsaris S, Zahn M, Keenlyside N (2011) The changing atmospheric water cycle in Polar Regions in a warmer climate. Tellus A 63:907–920. doi:10.1111/j.1600-0870.2011.00534.x

Cao L, Bala G, Caldeira K (2012) Climate response to changes in atmospheric carbon dioxide and solar irradiance on the time scale of days to weeks. Environ Res Lett 7:034015. doi:10.1088/1748-9326/7/3/034015

Chou C, Neelin JD, abd J-Y Tu CAC (2009) Evaluating the "rich get richer" mechanism in tropical precipitation change under global warming. J Clim 22:1982–2005

Chung ES, Soden BJ (2010) Radiative signature of increasing atmospheric carbon dioxide in HIRS satellite observations. Geophys Res Lett 37:L07707. doi:10.1029/2010GL042698

Collins WJ, Bellouin N, Doutriaux-Boucher M, Gedney N, Halloran P, Hinton T, Hughes J, Jones CD, Joshi M, Liddicoat S, Martin G, O'Connor F, Rae J, Senior C, Sitch S, Totterdell I, Wiltshire A, Woodward S (2011) Development and evaluation of an earth-system model—HadGEM2. Geosci Model Dev Discuss 4:997–1062. doi:10.5194/gmdd-4-997-2011

Dee DP, Uppala SM, Simmons AJ, Berrisford P, Poli P, Kobayashi S, Andrae U, Balmaseda MA, Balsamo G, Bauer P, Bechtold P, Beljaars ACM, van de Berg L, Bidlot J, Bormann N, Delsol C, Dragani R, Fuentes M, Geer AJ, Haimberger L, Healy SB, Hersbach H, Hólm EV, Isaksen L, Kållberg P, Köhler M, Matricardi M, McNally AP, Monge-Sanz BM, Morcrette JJ, Park BK, Peubey C, de Rosnay P, Tavolato C, Thépaut JN, Vitart F (2011) The ERA-Interim reanalysis: configuration and performance of the data assimilation system. Q J Roy Meteorol Soc 137:553–597. doi:10.1002/qj.828

Dettinger MD, Ralph FM, Das T, Neiman PJ, Cayan DR (2011) Atmospheric rivers, floods and the water resources of California. Water 3:445–478. doi:10.3390/w3020445

Gent PR, Danabasoglu G, Donner LJ, Holland MM, Hunke EC, Jayne SR, Lawrence DM, Neale RB, Rasch PJ, Vertenstein M, Worley PH, Yang ZL, Zhang M (2011) The community climate system model version 4. J Clim 24:4973–4991. doi:10.1175/2011JCLI4083.1

Gu G, Adler RF (2012) Interdecadal variability/long-term changes in global precipitation patterns during the past three decades: global warming and/or pacific decadal variability? Clim Dyn. doi:10.1007/s00382-012-1443-8

Gu G, Adler RF, Huffman GJ, Curtis S (2007) Tropical rainfall variability on interannual-to-interdecadal and longer time scales derived from the GPCP monthly product. J Clim 20:4033–4046

Haerter JO, Berg P, Hagemann S (2010) Heavy rain intensity distributions on varying time scales and at different temperatures. J Geophys Res 115:D17102

Hansen J, Johnson D, Lacis A, Lebedeff S, Lee P, Rind D, Russell G (1981) Climate impact of increasing atmospheric carbon dioxide. Science 213:957–966. doi:10.1126/science.213.4511.957

Hansen J, Sato M, Ruedy R, Kharecha P, Lacis A, Miller R, Nazarenko L, Lo K, Schmidt GA, Russell G, Aleinov I, Bauer S, Baum E, Cairns B, Canuto V, Chandler M, Cheng Y, Cohen A, Del Genio A, Faluvegi G, Fleming E, Friend A, Hall T, Jackman C, Jonas J, Kelley M, Kiang NY, Koch D, Labow G, Lerner J, Menon S, Novakov T, Oinas V, Perlwitz J, Perlwitz J, Rind D, Romanou A, Schmunk R, Shindell D, Stone P, Sun S, Streets D, Tausnev N, Thresher D, Unger N, Yao M, Zhang S (2007) Climate simulations for 1880-2003 with GISS modelE. Clim Dyn 29:661–696. doi:10.1007/s00382-007-0255-8

Hansen J, Sato M, Kharecha P, von Schuckmann K (2011) Earth's energy imbalance and implications. Atmos Chem Phys 11:13421–13449. doi:10.5194/acp-11-13421-2011

Hawkins E, Sutton R (2011) The potential to narrow uncertainty in projections of regional precipitation change. Clim Dyn 37:407–418. doi:10.1007/s00382-010-0810-6

Held IM, Soden BJ (2006) Robust responses of the hydrological cycle to global warming. J Clim 19:5686–5699

Hourdin F, Grandpeix JY, Rio C, Bony S, Jam A, Cheruy F, Rochetin N, Fairhead L, Idelkadi A, Musat I, Dufresne JL, Lahellec A, Lefebvre MP, Roehrig R (2012) LMDZ5B: the atmospheric component of the IPSL climate model with revisited parameterizations for clouds and convection. Clim Dyn. doi:10.1007/s00382-012-1343-y

Huffman GJ, Adler RF, Bolvin DT, Gu G (2009) Improving the global precipitation record: GPCP version 2.1. Geophys Res Lett 36:L17808. doi:10.1029/2009GL040000

Ingram W (2010) A very simple model for the water vapour feedback on climate change. Q J R Meteorol Soc 136:30–40. doi:10.1002/qj.546

John VO, Allan RP, Soden BJ (2009) How robust are observed and simulated precipitation responses to tropical warming. Geophys Res Lett 36:L14702. doi:10.1029/2009GL038276

Lambert FH, Webb MJ (2008) Dependency of global mean precipitation on surface temperature. Geophys Res Lett 35:L16706. doi:10.1029/2008GL034838

Lavers DA, Allan RP, Wood EF, Villarini G, Brayshaw DJ, Wade AJ (2011) Winter floods in Britain are connected to atmospheric rivers. Geophys Res Lett 38:L23803. doi:10.1029/2011GL049783

Lenderink G, Mok HY, Lee TC, van Oldenborgh GJ (2011) Scaling and trends of hourly precipitation extremes in two different climate zones—Hong Kong and The Netherlands. Hydrol Earth Syst Sci 15:3033–3041. doi:10.5194/hess-15-3033-2011

Liu C, Allan RP (2012) Multisatellite observed responses of precipitation and its extremes to interannual climate variability. J Geophys Res 117:D03101. doi:10.1029/2011JD016568

Liu C, Allan RP, Huffman GJ (2012) Co-variation of temperature and precipitation in CMIP5 models and satellite observations. Geophys Res Lett 39:L13803. doi:10.1029/2012GL052093

Loeb NG, Lyman JM, Johnson GC, Allan RP, Doelling DR, Wong T, Soden BJ, Stephens GL (2012) Observed changes in top-of-the-atmosphere radiation and upper-ocean heating consistent within uncertainty. Nat Geosci 5:110–113. doi:10.1038/ngeo1375

Lu J, Cai M (2009) Stabilization of the atmospheric boundary layer and the muted global hydrological cycle response to global warming. J Hydrometeorol 10:347–352. doi:10.1175/2008JHM1058.1

Manabe S, Wetherald RT (1967) Thermal equilibrium of the atmosphere with a given distribution of relative humidity. J Atmos Sci 24:241–259

Manabe S, Wetherald RT (1975) The effects of doubling the CO_2 concentration on the climate of a general circulation model. J Atmos Sci 32:3–15

Manabe S, Wetherald RT (1980) On the distribution of climate change resulting from an increase in CO_2 content in the atmosphere. J Atmos Sci 37:99–118

Martin G (2012) Quantifying and reducing uncertainty in the large-scale responses of the water cycle. Surv Geophys (accepted) doi:10.1007/s10712-012-9203-1

McInerney D, Moyer E (2012) Direct and disequilibrium effects on precipitation in transient climates. Atmos Chem Phys Discuss 12:19649–19681. doi:10.5194/acpd-12-19649-2012

Meehl G, Stocker T, Collins W, Friedlingstein P, Gaye A, Gregory J, Kitoh A, Knutti R, Murphy J, Noda A, Raper S, Watterson I, Weaver A, Zhao ZC (2007) Global climate projections. Climate change 2007:

the physical science basis. Contribution of working group I to the fourth assessment report of the Intergovernmental Panel on Climate Change, Cambridge University Press, Cambridge, pp 747–845
Merrifield MA (2011) A shift in western tropical Pacific Sea level trends during the 1990s. J Clim 24:4126–4138 doi:10.1175/2011JCLI3932.1
Min S, Zhang X, Zwiers FW, Hegerl GC (2011) Human contribution to more-intense precipitation extremes. Nature 470:378–381
Ming Y, Ramaswamy V, Persad G (2010) Two opposing effects of absorbing aerosols on global-mean precipitation. Geophys Res Lett 37:L13701
Mitchell J, Wilson CA, Cunnington WM (1987) On CO_2 climate sensitivity and model dependence of results. Q J Roy Meteorol Soc 113:293–322
Morice CP, Kennedy JJ, Rayner NA, Jones PD (2012) Quantifying uncertainties in global and regional temperature change using an ensemble of observational estimates: the HadCRUT4 data set. J Geophys Res 117:D08101. doi:10.1029/2011JD017187
O'Gorman PA (2012) Sensitivity of tropical precipitation extremes to climate change. Nat Geosci 5:697–700 doi:10.1038/ngeo1568
O'Gorman PA, Schneider T (2009) The physical basis for increases in precipitation extremes in simulations of 21st-century climate change. Proc Nat Acad Sci 106:14773–14777
O'Gorman PA, Allan RP, Byrne MP, Previdi M (2012) Energetic constraints on precipitation under climate change. Surv Geophys 33:585–608. doi:10.1007/s10712-011-9159-6
Peterson TC, Stott PA, Herring S (2012) Explaining extreme events of 2011 from a climate perspective. Bull Am Meteorol Soc 93:1041–1067. doi:10.1175/BAMS-D-12-00021.1
Power SB, Delage F, Colman R, Moise A (2011) Consensus on twenty-first-century rainfall projections in climate models more widespread than previously thought. J Clim 25:3792–3809
Prata F (2008) The climatological record of clear-sky longwave radiation at the earth's surface: evidence for water vapour feedback? Int J Remote Sens 29:5247–5263. doi:10.1080/01431160802036508
Raddatz TJ, Reick CH, Knorr W, Kattge J, Roeckner E, Schnur R, Schnitzler KG, Wetzel P, Jungclaus J (2007) Will the tropical land biosphere dominate the climate-carbon cycle feedback during the twenty-first century? Clim Dyn 29:565–574. doi:10.1007/s00382-007-0247-8
Ramanathan V (1981) The role of ocean–atmosphere interactions in the CO_2 climate problem. J Atmos Sci 38:918–930
Richter I, Xie SP (2008) The muted precipitation increase in global warming simulations: a surface evaporation perspective. J Geophys Res 113:D24118. doi:10.1029/2008JD010561
Roeckner E, Bäuml G, Bonaventura L, Brokopf R, Esch M, Giorgetta M, Hagemann S, Kirchner I, Kornblueh L, Manzini E, Rhodin A, Schlese U, Schulzweida U, Tompkins A (2003) The atmospheric general circulation model ECHAM 5. Part I: model description. Technical report 349, 140 pp, Max-Plank institüte für Meteorologie, Hamburg
Simmons AJ, Willett KM, Jones PD, Thorne PW, Dee DP (2010) Low-frequency variations in surface atmospheric humidity, temperature, and precipitation: inferences from reanalyses and monthly gridded observational data sets. J Geophys Res 115:D01110. doi:10.1029/2009JD012442
Soden BJ, Jackson DL, Ramaswamy V, Schwarzkopf MD, Huang X (2005) The radiative signature of upper tropospheric moistening. Science 310:841–844
Sohn BJ, Yeh SW, Schmetz J, Song HJ (2012) Observational evidences of Walker circulation change over the last 30years contrasting with GCM results. Clim Dyn 1–12. doi:10.1007/s00382-012-1484-z
Stackhouse PW Jr, Gupta SK, Cox SJ, Zhang T, Mikovitz JC, Hinkelman LM (2011) 24.5-year srb data set released. GEWEX News 21:10–12
Stephens GL, Ellis TD (2008) Controls of global-mean precipitation increases in global warming GCM experiments. J Clim 21:6141–6155
Sugiyama M, Shiogama H, Emori S (2010) Precipitation extreme changes exceeding moisture content increases in MIROC and IPCC climate models. Proc Natl Acad Sci 107:571–575
Taylor KE, Stouffer RJ, Meehl GA (2011) An overview of CMIP5 and the experiment design. Bull Am Meteorol Soc 93:485–498. doi:10.1175/BAMS-D-11-00094.1
Trenberth KE (2011) Changes in precipitation with climate change. Clim Res 47:123–138
Trenberth KE, Fasullo JT, Mackaro J (2011) Atmospheric moisture transports from ocean to land and global energy flows in reanalyses. J Clim 24:4907–4924. doi:10.1175/2011JCLI4171.1
Turner AG, Slingo JM (2009) Uncertainties in future projections of extreme precipitation in the indian monsoon region. Atmos Sci Lett 10:152–158
Vecchi GA, Soden BJ, Wittenberg AT, Held IM, Leetmaa A, Harrison MJ (2006) Weakening of tropical pacific atmospheric circulation due to anthropogenic forcing. Nature 441:73–76

Volodin EM, Dianskii NA, Gusev AV (2010) Simulating present-day climate with the INMCM4.0 coupled model of the atmospheric and oceanic general circulations. Izvestiya Atmos Ocean Phys 46:414–431. doi:10.1134/S000143381004002X

Voldoire A, Sanchez-Gomez E, Salas y Mélia D, Decharme B, Cassou C, Sénési S, Valcke S, Beau I, Alias A, Chevallier M, Déqué M, Deshayes J, Douville H, Fernandez E, Madec G, Maisonnave E, Moine M-P, Planton S, Saint-Martin D, Szopa S, Tyteca S, Alkama R, Belamari S, Braun A, Coquart L, Chauvin F (2012) The CNRM-CM5.1 global climate model: description and basic evaluation. Clim Dyn. doi:10.1007/s00382-011-1259-y

Watanabe M, Suzuki T, O'ishi R, Komuro Y, Watanabe S, Emori S, Takemura T, Chikira M, Ogura T, Sekiguchi M, Takata K, Yamazaki D, Yokohata T, Nozawa T, Hasumi H, Tatebe H, Kimoto M (2010) Improved climate simulation by MIROC5: mean states, variability, and climate sensitivity. J Clim 23:6312–6335. doi:10.1175/2010JCLI3679.1

Wentz FJ, Schabel M (2000) Precise climate monitoring using complementary satellite data sets. Nature 403:414–416

Wentz FJ, Ricciardulli L, Hilburn K, Mears C (2007) How much more rain will global warming bring? Science 317:233–235

Willett KM, Jones PD, Gillett NP, Thorne PW (2008) Recent changes in surface humidity: Development of the HadCRUH dataset. J Clim 21(20):5364–5383

Wu P, Wood R, Ridley J, Lowe J (2010) Temporary acceleration of the hydrological cycle in response to a CO2 rampdown. Geophys Res Lett 37:L12705. doi:10.1029/2010GL043730

Wu T et al. (2012) The 20th century global carbon cycle from the Beijing Climate Center Climate System Model (BCC CSM). J Clim (in press)

Yukimoto S, Y A, Hosaka M, Sakami T, Yoshimura H, Hirabara M, Tanaka TY, Shindo E, Tsujino H, Deushi M, Mizuta R, Yabu S, Obata A, Nakano H, Ose T, Kitoh A (2012) A new global climate model of meteorological research institute: MRI-CGCM3—model description and basic performance. J Meteorol Soc Jpn (under preparation)

Zahn M, Allan RP (2011) Changes in water vapor transports of the ascending branch of the tropical circulation. J Geophys Res 116:D18111. doi:10.1029/2011JD016206

Zahn M, Allan RP (2012) Climate Warming related strengthening of the tropical hydrological cycle. J Clim. doi:10.1175/JCLI-D-12-00222.1

Zhang Y, Rossow WB, Lacis AA, Oinas V, Mishchenko MI (2004) Calculation of radiative fluxes from the surface to top of atmosphere based on ISCCP and other global data sets: refinements of the radiative transfer model and the input data. J Geophys Res 109:D19105. doi:10.1029/2003JD004457

Zhang ZS, Nisancioglu K, Bentsen M, Tjiputra J, Bethke I, Yan Q, Risebrobakken B, Andersson C, Jansen E (2012) Pre-industrial and mid-Pliocene simulations with NorESM-L. Geosci Model Dev Discuss 5:119–148. doi:10.5194/gmdd-5-119-2012

DOI 10.1007/978-94-017-8789-5_5
Reprinted from *Surveys in Geophysics* Journal, DOI 10.1007/s10712-012-9203-1

Quantifying and Reducing Uncertainty in the Large-Scale Response of the Water Cycle

Gill M. Martin

Received: 31 May 2012 / Accepted: 4 September 2012 / Published online: 5 October 2012

Abstract Despite their obvious environmental, societal and economic importance, our understanding of the causes and magnitude of the variations in the global water cycle is still unsatisfactory. Uncertainties in hydrological predictions from the current generation of models pose a serious challenge to the reliability of forecasts and projections across time and space scales. This paper provides an overview of the current issues and challenges in modelling various aspects of the Earth's hydrological cycle. These include: the global water budget and water conservation, the role of model resolution and parametrisation of precipitation-generating processes on the representation of the global and regional hydrological cycle, representation of clouds and microphysical processes, rainfall variability, the influence of land–atmosphere coupling on rainfall patterns and their variability, monsoon processes and teleconnections, and ocean and cryosphere modelling. We conclude that continued collaborative activity in the areas of model development across timescales, process studies and climate change studies will provide better understanding of how and why the hydrological cycle may change, and better estimation of uncertainty in model projections of changes in the global water cycle.

Keywords Hydrological cycle · Moisture · Precipitation · Modelling

1 Introduction

The hydrological cycle is the process by which water travels from the Earth's surface to the atmosphere and then back to the surface again. The sun provides the energy for a continuous exchange of moisture between the oceans, the land and the atmosphere. The atmospheric water cycle is the driving force of weather and climate, and the spatial and temporal characteristics of precipitation—too much, too little, at the wrong time, in the wrong place—have profound effects on all aspects of life. Substantial changes in the global water cycle (GWC) are an expected consequence of a warming climate. Such changes could prove a significant

G. M. Martin (✉)
Met Office Hadley Centre, FitzRoy Road, Exeter, UK
e-mail: Gill.martin@metoffice.gov.uk

challenge for societies and ecosystems and impose substantial pressures on water management, urban planning, agricultural production and tourism. Despite their obvious environmental, societal and economic importance, our understanding of the causes and magnitude of the variations in the GWC is still unsatisfactory. Key issues for climate change that hinge on the global water cycle include: (1) the strength and variability of global and regional hydrological cycles in a warmer world; (2) freshwater forcing and salinity budget of the global oceans; (3) terrestrial ecosystems and their dependence on water availability; (4) the fate of polar ice caps and glaciers with consequent sea level rise.

Uncertainties in hydrological predictions from the current generation of models pose a serious challenge to the reliability of forecasts across timescale and space scale. On climate timescales, the GWC is highlighted as a key source of uncertainty in the latest Intergovernmental Panel on Climate Change (IPCC) report (Randall et al. 2007), and problems with forecasting precipitation continue to affect measures of forecast accuracy. We require a more holistic approach to understanding, modelling and predicting the global and regional terrestrial water cycle and its role in the impacts of hazardous weather, climate variability and climate change. This must extend from the prediction of hydrological extremes (floods and droughts) to an integrated assessment of water, food and natural resources.

The growth of systematic errors in models remains one of the central problems in producing accurate climate change projections for the next 50–100 years. Although great advances have been made in global modelling in recent decades, there are still large uncertainties in many processes, such as clouds, convection, and coupling to oceans and the land surface. Such uncertainties are related to different physical parametrisations and to model horizontal and vertical resolution. Global precipitation distribution, intensity and variation on a range of timescales are also not well observed, particularly over oceans.

Similar systematic errors are seen on a range of timescales from Numerical Weather Prediction (NWP)/seasonal/decadal/climate. "Seamless" model development, where the same model configuration is developed and tested across all timescales (e.g., Martin et al. 2010), has been instrumental in diagnosing and reducing errors in tropical circulation and precipitation. Increasing model resolution has also been shown to improve the distribution of tropical precipitation. This will ultimately lead to better predictions and projections for the GWC.

This paper provides an overview of the current issues and challenges in modelling various aspects of the Earth's hydrological cycle. These include: the global water budget and water conservation, the role of model resolution and parametrisation of precipitation-generating processes on the representation of the global and regional hydrological cycle, representation of clouds and microphysical processes, rainfall variability, the influence of land–atmosphere coupling on rainfall patterns and their variability, monsoon processes and teleconnections, and ocean and cryosphere modelling. We include examples of studies being done around the world which shed light on the relevant processes and limitations in their modelling. We conclude that continued collaborative activity in the areas of model development across timescales, process studies and climate change studies will provide better understanding of how and why the hydrological cycle may change, and better estimation of uncertainty in model projections of changes in the global water cycle.

2 Scientific Challenges in Modelling the Global Water Cycle

The main scientific challenges in modelling the global water cycle are now discussed, with reference to research currently underway in a number of institutes around the world.

2.1 Global Water Budget

A long-standing on-going challenge is to provide a reliable estimate of the annual mean global water cycle. An estimate made by Trenberth et al. (2007) using observational data is shown in Fig. 1. However, there are large uncertainties in many of the estimated numbers, not least because quantitative knowledge of the various components and their variability is still fairly limited because of a lack of reliable data for surface evaporation, oceanic precipitation, terrestrial runoff and several other fields. A further challenge is to determine the interannual and longer-term variability of this cycle, particularly those aspects that may be associated with climate change.

Global warming is likely to intensify the global hydrological cycle and bring changes to the water vapour transport, changing the magnitude and spatial distribution of freshwater fluxes between atmosphere, land and sea ice into the ocean (e.g., Held and Soden 2006). This, in turn, will modify the salinity distribution at the surface of the ocean and could potentially alter the ocean circulation through effects on its density (Durack et al. 2012). An appropriate representation of these fluxes by general circulation models (GCMs) is essential in order to make confident projections of future characteristics of the thermohaline circulation, in particular its stability, under global warming conditions. It is therefore very important to investigate the skill of climate models to simulate large-scale water vapour transports and their mechanisms.

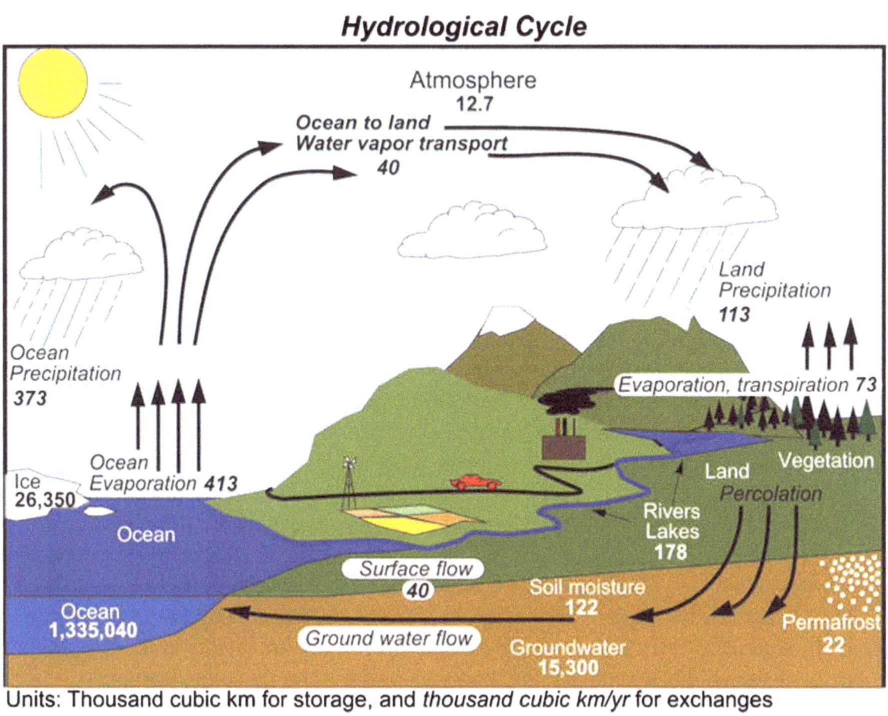

Fig. 1 The hydrological cycle. Estimates of the main water reservoirs, given in plain font in 10^3 km^3, and the flow of moisture through the system, given in slant font (10^3 km^3 $year^{-1}$). From Trenberth et al. (2007), their Fig. 1. © Copyright 2007 AMS (http://www.ametsoc.org/pubs/cr_2005.html)

A water budget quantifies systematically the flows and reservoirs of water in the water cycle based on the principle of the conservation of mass. That principle assumes that water is neither created nor destroyed in the system. However, GCMs may be subject to parametrisation errors which affect water conservation and thus the closure of the water budget. It is essential that such problems are addressed in order that reliable estimates of future changes in the water cycle can be made. For example, Collins et al. (2011) showed a budget diagram of long-term means of global water fluxes from the various sub-models in a climate configuration of the Met Office Unified Model (MetUM), HadGEM2-AO. The fluxes were shown to be close to equilibrium, with the exception of the atmosphere. This apparent imbalance has been traced to a lack of water conservation in two areas of the atmospheric model and has since been corrected (J. Rodriguez, personal communication, May 2012). In addition, Collins et al. (2011) showed that a net freshwater flux field must be applied to the ocean in order to balance the lack of iceberg-calving processes in the GCM (see Sect. 2.3.4), a problem which is common to many climate models.

Global water conservation is usually represented with an equation that involves vertically integrated quantities:

$$\frac{\partial W}{\partial t} + \nabla_{\mathrm{H}} \cdot \mathbf{Q} = E - P$$

where the first term on the left-hand side represents the precipitable water tendency, the second term represents the moisture flux divergence, E is the evapotranspiration from the surface and P the precipitation at the surface. Excess of evaporation over precipitation is balanced by the local rate of change of moisture storage and the loss through horizontal advection. Budget equations like the one above have been used before to estimate quantities like moisture, energy and momentum, deducing the fluxes as a residual, after computing the other variables in the budget equation (e.g., Trenberth et al. 2007).

Rodríguez et al. (2011) used this method to estimate the freshwater fluxes from the atmosphere and land surface over large-scale ocean basins, in both a range of MetUM models and in reanalyses. This showed an excess of evaporation in the tropical-subtropical Atlantic in the models, which is manifested mainly as a lack of water vapour imported across the African boundary. Models also have difficulties in representing the surface fluxes over the tropical-subtropical Pacific. The freshwater budget in this region suffers from too much import of water vapour across the Indonesia boundary and an excessive export of water vapour to the mid-latitude regions. This surplus of water vapour from the tropical-subtropical regions makes, in turn, the surface fluxes over the oceans too fresh at mid-latitudes and can also produce a fresh bias over the Southern Ocean.

The above errors are consistent with an excessively strong hydrological cycle. Water vapour recycles between 12 and 24 % faster in climate models, compared with estimates from observations (Rodriguez et al. 2011). The shorter water vapour residence times in the GCMs are the effect of a combination of an enhanced global precipitation rate and a deficit in the water content of the atmosphere (e.g., Demory et al. 2012). These characteristics are related to the tendency for models to rain more frequently and with less intensity than observed (Stephens et al. 2010).

The characteristics of the water cycle in various regions in the globe are now being studied. Regional water budgets on different timescales are being evaluated, and the capacity of the GCMs to simulate the workings of the water cycle in those areas is being assessed. Preliminary analysis of the monthly mean water budget components in Central Africa in a recent climate configuration of the MetUM (J. Rodriguez, personal communication, May 2012) suggests that, although the seasonal variability of rainfall is well represented, the moisture convergence is underestimated for most of the year. This makes local evaporation the dominant source of model precipitation for most of the year, instead of moisture convergence from large-scale circulation as suggested by reanalyses. This may be related to model resolution, as discussed in the next section.

2.2 The Role of Model Resolution

It is widely believed that scale interactions (between short/small and long/large scales in time and space, and vice versa) are an important feature of the climate system, particularly in the tropics (e.g., Slingo et al. 2003). Consequently, a change in the resolution of a climate model can be expected to affect the simulated climate, not only at the scale of the grid point/time-step but also at much larger scales. Examples where such impacts have been documented include the simulation of ocean eddies, and associated improvements in the mean ocean state, in the Hadley Centre Coupled Eddy-permitting model (HadCEM; Roberts et al. 2004), improvements in the simulated tropical climate due to the representation of small-scale atmosphere–ocean interactions in the High-resolution Global Environment Model (HiGEM; Shaffrey et al. 2009), and improvements in mid-latitude variability through reduced North Atlantic SST biases (Scaife et al. 2011).

These and many other studies have resulted in several modelling centres actively pursuing increased horizontal and vertical resolution in various components of both regional and, increasingly, global models for a range of timescales. The results of some of these advances are mentioned in the sub-sections of Sect. 2.3, while, in the next sub-sections, we focus specifically on the relationship between model resolution and global precipitation, and the issue of resolved versus parametrised processes.

2.2.1 Influence of Resolution on Global Precipitation

In order to understand the systematic impact of model resolution in the simulation of the hydrological cycle, it is useful to use hierarchy of models with the same physical configuration. Demory et al. (2012) used a series of simulations of the MetUM climate configuration HadGEM1 (Martin et al. 2006) and a prototype configuration of HadGEM3 (Hewitt et al. 2011), with a wide range of horizontal resolutions, to investigate the impacts of resolution on the mean simulated model precipitation at large spatial scales. Demory et al. (2012) found that there is a redistribution of precipitation from ocean to land with resolution and that the ratio of E to *P* over land decreases with resolution (Fig. 2, black dots and circles), bringing the ratio closer to the range of observed values estimated by Trenberth et al. (2007, 2011). This suggests that higher resolution models decrease moisture recycling over land.

These changes are associated with an increase in the transport of moisture from ocean to land at higher resolution (red squares on Fig. 2). By decomposing the moisture convergence into mean and transient (which is related to eddies) terms, it has been found that the increased moisture convergence mainly comes from the mean flow, but that the transient term, associated with storminess, becomes more important with increased resolution (M.-E. Demory, personal communication, May 2012). Demory et al. (2012) conclude that a resolution of at least 50 km is necessary to simulate the mean global hydrological cycle.

2.2.2 Resolved Versus Parametrised Processes in Models

With the increase in computing power in recent years, the influence of resolving hydrological processes, compared with using physical parametrisations, on the distribution and variability of precipitation in global models has become a key area of interest among

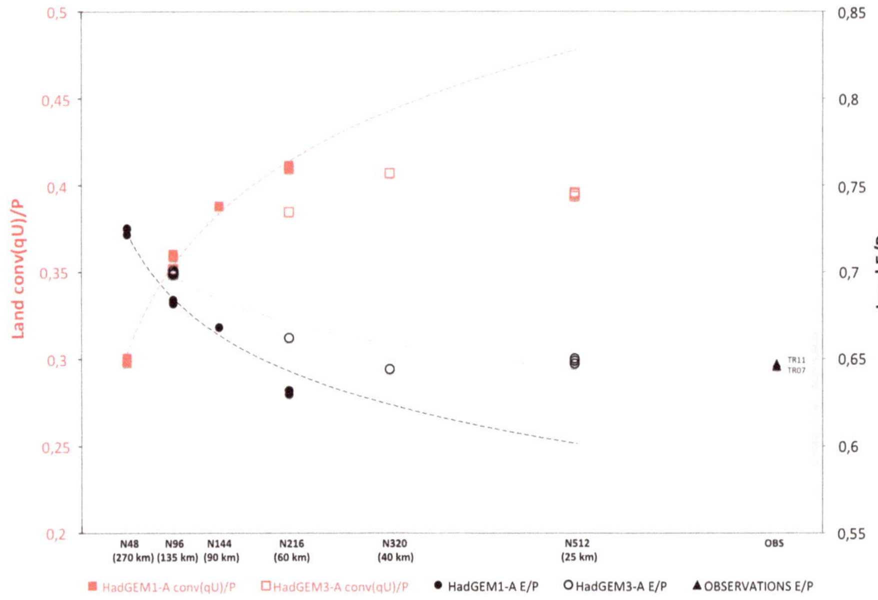

Fig. 2 Evaporation to precipitation ratio (*black dots*) and moisture convergence to precipitation ratio (*red squares*) over land for each member of HadGEM1-A (*solid*) and HadGEM3-A (*clear*) models with various resolutions. The *lines* are extrapolated polynomial fits based on N48, N96 and N144 for HadGEM1-A and N96, N216, N320 and N512 for HadGEM3-A. OBS corresponds to E/P estimates from Trenberth et al. (2007, 2011). The *grey bar* includes estimates from previous studies. See Demory et al. (2012) for more details. Figure provided by M-E Demory, May 2012

modelling centres. Many groups are now using very high resolution (~km-scale) simulations, in which convection is represented explicitly without the need for a convective parametrisation scheme, in order to evaluate and understand the deficiencies in the simulations which use parametrised convection. Such studies reveal that convection parametrisations are often associated with poor representation of the diurnal cycle of precipitation over land (e.g., Bechtold et al. 2004; Wang et al. 2007; Hohenegger et al. 2009; and see Sect. 2.3.1.2), over-persistence of light rain and too few dry days (e.g., Stephens et al. 2010; Kendon et al. 2012) and differences in the sign of simulated soil moisture–precipitation feedback (e.g., Hohenegger et al. 2009; and Sect. 2.3.2).

Variability on longer timescales (e.g., 20–50 day variability associated with the Madden Julian Oscillation (MJO), monsoon active/break cycles, etc.) is also often poorly represented in models (e.g., Kim et al. 2009), both in terms of total variance and also the direction, speed and extent of propagation. Predicting such variations, and understanding how they may change in a future climate, is crucially important to many tropical regions, and is therefore a key area of research among modelling groups. 'Cascade', a UK consortium project funded by the Natural Environment Research Council (NERC), seeks to better understand the interaction between tropical convection at the cloud-system scale and larger-scale processes including the MJO. Holloway et al. (2012) showed 'Cascade' simulations of the tropical atmosphere over a very large domain at several different horizontal resolutions and with both parametrised and explicit convection versions for a 10-day MJO case study in April 2009. They found that the explicit convection simulations had precipitation distributions that were much more similar to observations than the simulations with parametrised convection, although they found that this did not necessarily result in a better MJO simulation.

Comparing precipitation from a 1.5 km resolution regional model simulation over the UK with gridded hourly radar rainfall, Kendon et al. (2012) showed that, although the 1.5 km model appears to have a tendency for heavy rain to be too intense, it gives a much better representation of its spatial and temporal structure. With parametrised convection, heavy rain events are not heavy enough and tend to be too persistent and widespread. There are also not enough short-duration high-peak intensity events. These biases are significantly reduced in the 1.5 km model. In addition, the 1.5 km simulation shows a much better representation of the occurrence of dry days.

Similar discussions regarding model horizontal resolution concern the storage, movement, and quality of water at and near the land surface. A recent opinion paper by Wood et al. (2011) discusses the need for much higher horizontal resolution in continental- or even global-scale models in order to include detailed information about these processes. They suggest that "Adequately addressing critical water cycle science questions and applications requires systems that are implemented globally at much higher resolutions, on the order of 1 km, resolutions referred to as hyperresolution in the context of global land surface models". However, they note the need for improvements in satellite remote sensor resolution and the development of advanced downscaling methodologies, as well as improvements to in situ observation networks, in order to support such a modelling effort.

We have described research which indicates that improving model horizontal resolution can have significant benefits for modelling the hydrological cycle. However, improved representation of hydrological processes through better physical parametrisations is also crucial. This is discussed in the next section.

2.3 Representing Hydrological Processes and Phenomena

As mentioned in the introduction, although great advances have been made in global modelling in recent decades, there are still large uncertainties in many processes, such as clouds, convection, and coupling to oceans and the land surface. In the previous section, we discussed the influence of model resolution on these uncertainties. In the next subsections, we highlight the role of model physical parametrisations in the representation of cloud microphysics, convection and precipitation processes, in the interactions between the atmosphere and the land, the oceans and the cryosphere, and in the representation of one of the largest seasonal phenomena in the hydrological cycle: the monsoons.

2.3.1 Clouds, Humidity and Radiation

The surface radiation balance is a key determinant of the intensity of the GWC. Decadal variations in the surface radiation balance during the 20th century are reflected in the variations in intensity of the GWC, but these are not captured in models. The simulation of cloud processes and feedbacks by GCMs remains one of the most critical aspects of climate modelling. In particular, cloud radiative feedbacks remain the primary source of uncertainty for transient and equilibrium climate sensitivity estimates (e.g., Randall et al. 2007; Dufresne and Bony 2008) and play a critical role in anthropogenic aerosol-induced climate forcing (Lohmann and Feichter 2005).

In addition, clouds play a key role in the hydrological cycle and in the large-scale atmospheric circulation, at both planetary and regional scales. By affecting precipitation and atmospheric dynamics, uncertainties in cloud and moist processes remain a major concern for virtually all aspects of climate modelling and climate change research. In a context where the climate modelling community is increasingly focussing its efforts on regional climate change impacts and biogeochemical (e.g., carbon and aerosols) climate feedbacks, improving our understanding of cloud-climate interactions and assessing our confidence in the simulation of cloud processes and feedbacks in climate models is imperative.

In the following sub-subsections, we discuss current research on the parametrisation of cloud and precipitation processes, we highlight common systematic errors such as failure to reproduce the diurnal cycle of clouds and precipitation over land, and we discuss issues surrounding the representation of hydrological processes in the mid-latitudes.

2.3.1.1 Model Representations of Cloud and Precipitation and Their Evaluation The formulation of cloud microphysical parametrisations is very important for simulation of the hydrological cycle and for model evolution because they modify the three-dimensional structure of temperature and humidity directly (e.g., condensation/evaporation) or indirectly by interacting with other parametrisations (e.g., radiation) and the large-scale dynamics. Therefore, the evaluation and improvement of these parametrisations is crucial to improving our weather forecasts or increasing our confidence in climate projections.

Improvements to the representation of clouds, humidity and radiation in models have been a focus in several modelling groups in recent years (e.g., Collins et al. 2006; Wilson et al. 2008; Salzmann et al. 2010). For example, in the MetUM, the new PC2 cloud scheme (with prognostic cloud and condensate) improves cloud and humidity distributions and, in combination with more advanced aerosol schemes, results in improved radiation balance (Walters et al. 2011).

Model Intercomparison projects, including the Third Climate Model Intercomparison Project (CMIP3), have always exhibited a large range of cloud-climate feedbacks (Webb et al. 2006; Dufresne and Bony 2008). There are so many factors or physical processes that may potentially contribute to this spread, that interpreting the origin of inter-model differences has turned out to be difficult, and that designing specific observational tests to assess the different feedbacks has remained elusive.

Satellites have proven to be very helpful tools for model evaluation because they provide global or near-global coverage, thereby giving a representative sample of all meteorological conditions. However, satellites do not measure directly those geophysical quantities of interest, such as the amount or phase of cloud condensate. They measure the intensity of radiation coming from a particular area and direction in a particular wavelength range (Bodas-Salcedo et al. 2011). A great deal of research has been conducted into producing satellite retrievals of many different geophysical variables, such as water vapour, atmospheric temperature, cloud properties and land surface products (e.g., Chahine et al. 2006; Wylie et al. 2005; Schaaf et al. 2002). Satellite retrievals have been used in numerous studies to analyse the performance of NWP and climate models (e.g., Allan et al. 2007; Pincus et al. 2008).

In the last two decades, a different avenue has been followed to exploit satellite data in model evaluation: the use of forward modelling of basic satellite measurements from model fields (Bodas-Salcedo et al. 2011). Simulators have been developed that mimic the observational process and essentially acknowledge the issue that a retrieval produced by a satellite might not be directly comparable to a model variable, giving rise to multiple values of geophysical quantities from different sensors and retrieval algorithms. A pilot model intercomparison using the CFMIP Observation Simulator Package (COSP;

Bodas-Salcedo et al. 2011) was carried out using four climate models. The study focussed on a region in the north Pacific, characterised by a rich mixture of different cloud types, and showed that, while all models capture the basic features of these diagnostics, they also show a lack of low-level, non-precipitating cloud (cumulus congestus and mixed-phase layered clouds). In future, this analysis will be extended to other models as part of the second Cloud Feedback Intercomparison Project (CFMIP-2) which is part of CMIP5 (see http://cfmip.metoffice.com/CFMIP2.html).

2.3.1.2 Diurnal Cycle of Clouds and Precipitation There is a vast amount of evidence that shows that the diurnal cycle of precipitation is misrepresented in models (e.g., Dai and Trenberth 2004; Bechtold et al. 2004; Stratton and Stirling 2011). All show that the peak in precipitation and outgoing longwave radiation over land occurs several hours too early. Realistic representation of the diurnal cycle of clouds and precipitation in models is important because the diurnal cloud–sun correlation rectifies onto the mean radiation balance, affecting climate simulation and weather prediction (Wang et al. 2007, and references therein). Studies with high-resolution models which explicitly represent convection show improved representation of the diurnal cycle (e.g., Clark et al. 2007; Hohenegger et al. 2009; Love et al. 2011; Kendon et al. 2012; and see Fig. 3), with convection peaking later in the day. Studies with global climate models which use "superparametrisation" techniques (where thousands of embedded cloud-resolving models (CRMs) are used to handle sub-grid cloud physics instead of parametrisations) show similar improvements in the diurnal cycle of rainfall over land (e.g., Pritchard 2011).

Wang et al. (2007) attempted to address this shortcoming of convective parametrisations by changing the entrainment/detrainment rates for deep and shallow convection. They found that this prolonged the development and reduced the strength of deep convection,

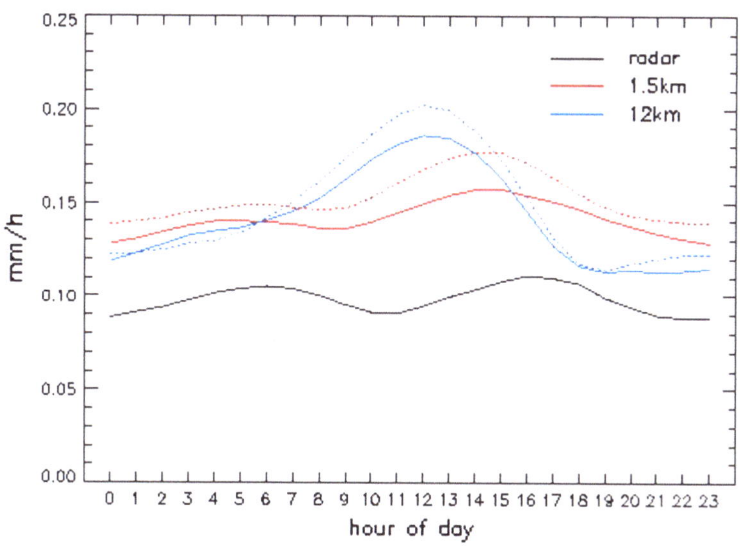

Fig. 3 Diurnal cycle of rainfall (mm/h) in radar and 1.5 and 12 km Regional Climate Models. Results correspond to years 2003–2010 in the radar, and years 1990–1997 (*solid*) and 2000–2007 (*dotted*) in the models. Plotted is the mean rainfall at each hour of the day, across all southern UK grid boxes and all 8 years in the respective data sets. From Kendon et al. (2012). © Copyright 2012 AMS (http://www.ametsoc.org/pubs/cr_2005.html)

thus delaying the mature phase and reducing the amplitude of the convective precipitation diurnal cycle over the land. In addition, they also found changes in the temporal variability of daily mean precipitation and the partitioning of stratiform and convective rainfall in the model. Stratton and Stirling (2011) made a similar attempt by changing the deep convective parametrisation over land in the MetUM to make the entrainment vary with the height of the lifting condensation level. This reduced the depth of the convection early in the day and delayed the development of deep convection by around 2 h, thereby improving the timing of maximum precipitation. The amplitude of the diurnal harmonic of precipitation was also improved. However, although the amount of precipitation in the evening was increased, Stratton and Stirling (2011) noted that further improvements are required for the decay phase of the diurnal cycle.

2.3.1.3 Precipitation in Mid-latitude Cyclones Mid-latitude cyclones are the primary means of transporting energy and moisture poleward at mid-latitudes and can bring severe weather in terms of heavy rain, leading to flooding, and damaging strong winds to major population centres. Therefore, determining whether the physical parametrisation of clouds and cloud properties used in models are providing realistic representations of these phenomena is important. Recent studies have shown a response of the storm tracks in climate change experiments, with the storm track moving poleward in some models (e.g., Meehl et al. 2007). This seems to be accompanied by a consistent poleward shift of mid-latitude precipitation (e.g., Held and Soden 2006). However, the poleward shift in the storm track is not seen in all models or in all locations. Also, the change in the distribution of the intensity of cyclones under climate change is not consistent across models, and the mechanisms causing these changes are still unclear.

A good simulation of mid-latitude cyclones is a prerequisite to capturing the main characteristics of extreme events in the present-day climate and to having confidence in the response of cyclones to climate change. Field et al. (2011) assessed how well the cloud and precipitation is predicted in the MetUM by physical parametrisations in the presence of an accurate thermodynamic and dynamic representation of the atmosphere. The main differences seen between the model and the satellite data were a lack of cloud above 2 km in the model associated with the main precipitation region.

Representing mid-level layer cloud is a challenge for global models (e.g., Webb et al. 2001). One significant obstacle is that the thickness of model levels at altitudes of $\sim$15 km is typically around 600 m, which is similar to the thickness of altostratus/cumulus layers (Field et al. 2011). The second challenge is that even if thin mid-level supercooled liquid cloud can be produced by a model, it will be efficiently transformed to ice by the ice nucleation scheme and fall out. This formation of ice reduces both the lifetime and radiative effect of these clouds. This has possible implications for the radiative effect of mid-latitude cyclones on the climate system (Field et al. 2011). Increased vertical resolution is now being implemented in many models, and studies such as Field et al. (2011) are informing the development of microphysical parametrisation schemes.

2.3.2 Land–Atmosphere Interactions

The land surface is an important component of the climate system. It controls the partitioning of available energy at the surface between sensible and latent heat, and the partitioning of available water between evaporation and runoff. The state of the land surface has the potential to influence large-scale circulations (Taylor 2008). Variations in rainfall

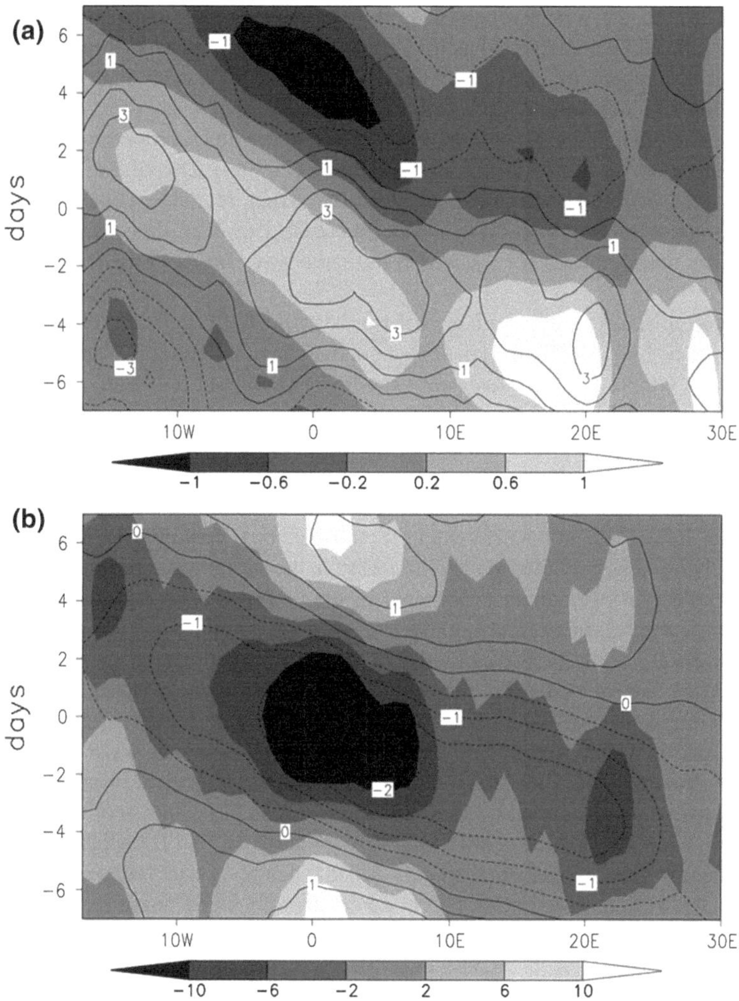

Fig. 4 Evolution in time and longitude of the differences (wet minus dry) in intraseasonal composites (defined by minima (wet) and maxima (dry) in filtered surface heat fluxes averaged over a 12.5°–17.5°N, 2.5°W–2.5°E subdomain), averaged between 12.5° and 17.5°N. **a** Meridional wind at 925 hPa (*shaded*; m s^{-1}) and cold cloud coverage at a threshold of −40 °C (contour; %). **b** Estimated surface heating (*shaded*; W m^{-2}) and 925-hPa temperature (contours; °C). From Taylor (2008). © Copyright 2008 AMS (http://www.ametsoc.org/pubs/cr_2005.html)

influence soil moisture and vegetation, which in turn affect the partition of surface heat and moisture fluxes into the atmosphere. If the large-scale circulation is sensitive to surface fluxes, and the surface responds to climate anomalies with sufficient amplitude and spatial coherence, conditions for feedback may be met. Monsoon circulations potentially provide these conditions, driven by gradients in ocean and land surface fluxes, with the transport of oceanic moisture into the continent being sensitive to land surface conditions.

Taylor (2008) demonstrated that intraseasonal feedbacks can occur between soil moisture and rainfall in monsoon regions such as West Africa (Fig. 4). Increases in soil moisture associated with a wet spell induce a cool high pressure anomaly, favouring southerly flow to the west and northerly flow to the east. The induced meridional flow influences rainfall via moisture advection, triggering a westward shift in rain, soil moisture, and low-level temperature. Thus, maxima in cold cloud (indicating deep convection) tend to follow maxima in southerly flow with a lag of the order of 1 day (Fig. 4a), and negative surface heating anomalies due to increased convection, with associated negative low-level temperature anomalies, also propagate westwards (Fig. 4b). Douville et al. (2001) demonstrated that increased soil moisture over the Sahel region is also associated with increased monsoon rainfall on the seasonal timescale. However, they also showed that the South Asian monsoon does not respond in the same way. This is discussed further in Sect. 2.3.5.

Koster et al. (2004) queried whether there exist specific locations ("hot spots") on the Earth's surface for which soil moisture anomalies have a substantial impact on precipitation. In the Global Land Atmosphere Coupling Experiment (GLACE), a dozen climate modelling groups performed the same highly controlled numerical experiment as part of a coordinated comparison project, allowing a multi-model estimation of the regions on Earth where precipitation is affected by soil moisture anomalies during Northern Hemisphere summer. These included the central Great Plains of North America, the Sahel, equatorial Africa, and India, and, to a lesser extent, South America, central Asia, and China. However, there was extensive inter-model variability in the strength and positioning of the "hot spots", a reflection of ongoing uncertainty in the proper way to represent in models the physical processes defining land–atmosphere coupling strength.

Comer and Best (2012) applied the GLACE approach to a recent climate configuration of the MetUM and showed a significant strengthening of the atmosphere's response to soil moisture compared with the earlier version, HadAM3 (the third Hadley Centre Atmospheric Model), which was one of the weakest of the models assessed in the GLACE intercomparison project. This strengthening was particularly evident over the Sahel region of Africa, which was identified by the GLACE project as a key "hot-spot" region for land–atmosphere coupling (Koster et al. 2004). Comer and Best (2012) could not attribute this improvement to developments in the land surface scheme; instead, it appeared to be mainly due to improvements in the atmospheric parametrisations and possibly also the model's vertical resolution. In addition to testing the coupling strength in the updated model, Comer and Best (2012) also demonstrated a dependency of the coupling strength on the way in which soil parameters are specified.

Understanding the nature and timescale of land–atmosphere coupling is an area of active research around the world. Various studies have shown that there may be both positive and negative feedbacks between soil moisture and precipitation depending on the time and space scales considered (e.g., Cook et al. 2006; Taylor 2008; Hohenegger et al. 2009; Schlemmer et al. 2012; and references therein). Observations and idealised modelling studies are being used to improve our understanding of land–atmosphere interactions. For example, Schlemmer et al. (2012) investigated the influence of soil moisture and atmospheric stability on mid-latitude diurnal convection and land–atmosphere exchange using an idealised cloud-resolving modelling framework and demonstrated a positive soil moisture–precipitation feedback on a scale of 10–50 km. In contrast, Cook et al. (2006) showed a negative soil moisture–precipitation feedback in southern Africa that was related to stronger atmospheric stratification over wet soils and to the formation of anticyclonic circulations that induced subsidence and divergence at the surface. Hohenegger et al. (2009) found a similar stabilization of the atmospheric profile over wet soil in summertime Europe. Furthermore, they found different signs of the soil moisture–precipitation feedback in simulations using parametrised convection and simulations using explicitly resolved convection.

Increasing greenhouse gas concentrations are expected to enhance the interannual variability of summer climate in Europe and other mid-latitude regions, potentially causing more frequent heatwaves (e.g., Schär et al. 2004; Clark et al. 2010). Seneviratne et al. (2006) showed that the increase in summer temperature variability predicted in central and eastern Europe is mainly due to feedbacks between the land surface and the atmosphere. Furthermore, they suggested that land–atmosphere interactions increase climate variability in this region because climatic regimes in Europe shift northwards in response to increasing greenhouse gas concentrations, creating a new transitional climate zone with strong land–atmosphere coupling in central and eastern Europe. This highlights the crucial role of land–atmosphere interactions in future climate change.

Several of the climate models participating in CMIP5 include dynamic vegetation models (e.g., Collins et al. 2011). Systematic biases in the distribution and variability of vegetation types in such models may affect surface fluxes of heat and moisture, which themselves may influence atmospheric temperature and humidity profiles, thereby affecting cloud formation. Large-scale circulation may also be affected, which could have an impact on non-local sources of precipitation. Martin and Levine (2012) showed how South Asian summer monsoon rainfall is affected by the inclusion of a dynamic vegetation scheme in a member of the HadGEM2 model family. They found that systematic increases in the bare soil fraction in key dust-producing regions, arising due to systematic dry rainfall biases particularly over India, affected the monsoon circulation through changes in the radiative balance, while changes in the needleleaf tree fraction over northern Eurasia also affected the monsoon through changes in winter snow cover. Other work is also ongoing to understand how such changes in the representation of the land surface during current and past climate conditions may affect the projected changes in temperature and precipitation in future.

2.3.3 Ocean–Atmosphere Processes and Interactions

The importance of coupled ocean–atmosphere interactions in climate variability and change has long been established. Of particular importance is the simulation of the El Nino–Southern Oscillation (ENSO), a coupled phenomenon whereby warming/cooling in the tropical Pacific Ocean that takes place at intervals of 2–7 years is associated with a large-scale tropical east–west seesaw in southern Pacific sea level surface pressure. Although ENSO originates in the tropical Pacific, it affects global climate and weather events such as drought/flooding and tropical storms. Therefore, understanding and predicting ENSO are crucial to both the scientific community and the public. Simulating the time-mean properties in the tropics has continually been a challenge for coupled GCMs. Though most models can internally generate the fundamental mechanisms that drive El Niño properties, most models simulate a mean zonal equatorial wind stress that is too strong and that has an annual amplitude that is also too strong (Guilyardi et al. 2009). This has profound effects on ENSO behaviour in that it limits the regimes in which interannual anomalies can develop.

Persistent systematic errors have been noted in several generations of global coupled models (Randall et al. 2007). These include a double intertropical convergence zone (ITCZ) pattern with excessive precipitation off the equator but insufficient precipitation on the equator, which is often associated with an excessive and overly narrow sea surface temperature (SST) cold tongue that extends too far west into the western Pacific. SST

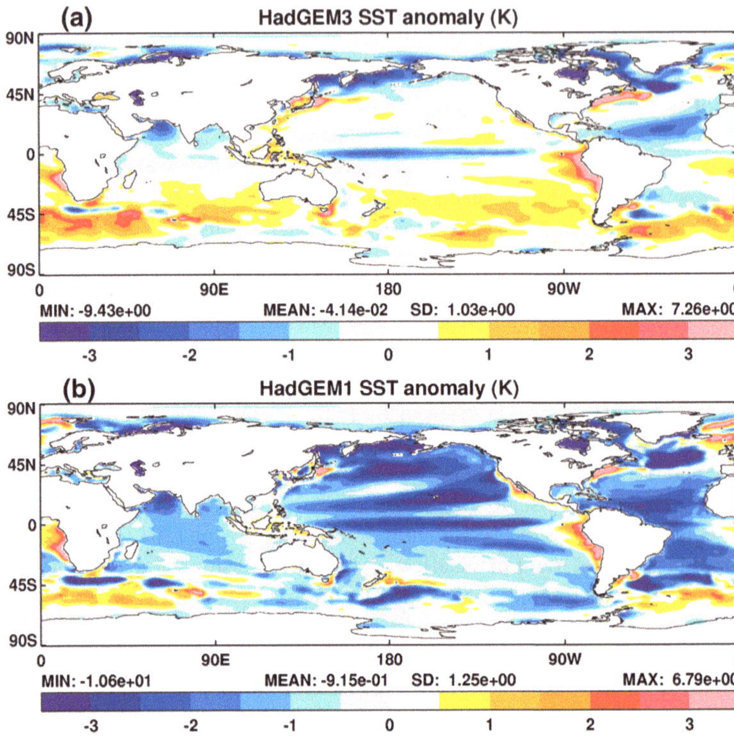

Fig. 5 Sea surface temperature (K) anomalies with respect to initial conditions for **a** HadGEM3 r1.1 and **b** HadGEM1. From Hewitt et al. (2011). © Hewitt et al. (2011). This work is distributed under the Creative Commons Attribution 3.0 License

biases and poor variability are also seen in the tropical Atlantic and Indian Oceans, affecting the simulation of monsoons in these regions, and also in the midlatitude regions of the Atlantic and the Southern Ocean, affecting the storm tracks (e.g., Scaife et al. 2011).

Development of ocean models with improved vertical and horizontal resolution and physical parametrisations, in parallel with similar development of their atmospheric counterparts, is helping to reduce SST biases and improve variability (see Sect. 2.2). The HadGEM2 Development Team (2011) showed how improvements in both the atmospheric physical parametrisations (which resulted in improved near-surface winds across the equatorial Pacific) and changes to the background vertical tracer diffusivity in the upper 500 m of the ocean led to improved SSTs and better simulation of ENSO in the HadGEM2 model family. Several recent climate configurations use the NEMO (Madec 2008) ocean model (e.g., the MetUM (Hewitt et al. 2011), EC-Earth (Hazeleger et al. 2011) and CNRM-CM5.1 (Voldoire et al. 2012)). This can also be associated with improved SSTs (e.g., Fig. 5).

Ocean circulations are a key part of the global water cycle. The climate of Europe is strongly influenced by the North Atlantic ocean circulation. Variations of the strength of the Thermohaline Circulation (THC) or the Meridional Overturning Circulation (MOC) are in several studies implicated as a main driver for decadal and longer timescale changes for European and Northern hemisphere climate (e.g., Mignot et al. 2007, and references therein). Likewise, variation in the THC is a commonly attributed mechanism for nonlinear and abrupt (i.e. decadal scale) climate changes. Yet the observational and model underpinning of these hypotheses are at best sketchy making it very difficult to come to firm conclusions. Reliable quantification of the variability and stability of the THC and its atmospheric implications in the current and future climate are therefore a major challenge in climate research.

Projects such as "Thermohaline Overturning—at Risk?" (THOR; http://www.eu-thor.eu/THOR-in-short.532.0.html) and RAPID-WATCH: Monitoring the Atlantic Meridional Overturning Circulation (http://www.noc.soton.ac.uk/rapid/rw/index.php) aim to improve our understanding of the Atlantic MOC. Areas of current work include quantification of the risk of future shutdown of the MOC driven by salinity changes in the North Atlantic and Arctic Oceans under climate change. Under climate change, there may be an increase in the amount of freshwater input into the North Atlantic Ocean due to increased precipitation and melt water from the Greenland ice sheets. By reducing the density of the surface waters, this could stop the sinking of dense water in the North Atlantic and lead to a slow down or even a shutdown of the MOC (e.g., Mignot et al. 2007, and references therein).

Mechanisms of natural variability are also being investigated including interactions between the MOC, extratropical salinity anomalies and the position of the inter-tropical convergence zone on centennial timescales. As part of the RAPID-WATCH project, work is underway using models to find variables that could be used to detect a significant trend in the strength of the MOC earlier than is possible by observing it directly. Examples could include temperatures and salinity in certain areas of the Nordic Seas.

2.3.4 The Role of the Cryosphere

The cryosphere encompasses the regions of the Earth's surface where water is in solid form, including sea ice, lake ice, river ice, snow cover, glaciers, ice caps and ice sheets, and frozen ground (which includes permafrost). Although the more spectacular parts of the hydrological cycle may be apparent in the tropics and subtropics, the cryosphere is an integral part of the global climate system, with important linkages and feedbacks generated through its influence on surface energy and moisture fluxes, clouds, precipitation, hydrology, atmospheric and oceanic circulation (see papers included in Bengtsson et al. 2011).

Sea level rise is arguably the most certain consequence of a warming climate, due to thermal expansion of water and the dependence of ice sheets and glaciers on low temperatures. It is currently thought that melting of the major ice sheets contribute around 1.2 mm of the approx. 3 mm of annual sea level rise (Lemke et al. 2007). An understanding of how this will change in the future is a policy driver. The key issue is not whether sea level will rise, but by how much and how fast. The West Antarctic ice sheet and Greenland ice sheets each contain enough land-based ice to raise sea level directly by several metres. The rapid disintegration of either of these ice sheets could cause sea level rise that is too great, or too fast for many coastal populations and ecosystems to adapt to. In addition to sea level rise, fresh water from melting of the Greenland ice sheet may contribute to a slowdown of the North Atlantic overturning circulation.

The IPCC Fourth Assessment Report (Meehl et al. 2007) acknowledged that current models do not adequately treat the dynamic response of ice sheets to climate change, and that this is the largest uncertainty in assessing potential rapid sea level rise. Many current climate models do not include an ice sheet model; in order to close the global water budget, the accumulation of frozen water on the permanent ice sheets is often returned to the freshwater cycle via a freshwater flux field (invariant in time) that is applied to the ocean (e.g., HadGEM2 Development Team 2011). This is necessary because there is no explicit representation in the model of the flow of ice from accumulation regions to iceberg calving at the coast.

The challenge for inclusion of ice sheets in GCMs is to develop the physical processes of interaction between ice sheets and oceans. It is this interaction that has caused the thinning of ice shelves and the speed-up of peripheral glaciers in both Greenland and Antarctica. The West Antarctic ice sheet is grounded below sea level and 3 m of sea level rise could result with glacial calving exceeding the supply of ice from the interior, as the coastline retreats poleward. Ice sheet models (e.g., Rutt et al. 2009) are being used to investigate the contribution of glacier speed-up and increased ice discharge as icebergs to sea level rise.

Mountain glaciers and ice caps include only a minor fraction of all water on Earth bound in glacier ice (<1 %) compared with the Antarctic and Greenland ice sheets (>99 %), but their retreat has dominated the eustatic sea level contribution in the past century (Meier et al. 2007). Mountain glacier schemes are being developed and used alongside climate change models to make projections of future changes in sea level due to volume changes in mountain glaciers and ice caps (e.g., Radić and Hock 2011).

Sea ice plays an important role in the climate system through high surface albedo, insulating the ocean, and influencing the ocean salinity through brine rejection when ice forms and surface freshening when ice melts. Changes in sea ice under a warming scenario significantly influence the local water cycle. The Los Alamos sea ice model, CICE, has been introduced into a number of climate models (e.g., Hewitt et al. 2011; Holland et al. 2012). The use of this community model means that implementations of new physics can be shared but still allows different modelling centres to use different dynamics and parametrisations, maintaining model diversity.

2.3.5 Global Monsoons

The monsoons in India, Africa, East Asia represent the largest seasonal redistribution of water within the hydrological cycle. They are also crucial to the economies of those countries and the livelihoods of the local population. Accurate predictions of monsoon onset, variations within the season and the overall seasonal rainfall amount and its regional distribution, and how these may change in the future, are required for effective agricultural and water resource management. We therefore focus this sub-section on the wide-ranging studies of monsoon processes that are underway around the world.

The monsoons are complex large-scale climate phenomena whose simulation has proved a challenge for modellers for several decades. Many studies have shown sensitivity to convection and boundary layer parametrisation (e.g., Mukhopadhyay et al. 2010 and references therein; Hong 2010 and references therein), cloud microphysics and land surface properties (e.g., Douville et al. 2001; Yasunari et al. 2006), as well as model resolution (e.g., Kim et al. 2008). Idealised sensitivity experiments have shed light on the major forcing regions, such as orographic forcing from the Himalayas and the Tibetan plateau (e.g., Boos and Kuang 2010) and the western Ghats (e.g., Wang and Chang 2012), land–sea contrasts between the Indian peninsula and the surrounding ocean, and sea surface temperature forcing from the Arabian Sea, Bay of Bengal and the equatorial Indian Ocean (Levine and Turner 2011).

Analysis of the impact of changes to convection parametrisation has provided insight into how problems with the distribution of precipitation intensity affect mean monsoon rainfall biases. For example, Mukhopadhyay et al. (2010) showed that different seasonal

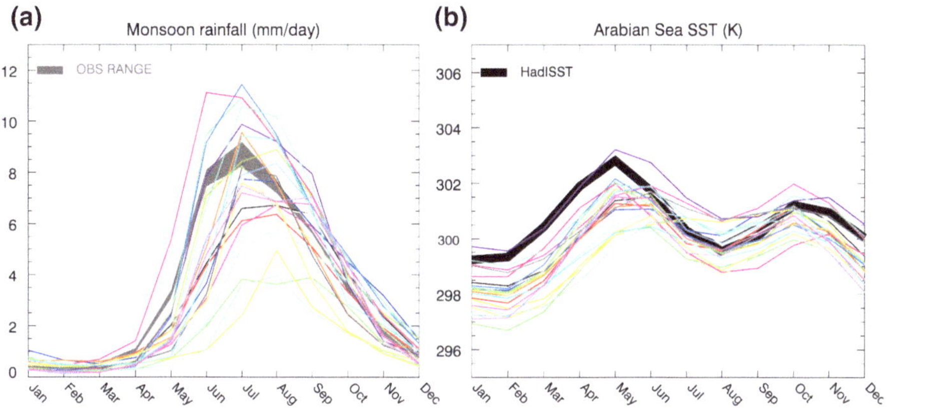

Fig. 6 Monthly mean seasonal cycle of **a** monsoon rainfall (65–95°E, 10–30°N) in mm/day for 24 CMIP5 models (*colours*) compared with a range of observational estimates (OBS; *thick black/grey lines*), and **b** Arabian Sea surface temperature (sea points within 50–75°E, 7.5–30°N) in K for CMIP5 models (*colours*) and HadISST (Rayner et al. 2003; *black line*). See Levine et al. (2012) for more details. Figure provided by R. C. Levine, May 2012

mean rainfall biases produced by different convection schemes applied within the same mesoscale model could be attributed to differences in the spread of rainfall rates produced by the schemes, which are related to differences in the amount and vertical distribution of diabatic heating.

Boos and Kuang (2010) showed that the strength of the South Asian monsoon is dependent on Himalayan orography, rather than elevated heating over the Tibetan Plateau. Their model study suggested that, although Tibetan plateau heating locally enhances rainfall along its southern edge, the large-scale South Asian summer monsoon circulation is otherwise unaffected by removal of the plateau, provided that the narrow orography of the Himalayas and adjacent mountain ranges is preserved. Additional observational and model results presented by Boos and Kuang (2010) suggested that these mountains produce a strong monsoon by insulating warm, moist air over continental India from the cold and dry extratropics.

Strong South Asian summer monsoons depend on moisture fluxes across the Arabian Sea. However, Arabian Sea cold SST biases are common in coupled models (Levine et al. 2012; and Fig. 6), and it has been shown that these significantly reduce monsoon rainfall (Levine and Turner 2011). Model projections for future monsoons indicate generally small positive changes in monsoon rainfall, while models also predict warming of Arabian Sea surface temperatures of a similar order to the magnitude of typical climate model biases in this region (Levine and Turner 2011). Therefore, it is possible that climate models with relatively large cold biases in the Arabian Sea are potentially underestimating the impact of greenhouse gas forcing and associated surface warming on the monsoon (Levine et al. 2012). Similarly, any models with warm Arabian Sea SST biases may overestimate increases of monsoon rainfall in future climate scenarios as a result of excessive acceleration of evaporation.

As mentioned in Sect. 2.3.2, although the basic drivers of the South Asian and West African monsoon systems are similar (i.e. the seasonal reversal of the land-sea temperature contrast and the resulting inflow of warm, moist air from the tropical ocean onto the landmass), the West African monsoon shows an additional significant dependence of rainfall on soil moisture on the seasonal timescale. Douville et al. (2001) showed that water recycling (where increased evaporation from wetter soil enhances precipitation) has a stronger influence on the West African monsoon than the South Asian monsoon because the moisture convergence in the latter region is much larger and itself responds to increased soil moisture in such a way as to counteract the local enhancement of precipitation. This illustrates that the soil–precipitation feedback on the monsoons is regionally dependent.

Studies such as that of Turner and Slingo (2011) have indicated that the cooling associated with increased Eurasian snow cover in boreal spring can be associated with weakening of the Tibetan anticyclone and weaker upper level easterly winds over the monsoon region in summer, as a result of the reduction in the tropospheric meridional temperature gradient. However, Turner and Slingo (2011) noted that this response could be reversed if the opposing changes over the Himalayas and Tibetan Plateau were dominant. Martin and Levine (2012) demonstrated that the inclusion of dynamic vegetation in the HadGEM2 model family generated a similar response through changes in the distribution of needleleaf trees, and the resulting impact on snow cover, over northeast Eurasia.

Simulating monsoons, therefore, requires adequate representation of hydrological processes in nearly all components of the climate system. Despite many decades of research focused on these phenomena, they remain a challenge for model development.

3 Summary

The challenge to quantify and reduce uncertainty in the large-scale response of the global water cycle is immense, not least because the hydrological cycle involves almost every component of the climate system. In this review, we have attempted to demonstrate the vast amount of research which is currently underway to improve our understanding of hydrological cycle processes and their representation in models. This includes observational analyses, systematic increases in model resolution, parametrisation development, inclusion of Earth system processes and idealised modeling studies.

The role of resolution in the representation of the hydrological cycle is complex: there is evidence that global models require a horizontal resolution of at least 50 km in order to represent the global hydrological cycle, while km-scale resolution facilitates realistic representation of rainfall distribution and variability on a local scale, and may be required for other areas of the hydrological cycle such as the storage and movement of water through the land surface. Increased vertical resolution may also be required in order that layer cloud processes can be represented.

The realism of rainfall in a model is a key indicator of its skill in representing the underlying physical processes, and hence for projecting future changes in rainfall. In particular, the spatial and temporal structure of rainfall is arguably more important than the absolute rainfall amount, which is typically used to assess model skill. We have highlighted studies which indicate that explicit representation of convective processes can improve the spatial and temporal structure of rainfall, but such models are still very expensive to run. Therefore, there is a need for such experiments to inform the development of improved parametrisations. Of particular interest in future will be whether models with explicit treatment of convection show a change in the spatial and temporal characteristics of heavy rainfall in a warmer climate, since this is unlikely to be captured by coarser resolution models. Accurately predicting such changes is essential for estimating changes in flood risk.

Development and evaluation of new physical parametrisations across timescales from daily/monthly/seasonal/decadal/centennial has been shown to be beneficial in several modelling centres. Seamless model development improves confidence in predictions across all timescales and enables separation of local versus remote influences. Idealised modelling studies are also useful for understanding hydrological cycle processes. As models become increasing complex, it is essential that the interactions between model processes and their systematic errors are understood and, where necessary, reduced, in order to facilitate increase confidence in, and better interpretation of, model projections.

We have highlighted issues in understanding and modelling processes within all of the components of the hydrological cycle, including atmosphere, land surface, ocean and cryosphere, and in the interactions between them. Although there is a large amount of research effort currently being expended in these different areas, there is a need for more coordinated collaborative activity in order that all may benefit. Studies of phenomena such as monsoons bring together several of these areas, as well as encompassing a wide range of timescales.

Multi-model studies such as CMIP5 permit evaluation and quantification of uncertainty in the response of the global water cycle to climate change. Systematic evaluation of the sensitivity of climate change projections in CMIP5 models to particular physical parametrisations (e.g., convection), processes (e.g., Earth system) and resolution (horizontal and vertical), along with the use of idealised climate change scenarios, will provide better understanding of how and why the hydrological cycle may change, and better estimation of uncertainty in model projections of changes in the global water cycle.

Acknowledgments The author gratefully acknowledges the contribution to this overview from members of the Improving Hydrological Predictions Working Group, which is supported by the Joint DECC/Defra Met Office Hadley Centre Climate Programme (GA01101). Particular thanks go to Marie-Estelle Demory and Richard Levine for providing Figs. 2 and 6.

References

Allan RP, Slingo A, Milton SF, Brooks ME (2007) Evaluation of the Met Office global forecast model using Geostationary Earth Radiation Budget (GERB) data. Q J R Meteorol Soc 133:1993–2010. doi:10.1002/qj.166

Bechtold P, Chaboureau J-P, Beljaars A, Betts AK, Kohler M, Miller M, Redelsperger J-L (2004) The simulation of the diurnal cycle of convective precipitation over land in a global model. Q J R Meteorol Soc 130:3119–3137

Bengtsson L, Koumoutsaris S, Bonnet R, Herland E-A, Huybrechts P, Johannessen O, Milne G, Oerlemans H, Ohmura A, Ramstein G, Woodworth P (eds) (2011) Special issue: ISSI workshop on the earth's cryosphere and sea level change. Surv Geophys 32:315–657. doi:10.1007/s10712-011-9136-0

Bodas-Salcedo A, Webb MJ, Bony S, Chepfer H, Dufresne J-L, Klein SA, Zhang Y, Marchand R, Haynes JM, Pincus R, John VO (2011) COSP: satellite simulation software for model assessment. Bull Am Meteorol Soc. doi:10.1175/2011BAMS2856.1

Boos WR, Kuang Z (2010) Dominant control of the South Asian monsoon by orographic insulation versus plateau heating. Nature 463:218–222. doi:10.1038/nature08707

Chahine MT, Pagano TS, Aumann HH, Atlas R, Barnet C, Chen L, Divakarla M, Fetzer EJ, Goldberg M, Gautier C, Granger S, Irion FW, Kakar R, Kalnay E, Lambrigtsen BH, Lee SY, Le Marshall J, McMillan W, McMillin L, Olsen ET, Revercomb H, Rosenkranz P, Smith WL, Staelin D, Strow LL, Susskind J, Tobin D, Wolf W (2006) The Atmospheric InfraRed Sounder (AIRS): improving weather forecasting and providing new insights into climate. Bull Am Meteorol Soc 87:911–926. doi:10.1175/BAMS-87-7-911

Clark AJ, Gallus WA, Chen T-C (2007) Comparison of the diurnal precipitation cycle in convection-resolving and non-convection-resolving mesoscale models. Mon Weath Rev 135:3456–3473. doi:10.1175/MWR3467.1

Clark RT, Murphy JM, Brown SJ (2010) Do global warming targets limit heatwave risk? Geophys Res Lett 37:L17703. doi:10.1029/2010GL043898

Collins WD, Rasch PJ, Boville BA, Hack JJ, McCaa JR, Williamson DL, Briegleb BP, Bitz CM, Lin S-J, Zhang M (2006) The formulation and atmospheric simulation of the Community Atmosphere Model version 3 (CAM3). J Clim 19:2144–2161

Collins WJ, Bellouin N, Doutriaux-Boucher M, Gedney N, Halloran P, Hinton T, Hughes J, Jones CD, Joshi M, Liddicoat S, Martin G, O'Connor F, Rae J, Senior C, Sitch S, Totterdell I, Wiltshire A, Woodward S (2011) Development and evaluation of an Earth-system model: HadGEM2. Geosci Model Dev 4:1051–1075. http://www.geosci-model-dev.net/4/1051/2011/gmd-4-1051-2011.html

Comer R, Best M (2012). Revisiting GLACE: understanding the role of the land surface in land-atmosphere coupling. J Hydrometeorol (accepted)

Cook BI, Bonan GB, Levis S (2006) Soil moisture feedbacks to precipitation in southern Africa. J Clim 19:4198–4206

Dai A, Trenberth KE (2004) The diurnal cycle and its depiction in the Community Climate System Model. J Clim 17:930–951. doi:10.1175/1520-0442(2004)017<0930:TDCAID>2.0.CO;2

Demory M-E, Vidale PL, Roberts MJ, Berrisford P, Strachan J (2012) The role of horizontal resolution in representing processes that drive the global hydrological cycle. Geophys Res Lett (submitted)

Douville H, Chauvin F, Broqua H (2001) Influence of soil moisture on the Asian and African monsoons. Part I: mean monsoon and daily precipitation. J Clim 14:2381–2403

Dufresne JL, Bony S (2008) An assessment of the primary sources of spread of global warming estimates from coupled ocean-atmosphere models. J Clim 21:5135–5144. doi:10.1175/2008JCLI2239.1

Durack PJ, Wijffels SE, Matear RJ (2012) Ocean salinities reveal strong global water cycle intensification during 1950 to 2000. Science 336:455–458. doi:10.1126/science.1212222

Field PR, Bodas-Salcedo A, Brooks ME (2011) Using model analysis and satellite data to assess cloud and precipitation in midlatitude cyclones. Q J R Meteorol Soc 137:1501–1515. doi:10.1002/qj.858

Guilyardi E, Wittenberg A, Fedorov A, Collins M, Wang C, Capotondi A, van Oldenborgh GJ, Stockdale T (2009) Understanding El Niño in ocean-atmosphere general circulation models: progress and challenges. Bull Am Meteorol Soc 90:325–340

HadGEM2 Development Team: Martin GM, Bellouin N, Collins WJ, Culverwell ID, Halloran PR, Hardiman SC, Hinton TJ, Jones CD, McDonald RE, McLaren AJ, O'Connor FM, Roberts MJ, Rodriguez, JM, Woodward S, Best MJ, Brooks ME, Brown AR, Butchart N, Dearden C, Derbyshire SH, Dharssi I, Doutriaux-Boucher M, Edwards JM, Falloon PD, Gedney N, Gray LJ, Hewitt HT, Hobson M, Huddleston MR, Hughes J, Ineson S, Ingram WJ, James PM, Johns TC, Johnson CE, Jones A, Jones CP, Joshi MM, Keen AB, Liddicoat S, Lock AP, Maidens AV, Manners JC, Milton SF, Rae JGL, Ridley JK, Sellar A, Senior CA, Totterdell IJ, Verhoef A, Vidale PL, Wiltshire A (2011) The HadGEM2 family of Met Office Unified Model climate configurations. Geosci Model Dev 4:723–757. doi:10.5194/gmd-4-723-2011, http://www.geosci-model-dev.net/4/723/2011/gmd-4-723-2011.html

Hazeleger W, Wang X, Severijns C, Stefanescu S, Bintanja R, Sterl A, Wyser K, Semmler T, Yang S, van den Hurk B, van Noije T, van der Linden E, van der Wiel K (2011) EC-Earth V2.2: description and validation of a new seamless Earth system prediction model. Clim Dyn (online first 2011). doi:10.1007/s00382-011-1228-5

Held IM, Soden BJ (2006) Robust responses of the hydrological cycle to global warming. J Clim 19:5686–5699. doi:10.1175/JCLI3990.1

Hewitt HT, Copsey D, Culverwell ID, Harris CM, Hill RSR, Keen AB, McLaren AJ, Hunke EC (2011) Design and implementation of the infrastructure of HadGEM3: the next-generation Met Office climate modelling system. Geosci Model Dev 4:223–253

Hohenegger C, Brockhaus P, Bretherton CS, Schär C (2009) The soil moisture–precipitation feedback in simulations with explicit and parameterized convection. J Clim 22:5003–5020. doi:10.1175/2009JCLI2604.1

Holland MM, Bailey DA, Briegleb BP, Light D, Hunke E (2012) Improved sea ice shortwave radiation physics in CCSM4: the impact of melt ponds and aerosols on Arctic Sea ice. J Clim 25:1413–1430. doi:10.1175/JCLI-D-11-00078.1

Holloway CE, Woolnough SJ, Lister GMS (2012) Precipitation distributions for explicit versus parametrized convection in a large-domain high-resolution tropical case study. Q J R Meteorol Soc. doi:10.1002/qj.1903

Hong S-Y (2010) A new stable boundary-layer mixing scheme and its impact on the simulated East Asian summer monsoon. Q J R Meteorol Soc 136:1481–1496. doi:10.1002/qj.665

Kendon EJ, Roberts NM, Senior CA, Roberts MJ (2012) Realism of rainfall in a very high resolution regional climate model. J Clim 25:5791–5806. doi: 10.1175/JCLI-D-11-00562.1

Kim D, Sperber K, Stern W, Waliser D, Kang I-S, Maloney E, Wang, Weickmann K, Benedict J, Khairoutdinov M, Lee M-I, Neale R, Suarez M, Thayer-Calder K, Zhang G (2009) Application of MJO simulation diagnostics to climate models. J Clim 22:6413–6436. doi:10.1175/2009JCLI3063.1

Kim H-J, Wang B, Ding Q (2008) The global monsoon variability simulated by CMIP3 coupled climate models. J Clim 21:5271–5294. doi:10.1175/2008JCLI2041.1
Koster RD, Dirmeyer PA, Guo Z, Bonan G, Chan E, Cox P, Davies H, Gordon T, Kanae S, Kowalczyk E, Lawrence D, Liu P, Lu S, Malyshev S, McAvaney B, Mitchell K, Oki T, Oleson K, Pitman A, Sud Y, Taylor C, Verseghy D, Vasic R, Xue Y, Yamada T (2004) Regions of strong coupling between soil moisture and precipitation. Science 305:1138–1140
Lemke P, Ren J, Alley RB, Allison I, Carrasco J, Flato G, Fujii Y, Kaser G, Mote P, Thomas RH, Zhang T (2007) Observations: changes in snow, ice and frozen ground. In: Solomon S, Qin D, Manning M, Chen Z, Marquis M, Averyt KB, Tignor M, Miller HL (eds) Climate change 2007: the physical science basis. Contribution of Working Group I to the fourth assessment report of the intergovernmental panel on climate change. Cambridge University Press, Cambridge, New York
Levine RC, Turner AG (2011) Dependence of Indian monsoon rainfall on moisture fluxes across the Arabian Sea and the impact of coupled model sea surface temperature biases. Clim Dyn (online first). doi:10.1007/s00382-011-1096-z
Levine RC, Turner AG, Marathayil D, Martin GM (2012) The role of northern Arabian Sea surface temperature biases in CMIP5 model simulations and future projections of Indian summer monsoon rainfall. Clim Dyn (submitted)
Lohmann U, Feichter J (2005) Global indirect aerosol effects: a review. Atmos Chem Phys 5:715737. SRefID: 16807324/acp/20055715
Love BS, Matthews AJ, Lister GMS (2011) The diurnal cycle of precipitation over the maritime continent in a high-resolution atmospheric model. Q J R Meteorol Soc 137:934–947. doi:10.1002/qj.809
Madec G (2008) NEMO ocean engine. Note du Pole de modélisation, Institut Pierre-Simon Laplace (IPSL), France, No 27 ISSN No 1288-1619
Martin GM, Levine RC (2012) The influence of dynamic vegetation on the present-day simulation and future projections of the South Asian summer monsoon in the HadGEM2 family. Earth Syst Dyn Discuss 3:759–799. doi:10.5194/esdd-3-759-2012
Martin GM, Ringer MA, Pope VD, Jones A, Dearden C, Hinton TJ (2006) The physical properties of the atmosphere in the new Hadley Centre Global Environment Model, HadGEM1. Part I: model description and global climatology. J Clim 19:1274–1301. http://journals.ametsoc.org/doi/abs/10.1175/JCLI3636.1
Martin GM, Milton SF, Senior CA, Brooks ME, Ineson S, Reichler T, Kim J (2010) Analysis and reduction of systematic errors through a seamless approach to modelling weather and climate. J Clim 23:5933–5957. doi:10.1175/2010JCLI3541.1
Meehl GA, Stocker TF, Collins WD, Friedlingstein P, Gaye AT, Gregory JM, Kitoh A, Knutti R, Murphy JM, Noda A, Raper SCB, Watterson IG, Weaver AJ, Zhao Z-C (2007) Global climate projections. In: Solomon S (ed) Climate change 2007, the physical science basis, Chap 10. Cambridge Univ. Press, Cambridge, pp 747–845
Meier MF, Dyurgerov MB, Rick UK, O'Neel S, Pfeffer WT, Anderson RS, Anderson SP, Glazovsky AF (2007) Glaciers dominate eustatic sea-level rise in the 21st century. Science 317:1064–1067
Mignot J, Ganopolski A, Levermann A (2007) Atlantic subsurface temperatures: response to a shutdown of the overturning circulation and consequences for its recovery. J Clim 20:4884–4898. doi:10.1175/JCLI4280.1
Mukhopadhyay P, Taraphdar S, Goswami, BN, Krishnakumar K (2010) Indian summer monsoon precipitation climatology in a high-resolution regional climate model: impacts of convective parameterization on systematic biases. Weather Forecast 25:369–387. doi:10.1175/2009WAF2222320.1
Pincus R, Batstone CP, Patrick-Hofmann RJ, Taylor KE, Gleckler PE (2008) Evaluating the present-day simulation of clouds, precipitation and radiation in climate models. J Geophys Res 133:D14209. doi:10.1029/2007JD009334
Pritchard MS (2011) Simulated diurnal rainfall physics in a multi-scale global climate model with embedded explicit convection, Ph.D. Thesis, University of California, San Diego, 2011. Available at http://gradworks.umi.com/34/81/3481689.html
Radić V, Hock R (2011) Regionally differentiated contribution of mountain glaciers and ice caps to future sea-level rise. Nat Geosci 4:91–94. doi:10.1038/NGEO1052
Randall DA, Wood RA, Bony S, Colman R et al. (2007) Climate models and their evaluation. In: Solomon S, Qin D, Manning M, Chen Z, Marquis M, Averyt KB, Tignor M, Miller HL (eds) Climate change 2007: the physical science basis. Contribution of Working Group I to the fourth assessment report of the intergovernmental panel on climate change. Cambridge University Press, Cambridge, New York
Rayner NA, Parker DE, Horton EB, Folland CK, Alexander LV, Rowell DP, Kent EC, Kaplan A (2003) Global analyses of sea surface temperature, sea ice, and night marine air temperature since the late nineteenth century. J Geophys Res 108:4407. doi:10.1029/2002JD002670

Roberts MJ, Banks H, Gedney N, Gregory J, Hill R, Mullerworth S, Pardaens A, Rickard G, Thorpe R, Wood R (2004) Impact of an eddy-permitting ocean resolution on control and climate change simulations with a global coupled GCM. J Clim 17:3–20. doi:10.1175/1520-0442(2004)017<0003:IOAEOR>2.0.CO;2

Rodríguez JM, Johns TC, Thorpe RB, Wiltshire A (2011) Using moisture conservation to evaluate oceanic surface freshwater fluxes in climate models. Clim Dyn (online first). doi:10.1007/s00382-010-0899-7

Rutt IC, Hagdorn M, Hulton NRJ, Payne AJ (2009) The 'Glimmer' community ice sheet model. J Geophys Res 114(F2): F02004. doi:10.1029/2008JF001015

Salzmann M, Ming Y, Golaz J-C, Ginoux PA, Morrison H, Gettelman A, Krämer M, Donner LJ (2010) Two-moment bulk stratiform cloud microphysics in the GFDL AM3 GCM: description, evaluation, and sensitivity tests. Atmos Chem Phys 10:8037–8064. doi:10.5194/acp-10-8037-2010

Scaife AA, Copsey D, Gordon G, Harris C, Hinton T, Keeley SJ, O'Neill A, Roberts M, Williams K (2011) Improved Atlantic blocking in a climate model. Geophys Res Lett 38:L23703. doi:10.1029/2011GL049573

Schaaf CB, Gao F, Strahler AH, Lucht W, Li XW, Tsang T, Strugnell NC et al (2002) First operational BRDF, albedo nadir reflectance products from MODIS. Remote Sens Env 83:135–148

Schär C, Vidale PL, Lüthi D, Frei C, Häberli CA, Liniger MA, Appenzeller C (2004) The role of increasing temperature variability in European summer heatwaves. Nature 427:332–336

Schlemmer L, Hohenegger C, Schmidli J, Schär C (2012) Diurnal equilibrium convection and land surface–atmosphere interactions in an idealized cloud-resolving model. Q J R Meteorol Soc. doi:10.1002/qj.1892

Seneviratne SI, Lüthi D, Litschi M, Schär C (2006) Land–atmosphere coupling and climate change in Europe. Nature 443:205–209. doi:10.1038/nature05095

Shaffrey LC, Stevens I, Norton W, Roberts MJ, Vidale PL, Harle JD, Jrrar A, Stevens DP, Woodage MJ, Demory M-E, Donners J, Clark DB, Clayton A, Cole JW, Wilson SS, Connolley WM, Davies TM, Iwi AM, Johns TC, King JC, New AL, Slingo JM, Slingo A, Steenman-Clark L, Martin GM (2009) U.K. HiGEM: the new U.K. High-resolution Global Environment Model—model description and basic evaluation. J Clim 22:1861–1896. doi:10.1175/2008JCLI2508.1

Slingo J, Inness P, Neale R, Woolnough S, Yang GY (2003) Scale interactions on diurnal to seasonal timescales and their relevance to model systematic errors. Ann Geophys 46:139–155

Stephens GL, L'Ecuyer T, Forbes R, Gettlemen A, Golaz J-C, Bodas-Salcedo A, Suzuki K, Gabriel P, Haynes J (2010) Dreary state of precipitation in global models. J Geophys Res 115:D24211. doi:10.1029/2010JD014532

Stratton RA, Stirling AJ (2011) Improving the diurnal cycle of convection in GCMs. Q J R Meteorol Soc. doi:10.1002/qj.991

Taylor CM (2008) Intraseasonal land–atmosphere coupling in the West African Monsoon. J Clim 21:6636–6648. doi:10.1175/2008JCLI2475.1

Trenberth KE, Smith L, Qian T, Dai A, Fasullo J (2007) Estimates of the global water budget and its annual cycle using observational and model data. J. Hydrometeorol 8:758–769. doi:10.1175/JHM600.1

Trenberth KE, Fasullo JT, Mackaro J (2011) Atmospheric moisture transports from ocean to land and global energy flows in reanalyses. J Clim 24:4907–4924. doi:10.1175/2011JCLI4171.1

Turner AG, Slingo JM (2011) Using idealized snow forcing to test teleconnections with the Indian summer monsoon in the Hadley Centre GCM. Clim Dyn 36:1717–1735. doi:10.1007/s00382-010-0805-3

Voldoire A, Sanchez-Gomez E, Salas y Melia D, Decharme B, Cassou C, Sénési S, Valcke S, Beau I, Alias A, Chevallier M, Déqué M, Deshayes J, Douville H, Fernandez E, Madec G, Maisonnave E, Moine M-P, Planton S, Saint-Martin D, Szopa S, Tyteca S, Alkama R, Belamari S, Braun A, Coquart L, Chauvin F, (2012) The CNRM-CM5.1 global climate model: description and basic evaluation. Clim Dyn (online first). doi:10.1007/s00382-011-1259-y

Walters DN, Best MJ, Bushell AC, Copsey D, Edwards JM, Falloon PD, Harris CM, Lock AP, Manners JC, Morcrette CJ, Roberts MJ, Stratton RA, Webster S, Wilkinson JM, Willett MR, Boutle IA, Earnshaw PD, Hill PG, MacLachlan C, Martin GM, Moufouma-Okia W, Palmer MD, Petch JC, Rooney GG, Scaife AA, Williams KD (2011) The Met Office Unified Model Global Atmosphere 3.0/3.1 and JULES Global Land 3.0/3.1 configurations. Geosci Model Dev 4:919–941. doi:10.5194/gmd-4-919-2011

Wang Y, Zhou L, Hamilton K (2007) Effect of convective entrainment/detrainment on simulation of tropical precipitation diurnal cycle. Mon Weath Rev 135:567–585

Wang Z, Chang C-P (2012), A numerical study of the interaction between the large-scale monsoon circulation and orographic precipitation over South and Southeast Asia. J Clim 25:2440–2455. doi:10.1175/JCLI-D-11-00136.1

Webb M, Senior C, Bony S, Morcrette JJ (2001) Combining ERBE and ISCCP data to assess clouds in the Hadley Centre, ECMWF and LMD atmospheric climate models. Clim Dyn 17:905–922. doi:10.1007/s003820100157

Webb MJ, Senior CA, Sexton DMH, Ingram WJ, Williams KD, Ringer MA, McAvaney BJ, Colman R, Soden BJ, Gudgel R, Knutson T, Emori S, Ogura T, Tsushima Y, Andronova N, Li B, Musat I, Bony S, Taylor K (2006) On the contribution of local feedback mechanisms to the range of climate sensitivity in two GCM ensembles. Clim Dyn 27(1):1738. doi:10.1007/s0038200601112

Wilson DR, Bushell AC, Kerr-Munslow AM, Price JD, Morcrette CJ, Bodas-Salcedo A (2008) PC2: a prognostic cloud fraction and condensation scheme. II: climate model simulations. Q J R Meteorol Soc 134:2109–2125. doi:10.1002/qj.332

Wood EF, Roundy JK, Troy TJ, van Beek LPH, Bierkens MFP, Blyth E, de Roo A, Döll P, Ek M, Famiglietti J, Gochis D, van de Giesen N, Houser P, Jaffé PR, Kollet S, Lehner B, Lettenmaier DP, Peters-Lidard C, Sivapalan M, Sheffield J, Wade A, Whitehead P (2011) Hyperresolution global land surface modeling: meeting a grand challenge for monitoring Earth's terrestrial water. Water Resour Res 47:W05301. doi:101029/2010WR010090

Wylie D, Jackson DL, Menzel WP, Bates JJ (2005) Trends in global cloud cover in two decades of hirs observations. J Clim 18:3021–3031. doi:10.1175/JCLI3461.1

Yasunari T, Saito K, Takata K (2006) Relative roles of large-scale orography and land surface processes in the global hydroclimate. Part I: impacts on monsoon systems and the tropics. J Hydrometeorol 7:626–641. doi:10.1175/JHM515.1

DOI 10.1007/978-94-017-8789-5_6
Reprinted from *Surveys in Geophysics* Journal, DOI 10.1007/s10712-013-9220-8

Connecting Satellite Observations with Water Cycle Variables Through Land Data Assimilation: Examples Using the NASA GEOS-5 LDAS

Rolf H. Reichle · Gabriëlle J. M. De Lannoy · Barton A. Forman · Clara S. Draper · Qing Liu

Received: 7 September 2012 / Accepted: 22 January 2013 / Published online: 15 February 2013

Abstract A land data assimilation system (LDAS) can merge satellite observations (or retrievals) of land surface hydrological conditions, including soil moisture, snow, and terrestrial water storage (TWS), into a numerical model of land surface processes. In theory, the output from such a system is superior to estimates based on the observations or the model alone, thereby enhancing our ability to understand, monitor, and predict key elements of the terrestrial water cycle. In practice, however, satellite observations do not correspond directly to the water cycle variables of interest. The present paper addresses various aspects of this seeming mismatch using examples drawn from recent research with the ensemble-based NASA GEOS-5 LDAS. These aspects include (1) the assimilation of coarse-scale observations into higher-resolution land surface models, (2) the partitioning of satellite observations (such as TWS retrievals) into their constituent water cycle components, (3) the forward modeling of microwave brightness temperatures over land for radiance-based soil moisture and snow assimilation, and (4) the selection of the most relevant types of observations for the analysis of a specific water cycle variable that is not observed (such as root zone soil moisture). The solution to these challenges involves the careful construction of an observation operator that maps from the land surface model variables of interest to the space of the assimilated observations.

Keywords Land data assimilation · Land surface modeling · Satellite remote sensing · Soil moisture · Snow · Terrestrial water storage · Ensemble Kalman filter

R. H. Reichle (✉) · G. J. M. De Lannoy · C. S. Draper · Q. Liu
Global Modeling and Assimilation Office (Code 610.1), NASA Goddard Space Flight Center, 8800 Greenbelt Road, Greenbelt, MD 20771, USA
e-mail: rolf.reichle@nasa.gov

G. J. M. De Lannoy · C. S. Draper
Universities Space Research Association, Columbia, MD, USA

B. A. Forman
Department of Civil and Environmental Engineering, University of Maryland, College Park, MD, USA

Q. Liu
Science Systems and Applications, Inc., Lanham, MD, USA

1 Introduction

The water cycle plays a crucial role in Earth's climate and environment, yet there are still large gaps in our understanding of its components, particularly at the land surface (Lahoz and De Lannoy 2013; Trenberth and Asrar 2013). Over the past decade, there has been a steady increase in the number and types of satellite observations (or retrievals) related to land surface hydrological conditions, including soil moisture, snow, and terrestrial water storage (TWS; Bartalis et al. 2007; Bruinsma et al. 2010; Clifford 2010; de Jeu et al. 2008; Entekhabi et al. 2010; Foster et al. 2005, 2011; Gao et al. 2010; Hall and Riggs 2007; Hall et al. 2010; Horwath et al. 2011; Kelly 2009; Kerr et al. 2010; Li et al. 2007; Liu et al. 2011b; Njoku et al. 2003; Parinussa et al. 2012; Pulliainen 2006; Rowlands et al. 2005, 2010; Swenson and Wahr 2006; Tedesco and Narvekar 2010; Tedesco et al. 2010; Wahr et al. 2004).

These observations can be assimilated into land surface models to provide land surface hydrological estimates that are generally superior to the satellite observations or model estimates alone (Andreadis and Lettenmaier 2006; Crow and Wood 2003; De Lannoy et al. 2012; de Rosnay et al. 2012a, b; Draper et al. 2012; Drusch 2007; Dunne and Entekhabi 2006; Durand and Margulis 2008; Forman et al. 2012; Houborg et al. 2012; Li et al. 2012; Liu et al. 2011a; Margulis et al. 2002; Pan and Wood 2006; Pan et al. 2008; Reichle and Koster 2005; Reichle et al. 2007, 2009; Sahoo et al. 2012; Su et al. 2008, 2010; Zaitchik et al. 2008).

However, land data assimilation systems must be designed carefully such that a number of conceptual problems can be overcome and the potential improvements from data assimilation can be realized. Earlier work addressed the bias between the satellite observations and model estimates within the assimilation system (De Lannoy et al. 2007; Drusch et al. 2005; Kumar et al. 2012; Reichle and Koster 2004). Moreover, approaches to efficient error modeling within the assimilation system, including adaptive methods, needed to be developed (Crow and Reichle 2008; Crow and van den Berg 2010; Reichle et al. 2008a, b). An overview of some relevant earlier literature in the context of the ensemble-based Goddard Earth Observing System Model, Version 5 (GEOS-5) land data assimilation system (LDAS) developed at the NASA Global Modeling and Assimilation Office (GMAO) is provided by Reichle et al. (2009).

Despite the early successes, the design and application of land data assimilation systems still face additional conceptual problems. While land surface models are flexible in the design and choice of model variables, satellite observations do not necessarily correspond directly to the water cycle variables of interest. For example, space-borne microwave observations can be converted into estimates of snow amount or surface soil moisture, but the spatial resolution of such microwave-based retrievals is usually much coarser than desired. Moreover, satellites typically observe electromagnetic properties such as backscatter and/or radiances (or brightness temperatures) that are only indirectly related to snow amounts or soil moisture levels. Furthermore, satellite-observed backscatter and radiances are at best sensitive to moisture in the top few centimeters of the soil. Information on important water cycle components such as root zone soil moisture must therefore be gained through even more indirect pathways in the land data assimilation system.

The present paper addresses several major challenges that all relate to a seeming mismatch between the assimilated observations and the water cycle variables of interest. This mismatch can be overcome through the careful design of the land data assimilation system. The conceptual challenges discussed here can be summarized as follows:

1. How can coarse-scale satellite observations increase our knowledge of land surface conditions at finer scales (horizontal downscaling), and how can unobserved areas be updated using information from neighboring observations?
2. How can vertically integrated measurements (such as TWS) be partitioned into their component variables within the assimilation system?
3. How can satellite radiances (rather than geophysical retrievals) be assimilated to improve estimates of land surface hydrological conditions (e.g., soil moisture and snow)?
4. How can the most relevant types of observations be selected for the analysis of a water cycle component that is not observed (such as root zone soil moisture)?

The present paper illustrates each of these conceptual problems based on recent progress using the GEOS-5 system for land surface hydrological data assimilation. The examples use satellite observations of land surface water cycle components from the Advanced Microwave Scanning Radiometer for EOS (AMSR-E), the Moderate Resolution Imaging Spectroradiometer (MODIS), the Gravity Recovery and Climate Experiment (GRACE) mission, the Advanced Scatterometer (ASCAT), and the Soil Moisture Ocean Salinity (SMOS) mission for the analysis of soil moisture (AMSR-E, ASCAT, SMOS, GRACE), snow (AMSR-E, MODIS, GRACE), and TWS (GRACE). After a brief discussion of the GEOS-5 LDAS, Sect. 2 provides details and references for the various satellite observations used in the examples. Section 3 addresses each of the above-mentioned challenge questions in a separate subsection. Results are discussed and summarized in Sect. 4. Finally, Sect. 5 provides conclusions and a brief outlook on future research directions.

2 Data and Methods

2.1 GEOS-5 Land Data Assimilation System

The GEOS-5 LDAS consists of the NASA Catchment land surface model and an implementation of the ensemble Kalman filter (EnKF; Evensen 2003). The GEOS-5 EnKF has also been included in the NASA Land Information System, a comprehensive land surface modeling and assimilation software framework, so that it can be used with a variety of land surface models (Kumar et al. 2008a, b). A brief summary of the key characteristics of the system is provided below. For a more comprehensive discussion, see Reichle et al. (2009) and references therein.

The Catchment land surface model (hereinafter Catchment model; Ducharne et al. 2000; Koster et al. 2000) differs from traditional, layer-based land surface models by including an explicit treatment of the spatial variation within each hydrological catchment (or computational element) of the soil water and water table depth, as well as its effect on runoff and evaporation. Within each element, the vertical profile of soil water down to the bedrock is given by the equilibrium soil moisture profile and the deviations from the equilibrium profile. The deviations are described by excess and deficit variables for a 0–2 cm (or 0–5 cm) surface layer and for a “root zone” layer that extends from the surface to a depth z_R of 75 cm $\leq z_R \leq$ 100 cm depending on local soil conditions. The spatial variability of soil moisture is diagnosed at each time step from the bulk water prognostic variables and the statistics of the catchment topography. One key feature of the Catchment model is the groundwater component implicit in the modeling of the water table depth (through the modeling of the subsurface water profile down to the bedrock). This groundwater component is critically important for the assimilation of TWS retrievals (Sect. 3.2).

The Catchment model also includes a state-of-the-art, multi-layer, global snow model (Stieglitz et al. 2001). In each watershed, the evolution of the amount of water in the snow pack (or snow water equivalent; SWE), the snow depth, and the snow heat content in response to surface meteorological conditions and snow compaction is modeled using three layers. The soil, vegetation, and snow model parameters used in the Catchment model are from the NASA GEOS-5 global modeling system (Rienecker et al. 2008).

The EnKF is a Monte-Carlo variant of the Kalman filter, which sequentially updates model forecasts in response to observations based on the relative uncertainty of the model and the observations. The key idea behind the EnKF is that the relevant parts of the model error covariance structure can be captured by a small ensemble of model trajectories. Each member of the ensemble experiences perturbed instances of the observed forcing fields (representing errors in the forcing data) and/or randomly generated noise that is added to the model parameters and prognostic variables (representing errors in model physics and parameters). The model error covariance matrices that are required for the filter update can then be diagnosed from the ensemble at the update time. The EnKF is flexible in its treatment of errors in model dynamics and parameters. It is also very suitable for modestly nonlinear problems and has become a popular choice for land data assimilation (Andreadis and Lettenmaier 2006; Durand and Margulis 2008; Kumar et al. 2008a, b; Pan and Wood 2006; Reichle et al. 2002a, b; Su et al. 2008; Zhou et al. 2006).

To realize the potential benefits from data assimilation, the assimilation system must be supplied with appropriate input parameters for the description of model and observation errors. For an ensemble-based system such as the GEOS-5 LDAS, for example, standard deviations, spatial and temporal correlations, and cross-correlations must be specified for the perturbations that are applied to each ensemble member. A detailed discussion of the error parameters in the examples discussed here is beyond the scope of the paper. The reader is referred to the references provided with each example as well as the overview discussion of Reichle et al. (2009).

2.2 Assimilated Observations

The data assimilation examples discussed in this paper use various types of satellite observations from a number of polar orbiting sensors/platforms, including passive and active microwave observations (AMSR-E, SMOS, and ASCAT), visible and near-infrared observations (MODIS), and gravimetric observations (GRACE).

AMSR-E, which operated with nominal performance between 2002 and 2011, is a scanning, dual polarization radiometer that measured microwave emission from the Earth at six frequencies (6.9, 10.7, 18.7, 23.9, 36.5, and 89.0 GHz), ranging in resolution from ~50 km at 6.9 GHz to ~5 km at 89.0 GHz (Knowles et al. 2006). Its successor, AMSR2, was launched in May 2012 (http://www.jaxa.jp/projects/sat/gcom_w/index_e.html). The training and validation of the empirical microwave radiative transfer model for snow-covered land surfaces in Sect. 3.3.2 uses the 10.7, 18.7, and 36.5 GHz AMSR-E brightness temperatures, while the snow assimilation example in Sect. 3.1 uses SWE retrievals that are based on the difference between the 18.7 and the 36.5 GHz brightness temperatures (Kelly 2009). The soil moisture assimilation examples in Sect. 3.4 use surface (top 1 cm) soil moisture retrievals that are derived from the 6.9 and 10.7 GHz brightness temperatures (de Jeu et al. 2008; Njoku et al. 2003).

SMOS was launched in 2009 and its Microwave Imaging Radiometer with Aperture Synthesis (MIRAS) sensor provides multi-angular L-band (1.4 GHz) brightness temperature observations at horizontal and vertical polarization and a nominal spatial resolution of 43 km (Kerr et al. 2010). SMOS brightness temperatures are used in Sect. 3.3.1.

ASCAT is a 5.3 GHz radar system that illuminates the Earth's surface and measures the energy scattered back to the instrument. The ASCAT surface (top 1 cm) soil moisture retrievals used in Sect. 3.4.2 are derived from these backscatter measurements (Bartalis et al. 2007; Wagner et al. 1999) and are provided in units of degree of saturation.

MODIS (2000-present) provides visible and near-infrared observations from which snow cover fraction (SCF) can be retrieved under clear-sky conditions (Hall and Riggs 2007). High-resolution (500 m) MODIS SCF retrievals are in Sect. 3.1.

Through the measurement of gravitational anomalies associated with the accumulation (or loss) of mass near the Earth's surface, GRACE provides approximately monthly, basin-scale (>150,000 km^2) estimates of variations in TWS, which includes snow, ice, surface water, soil moisture, and groundwater (Bruinsma et al. 2010; Horwath et al. 2011; Rodell et al. 2009; Rowlands et al. 2005, 2010; Swenson and Wahr 2006; Tang et al. 2010; Wahr et al. 2004). The assimilation experiments of Sect. 3.2 use GRACE TWS retrievals.

2.3 Validation Data and Approach

For each of the examples presented in Sect. 3, the output from the assimilation system was evaluated against independent data from various sources. In Sect. 3.1, in situ SWE measurements from United States Department of Agriculture Snowpack Telemetry (SNOTEL; Schaefer et al. 2007) network sites in Colorado were used for evaluation, along with snow depth measurements from National Oceanic and Atmospheric Administration Cooperative Observer Program (COOP; http://www.ncdc.noaa.gov) sites.

SWE estimates for the Mackenzie River basin, used for evaluation in Sect. 3.2, were derived from the daily snow depth product of the Canadian Meteorological Centre (CMC) daily snow analysis (Brasnett 1999; Brown and Brasnett 2010) at a horizontal resolution of approximately 24 km. The CMC snow analysis is based on optimal interpolation of in situ daily snow depth observations and aviation reports with a first-guess field generated from a snow model driven by output from the CMC weather model. Using the snow class map shown in Sturm et al. (1995), SWE estimates were obtained by multiplying the CMC snow depths with the Sturm et al. (2010) snow densities. Furthermore, runoff estimates for the Mackenzie River basin and its major sub-basins provided by the Global Runoff Data Center (GRDC; http://www.bafg.de/GRDC) were used in Sect. 3.2.

The radiative transfer models of Sect. 3.3 were evaluated with AMSR-E and SMOS microwave brightness temperatures using a split sample approach in which one portion of the satellite brightness data was used for calibration or training and another, different portion was used for evaluation.

In situ profile soil moisture observations used for evaluation in Sect. 3.4 are from the United States Department of Agriculture Soil Climate Analysis Network (SCAN)/SNOTEL (Schaefer et al. 2007) network in the contiguous US and from the Murrumbidgee Soil Moisture Monitoring Network (Smith et al. 2012) in Australia. Both sets of measurements were subjected to extensive quality control steps, including automatic detection of problematic observations and a visual inspection of the time series prior to using the data for evaluation.

Metrics used for skill assessment include the bias, root mean square error (RMSE), and time series correlation coefficient (R). When specified, anomalies were computed by removing a seasonally varying climatology from the data before computing the metrics.

3 Results

3.1 Assimilation of Sparse and Coarse-Scale Observations

Snow is an important component of the land system because of its strong impact on the land surface water and energy balance, weather, climate, and water resources (Barnett et al. 2005). However, land surface models often represent snow processes poorly. Satellite observations of SWE can be retrieved from passive microwave sensors, but they are only available at relatively coarse resolution. Moreover, SWE retrievals, like most satellite observations, do not provide complete spatial and continuous temporal coverage due to orbit or sensor limitations. The challenge is therefore to design an assimilation system that can use coarse-scale satellite observations to provide enhanced model estimates at the finer scales of interest (horizontal downscaling) and that can also propagate the information to intermittently unobserved areas.

Using AMSR-E SWE retrievals and MODIS SCF observations, De Lannoy et al. (2010, 2012) developed a data assimilation and downscaling technique for estimating fine-scale (1 km) snow fields using coarse-scale (25 km) SWE retrievals and fine-scale (500 m) SCF retrievals for a domain in Northern Colorado, USA. In their study, the authors used the LIS version of the GEOS-5 EnKF together with the Noah land surface model (Ek et al. 2003) (rather than the GEOS-5 LDAS and the Catchment model used elsewhere in this paper). The Noah model simulates a single snow layer with two prognostic variables for SWE and snow depth. The default LIS soil, vegetation, and general parameter tables for Noah were used, including a Noah-specific maximum snow albedo.

Figure 1 shows schematically how the coarse-scale SWE retrievals are used. The fine-scale model grid is represented by the dashed lines in the figure. The coarse-scale grid of the SWE observations is represented by the solid lines and light/dark gray shading, and the center points of individual SWE retrievals are marked with crosses. Let us now consider the analysis update of the fine-scale model grid cell indicated by the solid black square. First, it is important to emphasize that the coarse-scale SWE retrievals are *not* compared directly to the SWE estimate at the fine-scale model grid cell. Rather, the model SWE is aggregated to the coarse grid of the retrievals, that is, the fine-scale model forecast is mapped into the coarse-scale *observation space*. This aggregation is part of the *observation operator* that maps the model states to the observations. Observation-minus-model-forecast residuals (or *innovations*) are then computed at the coarse scale of the observation space. The *Kalman gain* matrix transforms the (observation-space) innovations into the (model-space) *increments*. It is computed from error correlations between the model states at the fine scale and the model-predicted measurements at the coarse scale. Finally, the increments are added to the (fine-scale) model forecast in the *analysis update*. See De Lannoy et al. (2010) for a discussion based on equations.

Second, multiple coarse-scale SWE retrievals in the vicinity of the fine-scale model grid cell in question are used for the analysis update. Specifically, the update uses the three coarse-scale SWE retrievals marked by black crosses that are within a given radius (indicated by the white semi-circle) around the fine-scale model grid cell in question (Fig. 1). Note that this model grid cell would be updated even if the SWE retrieval directly

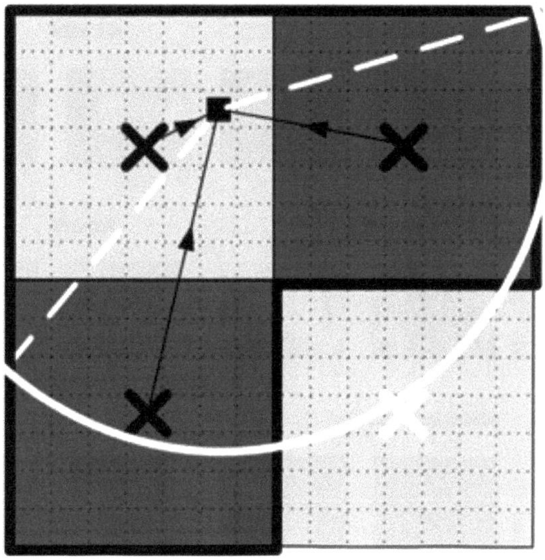

Fig. 1 Schematic of the distributed ("three-dimensional") EnKF update used for the assimilation of coarse-scale snow observations. See text for details. Adapted from De Lannoy et al. (2010)

covering it were unavailable—the two neighboring SWE retrievals (dark gray shading) would still contribute to the update. The connection between the neighboring SWE retrievals and the model grid cell in question relies on horizontal model *error* correlations that are due to, for example, errors in large-scale model forcing fields such as snowfall or air temperature.

To assimilate SCF, the Noah model snow depletion curve acts as the observation operator that converts fine-scale modeled SWE into SCF estimates. Unlike binary indicators of snow presence, the continuous SCF observations used here can thus be assimilated with an EnKF, taking advantage of the distribution of SCF values across the ensemble. Snow-free or fully snow-covered conditions in the model-forecast ensemble were addressed by supplementing the EnKF with rule-based update procedures (De Lannoy et al. 2012). If at a given time and location all members of the model-forecast ensemble are snow-free but the SCF observation indicates the presence of snow, then a nominal amount of snow is added to the model forecast. If all forecast ensemble members have full snow cover and the observed SCF indicates less than full cover, then the model-forecast SWE and snow depth are reduced by a fixed fraction.

Figure 2 shows several observed and modeled snow fields for one snow season. The top row shows the coarse-scale (25 km) AMSR-E SWE retrievals, with data missing when the satellite swath does not fully cover the study area. MODIS fine-scale estimates of SCF, shown in the second row, are available only for clear-sky conditions. The bottom four rows of Fig. 2 show that the assimilation of coarse-scale AMSR-E SWE and fine-scale MODIS SCF observations both result in realistic fine-scale spatial SWE patterns.

Through a quantitative validation of the assimilation results with independent measurements at individual SNOTEL and COOP sites over the course of 8 years, De Lannoy et al. (2012) demonstrate improvements from the assimilation of SWE and/or SCF retrievals in shallow snow packs, but not in deep snow packs (not shown). The validation also shows that joint assimilation of SWE and SCF retrievals yields significantly improved RMSE and correlation values. For example, the RMSE for SWE versus COOP site measurements was reduced by 21 % (from 78 to 62 mm) through the joint assimilation of satellite SWE and SCF retrievals. Furthermore, SCF assimilation was found to improve the

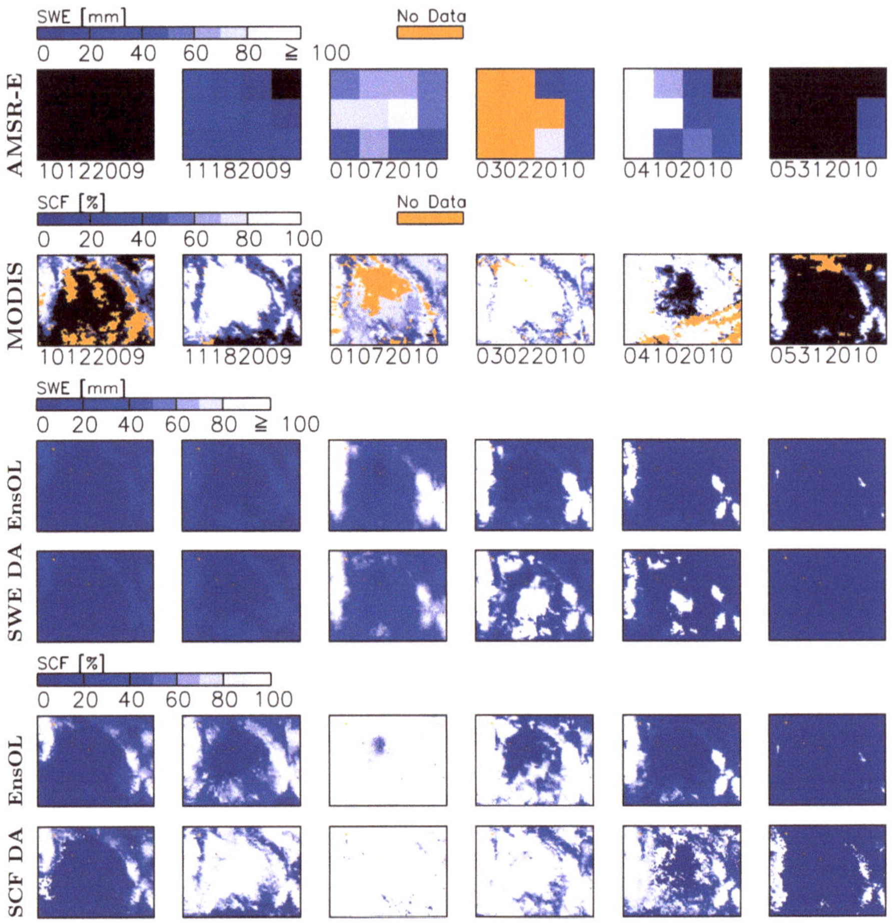

Fig. 2 SWE and SCF fields for 6 days (MMDDYYYY) in the winter of 2009–2010 for a 75 km by 100 km domain (1 km resolution) in northern Colorado. *Blue* (*white*) *colors* indicate low (high) SWE or SCF, *black shading* indicates no snow, and *orange shading* indicates no data. The *top two rows* show SWE and SCF satellite observations. The remaining *rows* show SWE (*rows 3* and *4*) and SCF (*rows 5* and *6*) for the ensemble Open Loop (EnsOL) forecast (no assimilation) and the analyses obtained through data assimilation (DA) of SWE or SCF. Adapted from De Lannoy et al. (2012)

timing of the onset of the snow season, albeit without a net improvement of SWE estimates. In areas of deep snow, however, AMSR-E retrievals are typically biased low and require bias correction (or scaling of the observations) prior to data assimilation. De Lannoy et al. (2012) also showed that the interannual SWE variations could not be improved through the assimilation of AMSR-E because the AMSR-E retrievals lack realistic interannual variability in deep snow packs. These deficiencies in the AMSR-E SWE retrievals motivated the development of the empirical microwave radiative transfer model (Sect. 3.3.2) toward a radiance-based snow analysis.

Of course, horizontal downscaling is not only important for snow assimilation. Low-frequency passive microwave brightness temperature observations such those from AMSR-E and SMOS (and the corresponding soil moisture retrievals) are at the coarse resolution of ~50 km. But for applications such as weather prediction, soil moisture estimates are needed at hydrometeorological scales of ~10 km or better. Examples of soil moisture downscaling based on data assimilation are provided by Reichle et al. (2001), Sahoo et al. (2012), and Zhou et al. (2006). Also, Reichle and Koster (2003) addressed the propagation of observational soil moisture information to unobserved regions.

3.2 Partitioning of Terrestrial Water Storage Observations

Passive microwave (e.g., AMSR-E) retrievals have been used in conjunction with land surface models to better characterize snow (Sect. 3.1) and soil moisture (Sect. 3.4). Gravimetric measurements such as from GRACE can provide monthly, basin-scale (>150,000 km^2) estimates of changes in TWS (Sect. 2.2). Since TWS is vertically integrated and includes groundwater, soil moisture, snow, and surface water, TWS retrievals offer significant insights into the regional- and continental-scale water balance and, through data assimilation, the potential to learn more about hydrological processes.

Besides the obvious spatial downscaling challenge presented by the basin-scale GRACE TWS retrievals, another challenge for the assimilation of GRACE-based TWS is the partitioning of the vertically integrated TWS retrievals into water cycle component variables. Like the horizontal downscaling of AMSR-E SWE retrievals discussed in the previous section, the partitioning of TWS retrievals can be accomplished through assimilation using an appropriate observation operator. In this case, the observation operator aggregates the fine-scale model estimates of soil moisture, groundwater, and snow to basin-scale TWS estimates. This observation operator enables the computation of the observation-minus-forecast residuals (or innovations) in the (basin-scale, TWS) space of the observations. The observation operator is also needed for the computation of the Kalman gain that transforms the innovations back into the space of the fine-scale model variables. Similarly, the required temporal aggregation of the model output to the monthly scale of the assimilated TWS retrievals is accomplished through the observation operator.

This concept was illustrated by Forman et al. (2012), who assimilated GRACE TWS retrievals over the Mackenzie River basin located in northwestern Canada (Fig. 3) using an updated version of the GEOS-5 LDAS developed by Zaitchik et al. (2008). The assimilation estimates were evaluated against independent SWE and river discharge observations (Sect. 2.3). Results suggest improved SWE estimates, including improved timing of the subsequent ablation and runoff of the snow pack. For example, Fig. 4 shows the improvements in SWE estimates resulting from the assimilation of GRACE TWS retrievals. The white bars represent model results without assimilation, whereas the gray bars represent results with assimilation. The labels on the y-axis of each subplot represent sub-basins of the Mackenzie River basin. As shown in Fig. 4, the assimilation of GRACE TWS retrievals generally reduced the mean difference and RMSE between the model and the independent CMC SWE estimates (Sect. 2.3). The reductions are greatest in the Liard basin, where the greatest amount of snow accumulation occurs. Here, the mean difference with the CMC estimates is reduced through GRACE data assimilation by 30 % (from 13.2 to 9.3 mm) and the RMSE is reduced by 18 % (from 24 to 19.6 mm). Smaller reductions occur in the other sub-basins. The correlation coefficient of the SWE anomalies (not shown) suggests a slight degree of degradation resulting from assimilation, but further analysis shows there is no statistically significant difference at the 5 % level. In summary, the assimilation of GRACE TWS information into the Catchment land surface model reduces the mean difference and RMSE in SWE estimates without adversely impacting estimates of interannual variability.

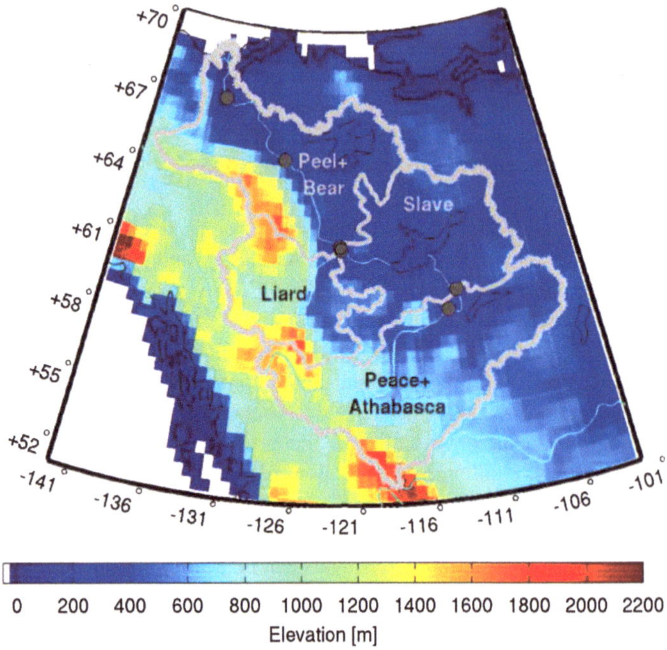

Fig. 3 Map of the 1,800,000 km^2 Mackenzie River Basin including GEOS-5 topography, sub-basin delineation, and GRDC observation locations (*solid dots*). Adapted from Forman et al. (2012)

Additional work was conducted to analyze modeled river discharge estimates against ground-based gauging stations. The findings (not shown) suggest that the assimilation of GRACE observations causes little or no change in the mean difference and RMSE of modeled river discharge, but that small, statistically significant improvements in the anomaly correlations were found. Improvements in the modeled river runoff anomalies are attributed to a redistribution of the water mass from the snow pack during the accumulation phase into the subsurface during the subsequent ablation and runoff phase. This redistribution of water by the assimilation framework effectively retains water within the hydrological basin for a longer period of time, which results in small but statistically significant improvements in modeled estimates of river discharge.

Investigation of the analysis increments can provide valuable insights into the behavior of the assimilation procedure and track how much and at what time water is being added to or removed from the individual TWS components. The thin, solid line in Fig. 5 shows the increments made to the subsurface water component. Averaged over the Mackenzie River basin and the 7-year experiment period, a total of 12.5 mm of water has been added into the subsurface by the assimilation procedure. This is most evident during the spring and summer. The thick, dashed line in Fig. 5 shows the increments for SWE. Averaged over time and space, SWE is removed during the accumulation phase with a small amount added back during the ablation and runoff phase for a total SWE increment of −45.1 mm. Acting together, the analysis increments to the subsurface water and SWE serve to reduce mass during snow accumulation and then increase the mass during ablation and runoff. These two phenomena essentially constrain the amplitude of the modeled TWS dynamics to achieve better agreement of the model estimates with the GRACE retrievals.

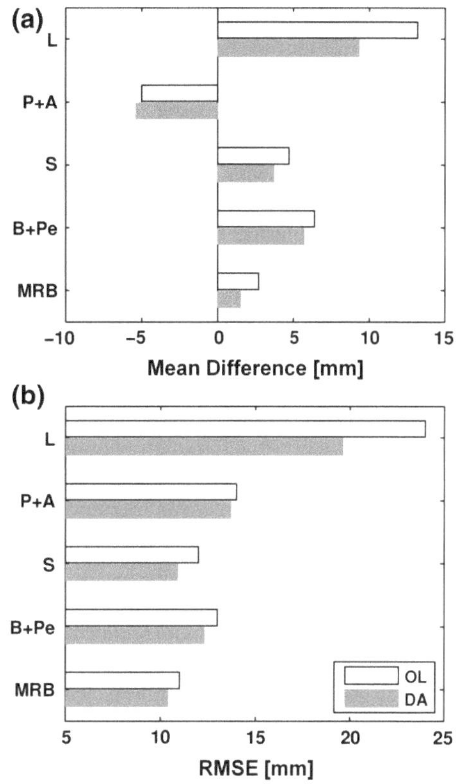

Fig. 4 SWE statistics of **a** mean difference and **b** RMSE for open loop (OL; *white*) and assimilation (DA; *light gray*) of GRACE TWS retrievals relative to CMC SWE estimates via Sturm et al. (2010). Statistics are for the Mackenzie River Basin (*MRB*) and its sub-basins Liard (*L*), Peace and Athabasca ($P + A$), Slave (*S*), and Bear and Peel ($B + Pe$) shown in Fig. 3. Adapted from Forman et al. (2012)

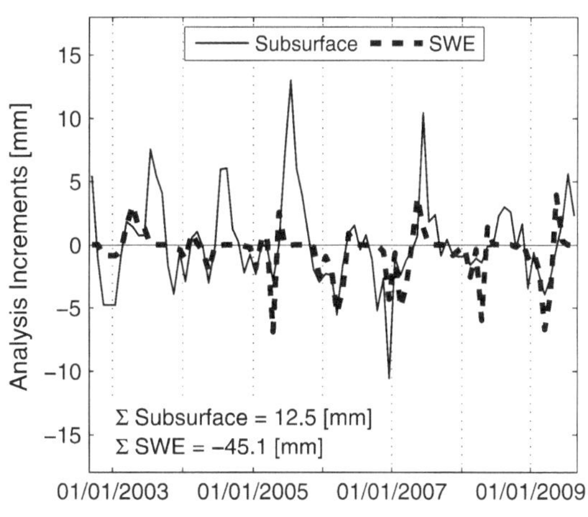

Fig. 5 Analysis increments for the entire Mackenzie River basin from GRACE TWS assimilation. The *thin, solid line* represents the subsurface water increments, whereas the thick, *dashed line* represents the SWE increments. Adapted from Forman et al. (2012)

The results shown in Figs. 4 and 5 imply that the assimilation procedure can effectively partition the vertically integrated GRACE TWS retrievals into their snow and subsurface water components. Houborg et al. (2012), Li et al. (2012), Su et al. (2010), and Zaitchik et al. (2008) further investigated the horizontal, vertical, and temporal disaggregation of GRACE TWS retrievals and reached similar conclusions for other basins in North America and Europe in different climate zones. Collectively, the growing body of research suggests that GRACE TWS assimilation can lead to better understanding of the hydrological cycle in remote regions of the globe where ground-based observation collection is difficult, if not impossible. This information could ultimately lead to improved freshwater resource management as well as reduced uncertainty in river discharge.

3.3 Microwave Radiative Transfer Models for Radiance Data Assimilation

It is well established for atmospheric data assimilation systems that the assimilation of satellite radiance observations is preferable to the assimilation of geophysical retrievals (Eyre et al. 1993; Joiner and Dee 2000). The former approach incorporates the radiative transfer model into the assimilation system and thereby avoids inconsistencies in the use of ancillary data between the assimilation system and the (pre-processed) geophysical retrievals. For land data assimilation, however, the vast majority of publications assimilate geophysical retrievals (Lahoz and De Lannoy 2013). In this section, we discuss the development of forward radiative transfer models (RTMs) that convert land surface model variables into microwave brightness temperatures. The first example presents such a model for warm-season microwave brightness temperatures (Sect. 3.3.1). The second example introduces a neural network approach to predict microwave brightness temperatures over snow-covered land (Sect. 3.3.2).

3.3.1 Warm-Season, L-Band Radiative Transfer Modeling

Global observations of brightness temperatures (Tb) at L-band (1.4 GHz) are available from the SMOS mission, and similar Tb observations are expected from the planned Soil Moisture Active Passive (SMAP; Entekhabi et al. 2010) mission. In preparation for the assimilation of Tb observations from SMOS and SMAP, De Lannoy et al. (2013) added a physically based, warm-season microwave RTM to the GEOS-5 Catchment model. The RTM is based on the commonly used, zero-order "tau-omega" approach that accounts for microwave emission by the soil and the vegetation canopy as well as attenuation by the vegetation. While the RTM is based on sound physical principles, determining the required parameter values for the microwave roughness, scattering albedo, and vegetation optical depth on a global scale is a serious challenge.

De Lannoy et al. (2013) collected three different sets of the literature values for the L-band RTM parameters. "Lit1" refers to parameters that are proposed for the future SMAP radiometer retrieval product, "Lit2" are parameters collected from the literature studies using the L-band Microwave Emission of the Biosphere model (Wigneron et al. 2007) and related models, and "Lit3" is the same as Lit2 except that the microwave roughness parameter is set to values used for SMOS monitoring in the European Centre for Medium-Range Weather Forecasts (ECMWF). The three sets of parameters are illustrated in Fig. 6, which shows the resulting microwave roughness (h), vegetation opacity (τ), and scattering albedo (ω) by vegetation class. As can be seen from the figure, there are large differences in h, τ, and ω between the three sets of the literature values. These differences translate into climatological differences in the simulated brightness temperatures.

For example, Fig. 7a–c shows the differences between 1-year mean (July 1, 2010–July 1, 2011) model simulations (using the three different literature-based sets of RTM parameters) and SMOS observations for H-polarized Tb at 42.5° incidence angle. Modeled brightness temperatures are at 36 km resolution, commensurate with the resolution of the SMOS observations. Brightness temperatures are screened for frozen soil conditions, snow on the ground, heavy precipitation, proximity to open water surfaces, and radio-frequency

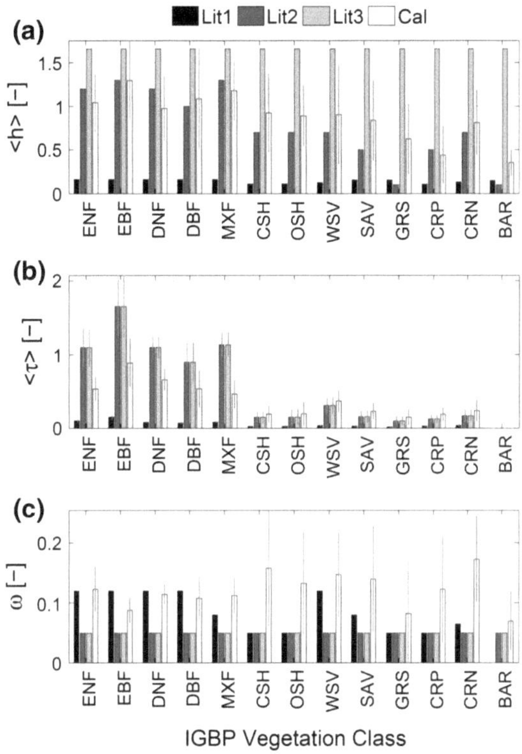

Fig. 6 **a** Time-mean <*h*> (July 1, 2010–July 1, 2011), **b** time-mean <*τ*>, and **c** time-invariant *ω*; (*Lit1*, *Lit2* and *Lit3*) before calibration, and (*Cal*) after calibration, spatially averaged by vegetation class. International Geosphere-Biosphere Program (*IGBP*) vegetation classes are (*ENF*) Evergreen Needleleaf Forest, (*EBF*) Evergreen Broadleaf Forest, (*DNF*) Deciduous Needleleaf Forest, (*DBF*) Deciduous Broadleaf Forest, (*MXF*) Mixed Forest, (*CSH*) Closed Shrublands, (*OSH*) Open Shrublands, (*WSV*) Woody Savannas, (*SAV*) Savannas, (*GRS*) Grasslands, (*CRP*) Croplands, (*CRN*) Cropland and Natural Vegetation, and (*BAR*) Barren or Sparsely Vegetated. *Thin gray lines* for *Cal* indicate the spatial standard deviation within each vegetation class. Adapted from De Lannoy et al. (2013)

interference. The figure shows that all three sets of the literature values for the RTM parameters lead to substantial biases against SMOS observations, with Lit1 being too cold (by 42.0 K on average) and Lit3 too warm (by 24.6 K on average). Even though Lit2 estimates are nearly unbiased in the global average, there are still significant regional biases in the simulated Tbs, with an average absolute bias of 12.7 K. Since such biases would interfere with the assimilation of satellite Tb, the RTM parameters need to be calibrated to achieve climatologically unbiased Tb simulations.

The most important RTM parameters determining h, τ, and ω have been calibrated, separately for each model grid cell, using multi-angular SMOS observations from July 1, 2011 to July 1, 2012. The calibration simultaneously minimizes, separately for each location, the difference between the modeled and observed climatological mean values, the difference between modeled and observed climatological standard deviations, and the deviations of the optimized parameters from prior guesses (that is, from Lit1, Lit2, or Lit3 values). Through investigating a number of calibration scenarios, De Lannoy et al. (2013) determined that it is best to simultaneously calibrate a subset of the RTM parameters that most directly determine h, τ, and ω.

After calibration, global Tb simulations for the validation year (July 1, 2010–July 1, 2011) are largely unbiased for multiple incidence angles and both H- and V-polarization. For example, Fig. 7d shows that the global average absolute bias is now just 2.7 K for H-polarized Tb at 42.5° incidence angle. It should be emphasized that an RMSE of approximately 10 K remains, which is partly due to seasonal biases and partly due to random errors. The former will be addressed in the assimilation system through bias estimation and correction, and the latter through the radiance-based soil moisture analysis.

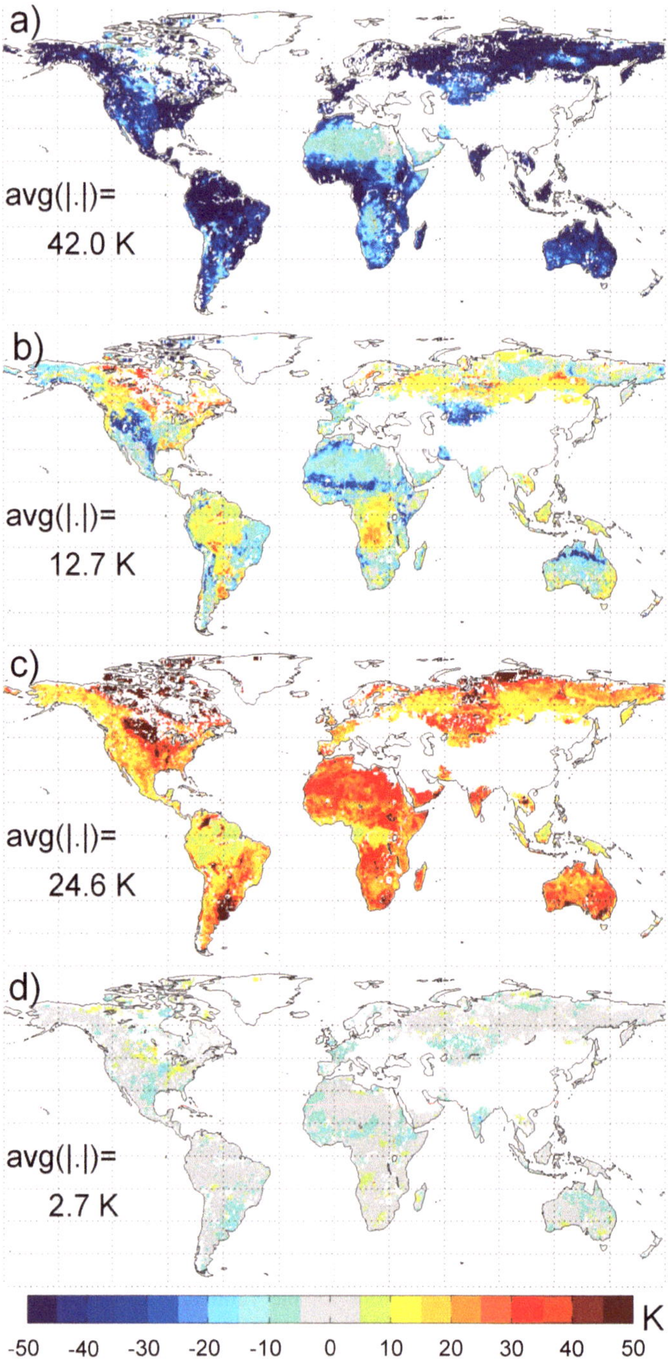

Fig. 7 Difference between 1-year (July 1, 2010–July 1, 2011) mean values of $Tb_H(42.5°)$ in Kelvin from GEOS-5 and SMOS observations for **a** Lit1, **b** Lit2, **c** Lit3, and **d** calibrated parameters. Within each subplot, "avg(|.|)" indicates the average absolute difference across the globe (excluding regions impacted by open water or radio-frequency interference that are shown in *white*). Adapted from De Lannoy et al. (2013)

The calibrated parameters are shown in Fig. 6. Results suggest, for example, that the roughness parameter (h) is too low in Lit1 and too high in Lit3. The calibrated vegetation opacity (τ) values distinguish clearly between high and low vegetation. The calibrated scattering albedo (ω) is increased over low vegetation, which reduces the vegetation effect in the simulated Tb. In summary, the climatological calibration generates plausible parameter values that are consistent with the underlying land modeling system.

3.3.2 Predicting Microwave Brightness Temperatures over Snow

As demonstrated in the previous section, the Catchment model (as do similar global land surface models) supports the application of a physically based microwave RTM for warm-season processes. However, the snow model components in global land surface models, including that in the Catchment model, are usually too simplistic to support physically based RTM modeling in the presence of snow. Specifically, global snow models lack reliable estimates of snow microphysical properties (such as grain size, ice layers, and depth hoar) which would be needed for physically based forward modeling of the microwave brightness temperatures. Forman et al. (2013) therefore constructed an empirical forward RTM for snow-covered land surfaces based on an Artificial Neural Network (ANN).

The Catchment model state variables used as input to the ANN include the density and temperature of the snowpack at multiple depths, the temperature of the underlying soil, the overlying air, and the vegetative canopy, and the total amount of water equivalent within the snowpack. In addition, a cumulative temperature gradient index (TGI) is used as a proxy for snow grain size evolution in the presence of a vapor pressure gradient. Using the above inputs, the ANN is trained and (independently) validated using 10.7, 18.7, and 36.5 GHz microwave brightness temperatures at H- and V-polarization from AMSR-E. The independent validation is accomplished as follows: From the 9-year AMSR-E data record, each single year is withheld in turn from the ANN training, and skill metrics for the resulting ANN predictions are computed only against the AMSR-E data that have been withheld from the ANN training.

Figure 8 demonstrates the performance of the ANN predictions relative to AMSR-E measurements that were not used during training. The figure illustrates the overall ability of the ANN to predict Tbs for the 10 GHz V-polarized channel. The ANN predictions are essentially unbiased (relative to the AMSR-E measurements) across the 9-year period (Fig. 8a). The RMSE is typically less than 5 K (Fig. 8b). In addition, the ANN demonstrates skill in predicting interannual variability, with anomaly R values well above 0.5 over large parts of North America (Fig. 8c). Relatively low skill can be seen in areas along the southern periphery, where the snowpack is relatively thin and ephemeral, as well as in areas north of the boreal forest, where sub-grid scale lake ice (which is not modeled in the land surface model) is common. In short, Fig. 8 suggests considerable skill by the ANN at predicting interannual variability in 10 GHz V-polarized Tbs across North America with negligible bias and a reasonable RMSE. The RMSE is somewhat higher but still reasonable (less than 10 K) for the higher frequencies and for H-polarization Tb (see Figures 4–6 of Forman et al. 2013).

Forman et al. (2013) also assessed the potential for using the ANN as a forward observation operator in radiance-based snow assimilation. For this demonstration, the observations are considered to be in the form of spectral differences in V-polarization brightness temperatures, $\Delta Tb \equiv Tb_V(18\ GHz) - Tb_V(36\ GHz)$. Since ΔTb typically increases with increasing SWE, this spectral difference is commonly used to estimate SWE

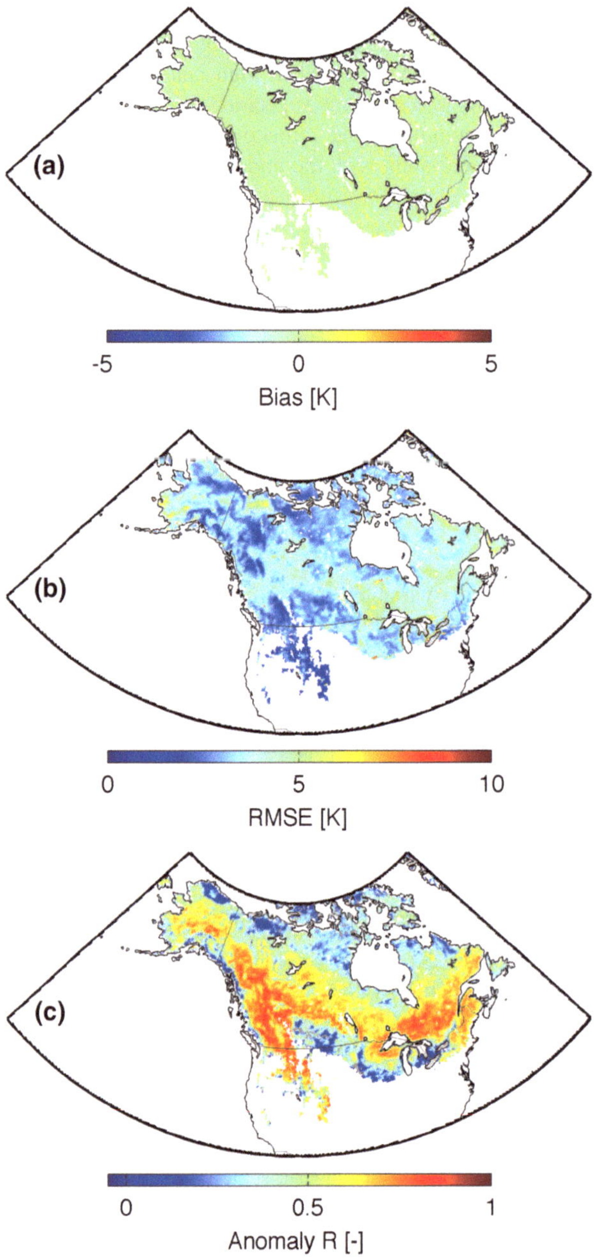

Fig. 8 **a** Bias, **b** RMSE, and **c** anomaly *R* for ANN simulated 10 GHz V-polarized Tb from September 1, 2002 to September 1, 2011 versus AMSR-E observations not used in training. Anomaly *R* values not statistically different from zero at the 95 % significance level based on a Fisher *Z* transform are shown in *gray*. Such non-significant *R* values occur in only a few very small regions

in retrieval algorithms (Kelly 2009). For the demonstration of the radiance-based assimilation considered here, observations of ΔTb imply that the resulting Kalman gain is proportional to error correlations between modeled SWE and ANN predictions of ΔTb. To obtain analysis increments, the Kalman gain would be multiplied with innovations in ΔTb (that is, the difference between actual AMSR-E observations of ΔTb and ANN predictions of ΔTb).

The Kalman gain computed for February 6, 2003 ranges from −10 to 15 mm K^{-1} as illustrated in Fig. 9. A gain of 1 mm K^{-1} equates to an increase of 1 mm in the posterior (updated) modeled SWE for a 1 K innovation (that is, for a difference of 1 K between AMSR-E ΔTb measurements and ANN ΔTb predictions). Similarly, a negative Kalman gain in the presence of a positive-valued innovation would equate to a reduction in modeled SWE. Most importantly, the results suggest that there is a nonzero error correlation between the model SWE forecasts and the simulated ΔTb measurements across much of the North American domain. Overall, the results suggest that the ANN could serve as a computationally efficient observation operator for radiance-based snow data assimilation at the continental scale.

3.4 Observation Selection for a Root Zone Soil Moisture Analysis

Knowledge of the amount of moisture stored in the root zone of the soil is important for many applications related to the transfer of water, energy, and carbon between the land and the atmosphere, including the assessment, monitoring, and prediction of drought (Seneviratne et al. 2010). At the global scale, soil moisture estimates are usually based on two sources of information: (1) direct observations of surface soil moisture from satellite and (2) observation-based precipitation forcing driving a numerical model of soil moisture dynamics. However, neither surface soil moisture retrievals nor precipitation observations provide direct measurements of soil moisture in the root zone. The selection of the most relevant types of observations for a root zone soil moisture analysis therefore presents an important conceptual problem.

A priori, it is not obvious whether the estimation of root zone soil moisture would benefit more from the use of precipitation observations (as, for example, in the Global Land Data Assimilation System; Rodell et al. 2003) or from the assimilation of surface soil moisture retrievals (as, for example, illustrated by Reichle et al. 2007). This section provides examples of both approaches. First, a land surface reanalysis that relies on observed precipitation is presented, followed by a root zone soil moisture analysis that is based on the assimilation of surface soil moisture retrievals. Finally, the two sources of soil moisture

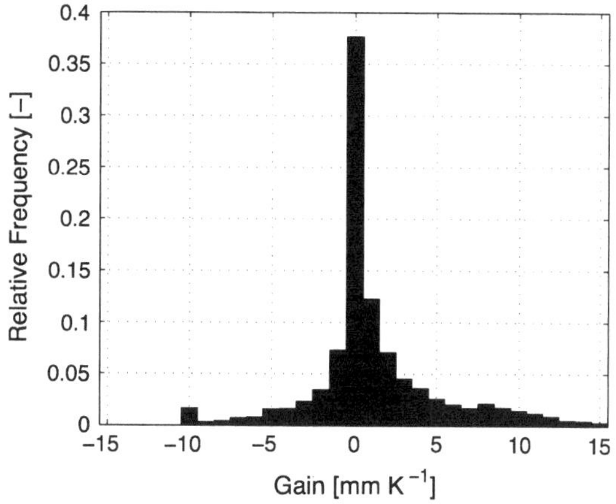

Fig. 9 Histogram of the Kalman gain on February 6, 2003 for SWE versus ΔTb = [Tb_V(18 GHz) − Tb_V(36 GHz)]

information are merged and compared directly in a single system, and their relative contributions to the skill of root zone soil moisture estimates are assessed.

3.4.1 Using Precipitation Observations

The Modern-Era Retrospective Analysis for Research and Applications (MERRA) is a state-of-the-art atmospheric reanalysis data product based on GEOS-5 that provides, in addition to atmospheric fields, global estimates of soil moisture, latent heat flux, snow, and runoff for 1979—present with a latency of about 1 month (Rienecker et al. 2011). A supplemental and improved set of land surface hydrological fields ("MERRA-Land") is generated routinely using an improved version of the land component of the MERRA system (Reichle et al. 2011; Reichle 2012). Specifically, the MERRA-Land estimates benefit from corrections to the MERRA precipitation forcing with the global gauge-based NOAA Climate Prediction Center "Unified" (CPCU) precipitation product and from revised parameter values in the rainfall interception model, changes that effectively correct for known limitations in the MERRA surface meteorological forcings.

With a few exceptions, the MERRA-Land data appear more accurate than the original MERRA estimates and are thus recommended for those interested in using MERRA output for land surface hydrological studies. As an example, Fig. 10 examines the drought conditions experienced across the western United States and along the East Coast. The MERRA and MERRA-Land drought indicator shown in the figure is derived by ranking, separately for each grid cell, the normalized, monthly mean root zone soil moisture anomalies for June, July, and August of 1980 through 2011 and converting the rank into percentile units. For comparison, the drought severity assessed independently by U.S. Drought Monitor is also shown. The figure clearly demonstrates that MERRA-Land data are more consistent with the Drought Monitor than MERRA data.

Reichle et al. (2011) and Reichle (2012) provide a more comprehensive and quantitative analysis of the skill (defined as the correlation coefficient of the anomaly time series with independent observations) in land surface hydrological fields from MERRA, MERRA-Land, and the latest global atmospheric reanalysis produced by ECWMF (ERA-I; Dee et al. 2011). Figure 11 shows that MERRA-Land and ERA-I root zone soil moisture skills (against in situ observations at 85 US stations) are comparable and significantly greater than that of MERRA. Furthermore, the runoff skill (against naturalized stream flow observations from 18 US basins) of MERRA-Land is typically higher than that of MERRA and ERA-I (not shown). Throughout the northern hemisphere, MERRA and MERRA-Land agree reasonably well with in situ snow depth measurements (from 583 stations) and with SWE from an independent analysis (not shown). In summary, through observations-based corrections of the MERRA precipitation forcing, MERRA-Land provides a supplemental and significantly improved land surface reanalysis product.

3.4.2 Assimilating Surface Soil Moisture Retrievals

Satellite retrievals of surface soil moisture are not used in MERRA-Land but would almost certainly have further improved the skill of root zone soil moisture estimates. Draper et al. (2012) illustrate the potential gains from assimilating ASCAT (Bartalis et al. 2007; Wagner et al. 1999) and 10.7 GHz AMSR-E Land Parameter Retrieval Model (LPRM; de Jeu et al. 2008) surface soil moisture retrievals. The retrievals are assimilated, both separately and jointly, over 3.5 years into the GEOS-5 LDAS, using MERRA forcing and initial conditions. Soil moisture skill is measured as the anomaly time series correlation coefficient

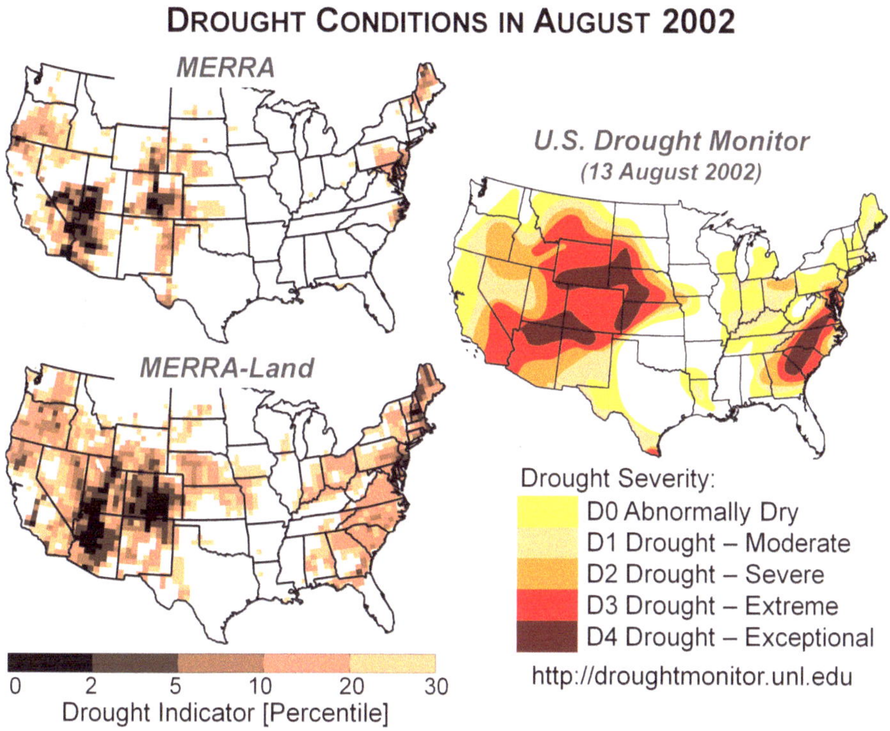

Fig. 10 Drought indicator derived from (*top left*) MERRA and (*bottom left*) MERRA-Land root zone soil moisture estimates for August 2002. *Darker colors* indicate more severe drought conditions. MERRA-Land estimates are more consistent than MERRA estimates with an independent drought assessment from the US Drought Monitor for August 13, 2002 (*right*)

Fig. 11 Skill (pentad anomaly R; dimensionless) of MERRA, MERRA-Land, and ERA-I estimates (2002–2009) versus SCAN in situ surface and root zone soil moisture measurements at 85 stations. *Error bars* indicate approximate 95 % confidence intervals. Adapted from Reichle (2012)

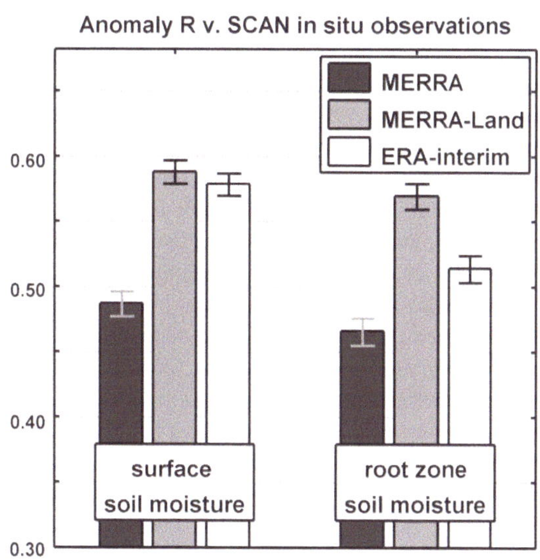

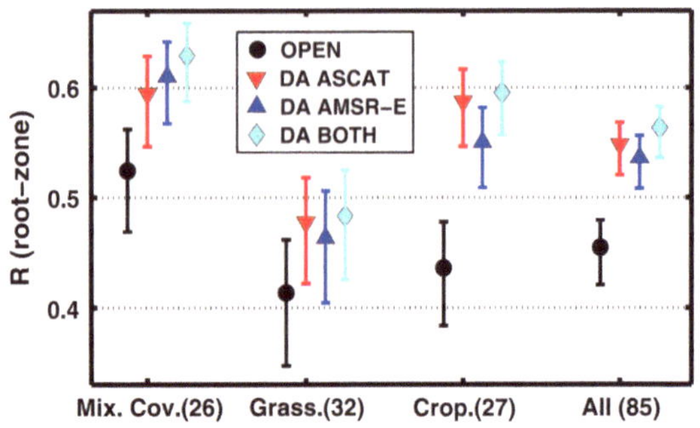

Fig. 12 Mean skill for root zone soil moisture from the open loop (ensemble mean, no assimilation), and the data assimilation (DA) of ASCAT, AMSR-E, and both surface soil moisture retrievals, averaged by land cover class, with 95 % confidence intervals. The number of sites in each land cover class is given in the axis labels. Skill is defined as the daily anomaly R value versus SCAN/SNOTEL and Murrumbidgee in situ observations. Adapted from Draper et al. (2012)

(R) with in situ soil moisture observations from the SCAN/SNOTEL network in the US (66 sites) and the Murrumbidgee Soil Moisture Monitoring Network in Australia (19 sites). These 85 sites are surrounded by terrain with low topographic complexity based on data provided with the ASCAT observations. Averaged over these sites, the ASCAT and AMSR-E surface soil moisture retrievals have similar skill (Draper et al. 2012).

Figure 12 shows the estimated R values and their 95 % confidence intervals for root zone soil moisture from the assimilation of ASCAT, AMSR-E, and both. The results are benchmarked against an open loop (no assimilation) model integration and have been averaged by land cover type (based on MODIS land cover classifications). Averaged across all 85 sites, assimilating ASCAT and/or AMSR-E surface soil moisture retrievals significantly improved the root zone soil moisture skill (at the 5 % level). The mean skill was increased from 0.45 for the open loop, to 0.55 for the assimilation of ASCAT, 0.54 for the assimilation of AMSR-E, and 0.56 for the assimilation of both.

Assimilating the ASCAT or AMSR-E retrievals also improved the mean R value over each individual land cover type, in most cases significantly. At the frequencies observed by AMSR-E and ASCAT, dense vegetation limits the accuracy of soil moisture observations, and so the improvements obtained over the mixed cover sites, which have 10–60 % trees or wooded vegetation, are very encouraging. For each land cover type, the skill obtained from the assimilation of ASCAT or AMSR-E retrievals was very similar. The combined assimilation of ASCAT and AMSR-E retrievals generally matched or slightly exceeded the mean R values from the single-sensor assimilation experiments.

Draper et al. (2012) also examined the contribution of the model skill and the observation skill to the skill of the assimilation estimates. The color surface in Fig. 13 shows the skill improvements (ΔR) in root zone soil moisture, where ΔR is defined as the skill (R) of the assimilation estimates (from the single-sensor assimilation of ASCAT or AMSR-E retrievals) minus that of the open loop model estimates. The skill improvements are shown as a function of the open loop model skill and the retrieval skill. Specifically, the ordinate measures the skill of the open loop root zone soil moisture estimates, and the abscissa measures the skill of the assimilated (ASCAT or AMSR-E) surface soil moisture retrievals. Where the skill of the assimilated retrievals is no more than 0.2 less than the open loop skill (below the dashed line), the assimilation improves the root zone soil moisture skill. The improvements increase (up to 0.4) as the observation skill increases relative to that of the

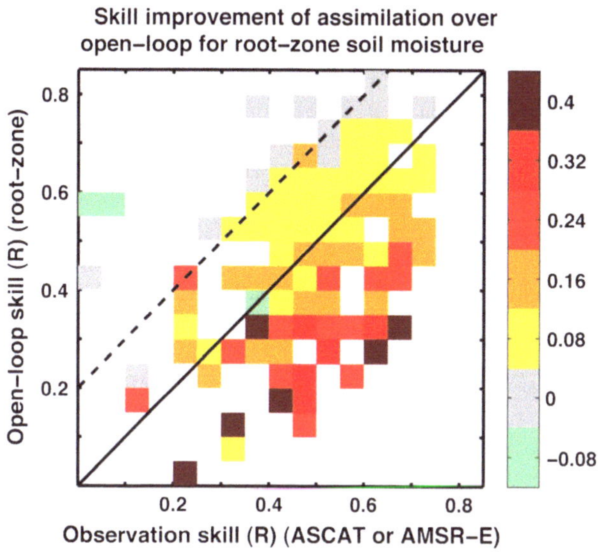

Fig. 13 Root zone soil moisture skill improvement (ΔR) from assimilating either ASCAT or AMSR-E surface soil moisture retrievals as a function of (ordinate) the open loop model skill and (abscissa) the observation skill. Skill improvement (ΔR) is defined as the skill of the assimilation product minus the open loop skill, with skill based only on days with data available from both satellites. Skill is assessed versus in situ measurements from the SCAN and Murrumbidgee networks. Significant improvements are found in the area below the dashed line where the skill of the retrievals may be lower than that of the open loop by up to 0.2. Adapted from Draper et al. (2012)

open loop (toward the bottom right hand corner). (The results are very similar if the ordinate measures surface soil moisture skill; not shown). Figure 13 thus provides a practical demonstration of the minimum skill required for soil moisture observations to be beneficial in a land data assimilation system and confirms the findings obtained by Reichle et al. (2008b) using synthetically generated observations. In summary, the assimilation of active or passive microwave data significantly improves the model root zone soil moisture estimates by a similar amount, even in cases where the assimilated surface soil moisture retrievals are less skillful than the open loop soil moisture estimates.

3.4.3 Combining Precipitation Observations and Surface Soil Moisture Retrievals

Liu et al. (2011a) used both precipitation observations and surface soil moisture retrievals within the GEOS-5 LDAS and investigated their relative contributions to the skill of root zone soil moisture estimates. Relative to baseline soil moisture estimates from MERRA, their study investigates soil moisture skill derived from (1) land model forcing corrections based on large-scale, gauge-, and satellite-based precipitation observations and (2) assimilation of surface soil moisture retrievals from AMSR-E. Three precipitation products were used (separately) to correct the MERRA precipitation toward gauge- and satellite-based observations: the NOAA Climate Prediction Center Merged Analysis of Precipitation (CMAP) pentad product ("standard" version), the Global Precipitation Climatology Project (GPCP) version 2.1 pentad product, and the NOAA Climate Prediction Center (CPC) daily unified precipitation analysis over the United States.

Two different surface soil moisture retrieval products were assimilated into the GEOS-5 LDAS: (1) the operational NASA Level-2B AMSR-E "AE-Land" product (version V09) archived at the National Snow and Ice Data Center (NSIDC; Njoku et al. 2003) and (2) the AMSR-E LPRM product (de Jeu et al. 2008). Soil moisture skill is assessed using in situ observations in the continental United States at the 37 single-profile sites within the SCAN network for which skillful AMSR-E retrievals are available. As in Sect. 3.4.2, skill is assessed in terms of the anomaly time series correlation coefficient *R*.

Figure 14 shows comparable average skill for surface soil moisture estimates from the two AMSR-E products and from the Catchment model with MERRA precipitation forcing without data assimilation. Consistent with the findings of Sect. 3.4.1, adding information from precipitation observations increases soil moisture skills for surface and root zone soil moisture. Consistent with the results of Sect. 3.4.2, assimilating satellite estimates of surface soil moisture also increases soil moisture skills, again for surface and root zone soil moisture. The salient result is that adding information from both sources (precipitation observations and surface soil moisture retrievals) increases soil moisture skills by almost the sum of the individual skill contributions, which demonstrates that precipitation corrections and assimilation of satellite soil moisture retrievals contribute important and largely independent amounts of information.

Liu et al. (2011a) also repeated their skill analysis against measurements from four USDA Agricultural Research Service ("CalVal") watersheds with high-quality distributed sensor networks that measure surface soil moisture at the scale of land model and satellite

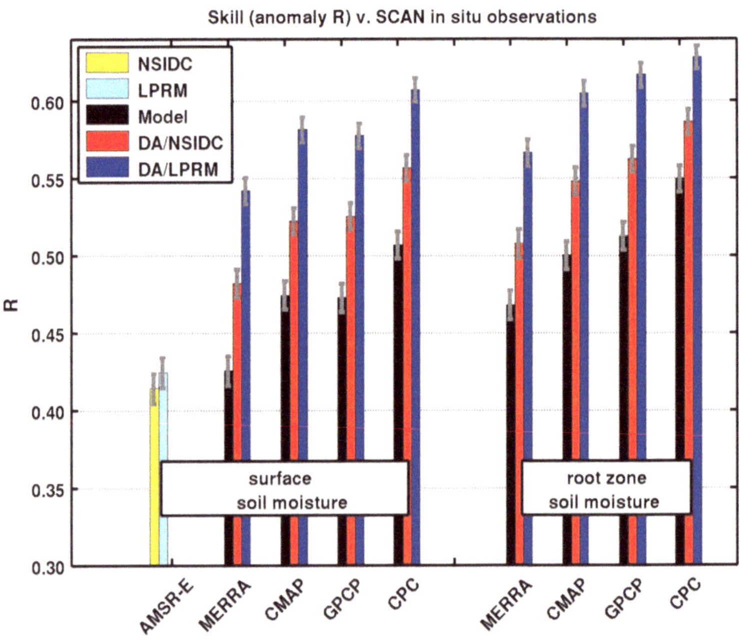

Fig. 14 Skill (daily anomaly *R*; dimensionless) versus SCAN in situ soil moisture measurements for estimates from two AMSR-E retrieval datasets (NSIDC and LPRM), the Catchment model forced with four different precipitation datasets (MERRA, CMAP, GPCP, and CPC), and the corresponding data assimilation integrations (*red bars*: DA/NSIDC and *blue bars*: DA/LPRM). Average is based on 37 SCAN sites for surface and 35 SCAN sites for root zone soil moisture. *Error bars* indicate approximate 95 % confidence. Adapted from Liu et al. (2011a)

estimates (Jackson et al. 2010). As expected, the skill of the satellite, model, and assimilation estimates is higher when it is assessed against the multi-sensor CalVal observations rather than against single-profile SCAN measurements (not shown). The relative skill contributions by precipitation corrections and soil moisture retrieval assimilation, however, remain unchanged (not shown). This corroborates the results shown in Fig. 14 which were obtained with a larger network of single-profile sensors.

Taken together, the results of this section strongly suggest that future land surface reanalysis efforts would benefit from the use of both precipitation observations and satellite retrievals of surface soil moisture because both types of observations contribute significant and largely independent amounts of information to the skill of root zone soil moisture in the analysis. Moreover, both active and passive surface soil moisture retrievals should be assimilated for maximum coverage and accuracy.

4 Summary and Discussion

The present study discussed several conceptual challenges in land surface hydrological data assimilation as part of an effort toward improving our understanding of the Earth's hydrological cycle (Trenberth and Asrar 2013). The challenges arise from a seeming mismatch between the assimilated observations and the water cycle variables of interest that can be overcome through the careful design of the assimilation system. This was illustrated with examples from recent research findings using the GEOS-5 LDAS.

The first challenge is the use of coarse-scale satellite observations to estimate land surface fields at finer scales of interest. Such horizontal downscaling can be accomplished by using a fine-scale land surface model and by defining an observation operator that maps from the fine-scale model space to the space of the coarse-scale observations (Sect. 3.1). In the presence of larger-scale model error correlations, the assimilation system can also spread observational information to unobserved locations.

The second challenge is the partitioning of satellite observations (such as TWS retrievals) into their component variables. This partitioning can again be accomplished through an observation operator. In the case of TWS assimilation, the observation operator maps from the fine-scale model estimates of soil moisture and snow to basin-scale TWS (Sect. 3.2). The observation operator therefore enables the computation of the observation-minus-forecast residuals (innovations). The observation operator is also needed for the computation of the Kalman gain matrix that transforms the observation-space (coarse-scale TWS) innovations into the model-space (fine-scale soil moisture and snow) analysis increments.

The third challenge is the development of microwave RTMs for use as observation operators in radiance-based data assimilation. Two examples were given. In the first example, a global microwave RTM for warm-season, L-band brightness temperatures was calibrated successfully using SMOS observations (Sect. 3.3.1). In the second example, an empirical approach based on an artificial neural network yielded robust model simulations of AMSR-E microwave brightness temperatures over snow-covered land at continental scales (Sect. 3.3.2). In both cases, the results are very encouraging and constitute progress toward replacing the commonly used assimilation of geophysical retrievals (such as SWE or surface soil moisture retrievals) with the direct assimilation of satellite radiances. Note that a radiance-based soil moisture analysis can partition the observational (brightness temperature) information into increments of model soil moisture, soil temperature, and vegetation water content (essentially, the model variables that most impact the brightness temperature). In other words, the microwave RTM, acting as the observation operator, takes on a role that is conceptually similar to that of the observation operator used for the partitioning of TWS information into its water cycle components (Sect. 3.2).

The fourth and final challenge addressed in the paper discusses the selection of the types of observations that are most relevant for the analysis of poorly observed variables. For the analysis of one such variable, root zone soil moisture, the use of gauge- and satellite-based precipitation observations along with active and passive surface soil moisture retrievals was investigated (Sect. 3.4). It was shown that the MERRA-Land surface reanalysis provides better estimates of root zone soil moisture than MERRA due to the use of gauge-based precipitation observations in MERRA-Land. Next, the potential skill gained from the assimilation of surface soil moisture retrievals was investigated. It was demonstrated that improved root zone soil moisture estimates can be obtained even where the skill of the assimilated surface soil moisture retrievals is somewhat poorer than that of the model estimates of surface soil moisture. For maximum coverage and accuracy, both active and passive retrievals should be assimilated. Finally, it was shown that the use of precipitation observations and the assimilation of surface soil moisture retrievals contribute significant and largely independent amounts of root zone soil moisture information. Therefore, future reanalyses should use both of these observation types. This finding is consistent with the general expectation that using more observations in a data assimilation system will improve its output.

In some cases (for example, Sects. 3.1 and 3.2), the appropriate observation operator and assimilation system configuration entail that neighboring grid cells (or land model tiles) are no longer computationally independent in the assimilation system, even if they are independent in the land model (Reichle and Koster 2003). These computational dependencies arise through spatially correlated perturbation fields or spatially distributed analysis update calculations. Such "three-dimensional" land data assimilation systems therefore necessitate greater computational resources than more simplistic, "one-dimensional" assimilation systems where all model grid cells (or tiles) are treated independently. It is assumed here that the purely technical challenge of computational demand can be overcome with sophisticated software engineering and the increasing availability of affordable and massively parallel computing architectures.

5 Conclusions and Outlook

The present paper focused on the seeming mismatch between satellite observations and the water cycle variables of interest, and how a mismatch can be overcome through careful design and application of a land data assimilation system. Responding to the challenge questions of Sect. 1, we find that, if designed properly, a land data assimilation system can enable

1. the horizontal downscaling of coarse-scale satellite observations,
2. the partitioning of vertically integrated satellite measurements such as TWS into their water cycle components,
3. the direct assimilation of satellite radiances for soil moisture or snow analyses, and
4. the propagation of information from observed fields such as precipitation and surface soil moisture into variables such as root zone soil moisture, that are of great interest but are not directly observed by satellites.

Naturally, many challenges still lie ahead. State-of-the-art land data assimilation algorithms are only now emerging in operational systems. Much of the recent progress has been achieved in so-called "off-line" (land-only) assimilation systems. These advances need to be incorporated into the coupled land–atmosphere systems used in atmospheric data assimilation and numerical weather prediction (NWP). Ground-breaking advances in coupled land–atmosphere data assimilation are being made, for example, at ECMWF (de Rosnay et al. 2012a, b). At the same time, the coupling of the GEOS-5 LDAS to the GEOS-5 atmospheric data assimilation system is underway at the NASA GMAO.

Moreover, much of the progress in land data assimilation has been with systems that assimilate only one type of observation, often surface soil moisture. In future, more emphasis will need to be placed on the assimilation of multiple types of observations within a single assimilation system, including observations of water cycle components such as soil moisture, SWE, snow cover fraction, TWS, and precipitation.

Future development should also address the addition or improvement of runoff routing and surface water storage model components in the global land surface models used in NWP. The planned NASA Surface Water and Ocean Topography (SWOT; Durand et al. 2010) mission, for instance, will provide high-resolution observations of surface water elevation. To improve our understanding of the global hydrological cycle, it will be crucial to incorporate these new observations into global land data assimilation systems, building on early studies such as those by Andreadis et al. (2007), Biancamaria et al. (2011), and Durand et al. (2008).

Finally, the existing global land data assimilation systems will need to consider the modeling of vegetation dynamics and the assimilation of current or planned satellite observations such as the Fraction of Absorbed Photosynthetically Active Radiation (FAPAR), the Leaf Area Index (LAI), or the multi-angular Photochemical Reflectance Index (PRI) (Albergel et al. 2010; Hilker et al. 2012; Kaminski et al. 2012; Knorr et al. 2010; Muñoz Sabater et al. 2008; Stöckli et al. 2011). Furthermore, current microwave sensors already provide observations of the freeze–thaw state of the landscape at coarse scales (Kim et al. 2010), and SMAP will provide much higher-resolution observations with continental coverage (Entekhabi et al. 2010). These vegetation and freeze–thaw observations link the hydrological and carbon cycles and should be used in global land data assimilation systems.

Acknowledgments The authors thank the organizers of the ISSI Workshop on "The Earth's Hydrological Cycle" held February 6–10, 2012 and two anonymous reviewers for their efforts. The research was supported by the NASA program on The Science of Terra and Aqua, the NASA Soil Moisture Active Passive mission, the NASA Postdoctoral Program, and the NASA High-End Computing program.

References

Albergel C, Calvet J-C, Mahfouf J-F, Rüdiger C, Barbu AL, Lafont S, Roujean J-L, Walker JP, Crapeau M, Wigneron J-P (2010) Monitoring of water and carbon fluxes using a land data assimilation system: a case study for southwestern France. Hydrol Earth Syst Sci 14:1109–1124. doi:10.5194/hess-14-1109-2010

Andreadis K, Lettenmaier D (2006) Assimilating remotely sensed snow observations into a macroscale hydrology model. Adv Water Resour 29:872–886

Andreadis KM, Clark EA, Lettenmaier DP, Alsdorf DE (2007) Prospects for river discharge and depth estimation through assimilation of swath-altimetry into a raster-based hydrodynamics model. Geophys Res Lett 34:L10403. doi:10.1029/2007GL029721

Barnett TP, Adam JC, Lettenmaier DP (2005) Potential impacts of a warming climate on water availability in snow-dominated regions. Nature 438:303–309. doi:10.1038/nature04141

Bartalis Z, Wagner W, Naeimi V, Hasenauer S, Scipal K, Bonekamp H, Figa J, Anderson C (2007) Initial soil moisture retrievals from the METOP-A Advanced Scatterometer (ASCAT). Geophys Res Lett 34:L20401. doi:10.1029/2007GL031088

Biancamaria S, Durand M, Andreadis KM, Bates PD, Boone A, Mognard NM, Rodríguez E, Alsdorf DE, Lettenmaier DP, Clark EA (2011) Assimilation of virtual wide swath altimetry to improve Arctic river modeling. Remote Sens Environ 115:373–381. doi:10.1016/j.rse.2010.09.008
Brasnett B (1999) A global analysis of snow depth for numerical weather prediction. J Appl Meteorol 38:726–740
Brown RD, Brasnett B (2010) Canadian Meteorological Centre (CMC) daily snow depth analysis data. Environment Canada, Boulder, Colorado, National Snow and Ice Data Center, available at http://nsidc.org/data/docs/daac/nsidc0447_CMC_snow_depth
Bruinsma S, Lemoine J-M, Biancale R, Vales N (2010) CNES/GRGS 10-day gravity field models (release 2) and the evaluation. Adv Space Res 45:587–601
Clifford D (2010) Global estimates of snow water equivalent from passive microwave instruments: history, challenges, and future developments. Int J Remote Sens 31:3707–3726
Crow WT, Reichle RH (2008) Comparison of adaptive filtering techniques for land surface data assimilation. Water Resour Res 44:W08423. doi:10.1029/2008WR006883
Crow WT, van den Berg MJ (2010) An improved approach for estimating observation and model error parameters for soil moisture data assimilation. Water Resour Res 46:W12519. doi:10.1029/2010WR009402
Crow WT, Wood EF (2003) The assimilation of remotely sensed soil brightness temperature imagery into a land surface model using ensemble Kalman filtering: a case study based on ESTAR measurements during SGP97. Adv Water Resour 26:137–149
de Jeu RAM, Wagner W, Holmes TRH, Dolman AJ, Giesen NC, Friesen J (2008) Global soil moisture patterns observed by space borne microwave radiometers and scatterometers. Surv Geophys 29:399–420. doi:10.1007/s10712-008-9044-0
De Lannoy GJM, Reichle RH, Houser PR, Pauwels VRN, Verhoest NEC (2007) Correcting for forecast bias in soil moisture assimilation with the ensemble Kalman filter. Water Resour Res 43:W09410. doi:10.1029/2006WR005449
De Lannoy GJM, Reichle RH, Houser PR, Arsenault KR, Verhoest NEC, Pauwels VRN (2010) Satellite-scale snow water equivalent assimilation into a high-resolution land surface model. J Hydrometeorol 11:352–369. doi:10.1175/2009JHM1192.1
De Lannoy GJM, Reichle RH, Arsenault KR, Houser PR, Kumar SV, Verhoest NEC, Pauwels VRN (2012) Multi-scale assimilation of AMSR-E snow water equivalent and MODIS snow cover fraction in northern Colorado. Water Resour Res 48:W01522. doi:10.1029/2011WR010588
De Lannoy GJM, Reichle RH, Pauwels VRN (2013) Global calibration of the GEOS-5 L-band microwave radiative transfer model over land using SMOS observations. J Hydrometeorol. doi:10.1175/JHM-D-12-092.1
de Rosnay P, Drusch M, Vasiljevic D, Balsamo G, Albergel C, Isaksen L (2012a) A simplified extended Kalman filter for the global operational soil moisture analysis at ECMWF. Q J R Meteorol Soc. doi:10.1002/qj.2023
de Rosnay P, Balsamo G, Albergel C, Munoz Sabater J, Isaksen L (2012b) Initialisation of land surface variables for numerical weather prediction. Surv Geophys. doi:10.1007/s10712-012-9207-x
Dee DP et al (2011) The ERA-Interim reanalysis: configuration and performance of the data assimilation system. Q J R Meteorol Soc 137:553–597. doi:10.1002/qj.828
Draper CS, Reichle RH, De Lannoy GJM, Liu Q (2012) Assimilation of passive and active microwave soil moisture retrievals. Geophys Res Lett 39:L04401. doi:10.1029/2011GL050655
Drusch M (2007) Initializing numerical weather prediction models with satellite derived surface soil moisture: data assimilation experiments with ECMWF's Integrated Forecast System and the TMI soil moisture data set. J Geophys Res 112:D03102. doi:10.1029/2006JD007478
Drusch M, Wood EF, Gao H (2005) Observation operators for the direct assimilation of TRMM microwave imager retrieved soil moisture. Geophys Res Let 32:L15403. doi:10.1029/2005GL023623
Ducharne A, Koster RD, Suarez MJ, Stieglitz M, Kumar P (2000) A catchment-based approach to modeling land surface processes in a general circulation model, 2: parameter estimation and model demonstration. J Geophys Res 105(20):24823–24838
Dunne S, Entekhabi D (2006) Land surface state and flux estimation using the ensemble Kalman smoother during the Southern Great Plains 1997 field experiment. Water Resour Res 42:W01407. doi:10.1029/2005WR004334
Durand M, Margulis S (2008) Effects of uncertainty magnitude and accuracy on assimilation of multi-scale measurements for snowpack characterization. J Geophys Res 113:D02105. doi:10.1029/2007JD008662
Durand M, Andreadis KM, Alsdorf DE, Lettenmaier DP, Moller D, Wilson MD (2008) Estimation of bathymetric depth and slope from data assimilation of swath altimetry into a hydrodynamic model. Geophys Res Lett 35:L20401. doi:10.1029/2008GL034150

Durand M, Fu LL, Lettenmaier DP, Alsdorf D, Rodríguez E, Esteban-Fernandez D (2010) The Surface Water and Ocean Topography mission: observing terrestrial surface water and oceanic submesoscale eddies. Proc IEEE 98:766–779

Ek M, Mitchell K, Yin L, Rogers P, Grunmann P, Koren V, Gayno G, Tarpley JD (2003) Implementation of Noah land surface model advances in the NCEP operational mesoscale Eta model. J Geophys Res 108(D22):8851. doi:10.1029/2002JD003296

Entekhabi D et al (2010) The Soil Moisture Active and Passive (SMAP) mission. Proc IEEE 98:704–716. doi:10.1109/JPROC.2010.2043918

Evensen G (2003) The ensemble Kalman filter: theoretical formulation and practical implementation. Ocean Dyn 53:343–367. doi:10.1007/s10236-003-0036-9

Eyre JR, Kelly GA, McNally AP, Andersson E, Persson A (1993) Assimilation of TOVS radiance information through one-dimensional variational analysis. Q J R Meteorol Soc 119:1427–1463. doi: 10.1002/qj.49711951411

Forman BA, Reichle RH, Rodell M (2012) Assimilation of terrestrial water storage from GRACE in a snow-dominated basin. Water Resour Res 48:W01507. doi:10.1029/2011WR011239

Forman BA, Reichle RH, Derksen C (2013) Estimating passive microwave brightness temperature over snow-covered land in North America using a land surface model and an artificial neural network. IEEE Trans Geosci Remote Sens (in press)

Foster JL, Sun C, Walker JP, Kelly R, Chang A, Dong J, Powell H (2005) Quantifying the uncertainty in passive microwave snow water equivalent observations. Remote Sens Environ 92(2):187–203

Foster JL et al (2011) A blended global snow product using visible, passive microwave and scatterometer satellite data. Int J Remote Sens 32(5):1371–1395. doi:10.1080/01431160903548013

Gao Y, Xie H, Lu N, Yao T, Liang T (2010) Toward advanced daily cloud-free snow cover and snow water equivalent products from Terra-Aqua MODIS and Aqua AMSR-E measurements. J Hydrol 385:23–35

Hall DK, Riggs GA (2007) Accuracy assessment of the MODIS snow products. Hydrol Process 21: 1534–1547

Hall DK, Riggs GA, Foster JL, Kumar SV (2010) Development and evaluation of a cloud-gap-filled MODIS daily snow-cover product. Remote Sens Environ 114:496–503

Hilker T, Hall FG, Tucker CJ, Coops NC, Black TA, Nichol CJ, Sellers PJ, Barr A, Hollinger DY, Munger JW (2012) Data assimilation of photosynthetic light-use efficiency using multi-angular satellite data: II. Model implementation and validation. Remote Sens Environ 121:287–300

Horwath M, Lemoine J-M, Biancale R, Bourgogne S (2011) Improved GRACE science results after adjustment of geometric biases in the Level-1B K-band ranging data. J Geod 85(1):23–38

Houborg R, Rodell M, Li B, Reichle RH, Zaitchik BF (2012) Drought indicators based on model assimilated GRACE terrestrial water storage observations. Water Resour Res 48:W07525. doi:10.1029/2011 WR011291

Jackson TJ et al (2010) Validation of Advanced Microwave Scanning Radiometer soil moisture products. IEEE Trans Geosci Remote Sens 48:4256–4272. doi:10.1109/TGRS.2010.2051035

Joiner J, Dee DP (2000) An error analysis of radiance and suboptimal retrieval assimilation. Q J R Meteorol Soc 126:1495–1514. doi:10.1002/qj.49712656514

Kaminski T, Knorr W, Scholze M, Gobron N, Pinty B, Giering R, Mathieu P-P (2012) Consistent assimilation of MERIS FAPAR and atmospheric CO_2 into a terrestrial vegetation model and interactive mission benefit analysis. Biogeosciences 9:3173–3184. doi:10.5194/bg-9-3173-2012

Kelly RE (2009) The AMSR-E snow depth algorithm: description and initial results. J Remote Sens Soc Jpn 29:307–317

Kerr Y et al (2010) The SMOS mission: new tool for monitoring key elements of the global water cycle. Proc IEEE 98(5):666–687

Kim Y, Kimball JS, McDonald KC, Glassy J (2010) Developing a global data record of daily landscape freeze/thaw status using satellite passive microwave remote sensing. IEEE Trans Geosci Remote Sens 49:949–960. doi:10.1109/TGRS.2010.2070515

Knorr W, Kaminski T, Scholze M, Gobron N, Pinty B, Giering R, Mathieu P-P (2010) Carbon cycle data assimilation with a generic phenology model. J Geophys Res 115:G04017. doi:10.1029/2009JG001119

Knowles KW, Savoie MH, Armstrong RL, Brodzik MJ (2006) Updated 2012. AMSR-E/Aqua daily EASE-grid brightness temperatures, September 2002 through May 2011. National Snow and Ice Data Center, Boulder, Colorado USA. Digital media

Koster RD, Suarez MJ, Ducharne A, Stieglitz M, Kumar P (2000) A catchment-based approach to modeling land surface processes in a general circulation model, 1: model structure. J Geophys Res 105(20): 24809–24822

Kumar SV, Reichle RH, Peters-Lidard CD, Koster RD, Zhan X, Crow WT, Eylander JB, Houser PR (2008a) A land surface data assimilation framework using the Land Information System: description and applications. Adv Water Resour 31:1419–1432. doi:10.1016/j.advwatres.2008.01.013

Kumar SV, Peters-Lidard C, Tian Y, Reichle R, Geiger J, Alonge C, Eylander J, Houser P (2008b) An integrated hydrologic modeling and data assimilation framework. IEEE Comput 41:52–59. doi:10.1109/MC.2008.511

Kumar SV, Reichle RH, Harrison KW, Peters-Lidard CD, Yatheendradas S, Santanello JA (2012) A comparison of methods for a priori bias correction in soil moisture data assimilation. Water Resour Res 48:W03515. doi:10.1029/2010WR010261

Lahoz W, De Lannoy GJM (2013) Closing the gaps in our knowledge of the hydrological cycle over land: conceptual problems. Surv Geophys. doi:10.1007/s10712-013-9221-7

Li L, Gaiser P, Jackson T, Bindlish R, Du J (2007) WindSat soil moisture algorithm and validation. In: IEEE international geoscience and remote sensing symposium 23–28 July 2007, pp 1188–1191. doi:10.1109/IGARSS.2007.4423017

Li B, Rodell M, Zaitchik BF, Reichle RH, Koster RD, van Dam TM (2012) Assimilation of GRACE terrestrial water storage into a land surface model: evaluation and potential value for drought monitoring in western and central Europe. J Hydrol 446–447:103–115. doi:10.1016/j.jhydrol.2012.04.035

Liu Q, Reichle RH, Bindlish R, Cosh MH, Crow WT, de Jeu R, De Lannoy GJM, Huffman GJ, Jackson TJ (2011a) The contributions of precipitation and soil moisture observations to the skill of soil moisture estimates in a land data assimilation system. J Hydrometeorol 12:750–765. doi:10.1175/JHM-D-10-05000.1

Liu YY, Parinussa RM, Dorigo WA, de Jeu RAM, Wagner W, van Dijk AIJM, McCabe MF, Evans JP (2011b) Developing an improved soil moisture dataset by blending passive and active microwave satellite-based retrievals. Hydrol Earth Syst Sci 15(2):425–436. doi:10.5194/hess-15-425-2011

Margulis SA, McLaughlin D, Entekhabi D, Dunne S (2002) Land data assimilation and estimation of soil moisture using measurements from the Southern Great Plains 1997 field experiment. Water Resour Res 38:1299. doi:10.1029/2001WR001114

Muñoz Sabater J, Rüdiger C, Calvet J-C, Fritz N, Jarlan L, Kerr Y (2008) Joint assimilation of surface soil moisture and LAI observations into a land surface model. Agric For Meteorol 148:1362–1373

Njoku EG, Jackson TJ, Lakshmi V, Chan TK, Nghiem SV (2003) Soil moisture retrieval from AMSR-E. IEEE Trans Geosci Remote Sens 41:215–229. doi:10.1109/TGRS.2002.808243

Pan M, Wood EF (2006) Data assimilation for estimating the terrestrial water budget using a constrained ensemble Kalman filter. J Hydrometeorol 7:534–547

Pan M, Wood EF, Wojcik R, McCabe MF (2008) Estimation of regional terrestrial water cycle using multi-sensor remote sensing observations and data assimilation. Remote Sens Environ 112:1282–1294. doi:10.1016/j.rse.2007.02.039

Parinussa RM, Holmes TRH, de Jeu RAM (2012) Soil moisture retrievals from the WindSat spaceborne polarimetric microwave radiometer. IEEE Trans Geosci Remote Sens. doi:10.1109/TGRS.2011.2174643

Pulliainen J (2006) Mapping of snow water equivalent and snow depth in boreal and subarctic zones by assimilating space-borne microwave radiometer data and ground-based observations. Remote Sens Environ 101:257–269

Reichle RH (2012) The MERRA-land data product (Version 1.1). NASA Global Modeling and Assimilation Office, Office Note No. 3. Available at http://gmao.gsfc.nasa.gov/pubs/

Reichle RH, Koster RD (2003) Assessing the impact of horizontal error correlations in background fields on soil moisture estimation. J Hydrometeorol 4:1229–1242

Reichle RH, Koster RD (2004) Bias reduction in short records of satellite soil moisture. Geophys Res Lett 31:L19501. doi:10.1029/2004GL020938

Reichle RH, Koster RD (2005) Global assimilation of satellite surface soil moisture retrievals into the NASA catchment land surface model. Geophys Res Lett 32:L02404. doi:10.1029/2004GL021700

Reichle RH, Entekhabi D, McLaughlin DB (2001) Downscaling of radiobrightness measurements for soil moisture estimation: a four-dimensional variational data assimilation approach. Water Resour Res 37:2353–2364. doi:10.1029/2001WR000475

Reichle RH, McLaughlin DB, Entekhabi D (2002a) Hydrologic data assimilation with the ensemble Kalman filter. Mon Weather Rev 130:103–114

Reichle RH, Walker JP, Koster RD, Houser PR (2002b) Extended vs. ensemble Kalman filtering for land data assimilation. J Hydrometeorol 3:728–740

Reichle RH, Koster RD, Liu P, Mahanama SPP, Njoku EG, Owe M (2007) Comparison and assimilation of global soil moisture retrievals from the Advanced Microwave Scanning Radiometer for the Earth

Observing System (AMSR-E) and the Scanning Multichannel Microwave Radiometer (SMMR). J Geophys Res 112:D09108. doi:10.1029/2006JD008033
Reichle RH, Crow WT, Koster RD, Sharif H, Mahanama SPP (2008a) Contribution of soil moisture retrievals to land data assimilation products. Geophys Res Lett 35:L01404. doi:10.1029/2007GL031986
Reichle RH, Crow WT, Keppenne CL (2008b) An adaptive ensemble Kalman filter for soil moisture data assimilation. Water Resour Res 44:W03423. doi:10.1029/2007WR006357
Reichle RH, Bosilovich MG, Crow WT, Koster RD, Kumar SV, Mahanama SPP, Zaitchik BF (2009) Recent advances in land data assimilation at the NASA global modeling and assimilation office. In: Park SK, Xu L (eds) Data assimilation for atmospheric, oceanic and hydrologic applications. Springer, New York, pp 407–428. doi:10.1007/978-3-540-71056-1
Reichle RH, Koster RD, De Lannoy GJM, Forman BA, Liu Q, Mahanama SPP, Toure A (2011) Assessment and enhancement of MERRA land surface hydrology estimates. J Clim 24:6322–6338. doi:10.1175/JCLI-D-10-05033.1
Rienecker MM et al (2008) The GEOS-5 data assimilation system—documentation of versions 5.0.1 and 5.1.0, and 5.2.0. NASA Tech. Rep. Series on Global Modeling and Data Assimilation, NASA/TM-2008-104606, vol 27, 92 pp. Available at http://gmao.gsfc.nasa.gov/pubs/
Rienecker MM et al (2011) MERRA—NASA's Modern-Era Retrospective analysis for Research and Applications. J Clim 24:3624–3648. doi:10.1175/JCLI-D-11-00015.1
Rodell M et al (2003) The global land data assimilation system. Bull Am Meteorol Soc 85:381–394. doi:10.1175/BAMS-85-3-381
Rodell M, Velicogna I, Famiglietti JS (2009) Satellite-based estimates of groundwater depletion in India. Nature 460:999–1002. doi:10.1038/nature08238
Rowlands DD, Luthcke SB, Klosko SM, Lemoine FGR, Chinn DS, McCarthy JJ, Cox CM, Anderson OB (2005) Resolving mass flux at high spatial and temporal resolution using GRACE intersatellite measurements. Geophys Res Lett 32:L04310. doi:10.1029/2004GL021908
Rowlands DD, Luthcke SB, McCarthy JJ, Klosko SM, Chinn DS, Lemoine FG, Boy J-P, Sabaka TJ (2010) Global mass flux solutions from GRACE: a comparison of parameter estimation strategies—mass concentrations versus Stokes coefficients. J Geophys Res 115:B01403. doi:10.1029/2009JB006546
Sahoo AK, De Lannoy GJM, Reichle RH, Houser PR (2012) Assimilation and downscaling of satellite observed soil moisture over the little river experimental watershed in Georgia, USA. Adv Water Resour 52:19–33. doi:10.1016/j.advwatres.2012.08.007
Schaefer GL, Cosh MH, Jackson TJ (2007) The USDA Natural Resources Conservation Service Soil Climate Analysis Network (SCAN). J Atmos Ocean Technol 24:2073–2077. doi:10.1175/2007JTECHA930.1
Seneviratne SI, Corti T, Davin EL, Hirschi M, Jaeger EB, Lehner I, Orlowsky B, Teuling AJ (2010) Investigating soil moisture–climate interactions in a changing climate: a review. Earth Sci Rev 99:125–161. doi:10.1016/j.earscirev.2010.02.004
Smith AB et al (2012) The Murrumbidgee soil moisture monitoring network data set. Water Resour Res 48:W07701. doi:10.1029/2012WR011976
Stieglitz M, Ducharne A, Koster RD, Suarez M (2001) The impact of detailed snow physics on the simulation of snow cover and subsurface thermodynamics at continental scales. J Hydrometeorol 2:228–242
Stöckli R, Rutishauser T, Baker I, Liniger MA, Denning AS (2011) A global reanalysis of vegetation phenology. J Geophys Res 116:G03020. doi:10.1029/2010JG001545
Sturm M, Holmgren J, Liston GE (1995) A seasonal snow cover classification system for local to global applications. J Clim 8:1261–1283
Sturm M, Taras B, Liston GE, Derksen C, Jonas T, Lea J (2010) Estimating snow water equivalent using snow depth data and climate classes. J Hydrometeorol 11:1380–1394
Su H, Yang Z-L, Niu G-Y, Dickinson RE (2008) Enhancing the estimation of continental-scale snow water equivalent by assimilating MODIS snow cover with the ensemble Kalman filter. J Geophys Res 113:D08120. doi:10.1029/2007JD009232
Su H, Yang Z-L, Dickinson RE, Wilson CR, Niu G-Y (2010) Multisensor snow data assimilation at the continental scale: the value of gravity recovery and climate experiment terrestrial water storage information. J Geophys Res 115:D10104. doi:10.1029/2009JD013035
Swenson S, Wahr J (2006) Post-processing removal of correlated errors in GRACE data. Geophys Res Lett 33:L08402. doi:10.1029/2005GL025285
Tang Q, Gao H, Yeh P, Oki T, Su F, Lettenmaier DP (2010) Dynamics of terrestrial water storage change from satellite and surface observations and modeling. J Hydrometeorol 11:156–170
Tedesco M, Narvekar PS (2010) Assessment of the NASA AMSR-E SWE product. IEEE J Sel Top Appl Earth Obs Remote Sens 3:141–159

Tedesco M, Reichle R, Löw A, Markus T, Foster JL (2010) Dynamic approaches for snow depth retrieval from spaceborne microwave brightness temperature. IEEE Trans Geosci Remote Sens 48(4): 1955–1967

Trenberth KE, Asrar G (2013) Challenges and opportunities in watercycle research: WCRP contributions. Surv Geophys. doi:10.1007/s10712-012-9214-y

Wagner W, Lemoine G, Rott H (1999) A method for estimating soil moisture from ERS scatterometer and soil data. Remote Sens Environ 70:191–207. doi:10.1016/S0034-4257(99)00036-X

Wahr J, Swenson S, Zlotnicki V, Velicogna I (2004) Time-variable gravity from GRACE: first results. Geophys Res Lett 31:L11501. doi:10.1029/2004GL019779

Wigneron J-P et al (2007) L-band microwave emission of the biosphere (L-MEB) model: description and calibration against experimental data sets over crop fields. Remote Sens Environ 107:639–655

Zaitchik BF, Rodell M, Reichle RH (2008) Assimilation of GRACE terrestrial water storage data into a land surface model: results for the Mississippi River basin. J Hydrometeorol 9:535–548. doi:10.1175/2007 JHM951.1

Zhou Y, McLaughlin D, Entekhabi D (2006) Assessing the performance of the ensemble Kalman filter for land surface data assimilation. Mon Weather Rev 134:2128–2142

DOI 10.1007/978-94-017-8789-5_7
Reprinted from *Surveys in Geophysics* Journal, DOI 10.1007/s10712-012-9207-x

Initialisation of Land Surface Variables for Numerical Weather Prediction

Patricia de Rosnay · Gianpaolo Balsamo · Clément Albergel · Joaquín Muñoz-Sabater · Lars Isaksen

Received: 1 June 2012 / Accepted: 8 October 2012 / Published online: 30 October 2012

Abstract Land surface processes and their initialisation are of crucial importance for Numerical Weather Prediction (NWP). Current land data assimilation systems used to initialise NWP models include snow depth analysis, soil moisture analysis, soil temperature and snow temperature analysis. This paper gives a review of different approaches used in NWP to initialise land surface variables. It discusses the observation availability and quality, and it addresses the combined use of conventional observations and satellite data. Based on results from the European Centre for Medium-Range Weather Forecasts (ECMWF), results from different soil moisture and snow depth data assimilation schemes are shown. Both surface fields and low-level atmospheric variables are highly sensitive to the soil moisture and snow initialisation methods. Recent developments of ECMWF in soil moisture and snow data assimilation improved surface and atmospheric forecast performance.

Keywords Land surface · Data assimilation · Numerical weather prediction · Soil moisture · Snow

1 Introduction

Land surface processes determine the lower boundary conditions of the atmosphere, and they represent a crucial component of the hydrological cycle (Mueller and Seneviratne 2012; Entekhabi et al 1999; Koster and Suarez 1992; Shukla and Mintz 1982). In Numerical Weather Prediction (NWP) and climate models, surface–atmosphere interaction processes are represented by Land Surface Models (LSMs). These models have been improved considerably during the last two decades and, nowadays, they represent exchanges of water and energy through the soil–plant–atmosphere continuum with a good

P. de Rosnay (✉) · G. Balsamo · C. Albergel · J. Muñoz-Sabater · L. Isaksen
European Centre for Medium-Range Weather Forecasts, Reading, Berkshire, UK
e-mail: Patricia.Rosnay@ecmwf.int

consistency between land surface fluxes and soil moisture (Best et al. 2011; Balsamo et al. 2009; Krinner et al. 2005; de Rosnay et al. 2002).

Land surface initialisation is of crucial importance for NWP. A number of studies have shown a significant impact of soil moisture conditions on weather forecast skill at short and medium range (van den Hurk et al. 2008; Drusch and Viterbo 2007; Douville et al. 2000; Mahfouf et al. 2000; Beljaars et al. 1996) as well as at seasonal range (Weisheimer et al. 2011; Koster et al. 2011, 2004). Cold processes are also a key component of the land–surface interactions. Snow is characterised by a very high albedo and a low thermal conductivity, and the snowpack constitutes a substantial water storage reservoir (De Lannoy et al. 2012; Brown and Mote 2009; Barnett et al. 2005). Snow has a strong influence on the summer water supply, and it affects the energy balance at the surface and the surface–atmosphere interactions (Gong et al. 2004; Walland and Simmonds 1997). So, initialisation of snow conditions has a large impact on the atmospheric forecast accuracy (Drusch et al 2004; Brasnett 1999).

In this paper, methods used in operational NWP models to analyse LSMs' prognostic variables are reviewed. Section 2 describes current snow analysis approaches used in NWP centres. It presents ground and satellite observations of snow that are relevant to operational applications and shows results of snow data assimilation experiments. Based on results from the European Centre for Medium-Range Weather Forecasts (ECMWF), the impact on the atmospheric forecasts is presented and compared for different snow data assimilation approaches. Section 3 reviews soil moisture analysis systems used for NWP applications. It includes a discussion on the use of satellite data to analyse soil moisture. ECMWF results are shown to illustrate the influence of different soil moisture analysis approaches on surface and low-level atmospheric fields. Concluding remarks are given in Sect. 4.

2 Snow Analysis

2.1 Snow Forecast Models

Snow processes are parameterised in LSMs to account for a range of processes, including snow accumulation on the ground, snow melting and snow compaction. The LSM used at ECMWF is H-TESSEL (Hydrology Tiled ECMWF Scheme for Surface Exchange over Land) (ECMWF 2012; Balsamo et al. 2009; Viterbo and Beljaars 1995). H-TESSEL snow parameterisation was revised in 2009 (Dutra et al. 2010). It now accounts for liquid water content in the snowpack, and it includes a new snow density formulation that expresses the fresh snow density as a function of wind speed and air temperature. Snow Cover Fraction (SCF) and Snow Water Equivalent (SWE) are related by a depletion curve which depends on snow density. So, H-TESSEL represents the SCF hysteresis between accumulation and depletion periods (Dutra et al. 2010).

H-TESSEL has an explicit treatment of the snowpack evolution, and it uses a single-layer snow model, in contrast to LSMs used at the United Kingdom Meteorological Office (UKMO) or at Météo France, which include a multi-layered snow scheme (Best et al. 2011; Dutra et al. 2010; Boone et al. 2004). Like most other LSMs, H-TESSEL represents the effects of snow on the surface roughness length and for sub-grid scale processes. Current LSMs represent well the duration of snow cover; however, they still have large uncertainties in terms of snow accumulation, due to inaccuracies in the meteorological forcing and to imperfect model parameterisations (Essery et al. 2009; Boone et al. 2004). Data assimilation approaches, by optimally combining models and observations, are expected to provide most accurate estimates of snow conditions (Pullen et al. 2011; Essery et al. 2009; Drusch et al. 2004; Brasnett 1999).

2.2 Snow Observations

Snow data assimilation strongly relies on snow depth ground observations (Drusch et al. 2004; Brasnett 1999). A major source of snow depth measurements is that provided by SYNOP stations (synoptic reports). These observations are available in near-real time (NRT) on the Global Telecommunication System (GTS), so they are suitable for NWP applications. In addition to SYNOP reports, most weather services maintain national snow depth measurements networks. For example, the SNOTEL (SNOwpack TELemetry) network provides snow depth measurements used in the NOAA (National Oceanic and Atmospheric Administration) National Weather Service's National Operational Hydrologic Remote Sensing Center (NOHRSC) SNOw Data Assimilation System (SNODAS). The NOAA COoperative Observer Program (COOP) also provides snow depth measurements over North America. However, data gathered from National Networks are not available on the GTS, and therefore, they are not suitable to be used in NWP snow analysis systems. In Europe, several countries are currently making available their snow depth measurements to the NWP community. The Swedish Meteorological and Hydrological Institute was the first to release its national network snow depth data on the GTS from December 2010. These data have been assimilated at ECMWF since March 2011 (de Rosnay et al. 2011a). Ground measurements of snow depth provide a very accurate local information, however, because of the variability of land surface and meteorological conditions, their representativeness can be limited, particularly in heterogeneous and in mountainous areas. Besides, many areas are sparsely observed (e.g., large areas in Siberia). Based on comparisons between pointwise SYNOP snow depth data and snow survey data sets, Takala et al. (2011) estimated the uncertainty of SYNOP snow depth data to be close to 0.12 m.

Satellite observations provide spatially integrated measurements with global coverage which makes them of high interest to provide consistent snow information for climate and NWP communities. SWE products based on passive microwave measurements, for example, from AMSR-E (Advanced Microwave Scanning Radiometer for Earth Observing System), product are available. However, retrieval algorithms are sensitive to many parameters such as snow grain size distribution and snow liquid water content, which are very difficult to estimate. Therefore, current satellite-based SWE products still have a limited accuracy, particularly for deep snow conditions (Takala et al. 2011). Future sensors, such as the proposed ESA (European Space Agency) Earth Explorer CoReH2O mission, are designed to accurately retrieve SWE, using dual polarisation measurements at frequencies optimal to separate grain size and SWE effects on the microwave emission (Rott et al. 2009).

While there are still high uncertainties in SWE retrievals from space-borne sensors, it is possible to estimate the Snow Cover Fraction with a good accuracy from Visible and Near infrared measurements in cloud-free conditions (Brubaker et al. 2009). The Moderate Resolution Imaging Spectroradiometer (MODIS) instruments provide high-resolution (0.05°) daily observations of snow cover. The MODIS snow cover product is used in the NASA (National Aeronautics and Space Administration)/NOAA Global Land Data Assimilation System (GLDAS, Rodell and Houser 2004). The NOAA/NESDIS (National Environmental Satellite, Data, and Information Service) Interactive Multi-sensor Snow and Ice Mapping System (IMS) combines ground observations and satellite data from microwave and visible sensors (using geostationary and polar orbiting satellites) to provide snow cover information in all weather conditions. It provides a binary information on snow cover. In other words, it indicates if there is snow or not on the ground, but if there is snow, it does not indicate the snow quantity on the ground. The IMS product is available daily for the northern hemisphere (Helfrich et al. 2007; Brubaker et al. 2009; Ramsay 1998). It is available from 1997 at a resolution of 24 km (Ramsay 1998) and from 2004 at 4 km resolution (Helfrich et al. 2007). The NOAA NESDIS IMS snow cover product has been used to analyse snow in NWP systems at ECMWF and the UKMO since 2004 and 2008, respectively (Pullen et al. 2011; de Rosnay et al. 2011b; Drusch et al. 2004). It is also used in the latest National Centers for Environmental Prediction (NCEP) Climate Forecast System Reanalysis (CFSR) (Meng et al. 2012).

2.3 Snow Analysis Methods

A number of NWP centres recently developed snow analysis approaches to improve the initialisation of snow variables, with expected impacts on the near surface weather parameters.

The UKMO snow analysis was implemented in operations 2008. It entirely relies on the NOAA NESDIS IMS 4 km Snow Cover information (Pullen et al. 2011). As part of the IMS pre-processing, the 4-km product is interpolated on the Unified Model grid, and a snow cover fraction is computed for each model grid point. To correct the model snow depth prognostic variable, a simple update approach is used, as described by Pullen et al. (2011). If the IMS product indicates snow-free conditions, the analysed snow depth is set to zero. Otherwise, the IMS snow cover is compared to the model background. If IMS indicates a region is snow covered and the model background agrees, the model is simply cycled, that is, the analysed snow depth is set to the background snow depth. If IMS indicates a region is covered by snow while the model background is snow free, the analysed snow depth is computed as a function of the observed snow cover using a logarithmic depletion curve.

The NASA/NOAA GLDAS snow analysis follows a similar approach along the same line, using the MODIS snow cover product (Rodell and Houser 2004). The NCEP CFSR reanalysis also relies on a simple update approach, with input data resulting from combined IMS and Air Force Weather Agency's snow depth analysis (SNODEP), as described by Meng et al. (2012).

Most of other NWP services use SYNOP snow depth reports available on the GTS. The snow analysis is a spatial interpolation of weighted background and observed snow depth. The DWD (Deutscher Wetterdienst) assimilates SYNOP reports of snow depth using the Cressman (1959) interpolation. The Cressman analysis accounts for weighting functions of vertical and horizontal distances between observations and model grid points. The Canadian Meteorological Center (CMC) uses a 2D Optimal Interpolation (OI) scheme developed by Brasnett (1999). Similarly to the Cressman interpolation, the OI expresses the observations weighting functions from vertical and horizontal structure functions. In addition, it accounts for covariance matrices of background and observations errors which enable to optimally combine model background and observations. At ECMWF, the latest ECMWF Re-Analysis, ERA-Interim, uses a Cressman interpolation for the snow analysis (Dee et al. 2011). The operational snow analysis relied on a Cressman interpolation for more than 20 years until it was replaced by a 2D Optimal Interpolation in November 2010 (de Rosnay et al. 2011b). The ECMWF operational snow analysis is a two-step algorithm. In the first step, a simple update scheme similar to the one used at NCEP or at the UKMO is used to account for the IMS snow cover information. Grid boxes, which are snow free in the model background but are snow covered in the satellite-derived product, are updated with a constant snow depth of 0.1 m of density 100 kg m^{-2}. In the second step, observations from ground stations reports and snow-free satellite observations (which enter the analysis with a snow depth equal to 0 m) are assimilated using an OI to produce the analysed snow field.

As described above for CMC, DWD, ECMWF, NASA/NOAA and NCEP, NWP systems generally use simple data assimilation approaches to initialise snow depth, ranging from simple update, Cressman Interpolation and Optimal Interpolation.

2.4 Results

Figure 1 shows snow depth analysis fields in north-east Asia on 30 October 2010, obtained from the ECMWF Integrated Forecasting System when using a Cressman snow analysis (top) and an OI snow analysis (bottom). A qualitative comparison shows that the Cressman analysis produces disc-shaped spurious patterns of snow in northern Asia related to the Cressman interpolation. The OI presents a smoother and more correct snow analysis without spurious patterns. The Optimal Interpolation analysis makes a better use of

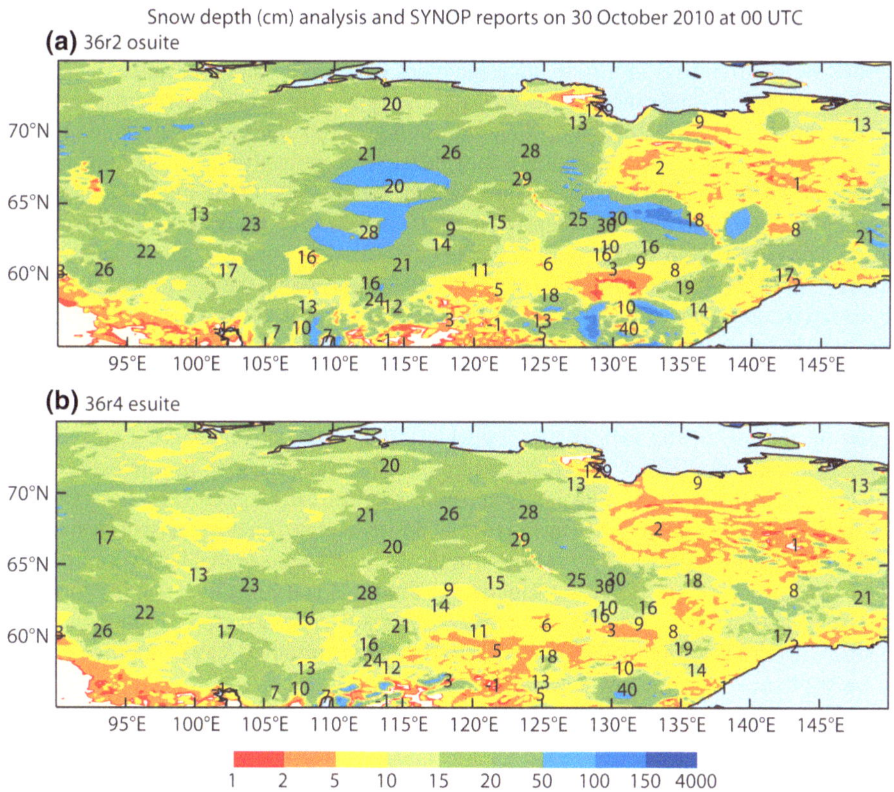

Fig. 1 Snow depth (cm) fields obtained using (*top*) a Cressman snow analysis from the operational ECMWF Integrated Forecasting System and (*bottom*) an OI snow analysis as tested at ECMWF before operational implementation, in northern Asia on 30 October 2010. SYNOP snow depth measurements are reported (cm) in *black* on the figure

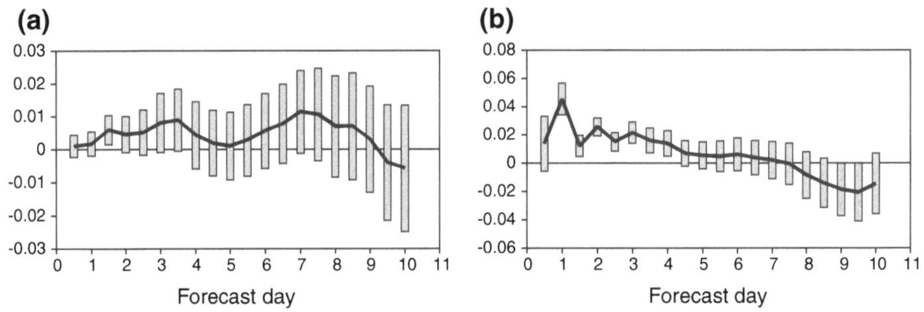

Fig. 2 Normalised root mean square forecast error difference for the ECMWF 1,000 hPa geopotential for (**a**) Cressman minus OI snow analyses, both using the SYNOP reports and the 24-km NOAA NESDIS IMS snow cover product and (**b**) Cressman minus OI, with OI using SYNOP and 4-km NOAA NESDIS IMS snow cover data (new ECMWF snow analysis) and Cressman using SYNOP and the 24-km IMS product (old ECMWF snow analysis). Statistics are computed based on daily analyses at 00 UTC from 01 December 2009 to 28 February 2010. *Vertical bars* show the 90 % confidence interval. The *x*-axis shows the forecast range from 0 to 10 days. Positive impact of the OI analysis compared to the Cressman analysis is shown by positive values

SYNOP snow depth data than Cressman. The difference between the two analyses mainly results from differences in the structure functions between OI and Cressman.

The 2009/2010 winter season, with cold and snowy conditions in the northern hemisphere (Cohen et al. 2010), highlighted the importance of good-quality snow analysis (de Rosnay et al. 2011b). Figure 2 shows the impact of different snow analysis configurations on the forecast 1000 hPa geopotential height error for the winter 2009–2010. Figure 2a shows the impact of the OI snow analysis compared to the Cressman snow analysis, with both schemes using SYNOP and IMS snow cover data at 24 km resolution. Figure 2b shows the impact of the revised ECMWF analysis implemented in 2010 compared to the previous analysis. The old analysis uses Cressman and observations from the SYNOP network and the IMS 24 km product. The new analysis relies on the OI and uses observations from SYNOP and the 4-km snow cover IMS product. The new analysis also accounts for an improved pre-processing and quality control of the IMS NESDIS data. In particular, based on an altitude threshold of 1500 m, the use of the IMS data is switched off in mountainous areas. Replacing the Cressman snow analysis by the OI has a relatively neutral impact on the atmospheric circulation, although a slight non-significant improvement can be seen (Fig. 2a). The new ECMWF snow analysis (Fig. 2b) has an overall positive impact on the atmospheric forecasts skill, with root mean square error forecast for the 1000 hPa geopotential height improved by 1–4 % in the short range (forecasts until day 4). Figures 1 and 2 illustrate that the combined improvements of the analysis approach (OI vs Cressman) and data pre-processing and quality control (IMS snow cover product resolution and altitude threshold) lead to improve both the snow depth fields and the low-level atmospheric forecast.

3 Soil Moisture Analysis

3.1 History of Soil Moisture Analysis for NWP

In the absence of a near-real-time global network for providing soil moisture information, using screen-level data has been the only source of information that has been continuously available from the SYNOP network for NWP soil moisture analysis systems. As shown by Mahfouf (1991), it provides indirect, but relevant information to analyse soil moisture. So, most of the current operational soil moisture analysis systems rely on analysed screen-level variables (2-m temperature and relative humidity).

In 1994, a nudging approach was implemented at ECMWF to analyse soil moisture, using the lowest atmospheric level specific humidity analysis increments. It was the first NWP centre to implement a soil moisture analysis scheme, mainly to prevent the soil moisture from drifting to unrealistic dry conditions in summer time. However, the nudging scheme was not accounting for processes that modulate the relation between soil moisture and specific humidity. So, soil moisture increments were affected by systematic biases with successive negative and positives increments at both the diurnal and seasonal scales (Mahfouf et al. 2000).

In 1999, the nudging soil moisture analysis was replaced by a 1D Optimal Interpolation analysis, as originally proposed by Mahfouf (1991). The OI soil moisture analysis implementation at ECMWF and evaluation are detailed in Douville et al. (2000) and Mahfouf et al. (2000). The OI soil moisture analysis relies on the relation between soil moisture and screen-level temperature and relative humidity. When soil moisture is underestimated, air temperature is expected to be overestimated and air humidity underestimated. In contrast, when soil moisture is overestimated, screen-level air temperature is too low and air humidity too large. Based on this relation between soil moisture and screen-level parameters, the soil moisture correction is computed as a function of the screen-level parameters correction. So, a dedicated screen-level parameters analysis was implemented, based on an OI approach, and its increments are then used as input of the soil moisture analysis (Mahfouf et al. 2000). The same method has been used to analyse soil temperature and snow temperature using the screen-level air temperature increments. The OI soil moisture and temperature analysis has been widely used for NWP applications in several NWP centres. It was used at ECMWF for operational NWP from July 1999 to November 2010. It has been used in ECMWF re-analyses ERA-40 (Uppala et al. 2005) and is still in use for ERA-Interim (Dee et al. 2011). An OI soil moisture analysis is currently used at Météo France (Giard and Bazile 2000), Environment Canada (Bélair et al. 2003) and in the High Resolution Limited Area Model (HIRLAM) (Rodríguez et al. 2003).

As shown by Drusch et al. (2009), the OI soil moisture analysis improves screen-level parameters forecasts, without, however, any positive impact on soil moisture. Furthermore, an important weakness of the OI approach is its lack of flexibility to easily account for new types of data including new generations of satellite data (Mahfouf et al. 2009). Also, it uses calibrated coefficients that would require to be updated for each change in the LSM. The OI, by using fixed coefficients, does not account for local processes such as cloud cover or soil moisture conditions that influence the coupling strength between soil moisture and screen-level parameters and therefore the magnitude of the increments (de Rosnay et al 2012).

Several NWP centres started to investigate the use of satellite data to analyse soil moisture, using a range of approaches based on simplified EKF (Draper et al. 2011) or the equivalent simplified 2D-Var (Balsamo et al. 2007), as well as EKF and an Ensemble Kalman Filter (Reichle et al. 2002, 2008). At Météo France, an offline EKF approach was evaluated, and the impact of ASCAT (Advanced SCATterometer) soil moisture data assimilation on the low-level atmospheric parameters was addressed (Mahfouf 2010). A limited impact of the EKF soil moisture analysis on relative humidity and air temperature was found. A simplified Extended Kalman Filter (EKF) soil moisture analysis, using screen-level parameters information, was implemented operationally at the German

Weather Service (Deutscher Wetterdienst) in 2000 (Hess 2001). Along the same line, ECMWF developed a point-scale simplified EKF soil moisture analysis (Seuffert et al. 2004). Preliminary investigations at local scale showed that the OI and the EKF soil moisture analyses give similar results when they both use screen-level parameters (Seuffert et al. 2004). The ECMWF simplified EKF soil moisture analysis was extended to be used at global scale (Drusch et al. 2009) and implemented in operations in 2010 (de Rosnay et al. 2012).

In the following section, differences between EKF and OI soil moisture analyses are presented in terms of soil moisture increments and low-level atmospheric parameters forecasts.

3.2 Comparison Between the OI and EKF Soil Moisture Analyses

de Rosnay et al. (2012) quantified monthly mean global soil moisture increments for both the OI and the EKF schemes for an entire annual cycle. They showed that the OI scheme systematically adds water to the soil. The global monthly mean value of the OI analysis increments was shown to be 5.5 mm, which represents a substantial and unrealistic contribution to the global water cycle. In contrast, the EKF global mean soil moisture analysis increments are much smaller, representing more reasonable global monthly mean increments of 0.5 mm. The reduction in increments between the EKF and the OI is mainly due to a systematic reduction in increments below the first layer. The OI increments computed for the first layer are amplified for deeper layers in proportion to the layer thickness, explaining the overestimation of the OI increments. In contrast, the EKF dynamical estimates, based on perturbed simulations, allow optimising soil moisture increments at different depths to match screen-level observations according to the strength of the local and current soil–vegetation–atmosphere coupling. The EKF accounts for additional controls due to meteorological forcing and soil moisture conditions. Thereby, it prevents undesirable and excessive soil moisture corrections (de Rosnay et al. 2012). Figure 3 illustrates monthly mean increments accumulated in the first metre of soil, for August 2009. In agreement with de Rosnay et al. (2012), it shows that larger increments are accumulated into the soil with the OI than with the EKF analysis scheme. It is, however, interesting to note that large increments remain with the EKF in the US Great Plains and in South America, showing these regions are affected by systematic soil moisture bias in the LSM. Further investigation will be carried out in the future to address this feature.

The impact of the soil moisture analysis scheme on analysed soil moisture was also studied using ground data from SMOSMANIA (Soil Moisture Observing System-Meteorological Automatic Network Integrated Application) (Calvet et al. 2007). de Rosnay et al. (2012) showed that ECMWF soil moisture is generally in good agreement with ground observations, with mean correlations higher than 0.78. Using the EKF instead of the OI scheme improves significantly the soil moisture analysis, with mean correlation between ECMWF and ground truth soil moisture higher than 0.84 when the EKF soil moisture analysis is used (de Rosnay et al. 2012).

Figure 4 shows the monthly mean impact of the EKF soil moisture analysis on the 36-h forecast of 2-m temperature at 0000 UTC for July-August-September 2009. It shows the difference in temperature error (in K) between the OI and EKF experiments. Positive values indicate that the EKF generally improves the 2-m temperature forecasts compared to the OI soil moisture analysis. It is consistent with the results shown by de Rosnay et al. (2012) to indicate that in most areas the 2-m temperature errors for OI are larger than the

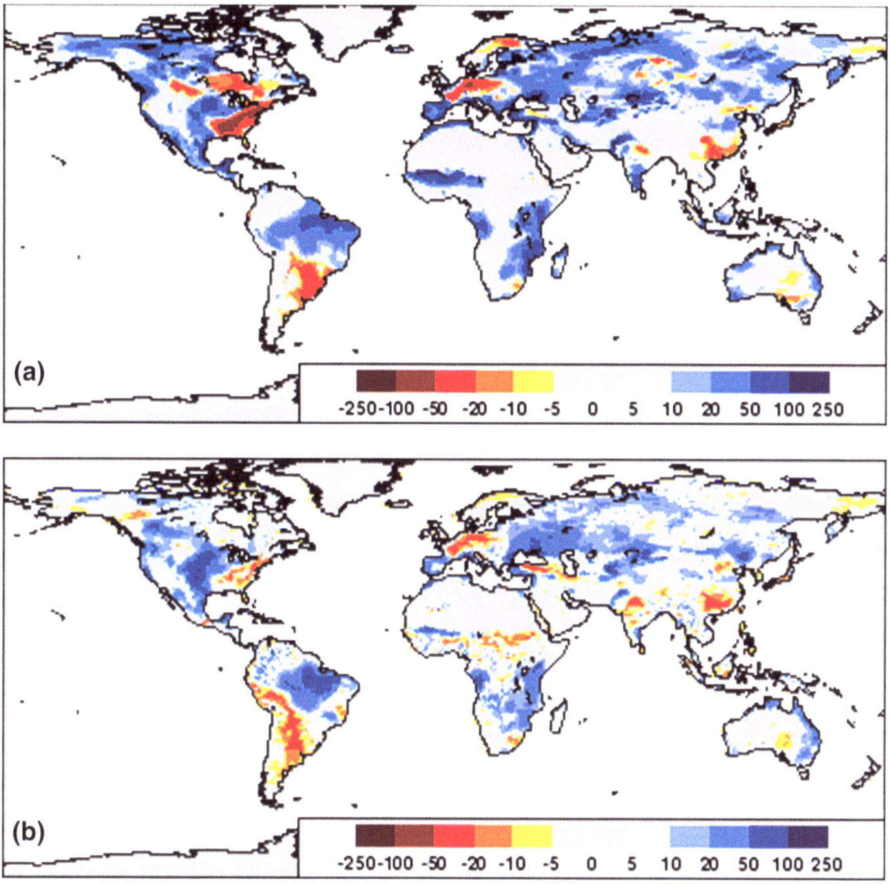

Fig. 3 Accumulated soil water increments (in mm) in the first metre of soil for August 2009, with the OI (*top*) and EKF (*bottom*) analyses

EKF errors, showing that the EKF soil moisture analysis has a slight positive impact on the 2-m temperature forecast.

3.3 Use of Satellite Data to Analyse Soil Moisture

In the past few years, several new space-borne microwave sensors have been developed to estimate soil moisture from space. They provide spatially integrated information on surface soil moisture at a scale relevant to NWP models. The active sensor ASCAT on MetOp was launched in 2006 (Bartalis et al. 2007). The EUMETSAT (European Organisation for the Exploitation of Meteorological Satellites) ASCAT surface soil moisture product is the first operational soil moisture product. It is available in near-real time, and it has been monitored operationally at ECMWF since September 2009 (de Rosnay et al. 2012). Scipal et al. (2008) evaluated the impact of scatterometer soil moisture products (from the European Remote-Sensing ERS) data assimilation in a simple nudging scheme. They showed that, compared to the model "open-loop" (without data assimilation), ASCAT soil moisture data

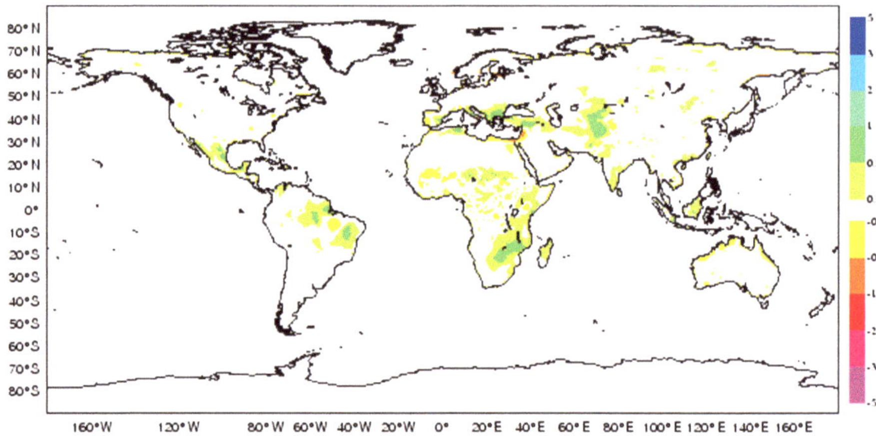

Fig. 4 Mean difference of monthly mean 36-h forecasts (12 UTC) error in 2-m temperature (in K) between the OI and the EKF soil moisture analysis schemes for July–August–September. *Greenish colours* indicate an improvement of the EKF compared to the OI, while *reddish colours* indicate a degradation

assimilation improves the model soil moisture and screen-level parameters. However, they found that compared to the OI soil moisture analysis, ASCAT soil moisture nudging scheme has a slightly negative impact on the atmospheric forecasts. De Rosnay et al. (2012) evaluated the use of ASCAT soil moisture data in the EKF soil moisture analysis, showing a neutral impact on both soil moisture and screen-level parameters forecasts. At the UKMO, Dharssi et al. (2011) investigated ASCAT surface soil moisture data assimilation using a simple nudging scheme, as already used at the UKMO to analyse soil moisture from screen-level parameter information. They showed that assimilating ASCAT data, in addition to screen-level information in their nudging scheme, improves soil moisture analysis and forecasts scores of screen-level parameters in the tropics, in Australia and in North America. Based on their positive evaluation results, ASCAT soil moisture nudging was implemented in operations in July 2010 at the UKMO.

The ESA SMOS (Soil Moisture and Ocean Salinity) mission was launched in 2009 (Kerr et al. 2007). Based on L-band passive microwave measurements, SMOS is the first mission dedicated to soil moisture remote sensing. The future NASA (National Aeronautics and Space Administration) SMAP (Soil Moisture Active and Passive) mission, planned to be launched in 2015, will be a soil moisture mission that combines active and passive microwave measurements to provide global soil moisture and freeze/thaw state (Entekhabi et al. 2010). ECMWF plays a major role in developing and investigating the use of new satellite data for soil moisture analysis. SMOS brightness temperature product has been monitored in near-real time since November 2010, as described in Sabater et al. (2011). Work toward assimilation of SMOS data over land is ongoing.

An extensive evaluation and comparison between SMOS, ASCAT and ECMWF soil moisture was conducted by Albergel et al. (2012), using existing soil moisture networks in Europe, Africa, Australia and the USA. Using more than 200 stations to evaluate the remotely sensed and analysed soil moisture products in contrasted climate conditions, the authors showed that (1) SMOS and ASCAT soil moisture products are of similar quality, with annual mean correlation 0.53 for ASCAT and 0.54 for SMOS, and (2) the analysed product is of better quality than the satellite products with 0.70 averaged correlation value.

Draper et al. (2012) recently investigated combined data assimilation, using an Ensemble Kalman Filter, of active and passive soil moisture satellite data. They evaluated the impact of soil moisture products data assimilation on the analysed soil moisture. They showed that, although correlation with ground data was better for the LSM than for the satellite data, data assimilation still has a positive impact on the analysed soil moisture. Their study confirms the potential of satellite-based soil moisture data for NWP applications.

4 Conclusion

This paper presented the current status of data assimilation systems used to initialise land surface variables for Numerical Weather Prediction. Different approaches used to analyse soil moisture and snow depth in Numerical Weather Prediction (NWP) systems were reviewed. Based on ECMWF experiments, analysis results and atmospheric forecast impact were presented for different land surface data assimilation approaches.

Snow processes strongly influence the hydrological cycle, and they have a large impact on the energy budget. So, accurate initialisation of snow depth for NWP applications is highly relevant. Snow analysis schemes currently used for operational NWP rely on simple approaches. Using the NOAA IMS snow cover product, a simple update approach is used at the UKMO. The NCEP latest reanalysis and the GLDAS also use a simple update approach using combined IMS/SNODEP products and the MODIS snow cover product, respectively. The German meteorological service relies on a Cressman interpolation and used the SYNOP snow depth reports. Since 1999 the Canadian Meteorological Center uses a 2D Optimal Interpolation to assimilate SYNOP snow depth observations. At ECMWF both SYNOP observations and the IMS snow cover data are assimilated. A Cressman Interpolation was used for more than 20 years before it was recently replaced by an Optimal Interpolation in 2010.

A qualitative comparison between snow depth fields illustrated differences between the Cressman and the Optimal Interpolation snow depth analyses. In contrast to Cressman, the Optimal Interpolation accounts for the model background and the observations errors, which allows to optimally combine model background and observations. So, by improving the structure functions, the Optimal Interpolation makes a better use of the observations than the Cressman interpolation. The quantitative impact of the snow analysis scheme on the atmospheric forecast was illustrated using ECMWF results. The revised ECMWF snow analysis, using an Optimal Interpolation and SYNOP observations combined with the 4-km IMS snow cover data and improved observation pre-processing and quality control, was compared to the old ECMWF snow analysis based on a Cressman interpolation that uses SYNOP and the 24-km IMS snow cover products without quality control. Results showed that the root mean square error forecast of the 1,000 hPa geopotential height in the northern hemisphere is reduced by 1–4 % in the short range (until forecast day 4) for the winter 2009–2010. This significant and large-scale impact of the snow analysis on the atmospheric forecast illustrates the major importance of the snow analysis for Numerical Weather Prediction applications. SWE products from satellite sensors are not yet used in NWP although they have a potential to provide reliable and near-real-time information on snow mass. It is expected that Snow Water Equivalent products quality will be improved in the next few years. Potential future satellite missions such as the proposed ESA CoReH2O mission are expected to provide SWE estimates from space with an improved accuracy compared to current products.

Concerning soil moisture, most NWP centres use a 1-D Optimal Interpolation analysis to initialise soil moisture, based on a dedicated screen-level parameters analysis. Both the German meteorological service and ECMWF use an Extended Kalman Filter soil moisture analysis in operations. The Extended Kalman Filter soil moisture analysis is based on a dedicated screen-level parameters analysis. Whereas the 1D Optimal Interpolation soil moisture analysis uses screen-level analysis increments as input of the soil moisture analysis, the Extended Kalman Filter, as implemented at ECMWF, uses analysed screen-level fields as input observations of the soil moisture analysis. ECMWF experiments showed that the Extended Kalman Filter soil moisture analysis consistently reduces, by 5 mm per month at global scale, the soil moisture increments compared to the Optimal Interpolation, and it slightly improves both soil moisture and screen-level parameters analyses and forecasts. The ECMWF soil moisture analysis evaluation against ground measurements from the SMOSMANIA network showed an improved correlation from 0.78 for the Optimal Interpolation to 0.84 for the Extended Kalman Filter soil moisture analysis. The Extended Kalman Filter analysis also makes it possible to combine screen-level parameters and satellite data, such as ASCAT or SMOS, to analyse soil moisture. Previous results with ASCAT data assimilation were discussed in this paper. ECMWF results showed a neutral impact of ASCAT data assimilation in the Extended Kalman Filter on both soil moisture and screen-level parameters. However, recent improvements in the ASCAT soil moisture products and in bias correction are expected to improve the impact of using ASCAT soil moisture data. In contrast to other centres, which mainly use Optimal Interpolation or Extended Kalman Filter approaches, the UKMO, soil moisture analysis relies on a simple nudging scheme. ASCAT soil moisture data assimilation was shown to have a positive impact on the screen-level parameters forecast at the UKMO, leading to operational ASCAT soil moisture assimilation from 2010. Developments are ongoing at the UKMO to replace their current nudging scheme by an Extended Kalman Filter approach which will open possibilities to combine different types of observations in their soil moisture analysis.

Kalman Filter–based land surface analysis systems (Ensemble or Extended), used in several NWP centres, either in research or operationally, open a wide range of further development possibilities, including exploiting new satellite surface data and products for the assimilation of soil moisture (e.g., SMOS or the future SMAP). At ECMWF an extension of the Extended Kalman Filter to analyse additional variables, such as snow temperature, snow mass and vegetation parameters, is planned for investigation in the near future.

Acknowledgments The authors thank two anonymous reviewers for their careful review of the manuscript and their helpful suggestions.

References

Albergel C, de Rosnay P, Gruhier C, Sabater JM, Hasenauer S, Isaksen L, Kerr Y, Wagner W (2012) Evaluation of remotely sensed and modelled soil moisture products using global ground-based in-situ observations. Remote Sens Environ 18:215–226. doi:10.1016/j.rse.2011.11.017

Balsamo G, Mahfouf JF, Bélair S, Deblonde G (2007) A land data assimilation system for soil moisture and temperature: an information content study. J Hydrometeorol 8:1225–1242. doi:10.1175/2007JHM819.1

Balsamo G, Viterbo P, Beljaars A, van den Hurk B, Hirsch M, Betts A, Scipal K (2009) A revised hydrology for the ECMWF model: verification from field site to terrestrial water storage and impact in the Integrated Forecast System. J Hydrometeorol 10:623–643

Barnett T, Adam J, Lettenmaier D (2005) Potential impacts of a warming climate on water availability in snow-dominated regions. Nature 438:303–309

Bartalis Z, Wagner W, Naeimi V, Hasenauer S, Scipal K, Bonekamp H, Figa J, Anderson C (2007) Initial soil moisture retrievals from the METOP-A advanced scatterometer (ASCAT). Geophys Res Lett 34. doi:10.1029/2007GL031088

Bélair S, Crevier LP, Mailhot J, Bilodeau J, Delage Y (2003) Operational implementation of the ISBA land surface scheme in the Canadian regional weather forecast model. Part I: warm season results. J Hydrometeorol 4:352–470

Beljaars ACM, Viterbo P, Miller M, Betts A (1996) Sensitivity to land surface parameterization and soil anomalies. Mon Weather Rev 124:362–383

Best M, Pryor M, Clark D, Rooney G, Essery R, Ménard C, Edwards J, Hendry M, Porson A, Gedney N, Mercado L, Sitch S, Blyth E, Boucher O, Cox P, Grimmond C, Harding R (2011) The joint UK land environment simulator (JULES), model description Part 1: energy and water fluxes. Geosci Model Dev 4:677–699. doi:10.5194/gmd-4-677-2011

Boone A, Habets F, Noilhan J, Clark D, Dirmeyer P, Fox S, Gusev Y, Haddeland I, Koster R, Lohmann D, Mahanama S, Mitchell K, Nasonova O, Niu GY, Pitman A, Polcher J, Shmakin A, Tanaka K, van den Hurk B, Vérant S, Verseghy D, Viterbo P, Yang ZL (2004) The Rhone-Aggregation land surface scheme intercomparison project: an overview. J Clim 17:187–208

Brasnett B (1999) A global analysis of snow depth for numerical weather prediction. J Appl Meteorol 38:726–740

Brown R, Mote P (2009) The response of northern hemisphere snow cover to a changing climate. J Clim 22:2124–2144

Brubaker K, Pinker R, Deviatova E (2009) Evaluation and comparison of MODIS and IMS snow-cover estimates for the continental United States using station data. J Hydrometeorol 6:1002–1017

Calvet JC, Fritz N, Froissard F, Suquia D, Petitpa B, Piguet B (2007) In situ soil moisture observations for the CAL/VAL of SMOS: the SMOSMANIA network. International geoscience and remote sensing symposium, IGARSS, Barcelona, Spain. doi:10.1109/IGARSS.2007.4423019

Cohen J, Foster J, Barlow M, Saito K, Jones J (2010) Winter 2009–2010: a case study of an extreme arctic oscillation event. Geophys Res Lett 37:117707. doi:10.1029/2010GL044256

Cressman G (1959) An operational objective analysis system. Mon Weather Rev 87(10):367–374

De Lannoy G, Reichle R, Arsenault K, Houser P, Kumar S, Verhoest N, Pauwels V (2012) Multiscale assimilation of Advanced Microwave Scanning Radiometer EOS snow water equivalent and Moderate Resolution Imaging Spectroradiometer snow cover fraction observations in northern Colorado. Water Resour Res 48:w01522. doi:10.1029/2011WR010588

de Rosnay P, Polcher J, Bruen M, Laval K (2002) Impact of a physically based soil water flow and soil–plant interaction representation for modeling large scale land surface processes. J Geophys Res 107(11). doi:10.1029/2001JD000634

de Rosnay P, Dragosavac M, Isaksen L, Andersson E, Haseler J (2011a) Use of new snow data from Sweden in IFS cycle 36r4. ECMWF Res Memo R483/PdR/1139

de Rosnay P, Balsamo G, Isaksen L (2011b) Snow analysis for numerical weather prediction at ECMWF. IGARSS 2011

de Rosnay P, Drusch M, Vasiljevic D, Balsamo G, Albergel C, Isaksen L (2012) A simplified extended Kalman filter for the global operational soil moisture analysis at ECMWF. Q J R Meteorol Soc. doi:10.1002/qj.2023

Dee D, Uppala S, Simmons A, Berrisford P, Poli P, Kobayashi S, Andrae U, Balsameda M, Balsamo G, Bauer P, Bechtold P, Beljaars A, van de Berg L, Bidlot J, Bormann N, Delsol C, Dragani R, Fuentes M, Geer A, Haimberger L, Healy S, Hersbach H, Hólm E, Isaksen L, Kållberg P, Köhler M, Marticardi M, McNally A, Monge-Sanz B, Morcrette JJ, Park BK, Peubey C, de Rosnay P, Tavolato C, Thépaut JN, Vitart F (2011) The ERA-Interim reanalysis: configuration and performance of the data assimilation system. Q J R Meteorol Soc 137:553–597. doi:10.1002/qj.828

Dharssi I, Bovis K, Macpherson B, Jones C (2011) Operational assimilation of ASCAT surface soil wetness at the met office. Hydrol Earth Syst Sci 15:2729–2746. doi:10.5194/hess-15-2729-2011

Douville H, Mahfouf JF, Beljaars A (2000) Evaluation of optimal interpolation and nudging techniques for soil moisture analysis using FIFE data. Mon Weather Rev 128:1733–1756

Draper C, Mahfouf JF, Walker JP (2011) Root zone soil moisture from the assimilation of scree-level variables and remotely sensed soil moisture. J Geophys Res 116:d02127. doi:10.1029/2010JD013829

Draper C, Reichle R, De Lannoy G, Liu Q (2012) Assimilation of passive and active microwave soil moisture retrievals. Geophys Res Lett 39:l04401. doi:10.1029/2011GL050655

Drusch M, Viterbo P (2007) Assimilation of screen-level variables in ECMWF's Integrated Forecast System: a study on the impact on the forecast quality and analyzed soil moisture. Mon Weather Rev 135:300–314

Drusch M, Vasiljevic D, Viterbo P (2004) ECMWF s global snow analysis: assessment and revision based on satellite observations. J Appl Meteorol 43:1282–1294

Drusch M, Scipal K, de Rosnay P, Balsamo G, Andersson E, Bougeault P, Viterbo P (2009) Towards a Kalman filter based soil moisture analysis system for the operational ECMWF Integrated Forecast System. Geophys Res Lett 36:l10401. doi:10.1029/2009GL037716

Dutra E, Balsamo G, Viterbo P, Miranda P, Beljaars A, Schär C, Elder K (2010) An improved snow scheme for the ECMWF land surface model: description and offline validation. J Hydrometeorol 11:899–916. doi:10.1175/2010JHM1249.1

ECMWF (2012) IFS documentation Cy37r2 operational implementation 18 May 2011. available at http://wwwecmwfint/research/ifsdocs/CY37r2

Entekhabi D, Asrar G, Betts A, Beven K, Bras R, Duffy C, Dunne T, Koster R, Lettenmaier D, DB ML, Shuttleworth W, van Genuchten M, Wei MY, Wood E (1999) An agenda for land surface hydrology research and a call for the second international hydrological decade. Bull Am Meteorol Soc 10:2043–2058

Entekhabi D, Njoku E, O'Neill P, Kellog K, Crow W, Edelstein W, Entin J, Goodman S, Jackson T, Johnson J, Kimball J, Piepmeier J, Koster R, Martin N, McDonald K, Moghaddam M, Moran S, Reichle R, Shi J, Spencer M, Thurman S, Tsang L, Van Zyl J (2010) The soil moisture active passive (SMAP) mission. Proc IEEE 98(5):704–716

Essery RLH, Rutter N, Pomeroy J, Baxter R, Stähli M, Gustafsson D, Barr A, Bartlett P, Elder K (2009) SNOWMIP2: an evaluation of forest snow process simulations. Bull Am Meteorol Soc 90:1120–1135. doi:10.1175/2009BAMS2629.1

Giard D, Bazile E (2000) Implementation of a new assimilation scheme for soil and surface variables in a global NWP model. Mon Weather Rev 128:997–1015

Gong G, Entekhabi D, Cohen J, Robinson D (2004) Sensitivity of atmospheric response to modeled snow anomaly characteristics. J Geophys Res 109:d06107. doi:10.1029/2003JD004160

Helfrich SR, McNamara D, Ramsay B, Baldwin T, Kasheta T (2007) Enhancements to, and forthcoming developments in the interactive multisensor snow and ice mapping system, (IMS). Hydrol Process 21:1576–1586. doi:10.1002/hyp.6720

Hess R (2001) Assimilation of screen-level observations by variational soil moisture analysis. Meteorol Atmos Phys 77:145–154

Kerr YH, Waldteufel P, Wigneron JP, Delwart S, Cabot F, Boutin J, Escorihuela M, Font J, Reul N, Gruhier C, Juglea S, Drinkwater M, Hahne A, Martín-Neira M, Mecklenburg S (2007) The SMOS mission: new tool for monitoring key elements of the global water cycle. Proc IEEE 98(5):666–687

Koster R, Mahanama P, Yamada T, Balsamo G, Berg A, Boisserie M, Dirmeyer P, Doblas-Reyes F, Drewitt G, Gordon C, Guo Z, Jeong J, Lee W, Li Z, Luo L, Malyshev S, Merryfield W, Seneviratne S, Stanelle T, van den Hurk B, Vitart F, Wood E (2011) The second phase of the global land-atmosphere coupling experiment: soil moisture contributions to subseasonal forecast skill. J Hydrometeorol 12:805–822

Koster RD, Suarez MJ (1992) Modeling the land surface boundary in climate models as a composite of independent vegetation stands. J Geophys Res 97:2697–2715

Koster RD, Dirmeyer P, Guo Z, Bonan G, Cox P, Gordon C, Kanae S, Kowalczyk E, Lawrence D, Liu P, Lu C, Malyshev S, McAvaney B, Mitchell K, Mocko D, Oki T, Oleson K, Pitman A, Sud Y, Taylor C, Verseghy D, Vasic R, Xue Y, Yamada T (2004) Regions of strong coupling between soil moisture and precipitation. Sciences 305:1138–1140

Krinner G, Viovy N, de Noblet-Ducoudré N, Ogée J, Polcher J, Friedlingstein P, Ciais P, Sitch S, Prentice I (2005) A dynamic global vegetation model for studies of the coupled atmosphere–biosphere system. Global Biogeochem Cycles 19:GB1015, 33 pp. doi:10.1029/2003GB002199

Mahfouf JF (1991) Analysis of soil moisture from near-surface parameters: a feasibility study. J Appl Meteorol 30:1534–1547

Mahfouf JF (2010) Assimilation of satellite-derived soil moisture from ASCAT in a limited-area NWP model. Q J R Meteorol Soc 136:784–798. doi:10.1002/qj.602

Mahfouf JF, Viterbo P, Douville H, Beljaars A, Saarinen S (2000) A revised land-surface analysis scheme in the Integrated Forecasting System. ECMWF Newslett 88

Mahfouf JF, Bergaoui K, Draper C, Bouyssel F, Taillefer F, Taseva L (2009) A comparison of two offline soil analysis schemes for assimilation of screen level observations. J Geophys Res 114. doi:10.1029/2008JD011077

Meng J, Yang R, Wei H, Ek M, Gayno G, Xie P, Mitchell K (2012) The land surface analysis in the NCEP climate forecast system reanalysis. J Hydrometeorol. doi:10.1175/JHM-D-11-090.1

Mueller B, Seneviratne S (2012) Hot days induced by precipitation deficits at the global scale. Proc Nat Acad Sci USA 109(31):12398–12403. doi:10.1073/pnas.1204330109

Pullen S, Jones C, Rooney G (2011) Using satellite-derived snow cover data to implement a snow analysis in the met office NWP model. J Appl Meteorol 50:958–973. doi:10.1175/2010JAMC2527.1

Ramsay B (1998) The interactive multisensor snow and ice mapping system. Hydrol Process 12:1537–1546
Reichle RH, Walker JP, Koster RD, Houser PR (2002) Extended versus ensemble Kalman filtering for land data assimilation. J Hydrometeorol 3:728–740
Reichle RH, Crow WT, Keppenne CL (2008) An adaptive ensemble Kalman filter for soil moisture data assimilation. Water Resour Res 44:W03423. doi:10.1029/2007WR006357
Rodell M, Houser P (2004) Updating a land surface model with MODIS-derived snow cover. J Hydrometeorol 5:1064–1075
Rodríguez A, Navascues B, Ayuso J, Järvenoja S (2003) Analysis of surface variables and parameterization of surface processes in HIRLAM. Part I: approach and verification by parallel runs. HIRLAM technical report No 59, Norrköping, Sweden, 52pp
Rott H, Cline D, Duguay C, Essery R, Haas C, Kern M, Macelloni G, Malnes E, Pulliainen J, Rebhan H, et al (2009) CoReH2O—cold regions hydrology high-resolution observatory. 2009 IEEE radar conference. doi:10.1109/RADAR.2009.4977133
Sabater JM, Fouilloux A, de Rosnay P (2011) Technical implementation of SMOS data in the ECMWF Integrated Forecasting System. IEEE Trans Geosc Remote Sens. doi:10.1109/LGRS.2011.2164777
Scipal K, Drusch M, Wagner W (2008) Assimilation of a ers scatterometer derived soil moisture index in the ECMWF numerical weather prediction system. Adv Water Resour. doi:10.1016/j.advwatres.2008.04.013
Seuffert G, Wilker H, ViterboM P Drusch, Mahfouf JF (2004) The usage of screen-level parameters and microwave brightness temperature for soil moisture analysis. J Hydrometeorol 5:516–531
Shukla J, Mintz Y (1982) Influence of land-surface evaporation on the Earth's climate. Science 215: 1498–1501
Takala M, Luojus K, Pulliainen J, Derksen C, Lemmetyinen J, Kärnä JP, Koskinen J, Bojkov B (2011) Estimating northern hemisphere snow water equivalent for climate research through assimilation of space-borne radiometer data and ground-based measurements. Remote Sens Environ 115:3517–3529. doi:10.1016/j.rse.2011.08.014
Uppala SM, Kållberg PW, Simmons A, Andrae U, Da Costa Bechtold V, Fiorino M, Gibson J, Haseler J, Hernandez A, Kelly G, Li X, Onogi K, Saarinen S, Sokka N, Allan R, Andersson E, Arpe K, Balmaseda M, Beljaars A, Van De Berg L, Bidlot J, Bormann N, Caires S, Chevallier F, Dethof A, Dragosavac M, Fisher M, Fuentes M, Hagemann S, Hólm E, Hoskins B, Isaksen L, Janssen P, Jenne R, Mcnally A, Mahfouf JF, Morcrette JJ, Rayner N, Saunders R, Simon P, Sterl A, Trenberth K, Untch A, Vasiljevic D, Viterbo P, Woollen J (2005) The ERA-40 re-analysis. Q J R Meteorol Soc 131:2961–3012. doi: 10.1256/qj.04.176
van den Hurk B, Ettema J, Viterbo P (2008) Analysis of soil moisture changes in Europe during a single growing season in a new ECMWF soil moisture assimilation system. J Hydrometeorol 9:116–131. doi: 10.1175/2007JHM848.1
Viterbo P, Beljaars ACM (1995) An improved land surface parameterization scheme in the ECMWF model and its validation. Technical report 75, ECMWF
Walland DJ, Simmonds I (1997) Modelled atmospheric response to changes in northern hemisphere snow cover. Clim Dyn 13:25–34. doi:10.1007/s003820050150
Weisheimer A, Doblas-Reyes P, Jung T, Palmer T (2011) On the predictability of the extreme summer 2003 over Europe. Geophys Res Lett 38. doi:10.1029/2010GL046455

DOI 10.1007/978-94-017-8789-5_8
Reprinted from *Surveys in Geophysics* Journal, DOI 10.1007/s10712-013-9221-7

Closing the Gaps in Our Knowledge of the Hydrological Cycle over Land: Conceptual Problems

William A. Lahoz · Gabriëlle J. M. De Lannoy

Received: 9 October 2012 / Accepted: 25 January 2013 / Published online: 15 March 2013

Abstract This paper reviews the conceptual problems limiting our current knowledge of the hydrological cycle over land. We start from the premise that to understand the hydrological cycle we need to make observations and develop dynamic models that encapsulate our understanding. Yet, neither the observations nor the models could give a complete picture of the hydrological cycle. Data assimilation combines observational and model information and adds value to both the model and the observations, yielding increasingly consistent and complete estimates of hydrological components. In this review paper we provide a historical perspective of conceptual problems and discuss state-of-the-art hydrological observing, modelling and data assimilation systems.

Keywords Hydrological cycle · Earth observation · Land surface models · Data assimilation

1 Introduction

The water stored on land is a key variable controlling numerous processes and feedback loops within the climate system (see, e.g., Dirmeyer 2000; Koster et al. 2004a, b; Seneviratne et al. 2010). It constrains plant transpiration and photosynthesis and thus is of major relevance for the Earth's water and energy cycles and impacts the exchanges of trace gases on land, including carbon dioxide. Figure 1, from IPCC (2007), provides an overview of the main terrestrial components and exchanges within the climate system. This shows the complexity of land processes and feedbacks, to a large extent owing to the high spatial

W. A. Lahoz (✉)
NILU, Instituttveien 18, PO Box 100, 2027 Kjeller, Norway
e-mail: wal@nilu.no

W. A. Lahoz
Météo-France, CNRM/GMGEC/CARMA, 42 avenue Gaspard Coriolis, 31100 Toulouse, France

G. J. M. De Lannoy
Global Modeling and Assimilation Office (Code 610.1), NASA/GSFC, Greenbelt, MD, USA
e-mail: Gabrielle.Delannoy@nasa.gov

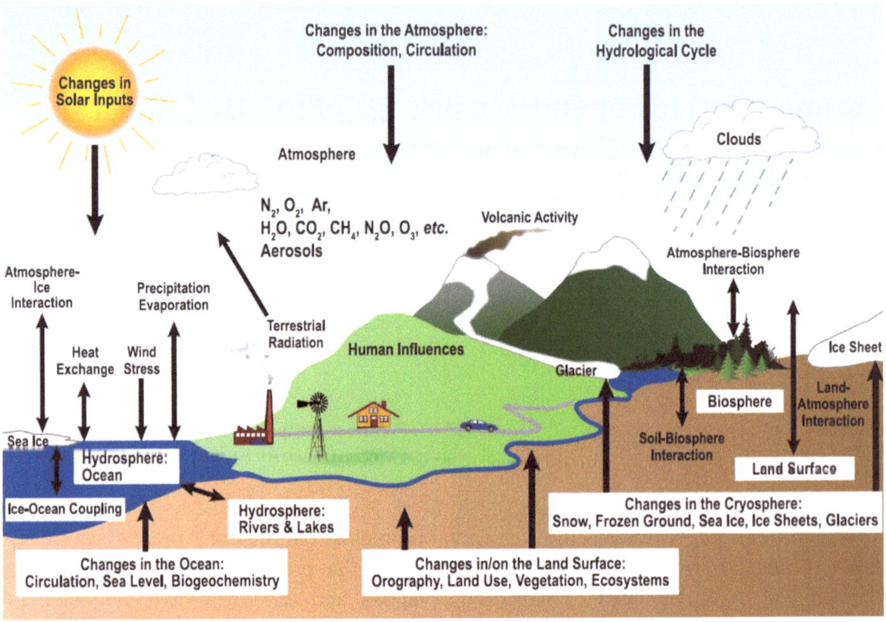

Fig. 1 Global climate system. Figure from IPCC (2007)

variability in soils, vegetation and topography (ranging from metres to kilometres). Processes affecting the amount of water stored on land, for example, precipitation and radiation, have spatial scales of kilometres (e.g., associated with weather fronts) and have high temporal variability (hours).

The amount of water stored in the unsaturated soil zone is generally referred to as *soil moisture*, although the exact definition can vary depending on the context. Soil moisture is one of the key geophysical variables for understanding the Earth's hydrological cycle. It is classed as an *essential climate variable* of the Global Climate Observing System (GCOS) (GCOS-107, 2006).

Soil moisture determines the partitioning of incoming water into infiltration and run-off. It directly affects plant growth and other organic processes and thus connects the water cycle to the carbon cycle. Run-off and base flow from the soil profile determine river flows and flooding, which connects hydrology with hydraulics. Soil moisture also has a significant impact on the partitioning of water and heat fluxes (latent and sensible heat), thereby connecting the hydrological (i.e. water) cycle with the energy cycle.

Soil moisture is a source of water for the atmosphere through evapotranspiration from land. Evapotranspiration is a major component of the continental water cycle, as it returns as much as 60 % of the whole land precipitation back to the atmosphere (e.g., Oki and Kanae 2006). Furthermore, evapotranspiration is also an important energy flux (Trenberth et al. 2009) and is connected to the *surface skin and soil temperature*, which make up other important state variables of the land surface system. Together, soil moisture, temperature and their impacts on the water, energy and carbon cycle play a major role in climate-change projections (IPCC 2007; Seneviratne et al. 2010). *Snow* on land is another important variable affecting the global energy and water budgets, because of its high albedo, low thermal conductivity, considerable spatial and temporal variation and medium-term capacity for water storage.

Quantifying the land state and fluxes and understanding soil moisture–temperature and soil moisture–precipitation couplings allow a better representation of hydrological processes in climate models and significantly help to reduce uncertainties in future climate scenarios, in particular regarding changes in climate variability and extreme events, and ecosystem/agricultural impacts (Seneviratne et al. 2010). This understanding is also crucially important for improving short-range numerical weather prediction (NWP) capabilities, in particularly regarding prediction of convective precipitation (Sherwood 1999; Adams et al. 2011, and references therein).

Hydrological observations are prone to *errors* and are *discrete in space and time* with the result that the information provided by these observations has *gaps*. Figure 2 shows an example of gaps in satellite observations. It is desirable to fill gaps in the observed information using additional information and computational techniques. Algorithms or models to fill in information gaps should organize, summarize and propagate the information from observations in an objective and consistent way. A simple approach such as linear interpolation could be a reasonably accurate "model", when observations are dense enough. However, linear interpolation may not be consistent with our advanced understanding of how the land surface behaves. A more realistic approach would be to fill in the gaps using a land surface model (LSM). While observations give an instantaneous view of the land surface, LSMs provide continuous estimates, based on physical laws that are derived from historical observations. These models are not perfect, and gaps in their structure, parametrization or initialization can be filled in with observations.

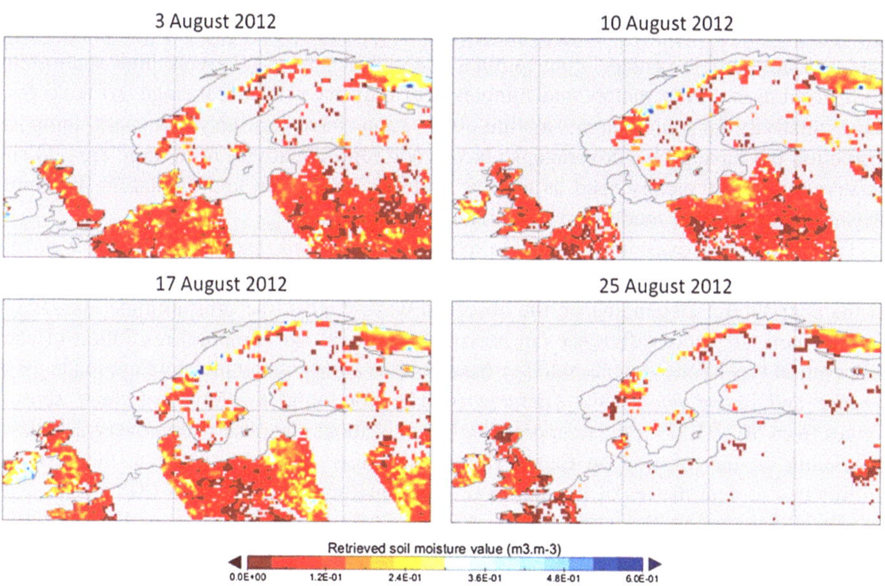

Fig. 2 Plot representing retrieved soil moisture data from the Soil Moisture Ocean Salinity (SMOS, Kerr et al. 2010) mission for August 3, 2012 (*top left panel*), August 10, 2012 (*top right panel*), August 17, 2012 (*bottom left panel*), and August 25, 2012 (*bottom right panel*), based on the observational geometry from ascending orbits from SMOS (units of $m^3\ m^{-3}$). *Blue* denotes relatively wet values; *red* denotes relatively dry values. The *uncoloured* (i.e. *grey*) areas over land represent gaps between the satellite orbits. Noteworthy are the sparse SMOS observations over Scandinavia, where retrievals from remotely sensed observations are particularly difficult, when the land is covered with snow, ice, forest, water bodies or rocks

Data assimilation (Kalnay 2003) provides an intelligent method to fill in the observational gaps using a model or to steer models using observations. By intelligent, it is meant an "objective" way which makes use of quantitative concepts (e.g., mathematical) for combining imperfect information. By combining observational and model information, data assimilation can be used to test the self-consistency and error characteristics of this information (Talagrand 2010b).

In this paper we focus on off-line land data assimilation, where the LSM is uncoupled from an atmospheric model. By using an uncoupled LSM, it can be forced with more observation-based forcings, rather than often inaccurate atmospheric analyses, and less computational resources are needed. The uncoupled approach can be regarded as a first step towards the land data assimilation goal of coupling an LSM to an atmospheric model to improve predictions at weather, seasonal and climate timescales (Palmer et al. 2008).

In this paper we discuss observations (Sect. 2), models (Sect. 3) and data assimilation methods (Sect. 4) used in the studies of the hydrological cycle and provide illustrative examples, with a focus on soil moisture. We pay special attention to the conceptual problems and key challenges associated with making use of observational and model information of the land surface in data assimilation systems (Sect. 5). We finish by providing conclusions (Sect. 6).

2 Observations of the Hydrological Cycle

Observations of the hydrological cycle are commonly divided into *conventional observations* (e.g., in situ ground-based measurements such as screen-level relative humidity) and *remotely sensed observations* (e.g., satellite or aircraft microwave observations). These data sets are complementary: conventional observations have relatively high spatio-temporal resolution (order metres and minutes) but only have local coverage, so have poor representativity for a large area; satellite observations have relatively low spatio-temporal resolution but have global coverage, so have good representativity for a large area. In situ observations are typically used as ground truth for calibration and validation of remote sensing products, and model and assimilation results.

Table 1 gives an overview of satellite sensors and missions that contribute to our current understanding of the hydrological cycle or may potentially contribute to this understanding in the near future. Depending on the observed wavelengths, the orbit altitude and design details, there are large differences in horizontal, vertical and temporal resolution of each observation type. For example, satellite-based observations of soil moisture are made using *passive* and *active microwave instruments*. The horizontal resolution of these sensors ranges from 50 to 10 km; the temporal resolution is about one observation every 2–3 days, depending on the location on Earth. These instruments typically penetrate the first few millimetres to centimetres of the soil: a few millimetres for the *X-band* (8–12 GHz, e.g., Advanced Microwave Sounding Radiometer for EOS, AMSR-E; Njoku and Chan 2006); ~1 cm for the *C-band* (4–8 GHz, e.g., AMSR-E; Advanced SCATterometer, ASCAT; Bartalis et al. 2007); and ~5 cm for the *L-band* (1–2 GHz, e.g., Soil Moisture Ocean Salinity, SMOS; Kerr et al. 2010). An immediate conceptual problem is to estimate soil moisture of actual interest in the root zone (1 m) at a finer resolution. For this, observational information needs to be transferred from the surface layer to the root zone (e.g., Calvet et al. 1998; Sabater et al. 2007; De Lannoy et al. 2007a; Draper et al. 2012) and downscaled from the coarse scale to finer scales (Reichle et al. 2001a; Pan et al. 2009;

Table 1 Characteristics of hydrological observations potentially available within the next decade (see "Appendix" for details of sensor acronyms)

Hydrological quantity	Remote sensing technique	Timescale	Spatial scale	Accuracy considerations	Examples of sensors
Precipitation	Thermal infrared	Hourly 1 day 15 days	4 km 1 km 60 m	Tropical convective clouds only	GOES, MODIS, AVHRR, Landsat, ASTER
	Passive microwave	3 h	10 km	Land calibration problems	TRMM, SSMI, AMSR-E, GPM
	Active microwave	Daily	10 m	Land calibration problems	TRMM, GPM
Surface soil moisture	Passive microwave	1–3 days	25–50 km	Limited to sparse vegetation, low topographic relief	AMSR-E, SMOS, Aquarius, SMAP
	Active microwave	3 days 30 days	3 km 10 m	Significant noise from vegetation and roughness	ERS, JERS, Radarsat, ASCAT
Surface skin temperature	Thermal infrared	1 h 1 day 15 days	4 km 1 km 60 m	Soil/vegetation average, cloud contamination	GOES, MODIS, AVHRR, Landsat, ASTER
Snow cover	Visible/thermal infrared	1 h 1 day 15 days	4 km 500 m–1 km 30–60 m	Cloud contamination, vegetation masking, bright soil problems	GOES, MODIS, AVHRR, Landsat, ASTER
Snow water equivalent (SWE)	Passive microwave	1–3 days	10 km	Limited depth penetration	AMSR-E
	Active microwave	30 days	100 m	Limited spatial coverage	SnoSat, SCLP, Cryosat-2, CoreH2O
Water level/velocity	Laser	10 days	100 m	Cloud penetration problems	ICESAT, ICESAT2, SWOT, DESDynl
	Radar	30 days	1 km	Limited to large rivers	TOPEX/POSEIDON
Total water storage changes	Gravity changes	30 days	1,000 km	Bulk water storage change	GRACE, GOCS, GRACEII
Evaporation		1 h 1 day 15 days	4 km 1 km 60 m	Significant assumptions	GOES, MODIS, AVHRR, Landsat, ASTER

Table updated from Houser et al. (2010)

De Lannoy et al. 2010; Sahoo et al. 2013) typically using a land surface model (discussed in Sect. 3) and/or land data assimilation (discussed in Sect. 4).

With the design of new sensors, one aims to gain resolution, increase the sensitivity to the variables of interest and reduce instrument errors (USGEO 2010). Examples of new missions for soil moisture observations are the SMOS and SMAP (Soil Moisture Active and Passive) missions, both using L-band sensors and designed with a target uncertainty lower than that of earlier missions, like AMSR-E and ASCAT. In November 2009, the ESA Earth Explorer mission SMOS was launched followed by another L-band mission, NASA/CONAE Aquarius (Le Vine et al. 2006), in June 2011. Aquarius measures various elements of the hydrological cycle, and its coarse resolution makes it less attractive for soil moisture estimation. The NASA mission SMAP is focused on soil moisture and freeze–thaw detection and is scheduled for launch in 2014 (Entekhabi et al. 2010a). To illustrate the importance of soil moisture information, Table 2 identifies key benefits from satellite soil moisture measurements.

A special issue on soil moisture from the SMOS mission has recently appeared in the *IEEE Transactions on Geoscience and Remote Sensing* (Kerr et al. 2012a). The papers in this special issue describe the SMOS mission (Mecklenburg et al. 2012); the radiometric performance (Kainulainen et al. 2012); the SMOS soil moisture retrieval algorithm (Kerr et al. 2012b; Mattar et al. 2012); the impact of radio frequency interference (RFI) on the SMOS soil moisture measurements (Castro et al. 2012; Misra and Ruf 2012; Oliva et al. 2012);

Table 2 Key benefits expected from satellite soil moisture observations

Area	Products	Comment
Meteorology	NWP models	Soil moisture plays a fundamental role in the transfer of water and energy between the surface and the atmosphere. Introduction of this variable in current NWP models will allow improving predictions, especially important under adverse meteorological conditions
Climatology	Models	Variability of the soil moisture time series with a long integration period may provide relevant information for the study of climate change
Risk Management	Flooding risk map	The soil's risk of flooding is significantly conditioned by the amount of water stored in the vadose zone. The generation of this type of products will require the inclusion of soil moisture data in hydrological and NWP models (precipitation predictions)
	Fire risk map	The risk of fire is determined by several factors, including meteorological, geophysical and biophysical factors. The information on soil moisture may be directly assimilated in drawing up fire risk maps as they provide direct information on evapotranspiration, water content assimilated by vegetation and quality of vegetation
	Famine risk map	The merging of geopolitical, meteorological/climatological information and data in the quality and estimates of agricultural and/or marine products (derived with the help of soil moisture data) may be of great use in early prediction of famine episodes in areas of Earth where resources are scarce
	Drought risk model	Analysing soil moisture trends in large areas may serve to generate drought models, along with data from other sensors
Agriculture	Agricultural production estimate	On the basis of soil moisture data and by means of the application of hydrological models, it is possible to determine the amount of water assimilated by the vegetation, a value that is very useful for estimating agricultural production
Hydrology	Models	The content of water stored in the soil is an important parameter to be taken into consideration in any hydrological model, as it is an indispensable variable in understanding the water cycle

Table adapted from http://www.cp34-smos.icm.csic.es/index.htm

soil processes in boreal regions (Rautiainen et al. 2012); disaggregation of SMOS data (Merlin et al. 2012); and various aspects of the validation of SMOS soil moisture data (Al Bitar et al. 2012; Bircher et al. 2012; dall'Amico et al. 2012; Jackson et al. 2012; Lacava et al. 2012; Mialon et al. 2012; Peischl et al. 2012b; Rowlandson et al. 2012; Sanchez et al. 2012; Schlenz et al. 2012; Schwank et al. 2012).

In view of the applications discussed later in this paper, we briefly mention that snow measurements are often provided by AMSR-E to measure snow water equivalent (SWE), and MODIS (MODerate resolution Imaging Spectroradiometer, Morisette et al. 2002) to give a picture of the snow-covered area. Finally, it is worth to mention GRACE (Gravity Recovery And Climate Experiment, Tapley et al. 2004) for its ability to measure an integrated water quantity of soil moisture and snow, as well as water in deeper layers.

Satellite instruments do not measure directly hydrological parameters. What they measure is photon counts (level 0 data). Algorithms then transform the level 0 data into radiances (level 1 data). Subsequently, using retrieval techniques (Rodgers 2000), retrievals of layer quantities (e.g., of soil moisture) or integrated amounts (e.g., total water storage) are derived (level 2 data). Fields derived from manipulation of level 2 data, for example, by interpolation to a common grid are termed level 3 data. Analyses derived from the assimilation of level 1 and/or 2 data are termed level 4 data.

Satellite observations (from level 0 and up) have associated with them a number of errors, including random and systematic errors in the measurement, and the error of representativeness (or representativity). Random errors (sometimes termed precision) have the property that averaging the data can reduce them. This is not the case of the systematic error or bias (sometimes termed accuracy). The error of representativeness is associated with the extent to which the measurement represents a point or volume in space. In land surface measurements, the error of representativeness is important to consider in comparisons of coarse-scale satellite data with point data.

Satellite-based hydrological data are becoming increasingly available, although little progress has been made in understanding their observational errors. Evaluation of the accuracy of land surface satellite data is a challenge, and novel methods to characterize their errors are being applied. Examples include *triple collocation* (e.g., Scipal et al. 2008; Dorigo et al. 2010; Parinussa et al. 2011); the *R-metrics* approach (Crow 2007; Crow and Zhan 2007; Crow et al. 2010); and *data assimilation* (Houser et al. 2010, and references therein).

A number of in situ network and airborne hydrological studies have been set up in the last decade for *evaluation of satellite data*. Examples of in situ networks include SMOSMANIA in France (Calvet et al. 2007; Albergel et al. 2009); NVE (Norges vassdrags-og energidirektorat, Norwegian Water Resources and Energy Directorate) in Norway (http://www.nve.no/en/); several large-scale (larger than 10,000 km^2) networks in the USA and elsewhere (see Table 1 in Crow et al. 2012); and several local- to regional-scale (larger than 100 km^2, smaller than 10,000 km^2) networks in the USA and elsewhere (see Table 2 in Crow et al. 2012). In situ soil moisture data from various networks across the world are consolidated in the International Soil Moisture Network (ISMN; http://www.ipf.tuwien.ac.at/insitu). As of January 2013, the ISMN includes data from 37 networks—Table 3 provides details. An example of an airborne study on evaluation of satellite data is the Australian Airborne Cal/Val Experiments for SMOS (AACE, Peischl et al. 2012a). An example of in situ ground-based station data used to evaluate satellite data is SMOSREX (de Rosnay et al. 2006). The temporal scale of in situ platforms ranges from minutes to hours; the spatial scale of in situ platforms ranges from tens of metres (individual stations) to thousands of kilometres (regional-scale networks). Along with the availability of dense

Table 3 Contributing networks to the International Soil Moisture Network (ISMN)

Name	Country	Stations	Website
AACES	Australia	49	http://www.moisturemap.monash.edu.au/
AMMA	Benin, Niger, Mali	7	http://amma-international.org/
ARM	USA	25	http://www.arm.gov
AWDN	USA	50	http://www.hprcc.unl.edu/awdn/
CALABRIA	Italy	5	http://www.cfcalabria.it
CAMPANIA	Italy	2	http://www.regione.campania.it/
CHINA	China	40	
COSMOS	USA, Germany, Switzerland, France, Brasil, Kenya, UK, Mexico	67	http://cosmos.hwr.arizona.edu/
FLUXNET-AMERIFLUX	USA	2	http://www.fluxnet.ornl.gov/fluxnet/index.cfm
FMI	Finland	1	http://fmiarc.fmi.fi/
GTK	Finland	7	
HOBE	Denmark	30	http://www.hobe.dk/
HSC_SELMACHEON	Korea	1	http://www.hsc.re.kr
HYDROL-NET_PERUGIA	Italy	1	http://www.dica.unipg.it/DICA
HYU_CHEONGMICHEON	Korea	1	http://wrrsl.hanyang.ac.kr/html/introduction.htm
ICN	USA	19	http://www.isws.illinois.edu/warm
IIT_KANPUR	India	1	http://www.iitk.ac.in
IOWA	USA	6	
MAQU	China	20	
MetEROBS	Italy	1	http://mistrals.sedoo.fr/HyMeX/Plateform-search?datsId=532
MOL-RAO	Germany	2	http://www.dwd.de/mol
MONGOLIA	Mongolia	44	
OZNET	Australia	52	http://www.oznet.org.au/
REMEDHUS	Spain	23	http://campus.usal.es/~hidrus/
RUSWET-AGRO	Former Soviet Union	78	
RUSWET-GRASS	Former Soviet Union	122	
RUSWET-VALDAI	Former Soviet Union	3	
SCAN	USA	182	http://www.wcc.nrcs.usda.gov/scan/
SMOSMANIA	France	21	http://www.hymex.org/
SNOTEL	USA	374	http://www.wcc.nrcs.usda.gov/snow/
SWEX_POLAND	Poland	6	
UDC_SMOS	Germany	11	http://www.geographie.uni-muenchen.de/department/fiona/forschung/projekte/index.php?projekt_id=103
UMBRIA	Italy	7	http://www.cfumbria.it/ http://hydrology.irpi.cnr.it/
UMSUOL	Italy	1	

Table 3 continued

Name	Country	Stations	Website
USCRN	USA	114	http://www.ncdc.noaa.gov/crn/
USDA-ARS	USA	4	
VAS	Spain	3	http://nimbus.uv.es/

Table adapted from http://www.ipf.tuwien.ac.at/insitu/index.php/insitu-networks.html

in situ data for validation, it is also important to select appropriate validation measures (Entekhabi et al. 2010b).

The assimilation of satellite data for land surface applications has only gained significance in the last decade; it started later than atmospheric and oceanographic data assimilation (see various chapters in Lahoz et al. 2010a). This can be attributed to: (1) a lack of dedicated land surface state (water and energy) remote sensing instruments; (2) inadequate retrieval algorithms for deriving global land surface information from remote sensing observations; and (3) a lack of mature techniques to objectively improve and constrain land surface model predictions using remote sensing data.

3 Models of the Hydrological Cycle

As discussed above, observational information has gaps in space and time. It is desirable to fill in these observational gaps using a model. Such models can range from simple linear interpolation to full land surface models (LSMs). Land surface processes are part of the global processes controlling the Earth, which are typically represented in global general circulation models (GCMs). The land component in these models is represented in (largely physically based) LSMs, which simulate the water and energy balance over land using simple algebraic equations or more complex systems of partial differential equations. The main *state variables* of these models include the water content and temperature of soil moisture, snow and vegetation. These variables are referred to as *prognostic* state variables. Changes in these state variables account for fluxes, for example, evapotranspiration and run-off, which are referred to as *diagnostic* state variables.

Most land surface models used in GCMs view the soil column as the fundamental hydrological unit, ignoring the role of, for example, topography on spatially variable processes (Stieglitz et al. 1997) to limit the complexity and computations for these coupled models. During the last decades, LSMs have become increasingly complex to accommodate for better understood processes, like snow and vegetation. Along with a more complex structure often comes a more complex parametrization, and several authors (Beven 1989; Duan et al. 1992) have stated that LSMs are over-parametrized given the data typically available for calibration. At larger scales, these models often rely on satellite-observed parameters, such as greenness and LAI (leaf area index). For field-scale studies, the LSMs are usually calibrated to specific circumstances to limit systematic prediction errors. Model calibration or parameter estimation relies on observed data and can be defined as a specific type of data assimilation (Nichols 2010).

Many LSMs have been developed and enhanced since the mid-1990s, with varying features, such as sub-grid variability, community-wide input, advanced physical representations and compatibility with atmospheric models (Houser et al. 2010). Some examples of widely used LSMs are the NCAR Community Land Model (CLM) (Oleson et al. 2010); the Variable Infiltration Capacity (VIC) Model (Liang et al. 1994); the Noah Model (Ek et al. 2003); the Catchment LSM (Koster et al. 2000); the TOPMODEL-based Land Atmosphere Transfer Scheme (TOPLATS) model (Famiglietti and Wood 1994); the Hydrology-Tiled European Centre for Medium-range Weather Forecasts (ECMWF) Scheme for Surface Exchange over Land (H-TESSEL) model (Balsamo et al. 2009); the SURFEX model (Le Moigne 2009); the Interaction between Soil Biosphere and Atmosphere (ISBA) model (Noilhan and Mahfouf 1996); and the Joint UK Land Environment Simulator (JULES) model (Best et al. 2011; Clark et al. 2011a). An example of an integrated system is the NASA Land Information System (LIS), which offers the capability to simulate with different models, observations and data assimilation techniques (Kumar et al. 2008).

An LSM has several elements, including a *soil moisture scheme*, a *snow scheme*, a *rainfall–run-off scheme* and a *routing/hydraulic scheme*. The *soil moisture scheme* can take several forms, such as explicit numerical solutions of Richards' equations over multiple discretized layers (e.g., in CLM), or using a force-restore method (e.g., Deardorff 1977, used in SURFEX), or other more non-traditional approaches, such as a soil moisture calculation as a deviation from the equilibrium soil moisture profile between the surface and the water table (Catchment LSM). The different profile structures involve different state variables, for example, describing soil moisture at the surface (*superficial volumetric water content*) or describing soil moisture over the root zone (*mean volumetric content of the root zone*). The coupling strength between the surface and deeper soil layers is a sensitive point for successful propagation of surface observations to deeper layers (Kumar et al. 2009).

The presence of snow covering the ground and vegetation can greatly influence the energy and mass transfers between the land surface and the atmosphere. Notably, the snow layer modifies the radiative balance by increasing the albedo. Furthermore, the amount of water stored in the snowpack has an important impact on water availability in the spring time. The prognostic variables in most snow schemes include variables related to *snow water equivalent* (SWE), including *snow depth* and *density*, and the *snow heat content*. These variables most often determine the diagnostics such as *snow area extent* and *albedo*.

The *snow scheme* can have one layer, or several layers. In a one-layer scheme, the evolution of the snow water equivalent of the snow reservoir depends on the precipitation of snow (a source) and the snow sublimation from the snow surface (a sink). Multi-layer schemes are often designed to have intermediate complexity, having simplified physical parametrizations based on those of highly detailed internal-process snow models, while having computational requirements resembling those of single-layer schemes (Loth et al. 1993; Lynch-Stieglitz 1994; Sun et al. 1999).

A number of approaches have been implemented for *rainfall–run-off schemes*. Water that cannot be stored in the soil profile either runs off over land (Horton run-off, e.g., Decharme and Douville 2006) or gravitationally drains out of the profile (Mahfouf and Noilhan 1996; Boone 1999). The TOPMODEL run-off approach combines key distributed effects of channel network topology and dynamic contributing areas for run-off generation (Beven and Kirkby 1979; Silvapalan et al. 1987). This formalism takes explicit account of topographic heterogeneities (Decharme et al. 2006; Decharme and Douville 2006, 2007). Run-off and drainage exiting from hydrological models can be used as a boundary to hydraulic models that predict river flow and potential flooding (Matgen et al. 2010).

A *hydraulic flood routing scheme* uses numerical methods to solve simultaneously the equations of continuity and momentum for a fluid (see, e.g., Guo 2006). It is often applied to a river network, typically in a hierarchy including hillslope routing, sub-network routing and main channel routing. An example of a routing scheme is the river transport model developed for the NCAR Community Land Model (CLM) (Branstetter and Erickson III 2003, and references therein). A river transport model is also useful because it can be used to evaluate the performance of an LSM against gauge station data.

Most LSMs are *soil–vegetation–atmosphere transfer* (SVAT) models, where the vegetation is not a truly dynamic component. Recently, coupling of hydrological or SVAT models with vegetation models has received some attention, to serve more specific ecological, biochemical or agricultural purposes. Dynamic vegetation models are used to simulate the evolution of vegetation cover, photosynthesis, carbon and nutrient inventories and the fluxes of water, CO_2, CH_4, N_2O, volatile organic carbon and fire-related emissions between the land surface and atmosphere. Illustrative examples of vegetation models are the Land biosphere Process and eXchange (LPX) model (Wania 2007; Spahni et al. 2010) and the CoupModel (Gustafsson et al. 2004; Jansson and Karlberg 2004; Jansson et al. 2005, 2008; Karlberg et al. 2006, 2007; Klemedtsson et al. 2008; Norman et al. 2008; Svensson et al. 2008).

There are a number of potential problems with LSMs that can cause errors in the forecast. These include components that cause "model error" and components that cause "predictability error". Components that cause "model error" are as follows: incomplete description of physical processes perhaps done for computational efficiency, perhaps a reflection of incomplete knowledge; inaccurate parameters; and inaccurate forcings. Components that cause "predictability error" are inaccurate initial states and boundaries. All these problems are the subject of research in the land surface modelling and assimilation community.

4 Data Assimilation of the Hydrological Cycle

4.1 Introduction

The only practical way to observe the land surface on continental to global scales is by satellite remote sensing. However, this cannot provide information on the entire system, and measurements only represent a snapshot in time. Land surface models can predict spatial/temporal land system variations, but these predictions are often poor, due to model initialization, parameter and forcing errors and inadequate model physics and/or resolution. A way forward is to merge the observational and model information through *data assimilation* (Kalnay 2003).

Mathematics provides rules for combining information objectively, based on principles which aim to maximize (or minimize) a quantity (e.g., a "penalty function") or on established *statistical concepts* (e.g., Bayesian methods) that relate *prior information* (understanding, which comes from prior combination of observations and models), with *new information* (e.g., an extra observation). The merged product, termed the posterior estimate or an *analysis*, *adds value* to both observational and model information. The data assimilation methodology takes account of the different nature (e.g., spatio-temporal resolution) of the observational and model information, using an *observation operator* (see, e.g., Talagrand 2010a).

Assimilation of land surface observations is at an earlier stage than, for example, assimilation of atmospheric observations (see various chapters in Lahoz et al. 2010a). However, during the past decade, land data assimilation has been a very active field of research. Land data assimilation considers both ground-based in situ data and satellite data. Often, satellite land surface data are assimilated and the process validated using in situ measurements. Assimilated satellite observations include retrievals of *land surface temperature*, *soil moisture*, *snow water equivalent* (SWE) and *snow cover area* (e.g., Van den Hurk et al. 2002; Andreadis and Lettenmaier 2006; Slater and Clark 2006; Bosilovich et al. 2007; Dong et al. 2007; Drusch 2007; Ni-Meister 2008; Reichle et al. 2008; Houser et al. 2010). Houser (2003) discusses the assimilation of land surface retrieved quantities and radiances. Early reviews of land data assimilation have been provided by McLaughlin (2002), Reichle (2008), Moradkhani (2008) and Houser et al. (2010).

Land data assimilation uses observations to constrain the physical parametrizations and initialization of land surface states critical for seasonal-to-interannual prediction. These constraints can be imposed in four ways: (1) by forcing the land surface primarily by observations (such as precipitation and radiation), often severe atmospheric NWP land surface forcing biases can be avoided (e.g., Saha et al. 2010; Reichle et al. 2011); (2) by employing innovative land surface data assimilation techniques, observations of land surface storages (such as snow, soil temperature and moisture) can be used to constrain unrealistic simulated storages (e.g., Houser et al. 2010; Reichle et al. 2013); (3) by tuning adjustable parameters (e.g., Pauwels et al. 2009; Vrugt et al. 2012); and (4) the land surface physical structure itself can be improved through the data assimilation process when the constant confrontation of model states against observations returns useful information about structural deficits. Integration of soil moisture information from satellite instruments, and ground-based and in situ observations of the land surface, using land data assimilation, provides a comprehensive picture of the state and variability of the land surface.

4.2 Data Assimilation Methods

Three methods are commonly used for land data assimilation (Houser et al. 2010): *variational* (3- and 4-dimensional, 3D-Var and 4D-Var); *sequential* (Kalman filter (KF) and Extended Kalman filter (EKF)); and *ensemble* (Ensemble Kalman filter, EnKF). Bouttier and Courtier (1999) provide details of these methods. Talagrand (2010a) and Kalnay (2010) discuss more recent developments in variational methods and ensemble methods, respectively.

In the *3-D variational* (3D-Var) method, a minimization algorithm is used to find a model state, $\mathbf{x}$, that minimizes the misfit between $\mathbf{x}$ and the background state $\mathbf{x}^b$, and also between the observation predictions $H(\mathbf{x})$ and the observations $\mathbf{y}$. The observation operator H maps the model state $\mathbf{x}$ to the measurement space, where $\mathbf{y}$ resides. In 3D-Var, we seek the minimum with respect to $\mathbf{x}$ of the penalty function, J, given by Eq. (1). The first term on the right hand side (J_b) quantifies the misfit to the background term, and the second term (J_o) is the misfit to the observations. If the observation operator is linear (written $\mathbf{H}$), the penalty function, J, is quadratic and is guaranteed to have a unique minimum.

$$J = \frac{1}{2}[x - x^b]^{\mathrm{T}}\mathbf{B}^{-1}[x - x^b] + \frac{1}{2}[y - H(x)]^{\mathrm{T}}\mathbf{R}^{-1}[y - H(x)] \quad (1)$$

4-D variational (4D-Var) assimilation is an extension of 3D-Var in which the temporal dimension is included, that is, 4D-Var is a *smoother*. In 4D-Var, observations are used at their correct time. 4D-Var has two new features compared to 3D-Var. First, it includes a model operator, M, that carries out the evolution forward in time. The first derivative, or differential, of M, $\mathbf{M}$, is the *tangent linear model* (if M is linear, represented by $\mathbf{M}$, its derivative is $\mathbf{M}$). The transpose of the tangent linear model operator, $\mathbf{M}^{\mathrm{T}}$, integrates the *adjoint variables* backward in time. The tangent linear model is only defined under the condition that the function J defined by Eq. (1) be differentiable—this is the *tangent linear hypothesis*. Second, J can include an extra term in which the model errors associated with the model's temporal evolution are accounted for. In the formulation of Zupanski (1997), an analogous term involving $\mathbf{Q}^{-1}$ is included in J, where $\mathbf{Q}$ is the model error covariance. Examples for the land surface using variational methods include Calvet et al. (1998) and Reichle et al. (2001a). Note that variational methods are very common for parameter estimation (e.g., Dumont et al. 2012), but with replacement of the misfit to the background with a misfit to prior parameter guesses.

In the *Kalman filter* (KF), a recursive sequential algorithm is applied to evolve a forecast, $\mathbf{x}^f$, and an analysis, $\mathbf{x}^a$, as well as their respective error covariance matrices, $\mathbf{P}^f$ and $\mathbf{P}^a$. The KF equations are (subscripts denote the time step) as follows:

$$x_n^f = \mathbf{M}_{n-1}x_{n-1}^a; \tag{2a}$$

$$\mathbf{P}_n^f = \mathbf{M}_{n-1}\mathbf{P}_{n-1}^a\mathbf{M}_{n-1}^{\mathrm{T}} + \mathbf{Q}_{n-1}; \tag{2b}$$

$$x_n^a = x_n^f + \mathbf{K}_n[y_n - \mathbf{H}_n x_n^f]; \tag{2c}$$

$$\mathbf{K}_n = \mathbf{P}_n^f\mathbf{H}_n^{\mathrm{T}}[\mathbf{R}_n + \mathbf{H}_n\mathbf{P}_n^f\mathbf{H}_n^{\mathrm{T}}]^{-1}; \tag{2d}$$

$$\mathbf{P}_n^a = [\mathbf{I} - \mathbf{K}_n\mathbf{H}_n]\mathbf{P}_n^f. \tag{2e}$$

Equation (2a) represents the forecast of the model fields from time step $n-1$ to n, while Eq. (2b) calculates the forecast error covariance from the analysis error covariance $\mathbf{P}^a$ and the model error covariance $\mathbf{Q}$. Equations (2c) and (2e) are the analysis steps, using the *Kalman gain* defined in Eq. (2d). $\mathbf{Q}$ and $\mathbf{P}^a$ are assumed to be uncorrelated. For optimality, all errors must be uncorrelated in time.

The KF can be generalized to nonlinear H and M operators, although in this case neither the optimality of the analysis nor the equivalence with 4D-Var holds. The resulting equations are known as the *Extended Kalman filter* (EKF) as, for example, used for the land surface by Boulet et al. (2002), Reichle et al. (2002b), Matgen et al. (2010), Rüdiger et al. (2010) and de Rosnay et al. (2012b).

The *Ensemble Kalman filter*, EnKF, uses a Monte Carlo ensemble of short-range forecasts to estimate $\mathbf{P}^f$. The estimation becomes more accurate as the ensemble size increases. The EnKF is more general than the EKF to the extent that it does not require validity of the tangent linear hypothesis. Evensen (2003) provides a comprehensive review of the theory and numerical implementation of the EnKF. Examples for the land surface are identified in Table 4 (see below).

The *Particle Filter* (PF) is also an ensemble method. It does not require a specific form for the state distribution but, typically, a re-sampling algorithm needs to be applied (van Leeuwen 2009). Because PF methods typically make no assumptions of linearity in the model equations or that model and observational errors are Gaussian, they are well suited to deal with the land surface where model evolution is highly nonlinear, and model and observational errors can be non-Gaussian. The PF has been applied in hydrology to estimate model parameters and state variables (e.g., Moradkhani et al. 2005a; Weerts and El Serafy 2006; Plaza et al. 2012; Vrugt et al. 2012).

4.3 Representation of Errors

Representation of *errors* is fundamental to data assimilation. One needs to consider errors in observations, background information and model (see Eqs. (1, 2a–e) above for identification of the error covariance matrices mentioned in the following). $\mathbf{R}$, the observational

Table 4 Selected studies on land surface data assimilation, sorted by assimilated observation type

Observation	State		Parameter/model
	EnKF/EnKS	Other	
Soil moisture, retrievals	Reichle and Koster (2005), Ni-Meister et al. (2006), Reichle et al. (2007, 2008), Kumar et al. (2009), Pan and Wood (2010), Liu et al. (2011), Han et al. (2012a), Draper et al. (2012), Sahoo et al. (2013)	Houser et al. (1998), Pauwels et al. (2002), Paniconi et al. (2003), Francois et al. (2003), Hurkmans et al. (2006), Parajka et al. (2006), Crow (2007), Crow and Bolten (2007), Parada and Liang (2008), Crow and van den Berg (2010), Draper et al. (2009), Mahfouf (2010), Dharssi et al. (2011), de Rosnay et al. (2012a, b)	Santanello et al. (2007), Ines and Mohanty (2009), Pauwels et al. (2009)
Soil moisture, in situ	Sabater et al. (2007), De Lannoy et al. (2007a, 2009), Camporese et al. (2009), Monsivais-Huerteroet et al. (2010), Han et al. (2012a)	Calvet et al. (1998), Wingeron et al. (1999), Walker et al. (2001b, 2002)	Boulet et al. (2002), De Lannoy et al. (2006), Vereecken et al. (2008), Loew and Mauser (2008), Nagarajanar et al. (2011)
Snow cover or albedo, retrievals	Clark et al. (2006), Su et al. (2008, 2010), De Lannoy et al. (2012), Arsenault et al. (2013)	Rodell and Houser (2004), Zaitchik and Rodell (2009), Saha et al. (2010), de Rosnay et al. (2012a, b)	Essery and Pomeroy (2004), Déry et al. (2005), Kolberg and Gottschalk (2010), Dumont et al. (2012)
Snow water equivalent, retrievals or in situ	Andreadis and Lettenmaier (2006), Slater and Clark (2006), Dong et al. (2007), De Lannoy et al. (2010, 2012), He et al. (2012)	Brasnett (1999), Sun et al. (2004), Drusch et al. (2004)	Clark and Vrugt (2006), Clark et al. (2011b), Su et al. (2011)
Backscatter, from soil or vegetation	Flores et al. (2012)	Hoeben and Troch (2000), Zhan et al. (2006)	Marzahn and Ludwig (2009), Nearing et al. (2010)
Brightness temperature, for soil or vegetation	Margulis et al. (2002), Reichle et al. (2002a), Crow (2003), Crow and Wood (2003), Dunne and Entekhabi (2006)	Entekhabi et al. (1994), Galantowicz et al. (1999), Crosson et al. (2002), Reichle et al. (2001a, b), Jones et al. (2003), Wilker et al. (2006), Balsamo et al. (2006), Loew et al. (2009), Dumedah et al. (2011)	Zhang et al. (2011), Montzka et al. (2012), De Lannoy et al. (2013)
Brightness temperature, for snow	Durand and Margulis (2007), Durand et al. (2009)	DeChant and Moradkhani (2010)	Tedesco et al. (2010), Vachon et al. (2010), Forman et al. (2012a)

Table 4 continued

Observation	State		Parameter/model
	EnKF/EnKS	Other	
Surface soil or skin temperature, evapotranspiration, retrievals or in situ	Pipunic et al. (2008), Ghent et al. (2010), Reichle et al. (2010), Xu et al. (2011)	Castelli et al. (1999), Lakshmi (2000), Boni et al. (2001), Schuurmans et al. (2003), Bosilovich et al. (2007), Renzullo et al. (2008), Sini et al. (2008), Meng et al. (2009), Barrett and Renzullo (2009), Mackaro et al. (2011)	Caparrini et al. (2004), Kalma et al. (2008), Gutmann and Small (2010)
Water stage, retrievals	Andreadis et al. (2007), Durand et al. (2008), Biancamaria et al. (2010)	Matgen et al. (2010), Giustarini et al. (2011)	Montanari et al. (2009)
Terrestrial water storage, retrievals	Zaitchik et al. (2008), Su et al. (2010), Li et al. (2012), Forman et al. (2012b)	–	Günter (2008), Lo et al. (2010)
Discharge, gauge	Weerts and El Serafy (2006), Vrugt et al. (2006), Pauwels and De Lannoy (2006, 2009)	Aubert et al. (2003), Moradkhani et al. (2005a), Seo et al. (2009), Lee et al. (2011), Vrugt et al. (2012)	Madsen (2003), Moradkhani et al. (2005b), Montanari and Toth (2007), Vrugt et al. (2008), Quets et al. (2010)
Leaf area index, remotely sensed	Pauwels et al. (2006), Nearing et al. (2012)	Jarlan et al. (2008), Albergel et al. (2010), Rüdiger et al. (2010)	Lewis et al. (2012)
Screen-level observations	–	Balsamo et al. (2004), Seuffert et al. (2004), Drusch and Viterbo (2007), Mahfouf et al. (2009), Draper et al. (2011), Mahfouf and Bliznak (2011)	–

Synthetic observation studies are classified by the observation type that is mirrored. For land surface (-coupled) state updating, the studies are divided into sets using either the EnKF or EnKS (Ensemble Kalman Smoother) and those using any other assimilation technique. For parameter and model structure updating, examples relate to either forward models or land surface(-coupled) models

error covariance matrix, is typically assumed to be diagonal, although this is not always justified. $\mathbf{R}$ includes errors of the measurements themselves, $\mathbf{E}$, and errors of representativeness, $\mathbf{F}$; $\mathbf{R} = \mathbf{E} + \mathbf{F}$. $\mathbf{B}$ is the background error covariance matrix in variational methods (the analogue in the KF and ensemble methods is $\mathbf{P}^f$); its diagonal elements determine the relative weight of the forecasts, and its off-diagonal elements determine how information is spread spatially. Estimating $\mathbf{B}$ or $\mathbf{P}^f$ is a key part of the data assimilation method (Bannister 2008a, b). Estimating model error $\mathbf{Q}$ is a research topic.

In the EnKF, the background (or forecast) errors are represented by the spread of the ensemble. This simplifies the computation of $\mathbf{P}^f$, implicitly accounts for the model error $\mathbf{Q}$ and avoids the calculation of Eq. (2b). For land data assimilation, the relative fraction of the observation error **R** and the model error **Q** (associated with the temporal evolution of the model) is often tuned or adaptively updated (e.g., Desroziers et al. 2005; Reichle et al. 2008).

In general, in data assimilation, errors are assumed to be *Gaussian*. The most fundamental justification for assuming Gaussian errors, which is entirely pragmatic, is the relative simplicity and ease of implementation of statistical linear estimation under these conditions. Because Gaussian probability distribution functions are fully determined by their mean and variance, the solution of the data assimilation problem becomes computationally practical. Note that the assumption of a Gaussian distribution is often not justified in land data assimilation applications.

Typically, there are biases between different observations, and between observations and model (see, e.g., Ménard 2010). These biases are spatially and temporally varying, and it is a major challenge to estimate and correct them. Despite this, and mainly for pragmatic reasons, in data assimilation it is often assumed that errors are unbiased. For NWP many assimilation schemes now incorporate a *bias correction*, and various techniques have been developed to correct observations to remove biases (e.g., Dee 2005); these methods are now being applied to land data assimilation (De Lannoy et al. 2007a, b).

4.4 Advantages and Disadvantages of Assimilation Methods

The feasibility of 4D-Var has been demonstrated in NWP systems (see, e.g., Simmons and Hollingsworth 2002). Its main advantage is that it considers observations over a time window that is generally much longer than the model time step, that is, it is a *smoothing algorithm*. This allows more observations to constrain the system and, considering satellite coverage, increases the geographical area influenced by the data. For nonlinear systems (as is generally the case for the land surface), this feature of 4D-Var, together with the non-diagonal nature of the adjoint operator which transfers information from observed regions to unobserved regions, reduces the weight of the background error covariance matrix in the final 4D-Var analysis compared to the KF analysis (for linear systems, the general equivalence between 4D-Var and the KF implies that the same weight is given to all data in both systems).

In contrast to the above advantages of 4D-Var, three weaknesses must be mentioned. First, its numerical cost is very high compared to approximate versions of the KF or ensemble methods. Second, its formalism cannot determine the analysis error directly; rather, it has to be computed from the inverse of the *Hessian matrix* (again, this procedure is prohibitive in both computation time and memory). Finally, its formalism requires the calculation of the adjoint model, which is time-consuming and may be difficult for a system such as the land surface which exhibits nonlinearities and on–off processes (e.g., presence or lack of snow).

The EKF is capable of handling some departure from Gaussian distributions of model errors and nonlinearity of the model operator. However, if the model becomes too nonlinear or the errors become highly skewed or non-Gaussian, the trajectories computed by the EKF will become inaccurate.

The EnKF is attractive as, for example, it requires no derivation of a tangent linear operator or adjoint equations and no integrations backward in time, as for 4D-Var (see Evensen 2003). The EnKF also provides a cost-effective representation of the background error covariance matrix, $\mathbf{P}^f$. Several issues need to be considered in developing the EnKF: (1) ensemble size; (2) ensemble collapse; (3) correlation model for $\mathbf{P}^f$, including localization (see, e.g., Kalnay 2010); and (4) specification of model errors.

The major drawback of the above techniques is the underlying assumption that the model states have a Gaussian distribution. The PF does not require a specific form for the state distribution, but its major drawback is that distribution of particle weights quickly becomes skewed, and a re-sampling algorithm needs to be applied.

The EnKF and PF are complementary. This complementarity makes a hybrid EnKF/PF version highly attractive for systems that can exhibit nonlinear and non-Gaussian features, an example being the land surface. For example, the EnKF could be used as an efficient sampling tool to create an ensemble of particles with optimal characteristics with respect to observations. The PF methodology could then be applied on that ensemble afterwards to resolve nonlinearity and non-Gaussianity in the system. This method is getting increased attention (see, e.g., Kotecha and Djurić 2003).

4.5 Example of a Land Data Assimilation System

For illustrative purposes, we describe the elements of the NILU SURFEX-EnKF land data assimilation system (Lahoz et al. 2010b). These elements are the following: (1) a data assimilation scheme (mainly variants of the EnKF, but also variants of the PF, and the EKF); (2) a land surface model (SURFEX model developed at Météo-France, Le Moigne 2009); (3) observations; (4) the observation operator; and (5) error characteristics for the model and the observations.

The SURFEX model used at NILU (and at Météo-France) can be run in uncoupled or coupled mode. It includes the following elements:

- A soil and vegetation scheme: ISBA and ISBA-A-gs;
- A water surface scheme: COARE/ECUME (Coupled Ocean–Atmosphere Response Experiment/Exchange Coefficients from Unified Multi-campaign Estimates) for the sea; FLAKE for inland water;
- Urban and artificial areas: Town Energy Balance—TEB model;
- A surface boundary layer (SBL) scheme;
- Chemistry and aerosols;
- A land use database: ECOCLIMAP.

Figure 3 illustrates how SURFEX works. During a model time step, each surface grid box receives from the atmosphere the following information: upper air temperature, specific humidity, horizontal wind components, pressure, total precipitation, long-wave radiation, short-wave direct and diffuse radiation and, possibly, concentrations of chemical species and dust. In return, SURFEX computes averaged fluxes of momentum, sensible and latent heat, and, possibly, chemical species and dust fluxes. These fluxes are then sent back to the atmosphere with the addition of radiative terms like surface temperature, surface direct and diffuse albedo, and surface emissivity.

The above information transferred to the atmosphere from the land surface provides the lower boundary conditions for the radiation and turbulent schemes in an atmospheric model coupled to SURFEX or forced by SURFEX output. In SURFEX, each grid box is made up of four adjacent surfaces: one for nature, one for urban areas, one for sea or ocean and one for lake, identified by the global ECOCLIMAP land database. The SURFEX fluxes are the average of the fluxes computed over nature, town, sea/ocean or lake, weighted by their respective fraction.

The assimilation system at NILU is illustrated in Fig. 4 with reference to the EnKF. It can assimilate the following data: (1) 2-m screen-level temperature (T_{2m}) and 2-m screen-level relative humidity (RH_{2m}) provided, for example, by the SYNOP/CANARI

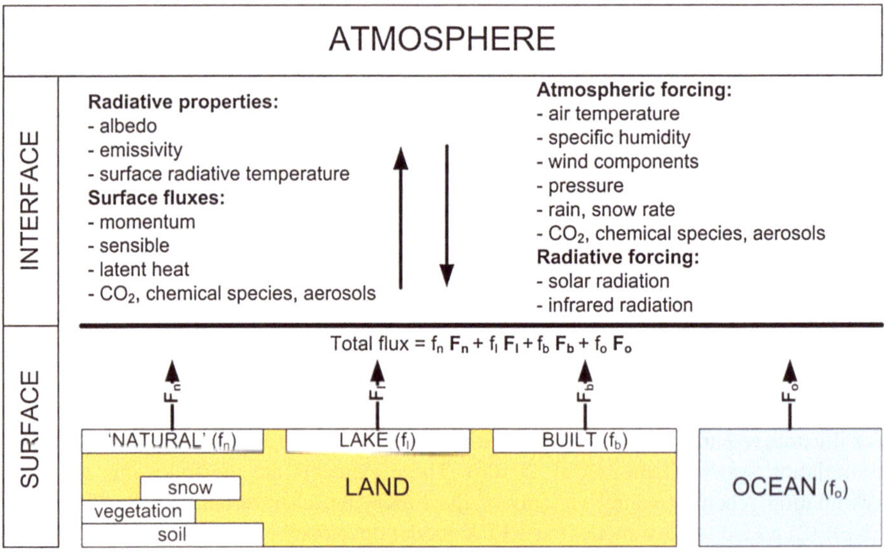

Fig. 3 Exchanges between the atmosphere and land surface implemented in the SURFEX LSM. See text. Based on Le Moigne (2009)

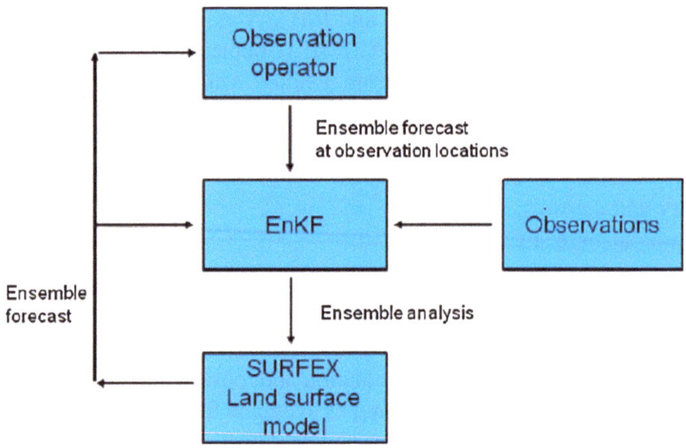

Fig. 4 Schematic of the NILU SURFEX-EnKF land DA system methodology. From Lahoz et al. (2010b)

(Code d'Analyse Nécessaire à Arpege pour ses Rejets et son Initialisation; Taillefer 2002) analysis; and (2) superficial soil moisture content data from satellites (e.g., from ASCAT, AMSR-E and SMOS). The *control variables* (Nichols 2010) of the NILU land DA system are the following:

- Surface temperature;
- Mean surface temperature;
- Superficial volumetric water content;
- Mean volumetric water content of the root zone.

4.6 Data Assimilation Research Applications

Table 4 shows a selection of studies using a variety of observation types to improve the land surface state or the state in a hydraulic, vegetation or snow model coupled to it. Because of its success in highly nonlinear land surface modelling (Reichle 2008), the EnKF has gained a lot of attention. Therefore, state estimation studies using an EnKF or EnKS (Ensemble Kalman Smoother, where the time integration is done forwards and backwards) are organized separately from those that use any other assimilation technique (e.g., variational, optimal interpolation). Also shown are a few examples on parameter estimation in land surface or forward models. While this review focuses on state estimation, parameter estimation and forcing correction are of utmost importance in land surface models. Land surface models are not chaotic and thus benefit less from state estimation than atmospheric or oceanic applications. By contrast, parameters and forcings determine the major part of the land surface model uncertainty, and great advances can be expected from combining state, bias, parameter and forcing estimation (Moradkhani et al. 2005b; De Lannoy et al. 2006; Vrugt et al. 2012). Here, we discuss a number of soil moisture and snow-related studies done mainly for state updating, with particular attention to the conceptual problems they address. Examples on evapotranspiration, surface or skin temperature, LAI (leaf area index), discharge and water stage assimilation are also provided in Table 4, but not discussed in detail.

4.6.1 Single-column Applications

To explore the possibilities and limitations of assimilation schemes, numerous studies have first explored single point-scale or grid cell-scale applications. For soil moisture assimilation, conceptual problems include the propagation of information from the surface to the entire soil profile; the optimization of assimilation techniques and update frequencies; and the identification of an allowable level of uncertainty in surface observations to be useful in a data assimilation scheme, mostly in view of satellite sensor design.

Georgakakos and Baumer (1996) performed a sensitivity study to document the impact of observation noise on Kalman filter (KF) results. Calvet et al. (1998) and Wingeron et al. (1999) assimilated surface soil moisture data from a soil profile in the highly instrumented field site of the Monitoring the Usable soil Reservoir EXperiment (MUREX) in France to update root zone soil moisture using variational approaches and investigated the importance of assimilation windows and observation frequencies. Similarly, Li and Islam (1999) studied the effect of assimilation frequency while directly inserting gravimetric measurements as surrogates for remote sensing data, and Aubert et al. (2003) suggested that a 1-week soil moisture update is sufficient. Walker et al. (2001a) showed in a synthetic profile study that the KF was superior to direct insertion. In a subsequent study with real data from the Nerrigundah catchment in Australia, Walker et al. (2001b) articulated the idea that soil moisture assimilation can solve issues with errors in forcings or initial conditions, but not errors caused by problems in the physics of the soil model.

De Lannoy et al. (2007a) used an EnKF to study vertical information propagation, and the effect of assimilation depth and frequency for an extensive set of soil profiles in an USDA field in Beltsville, USA. This study highlighted the effect of bias propagation through the profile and the need for bias estimation, a conceptual problem that was addressed with a two-stage forecast and bias filter (De Lannoy et al. 2007a, b). At the same time, Sabater et al. (2007) studied the concept of propagating surface observations to deeper model layers using different types of filtering, using ground data from the Surface

Monitoring of the Soil Reservoir EXperiment, SMOSREX. Camporese et al. (2009) set up synthetic soil profile assimilation experiments studying the effect of uncertainties, ensemble size, bias and other factors with an EnKF. Because of the large impact of parameters and forcings on soil moisture errors and biases, assimilation schemes have paid increasing attention to including parameter estimation along with state updating, as, for example, illustrated in Monsivais-Huerteroet et al. (2010). At present, EnKF filtering experiments are being conducted at point-scales to further identify and address conceptual problems with soil profile estimation, using surface observations (see, e.g., Han et al. 2012a).

Another important conceptual problem with soil moisture assimilation, initially addressed in a point-scale setting, is the direct assimilation of radiances or assimilation using an observation operator. This is done to avoid inconsistencies between auxiliary information that would be used in retrievals and that used in the land surface models. Entekhabi et al. (1994) estimated 1-m-deep bare soil moisture profiles using synthetic microwave brightness temperatures. This work was extended by Galantowicz et al. (1999) using eight days of L-band brightness temperature (T_b) data collected from a test plot in Beltsville, USA. Pathmathevan et al. (2003) assimilated microwave observations with a variational technique, but using a heuristic optimization, rather than an adjoint. Crosson et al. (2002) tested T_b assimilation at the point-scale with an EKF and showed that biases could not be overcome through assimilation. Crow (2003) successfully assimilated T_b for soil moisture and showed improvements at the plot-scale, using either synthetic or real field data. Crow analysed the EnKF performance in terms of the assumptions that underlie the KF. Crow and Wood (2003) also used the EnKF at two sites within the Southern Great Plains 1997 (SGP97) experimental domain and reported that T_b data assimilation was able to correct for rainfall errors. Wilker et al. (2006) highlighted the difficulty in mapping heterogeneous soil moisture into T_b using a forward operator and identified the representativeness errors associated with these data. Similar to the above studies, Hoeben and Troch (2000) used a KF including a forward backscatter model to explore the direct assimilation of radar microwave signals to estimate soil moisture profiles.

Snow data assimilation has conceptual problems inherent to the cumulative and temporary nature of this variable. Slater and Clark (2006) illustrated how a square root EnKF could improve the snow state at in situ sites in Colorado during the accumulation and melt phase. They also identified the temporal correlation in snowpacks and showed how it could limit the efficiency of filtering if not accounted for properly. In a synthetic study, Liston and Hiemstra (2008) proposed a technique to update snow retroactively, which would be useful for reanalysis applications, if observations would only be available at the end of the snow season. In situ snow data assimilation is performed operationally (see Sect. 4.7 below), usually with simple assimilation techniques. An example where both the snow state and parameters were estimated using an EnKF in a 1-D setting is given by Su et al. (2011).

A number of point- or single-grid-scale studies have tried to relate brightness temperature data to snowpack characteristics (Durand et al. 2008; Andreadis et al. 2008), in preparation for T_b assimilation. Many of these studies highlight the large sensitivity of snowpack estimates to model parameters (Davenport et al. 2012), which makes both forward simulation and inversion of T_b observations for SWE estimation a difficult task.

4.6.2 Distributed Applications

The most obvious advantage of remotely sensed observations is the possibility of performing large-scale and spatially distributed assimilation. It should be recognized, however, that despite the spatial coverage of data, for computational reasons assimilation is often performed per column, that is, using a 1-D filter. When the vertical columns (of snow or soil) are horizontally connected through the model physics or assimilation statistics, this is referred to as 3-D assimilation.

The assimilation of catchment-distributed soil moisture has often focused on the improvement of the state or initial conditions (Pauwels et al. 2001, 2002) and parameters in order to improve spatially integrated fluxes, such as discharge. However, it is also possible to use soil moisture assimilation to correct rainfall estimates (Crow and Ryu 2009). At the global scale, soil moisture assimilation will become increasingly important when coupled to the atmosphere for climate and seasonal predictions.

Spatially distributed studies initially focused on assimilation of retrievals with simple techniques and gradually developed towards more complex schemes, with the inclusion of forward models (observation operators) to directly assimilate, for example, microwave observations. Initial soil moisture retrieval studies explored the performance of different filter techniques, such as Newtonian nudging, statistical correction and statistical interpolation (Houser et al. 1998; Pauwels et al. 2001; Paniconi et al. 2003; Hurkmans et al. 2006), while during the last decade, variational and KF-based assimilation largely dominated this research field because of the proven robustness and flexibility of these latter techniques (Reichle et al. 2002a, b).

A typical conceptual problem with spatially distributed assimilation is the use of coarse-scale remotely sensed data to infer fine-scale information. There are many static disaggregation techniques that use auxiliary information to perform such a downscaling outside the assimilation scheme. Performing dynamic disaggregation within the assimilation scheme remains a research challenge. The latter concept consists of a 3-D filter with inclusion of spatially correlated (fine-scale) state and (coarse-scale) observation prediction errors and has been addressed in EnKF frameworks by Reichle et al. (2001b, 2013), Reichle and Koster (2003), Pan et al. (2009), De Lannoy et al. (2010) and Sahoo et al. (2013).

An important issue connected to 3-D filtering for disaggregation is the use of local observations to update neighbouring locations, for example, to propagate from observed swaths to unobserved locations. Often, this problem is solved with spatial interpolation or by relying on horizontal connections in the model equations (Walker et al. 2002). Alternatively, such horizontal information propagation can be done within an assimilation scheme that provides accurate error correlations between observed and non-observed observations and forecasts (Reichle and Koster 2003; De Lannoy et al. 2012). De Lannoy et al. (2009) used an adaptive KF to identify such spatial correlations, along with the magnitude of the forecast error, to optimize filter performance. Han et al. (2012b) studied the effect of spatial correlations in an OSSE (observing system simulation experiment) with a local ensemble transform Kalman filter. Filter technical issues such as update frequency (Walker and Houser 2004) and error estimation have also been addressed in a spatial context. Reichle and Koster (2005) demonstrated the validity of the concept that assimilation results should be better than either the model or observations alone. After re-scaling satellite observations from AMSR-E and SMMR to take bias out of the system, Reichle et al. (2007) showed that satellite observations can contribute valuable information, even if they are not accurate. Reichle et al. (2009) further assessed the quality of assimilation products as a function of retrieval and land surface model uncertainty in an OSSE and showed that soil moisture retrievals can have slightly less skill than the land surface model and still contribute to an overall higher skill in the assimilation product. This was confirmed in a real data assimilation study by Draper et al. (2012).

The importance of correctly specifying random errors and biases is a major conceptual challenge in the optimization of distributed assimilation systems. Bias mitigation has become a regular part of most soil moisture data assimilation systems (Reichle and Koster 2004; Drusch et al. 2005; Kumar et al. 2012; Sahoo et al. 2013), and random error specifications for soil moisture data assimilation have been studied through adaptive filtering (Crow and van Loon 2006; Reichle et al. 2008).

Another idea with potential benefit is multi-sensor assimilation for soil moisture estimation. As an example, Draper et al. (2012) showed how both active (ASCAT) and passive (AMSR-E) microwave retrievals can contribute to a similar improvement in assimilation results. Combining improved precipitation data with soil moisture retrieval assimilation (Liu et al. 2011) and combining discharge (Pauwels and De Lannoy 2006), temperature or LAI with soil moisture assimilation are other avenues that have been exploited for hydrological assimilation.

As already indicated for single-column applications, a major conceptual problem is the direct assimilation of brightness temperatures (T_b) or backscatter observations from satellite missions for soil moisture estimation. Reichle et al. (2001a, b) presented pioneering studies with a 3-D variational scheme to assimilate and disaggregate synthetic or real brightness temperatures over the SGP97 study area, while Margulis et al. (2002) used an EnKF and Dunne and Entekhabi (2006) compared an EnKF with an EnKS for the same T_b assimilation problem. Walker et al. (2002) also assimilated T_b directly, but from SMMR and using an EKF over Australia. Using a variational scheme, and with inclusion of both a land surface temperature and microwave brightness temperature observation operator, Barrett and Renzullo (2009) showed that both thermal (AVHRR) and microwave (AMSR-E) satellite observations can provide effective observational constraints on the modelled profile and on surface soil moisture. There are only a few studies on spatially distributed backscatter assimilation, but in a recent OSSE using an EnKF, Flores et al. (2012) showed the potential of the L-band radar information expected from the future SMAP mission.

For snow, spatially distributed assimilation studies include snow cover area (or snow cover fraction) and snow water equivalent (SWE) assimilation. A correct specification of the snow-covered area is important to represent feedbacks from the land to the atmosphere, while a good estimate of the actual amount of snow in the snowpack is of crucial importance for flood, drought and discharge predictions (He et al. 2012). Snow cover observations are typically fine-scale visible/near infrared observations that are only available in cloud-free areas, while SWE measurements are typically more inaccurate retrievals from T_b observations at a coarse scale (see Table 1). It can be expected that multi-sensor assimilation could help to further snow estimation (De Lannoy et al. 2012).

Because of its binary nature, snow cover in terms of the presence or absence of snow cannot be assimilated with filters that rely on continuous variables. Instead, rule-based algorithms have been proposed (Rodell and Houser 2004; Zaitchik and Rodell 2009; Roy et al. 2010). However, the snow cover fraction (SCF) is a more continuous variable that has been assimilated with KF-based algorithms (Clark et al. 2006; Su et al. 2008; De Lannoy et al. 2012). When assimilating SCF with a Kalman filter, there is a need to relate SCF to the actual SWE state variable through an observation operator, often defined as a snow depletion curve. It is also possible to use visible/near infrared snow albedo observations to update snow parameters such as grain size (Dumont et al. 2012).

The two dominant conceptual problems with satellite-based SWE assimilation are the coarse-scale nature and high uncertainty of the measurements. Initial attempts to assimilate SMMR or AMSR-E SWE retrievals only yielded marginal success (Andreadis and Lettenmaier 2006; Dong et al. 2007), because of retrieval errors due to signal saturation, presence of liquid water in the snowpack and multiple other factors. To address the coarse-scale issue, De Lannoy et al. (2010) proposed several 3-D filter options to disaggregate SWE data and propagate data from observed swaths to unobserved regions. These techniques showed great benefit in a synthetic data study. When using real AMSR-E retrievals (De Lannoy et al. 2012), and with bias mitigation through re-scaling added to the system, the assimilation analyses were affected by a lack of a realistic interannual signal in the retrievals.

To address the problems with SWE retrieval accuracy, the potential of direct radiance assimilation has been investigated (Durand and Margulis 2006; Andreadis et al. 2008; Durand et al. 2009; DeChant and Moradkhani 2010). However, these efforts rely on a good description of the snowpack in the land surface model, which is not always available for large-scale applications. To address this, Forman et al. (2013) developed an artificial neural network as a computationally attractive forward model in readiness for large-scale radiance assimilation. In preparation for the future SMAP mission, freeze–thaw assimilation (Bateni et al. 2013) has been investigated, because of its importance in understanding the carbon cycle.

The above studies update either snow or soil moisture separately. A major challenge for land data assimilation is making use of total water storage (TWS) observations from GRACE, which include soil moisture, snow and other water components at a very coarse scale (Table 1). Total water storage can be decomposed into soil and snow components and disaggregated to finer scales (Zaitchik et al. 2008; Su et al. 2010; Forman et al. 2012; Li et al. 2012; Reichle et al. 2013).

4.7 Towards Operational Land Data Assimilation

Land surface processes and their initialization are of crucial importance to address the challenge of seamless prediction from weather to seasonal and climate timescales (Palmer et al. 2008). It is well established that high skill in short- and medium-range forecasts of temperature and humidity over land requires proper initialization of soil moisture (Beljaars et al. 1996; Douville et al. 2000; Mahfouf et al. 2000; Drusch and Viterbo 2007; van den Hurk et al. 2008). A similar impact from soil moisture has been established for seasonal forecasts (Koster et al. 2004a, b, 2011; Weisheimer et al. 2011). Initialization of snow conditions also has a significant impact on forecast accuracy at weather timescales (Brasnett 1999; Drusch et al. 2004). Operational land data assimilation has initially focused on ingesting precipitation observations (e.g., Saha et al. 2010; Reichle et al. 2011), but improved snow and soil moisture state updates are now emerging, as documented, for example, for the ECMWF Integrated Forecasting System by de Rosnay et al. (2012a).

An unprecedented operational land data assimilation product will be provided by the Global Modeling and Assimilation Office (NASA GMAO) in the form of a level 4 satellite-based soil moisture product (Reichle et al. 2012; De Lannoy et al. 2013). The assimilation of SMAP brightness temperatures into the Goddard Earth Observing System land surface model will yield a global root zone soil moisture product.

5 Conceptual Problems and Key Challenges

To summarize, the conceptual problems in our understanding of the hydrological cycle over land can be grouped by observing, modelling and data assimilation systems. These are outlined below.

5.1 Assimilated Observations

- To be useful for model development and assimilation, the dominant modes (in space and time) of the land system must be sampled;
- To be efficient for state updating, observations need to be available at a reasonable time interval to capture short-term dynamical variations (cf. the importance of satellite overpass frequency; Walker and Houser 2004; Pan and Wood 2010);
- Observations must be collected in long enough historical records to identify long-term, climatological, statistics for bias mitigation (Reichle and Koster 2004) or trend identification;
- Observations need to be sampled at different spatial scales to capture both local and global processes;
- There is a need to have a reasonable signal-to-noise ratio (e.g., SMAP's target of brightness temperature uncertainty is 1.3 K; Entekhabi et al. 2010a), and an uncertainty in the error description appropriate for scientific studies;
- There is a need to relate observations to key system state variables, that is, there needs to be system observability.

5.2 Forward and Retrieval Models, with Particular Reference to Radiances and Backscatter Processes

- To achieve appropriate retrieval accuracy, there is a need to use advanced methods to describe physical processes in radiative transfer models (RTMs);
- When assimilating radiances at large scales (e.g., from microwave sensors), there is a need for calibration of RTMs (De Lannoy et al. 2013; Forman et al. 2013).

5.3 Land Surface Models

- There is a need to use advanced methods to describe physical processes (this limits structural uncertainty) and couple land surface models with models describing more specialized processes such as run-off routing, dynamic vegetation or snow (Pauwels et al. 2006);
- There is a need for consistent global parameter datasets to limit predictive uncertainty due to parameter uncertainty;
- There is a need for high-quality forcing data (this limits input uncertainty), mainly for precipitation (Maggioni et al. 2011; Reichle et al. 2011).

5.4 Data Assimilation Challenges

- There is a need to fill in the spatial and temporal gaps in observations (Reichle and Koster 2003; De Lannoy et al. 2012);
- There is a need to disaggregate data in space and time and into their individual components (Forman et al. 2012; Reichle et al. 2013);
- There is a need to ingest directly radiances or backscatter information (as opposed to retrievals) to avoid inconsistencies between auxiliary information in retrievals and land surface models (Crow and Wood 2003; Durand et al. 2009; Flores et al. 2012);

- There is a need to exploit the simultaneous use of multiple sensors (Pan et al. 2008; Draper et al. 2012) and explore the capabilities of new sensors (Andreadis et al. 2007; Durand et al. 2008);
- There is a need to combine state and input (forcing) information with parameter updates (Moradkhani et al. 2005b; Liu et al. 2011; Vrugt et al. 2012);
- There is a need to explore advanced filtering techniques, for example, the use of the particle filter to account for non-Gaussian errors (Plaza et al. 2012);
- There is a need to improve the representation of observation and forecast errors, and to specify biases in observational and model information (De Lannoy et al. 2007b; Crow and Reichle 2008; Reichle et al. 2008; De Lannoy et al. 2009; Crow and van den Berg 2010);
- There is a need to preserve water balance in the land system (Pan and Wood 2006; Yilmaz et al. 2011) and draw lessons from the information in the assimilation increments;
- There is a need to have access to adequate computational resources.

5.5 Validation

- Needs ground observations with substantial spatial and temporal coverage;
- Needs tools to address scaling and representativeness errors (Crow et al. 2012);
- Needs appropriate and effective validation metrics (Entekhabi et al. 2010b).

6 Conclusions

To understand the hydrological cycle over land, we need to make observations and develop models that encapsulate our understanding. These models have a basis on the information gathered from observations, as well as on previous experience, and are used to project our understanding into the future by making predictions. A crucial element in this procedure is confronting models with observations. *Data assimilation*, which combines observational and model information, provides an objective method to *confront* models against observations and *add value* to both the model and the observations. Data assimilation adds value to observations by filling the gaps between them and adds value to models by constraining them with observations. In this paper, we touch on the main conceptual problems that limit a full integration of land surface models and observations by reviewing progress in land surface data assimilation research over the last decade.

Collectively, the advent of new satellite missions, the increasing attention to forecast uncertainty due to errors in the land surface model structure, parameters and input, and the development of advanced assimilation techniques will eventually close the largest gaps in our understanding of the hydrological cycle over land.

Acknowledgments This paper arose from the International Space Science Institute (ISSI) workshop "The Earth's Hydrological Cycle", held at ISSI, Bern, Switzerland, on 6–10 February 2012. A NILU internal project supported WAL. Thanks to Alexandra Griesfeller for providing Fig. 2.

Appendix: Sensor acronyms

AACE	Australian Airborne Cal/Val Experiments (for SMOS)
AMSR	Advanced Microwave Sounding Radiometer on EOS Aqua
ASCAT	Advanced SCATterometer
ASTER	Advanced Spaceborne Thermal Emission and reflection Radiometer
AVHRR	Advanced Very High Resolution Radiometer
CONAE	COmisión Nacional de Actividades Espaciales (National Space Activities Commission)—Argentina Space Agency
CoReH2O	COld REgions Hydrology high-resolution Observatory
ERS	European Research Satellite
ESA	European Space Agency
GOES	Geostationary Operational Environmental Satellite
GPM	Global Precipitation Measurement
GRACE	Gravity Recovery And Climate Experiment
JERS	Japanese Earth Resources Satellite
MODIS	MODerate resolution Imaging Spectroradiometer
MUREX	Monitoring of the Usable Reservoir EXperiment
NASA	National Aeronautics and Space Administration
SCLP	Snow and Cold Land Process
SMAP	Soil Moisture Active and Passive
SMMR	Scanning Multichannel Microwave Radiometer
SMOS	Soil Moisture and Ocean Salinity
SMOSMANIA	Soil Moisture Observing System-Meteorological Automatic Network Integrated Application
SMOSREX	Surface MOnitoring of the Soil Reservoir EXperiment
SSM/I	Special Sensor Microwave Imager
SWOT	Surface Water Ocean Topography
TRMM	Tropical Rainfall Measuring Mission

References

Adams DKRM, Fernandes S, Kursinski ER, Maia JM, Sapucci LF, Machado LAT, Vitorello I, Galera Monico JF, Holub KL, Gutman S, Filizola N, Bennett RA (2011) A dense GNSS meteorological network for observing deep convection in the Amazon. Atmos Sci Lett 12. doi:10.1002/asl.312

Al Bitar A, Leroux D, Kerr YH, Merlin O, Richaume P, Sahoo A, Wood EF (2012) Evaluation of SMOS soil moisture products over continental U.S. using the SCAN/SNOTEL network. IEEE Trans Geosci Remote Sens 50:1572–1586

Albergel C, Rüdiger C, Carrer D, Calvet J-C, Fritz N, Naeimi V, Bartalis Z, Hasenauer S (2009) An evaluation of ASCAT surface soil moisture products with in situ observations in Southwestern France. Hydrol Earth Syst Sci 13:115–124

Albergel C, Calvet J-C, Mahfouf J-F, Rüdiger C, Barbu AL, Lafont S, Roujean J-L, Walker JP, Crapeau M, Wigneron J-P (2010) Monitoring of water and carbon fluxes using a land data assimilation system: a case study for southwestern France. Hydrol Earth Syst Sci 14:1109–1124

Andreadis KM, Lettenmaier DP (2006) Assimilating remotely sensed snow observation into a macroscale hydrology model. Adv Water Resour 29:872–886

Andreadis K, Clark E, Lettenmaier D, Alsdorf D (2007) Prospects for river discharge and depth estimation through assimilation of swath-altimetry into a raster-based hydrodynamics model. Geophys Res Lett 34:L10403

Andreadis KM, Liang D, Tsang L, Lettenmaier DP, Josberger EG (2008) Characterization of errors in a coupled snow hydrology microwave emission model. J Hydrometeorol 9:149–164
Arsenault K, Houser P, Dirmeyer P, De Lannoy G (2013) Impacts of snow cover fraction data assimilation on modeled energy and moisture budgets. J Geophys Res (in press)
Aubert D, Loumagne C, Oudin L (2003) Sequential assimilation of soil moisture and streamflow data into a conceptual rainfall-runoff model. J Hydrol 280:145–161
Balsamo G, Bouyssel F, Noilhan J (2004) A simplified bi-dimensional variational analysis of soil moisture from screen-level observations in a mesoscale numerical weather-prediction model. Q J R Meteorol Soc 130:895–915
Balsamo G, Bélair S, Deblonde G (2006) A global root-zone soil moisture analysis using simulated L-band brightness temperature in preparation for the hydros satellite mission. J Hydrometeorol 7:1126–1146
Balsamo G, Viterbo P, Beljaars A, van den Hurk B, Hirschi M, Betts AK, Scipal K (2009) A revised hydrology for the ECMWF model: verification from field site to terrestrial water storage and impact in the integrated forecast system. J Hydrometeorol 10. doi:10.1175/2008JHM1068.1
Bannister RN (2008a) A review of forecast error covariance statistics in atmospheric variational data assimilation. I: characteristics and measurements of forecast error covariances. Q J R Meteorol Soc 134:1951–1970
Bannister RN (2008b) A review of forecast error covariance statistics in atmospheric variational data assimilation. II: modelling the forecast error covariances. Q J R Meteorol Soc 134:1971–1996
Barrett D, Renzullo LJ (2009) On the efficacy of combining thermal and microwave satellite data as observational constraints for root-zone soil moisture estimation. J Hydrometeorol 10:1109–1127
Bartalis Z, Wagner W, Naeimi V, Hasenauer S, Scipal K, Bonekamp H, Figa J, Anderson C (2007) Initial soil moisture retrievals from the METOP-A advanced Scatterometer (ASCAT). Geophys Res Lett 34:L20401. doi:10.1029/2007GL031088
Bateni SM, Huang C, Margulis SA, Podest E, McDonald K (2013) Feasibility of characterizing snowpack and the freeze-thaw state of underlying soil using multifrequency active/passive microwave data. IEEE Trans Geosci Remote Sens. doi:10.1109/TGRS.2012.2229466
Beljaars A, Viterbo P, Miller M, Betts A (1996) The anomalous rainfall over the United States during July 1993: sensitivity to land surface parameterization and soil anomalies. Mon Weather Rev 124:362–383
Best MJ, Pryor M, Clark DB, Rooney GG, Essery RLH, Ménard CB, Edwards JM, Hendry MA, Porson A, Gedney N, Mercado LM, Sitch S, Blyth E, Boucher O, Cox PM, Grimmond CSB, Harding RJ (2011) The Joint UK Land Environment Simulator (JULES), model description—part 1: energy and water fluxes. Geosci Model Dev 4:677–699
Beven K (1989) Changing ideas in hydrology: the case of physically-based models. J Hydrol 105:157–172
Beven BJ, Kirkby MJ (1979) A physically-based variable contributing area model of basin hydrology. Hydrol Sci Bull 24:43–69
Biancamaria S, Durand M, Andreadis KM, Bates PD, Boone A, Mognard NM, Rodríguez E, Alsdorf DE, Lettenmaier DP, Clark EA (2010) Assimilation of virtual wide swath altimetry to improve Arctic river modelling. Remote Sens Environ. doi:10.1016/j.rse.2010.09.008
Bircher S, Balling JE, Skou N, Kerr YH (2012) Validation of SMOS brightness temperatures during the HOBE airborne campaign, Western Denmark. IEEE Trans Geosci Remote Sens 50:1468–1482
Boni G, Entekhabi D, Castelli F (2001) Land data assimilation with satellite measurements for the estimation of surface energy balance components and surface control on evaporation. Water Resour Res 37:1713–1722
Boone A (1999) Modelisation des processus hydrologiques dans le schema de surface ISBA: Inclusion d'un reservoir hydrologique, du gel et modelisation de la neige, PhD thesis, University Paul Sabatier, Toulouse, France, 2000, 252 pp
Bosilovich MG, Radakovich JD, Silva AD, Todling R, Verter F (2007) Skin temperature analysis and bias correction in a coupled land-atmosphere data assimilation system. J Meteorol Soc Jpn 85A:205–228
Boulet G, Kerr Y, Chehbouni A (2002) Deriving catchment-scale water and energy balance parameters using data assimilation based on extended Kalman filtering. Hydrol Sci J des Sci Hydrol 47:449–467
Bouttier F, Courtier P (1999) Data assimilation concepts and methods. ECMWF training notes, March 1999, available from http://www.ecmwf.int
Branstetter ML, Erickson DJ III (2003) Continental runoff dynamics in the Community Climate System Model 2 (CCSM2) control simulation. J Geophys Res 108:4550. doi:10.1029/2002JD003212
Brasnett B (1999) A global analysis of snow depth for numerical weather prediction. J Appl Meteorol 38:726–740
Calvet J-C, Noilhan J, Bessemoulin P (1998) Retrieving the root-zone soil moisture from surface soil moisture or temperature estimates: a feasibility study based on field measurements. J Appl Meteorol 37:371–386

Calvet J-C, Fritz N, Froissard F, Suquia D, Petitpa A, Piguet B (2007) In-situ soil moisture observations for the CAL/VAL of SMOS: the SMOSMANIA network. International geoscience and remote sensing symposium, IGARSS, 23–28 July 2007, Barcelona, Spain, pp 1196–1199. doi:10.1109/IGARSS.2007.4423019

Camporese M, Paniconi C, Putti M, Salandin P (2009) Ensemble Kalman filter data assimilation for a process-based catchment scale model of surface and subsurface flow. Water Resour Res 45:W10421

Caparrini F, Castelli F, Entekhabi D (2004) Variational estimation of soil and vegetation turbulent transfer and heat flux parameters from sequences of multisensor imagery. Water Resour Res 40:W12515.1–W12515.15

Castelli F, Entekhabi D, Caporali E (1999) Estimation of surface heat flux and an index of soil moisture using adjoint-state surface energy balance. Water Resour Res 35:3115–3125

Castro R, Gutierrez A, Barbosa J (2012) A first set of techniques to detect radio frequency interferences and mitigate their impact on SMOS data. IEEE Trans Geosci Remote Sens 50:1440–1447

Clark M, Vrugt J (2006) Unraveling uncertainties in hydrologic model calibration: addressing the problem of compensatory parameters. Geophys Res Lett 33:L06406.1–L06406.5

Clark MP, Slater AG, Barrett AP, Hay LE, McCabe GJ, Rajagopalan B, Leavesley GH (2006) Assimilation of snow covered area information into hydrologic and land-surface models. Adv Water Resour 29:1209–1221

Clark DB, Mercado LM, Sitch S, Jones CD, Gedney N, Best MJ, Pryor M, Rooney GG, Essery RLH, Blyth E, Boucher O, Harding RJ, Huntingford C, Cox PM (2011a) The Joint UK Land Environment Simulator (JULES), model description—part 2: carbon fluxes and vegetation dynamics. Geosci Model Dev 4:701–722

Clark MP, Hendrikx L, Slater AG, Kavetski D, Anderson B, Cullen NJ, Kerr T, Hreinsson EÖ, Woods RA (2011b) Representing spatial variability of snow water equivalent in hydrologic and land-surface models: a review. Water Resour Res 47:W07539

Crosson WL, Laymon CA, Inguva R, Schamschula MP (2002) Assimilating remote sensing data in a surface flux–soil moisture model. Hydrol Process 16:1645–1662

Crow W (2003) Correcting land surface model predictions for the impact of temporally sparse rainfall rate measurements using an ensemble Kalman filter and surface brightness temperature observations. J Hydrometeorol 4:960–973

Crow WT (2007) A novel method for quantifying value in spaceborne soil moisture retrievals. J Hydrometeorol 8:56–67. doi:10.1175/JHM553.1

Crow WT, Bolten JD (2007) Estimating precipitation errors using spaceborne surface soil moisture retrievals. Geophys Res Lett 34:L08403

Crow WT, Reichle RH (2008) Comparison of adaptive filtering techniques for land surface data assimilation. Water Resour Res 44:W08423. doi:10.1029/2008WR006883

Crow WT, Ryu D (2009) A new data assimilation approach for improving runoff prediction using remotely sensed soil moisture retrievals. Hydrol Earth Syst Sci 13:1–16

Crow WT, van den Berg J (2010) An improved approach for estimating observation and model error parameters in soil moisture data assimilation. Water Resour Res 46:W12519

Crow WT, van Loon E (2006) Impact of incorrect model error assessment on the sequential assimilation of remotely sensed surface soil moisture. J Hydrometeorol 7:421–432

Crow WT, Wood EF (2003) The assimilation of remotely sensed soil brightness temperature imagery into a land surface model using ensemble Kalman filtering: a case study based on ESTAR measurements during SGP97. Adv Water Resour 26:137–149

Crow WT, Zhan X (2007) Continental-scale evaluation of remotely sensed soil moisture products. IEEE Geosci Remote Sens Lett 4:451–455. doi:10.1109/LGRS.2007.896533

Crow WT, Miralles DG, Cosh MH (2010) A quasi-global evaluation system for satellite-based surface soil moisture retrievals. IEEE Geosci Remote Sens Lett 48:2516–2527

Crow WT, Berg AA, Cosh MH, Loew A, Mohanty BP, Panciera R, de Rosnay P, Ryu D, Walker JP (2012) Upscaling sparse ground-based soil moisture observations for the validation of coarse-resolution satellite soil moisture products. Rev Geophys Res 50:RG2002. doi:10.1029/2011RG000372

dall'Amico JT, Schlenz F, Loew A, Mauser W (2012) First results of SMOS soil moisture validation in the Upper Danube catchment. IEEE Trans Geosci Remote Sens 50:1507–1516

Davenport I, Sandells M, Gurney R (2012) The effects of variation in snow properties on passive microwave snow mass estimation. Remote Sens Environ 118:168–175

De Lannoy GJM, Houser PR, Pauwels VRN, Verhoest NEC (2006) Assessment of model uncertainty for soil moisture through ensemble verification. J Geophys Res 111:D10101.1–D10101.18. doi:10.1029/2005JD006367

De Lannoy GJM, Houser PR, Pauwels VRN, Verhoest NEC (2007a) State and bias estimation for soil moisture profiles by an ensemble Kalman filter: effect of assimilation depth and frequency. Water Resour Res 43:W06401. doi:10.1029/2006WR005100

De Lannoy GJM, Reichle RH, Houser PR, Pauwels VRN, Verhoest NEC (2007b) Correcting for forecast bias in soil moisture assimilation with the ensemble Kalman filter. Water Resour Res 43:W09410. doi:10.1029/2006WR00544

De Lannoy GJM, Houser PR, Verhoest NEC, Pauwels VRN (2009) Adaptive soil moisture profile filtering for horizontal information propagation in the independent column-based CLM2.0. J Hydrometeorol 10:766–779

De Lannoy GJM, Reichle RH, Houser PR, Arsenault KR, Pauwels VRN, Verhoest NEC (2010) Satellite-scale snow water equivalent assimilation into a high-resolution land surface model. J Hydrometeorol 11:352–369. doi:10.1175/2009JHM1194.1

De Lannoy GJM, Reichle R, Arsenault K, Houser P, Kumar S, Verhoest N, Pauwels V (2012) Multiscale assimilation of advanced microwave scanning radiometer-EOS snow water equivalent and moderate resolution imaging spectroradiometer snow cover fraction observations in northern Colorado. Water Resour Res 48:W01522

De Lannoy GJM, Reichle RH, Pauwels VRN (2013) Global calibration of the GEOS-5 L-band microwave radiative transfer model over land using SMOS observations. J Hydrometeorol (in press)

de Rosnay P, Calvet J-C, Kerr Y, Wigneron J-P, Lemaître F et al (2006) SMOSREX: a long term field campaign experiment for soil moisture and land surface processes remote sensing. Remote Sens Environ 102(377–389):2006

de Rosnay P, Balsamo G, Albergel C, Muñoz-Sabater J, Isaksen L (2012a) Initialisation of land surface variables for numerical weather prediction. Surv Geophys. doi:10.1007/s10712-012-9207-x

de Rosnay PD, Drusch M, Vasiljevic D, Balsamo G, Albergel C, Isaksen L (2012b) A simplified extended Kalman filter for the global operational soil moisture analysis at ECMWF. Q J R Meteorol Soc. doi:10.1002/qj.2023

Deardorff JW (1977) A parameterization of ground surface moisture content for use in atmospheric prediction models. J Appl Meteorol 16:1182–1185

DeChant C, Moradkhani H (2010) Radiance data assimilation for operational snow and streamflow forecasting. Adv Water Resour 34:351–364

Decharme B, Douville H (2006) Introduction of a sub-grid hydrology in the ISBA land surface model. Clim Dyn 26:65–78

Decharme B, Douville H (2007) Global validation of the ISBA sub-grid hydrology. Clim Dyn 29:21–37

Decharme B, Douville H, Boone A, Habets F, Noilhan J (2006) Impact of an exponential profile of saturated hydraulic conductivity within the ISBA LSM: simulations over the Rhône basin. J Hydrometeorol 7:61–80

Dee DP (2005) Bias and data assimilation. Q J R Meteorol Soc 131:3323–3343

Déry SJ, Salomonson VV, Stieglitz M, Hall DK, Appel I (2005) An approach to using snow areal depletion curves inferred from MODIS and its application to land surface modelling in Alaska. Hydrol Process 19:2755–2774

Desroziers G, Berre L, Chapnik B, Poli P (2005) A simple method to diagnose and adapt observation and background errors. Q J R Meteorol Soc 131:3385–3396

Dharssi I, Bovis KJ, Macpherson B, Jones CP (2011) Operational assimilation of ASCAT surface soil wetness at the Met Office. Hydrol Earth Syst Sci 15:2729–2746

Dirmeyer P (2000) Using a global soil wetness dataset to improve seasonal climate simulation. J Clim 13:2900–2921

Dong J, Walker JP, Houser PR, Sun C (2007) Scanning multichannel microwave radiometer snow water equivalent assimilation. J Geophys Res 112:D07108. doi:10.1029/2006JD007209

Dorigo W, Scipal K, Parinussa RM, Liu YY, Wagner W, de Jeu RAM, Naeimi V (2010) Error characterization of global active and passive microwave soil moisture data sets. Hydrol Earth Syst Sci 14:2605–2616

Douville H, Viterbo P, Mahfouf J-F, Beljaars ACM (2000) Evaluation of optimal interpolation and nudging techniques for soil moisture analysis using FIFE data. Mon Weather Rev 128:1733–1756

Draper CS, Mahfouf J-F, Walker J (2009) An EKF assimilation of AMSR-E soil moisture into the ISBA land surface scheme. J Geophys Res 114:D20104.1–D20104.13. doi:10.1029/2008JD011650

Draper C, Mahfouf J-F, Walker J (2011) Root zone soil moisture from the assimilation of screen-level variables and remotely sensed soil moisture. J Geophys Res 116:D02127

Draper CS, Reichle RH, De Lannoy GJM, Liu Q (2012) Assimilation of passive and active microwave soil moisture retrievals. Geophys Res Lett 39:L04401

Drusch M (2007) Initializing numerical weather prediction models with satellite surface soil moisture: Data assimilation experiments with ECMWF's integrated forecast system and the TMI soil moisture data set. J Geophys Res 112. doi:10.1029/2006JD007478
Drusch M, Viterbo P (2007) Assimilation of screen-level variables in ECMWF's integrated forecast system: a study on the impact on the forecast quality and analyzed soil moisture. Mon Weather Rev 135:300–314
Drusch M, Vasilievic D, Viterbo P (2004) ECMWF's global snow analysis: assessment and revision based on satellite observations. J Appl Meteorol 43:1282–1294
Drusch M, Wood EF, Gao H (2005) Observation operators for the direct assimilation of TRMM microwave imager retrieved soil moisture. J Geophys Res 32:L15403.1–L15403.4
Duan Q, Sorooshian S, Gupta VK (1992) Effective and efficient global optimization for conceptual rainfall-runoff models. Water Resour Res 28:1015–1031
Dumedah G, Berg AA, Wineberg M (2011) An integrated framework for a joint assimilation of brightness temperature and soil moisture using the nondominated sorting genetic algorithm II. J Hydrometeorol 12:1596–1609
Dumont M, Durand Y, Arnaud Y, Six D (2012) Variational assimilation of albedo in a snowpack model and reconstruction of the spatial mass-balance distribution of an alpine glacier. J Glaciol 58:151–164
Dunne S, Entekhabi D (2006) Land surface state and flux estimation using the ensemble Kalman smoother during the Southern Great Plains 1997 field experiment. Water Resour Res 42:W01407.1–W01407.15
Durand M, Margulis SA (2006) Feasibility test of multifrequency radiometric data assimilation to estimate snow water equivalent. J Hydrometeorol 7:443–457
Durand M, Margulis SA (2007) Correcting first-order errors in snow water equivalent estimates using a multifrequency, multiscale radiometric data assimilation scheme. J Geophys Res 112:D13121.1–D13121.15
Durand M, Andreadis K, Alsdorf D, Lettenmaier D, Moller D, Wilson M (2008) Estimation of bathymetric depth and slope from data assimilation of swath altimetry into a hydrodynamic model. Geophys Res Lett 35:L20401
Durand M, Kim E, Margulis SA (2009) Radiance assimilation shows promise for snowpack characterization. Geophys Res Lett 36:L02503.1–L02503.5
Ek MB, Mitchell KE, Lin Y, Rogers E, Grunmann P, Koren V, Gayno G, Tarpley JD (2003) Implementation of Noah land surface model advances in the National Centers for Environmental Prediction operational mesoscale Eta model. J Geophys Res 108:8851. doi:10.1029/2002JD003296.0
Entekhabi D, Nakamura H, Njoku EG (1994) Solving the inverse problem for soil moisture and temperature profiles by sequential assimilation of multifrequency remotely sensed observations. IEEE Trans Geosci Remote Sens 32:438–448
Entekhabi D, Njoku EG, O'Neill PE, Kellogg KH, Crow WT, Edelstein WN, Entin JK, Goodman SD, Jackson TJ, Johnson J, Kimball J, Piepmeier JR, Koster RD, Martin N, McDonald KC, Moghaddam M, Moran S, Reichle R, Shi JC, Spencer MW, Thurman SW, Tsang L, Van Zyl J (2010a) The Soil Moisture Active and Passive (SMAP) mission. Proc IEEE 98:704–716
Entekhabi D, Reichle RH, Koster RD, Crow WT (2010b) Performance metrics for soil moisture retrievals and application requirements. J Hydrometeorol 11:832–840
Essery R, Pomeroy J (2004) Implications of spatial distributions of snow mass and melt rate for snow-cover depletion: theoretical considerations. Ann Glaciol 38:261–265
Evensen G (2003) The ensemble Kalman filter: theoretical formulation and practical implementation. Ocean Dyn 53:343–367
Famiglietti J, Wood E (1994) Multiscale modeling of spatially variable water and energy balance processes. Water Resour Res 30:3061–3078
Flores A, Bras RL, Entekhabi D (2012) Hydrologic data assimilation with a hillslope-scale resolving model and L-band radar observations: synthetic experiments with the ensemble Kalman filter. Water Resour Res 48:W08509
Forman BA, Reichle RH, Rodell M (2012) Assimilation of terrestrial water storage from GRACE in a snow-dominated basin. Water Resour Res 48:W01507
Forman B, Reichle RH, Derksen C (2013) Estimating passive microwave brightness temperature over snowcovered land in North America using a land surface model and an artificial neural network. IEEE Trans Geosci Remote Sens (in press)
Francois C, Quesney A, Ottlé C (2003) Sequential assimilation of ERS-1 SAR data into a coupled land surface-hydrological model using an extended Kalman filter. J Hydrometeorol 4:473–487
Galantowicz JF, Entekhabi D, Njoku EG (1999) Tests of sequential data assimilation for retrieving profile soil moisture and temperature from observed L-band radiobrightness. IEEE Trans Geosci Remote Sens 37:1860–1870

GCOS-107 (2006) Systematic observation requirements for satellite-based products for climate. Supplemental details to the satellite-based component of the "implementation plan for the global observing system for climate in support of the UNFCCC", GCOS-107, WMO/TD No. 1338, Sept 2006

Georgakakos KP, Baumer OW (1996) Measurement and utilization of on-site soil moisture data. J Hydrol 184:131–152

Ghent D, Kaduk J, Remedios J, Ard J, Balzter H (2010) Assimilation of land surface temperature into the land surface model JULES with an ensemble Kalman filter. J Geophys Res 115:D19112

Giustarini L, Matgen P, Hostache R, Montanari M, Plaza D, Pauwels V, De Lannoy G, De Keyser R, Pfister L, Hoffmann L, Savenije H (2011) Assimilating SAR-derived water level data into a flood models: a case study. Hydrol Earth Syst Sci 15:2349–2365

Günter A (2008) Improvement of global hydrological models using GRACE data. Surv Geophys 2008:375–397

Guo JCY (2006) Urban hydrology and hydraulic design. Water Resources Publications, LLC, Colorado

Gustafsson D et al (2004) Modeling water and heat balance of the Boreal landscape—comparison of forest and arable land in Scandinavia. J Appl Meteorol 43:1750–1767

Gutmann E, Small E (2010) A method for the determination of the hydraulic properties of soil from MODIS surface temperature for use in land-surface models. Water Resour Res 46:W06520

Han E, Merwade V, Heathman G (2012a) Application of data assimilation with the root zone water quality model for soil moisture profile estimation in the upper Cedar Creek, Indiana. Hydrol Process 26:1707–1719

Han X, Li X, Hendricks H, Vereecken H, Montzka C (2012b) Spatial horizontal correlation characteristics in the land data assimilation of soil moisture. Hydrol Earth Syst Sci 26:1349–1363

He M, Hogue T, Margulis S, Franz K (2012) An integrated uncertainty and ensemble-based data assimilation approach for improved operational streamflow predictions. Hydrol Earth Syst Sci 16:815–831

Hoeben R, Troch PA (2000) Assimilation of active microwave observation data for soil moisture profile estimation. Water Resour Res 36:2805–2819

Houser PR (2003) Land data assimilation systems. In: Swinbank R, Shuytaev V, Lahoz WA (eds) Data assimilation for the earth system. NATO science series: IV: earth and environmental sciences, vol 26. Kluwer, Dordrecht, pp 345–360

Houser PR, Shuttleworth WJ, Famiglietti JS, Gupta HV, Syed KH, Goodrich DC (1998) Integration of soil moisture remote sensing and hydrologic modeling using data assimilation. Water Resour Res 34:3405–3420

Houser PR, De Lannoy GJM, Walker JP (2010) Land surface data assimilation. In: Lahoz WA, Khattatov B, Ménard R (eds) Data assimilation: making sense of observations. Springer, Berlin, pp 549–597

Hurkmans R, Paniconi C, Troch PA (2006) Numerical assessment of a dynamical relaxation data assimilation scheme for a catchment hydrological model. Hydrol Process 20:549–563

Ines A, Mohanty B (2009) Near-surface soil moisture assimilation for quantifying effective soil hydraulic properties using genetic algorithms: 2. Using airborne remote sensing during SGP97 and SMEX02. Water Resour Res 45:W01408

IPCC (2007) Climate change 2007: the physical science basis. Contribution of working group I to the fourth assessment report, Intergovernmental Panel on Climate Change, Cambridge University Press, Cambridge

Jackson TJ, Bindlish R, Cosh MH, Zhao T, Starks PJ, Bosch DD, Seyfried M, Moran MS, Goodrich DC, Kerr YH, Leroux D (2012) Validation of Soil Moisture and Ocean Salinity (SMOS) soil moisture over watershed networks in the U.S. IEEE Trans Geosci Remote Sens 50:1530–1543

Jansson P-E, Karlberg L (2004) CoupModel, coupled heat and mass transfer model for soil-plant-atmosphere system. Royal Institute of Technology (web-based on line documentation with details on use of model). http://www.lwr.kth.se/Vara%20Datorprogram/CoupModel/index.htm

Jansson C et al (2005) Preferential water flow in a glacial till soil. Nord Hydrol 36:1–11

Jansson P-E et al (2008) Simulated climate change impacts on fluxes of carbon in Norway spruce ecosystems along a climatic transect in Sweden. Biogeochemistry 89:81–94

Jarlan L, Balsamo G, Lafont S, Beljaars A, Calvet J-C, Mougin E (2008) Analysis of leaf area index in the ECMWF land surface model and impact on latent heat and carbon fluxes: application to West Africa. J Geophys Res 113:D24117

Jones A, Vukićević T, Vonder Haar T (2003) Variational data assimilation of soil moisture using 6 and 10 GHz passive microwave data. J Hydrometeorol 5:213–229

Kainulainen J, Colliander A, Closa J, Martin-Neira M, Oliva R, Buenadicha G, Rubiales Alcaine P, Hakkarainen A, Hallikainen MT (2012) Radiometric performance of the SMOS reference radiometers—assessment after one year of operation. IEEE Trans Geosci Remote Sens 50:1367–1383

Kalma J, McVicar T, McCabe M (2008) Estimating land surface evaporation: a review of methods using remotely sensed surface temperature data. Surv Geophys 29:421–469

Kalnay E (2003) Atmospheric modeling, data assimilation and predictability. Cambridge University Press, Cambridge

Kalnay E (2010) Ensemble Kalman filter: current status and potential. In: Lahoz WA, Khattatov B, Ménard R (eds) Data assimilation: making sense of observations. Springer, Berlin, pp 69–92

Karlberg L et al (2006) Modelling transpiration and growth in salinity-stressed tomato under different climatic conditions. Ecol Model 190:15–40

Karlberg L et al (2007) Modeling carbon turnover in five terrestrial ecosystems in the boreal zone using multiple criteria of acceptance. AMBIO J Hum Environ 35:448–458

Kerr YH, Waldteufel P, Wigneron J-P, Delwart S, Cabot F, Boutin J, Escorihuela M-J, Font J, Reul N, Gruhier C, Juglea SE, Drinkwater MR, Hahne A, Martín-Neira M, Mecklenburg S (2010) The SMOS mission: new tool for monitoring key elements of the global water cycle. Proc IEEE 98:666–687

Kerr YH, Font J, Martin-Neira M, Mecklenburg S (2012a) Introduction to the special issue on the ESA's Soil Moisture and Ocean Salinity Mission (SMOS)—instrument performance and first results. IEEE Trans Geosci Remote Sens 50:1351–1353

Kerr YH, Waldteufel P, Richaume P, Wigneron JP, Ferrazzoli P, Mahmoodi A, Al Bitar A, Cabot F, Gruhier C, Juglea SE, Leroux D, Mialon A, Delwart S (2012b) The SMOS soil moisture retrieval algorithm. IEEE Trans Geosci Remote Sens 50:1384–1403

Klemedtsson L et al (2008) Bayesian calibration method used to elucidate carbon turnover in forest on drained organic soil. Biogeochemistry 89:61–79

Kolberg S, Gottschalk L (2010) Interannual stability of grid cell snow depletion curves as estimated from MODIS images. Water Resour Res 46:W11555

Koster RD, Suarez MJ, Ducharne A, Stieglitz M, Kumar P (2000) A catchment-based approach to modeling land surface processes in a general circulation model 1. Model structure. J Geophys Res 105:24809–24822

Koster RD, Dirmeyer PA, Guo Z, Bonan G, Cox P, Gordon C, Kanae S, Kowalczyk E, Lawrence D, Liu P, Lu C, Malyshev S, McAvaney B, Mitchell K, Mocko D, Oki T, Oleson K, Pitman A, Sud Y, Taylor C, Verseghy D, Vasic R, Xue Y, Yamada T (2004a) Regions of strong coupling between soil moisture and precipitation. Science 305:1138–1140

Koster RD, Suarez M, Liu P, Jambor U, Berg A, Kistler M, Reichle R, Rodell M, Famiglietti J (2004b) Realistic initialization of land surface states: impacts on subseasonal forecast skill. J Hydrometeorol 5:1049–1063

Koster RD, Mahanama PP, Yamada TJ, Balsamo G, Berg AA, Boisserie M, Dirmeyer PA, Doblas-Reyes FJ, Drewitt G, Gordon CT, Guo Z, Jeong JH, Lee WS, Li Z, Luo L, Malyshev S, Merryfield WJ, Seneviratne SI, Stanelle T, van den Hurk BJJM, Vitart F, Wood EF (2011) The second phase of the global land-atmosphere coupling experiment: soil moisture contributions to subseasonal forecast skill. J Hydrometeorol 12:805–822

Kotecha JH, Djurić PM (2003) Gaussian particle filtering. IEEE Trans Signal Process 51:2592–2601

Kumar SV, Reichle RH, Peters-Lidard CD, Koster RD, Zhan X, Crow WT, Eylander JB, Houser PR (2008) A land surface data assimilation framework using the Land Information System: description and applications. Adv Water Resour 31:1419–1432

Kumar S, Reichle RH, Koster RD, Crow WT, Peters-Lidard CD (2009) Role of subsurface physics in the assimilation of surface soil moisture observations. J Hydrometeorol 10:1534–1547. doi: 10.1109/MC.2008.511

Kumar S, Reichle R, Harrison K, Peters-Lidard C, Yatheendradas S, Santanello J (2012) A comparison of methods for a priori bias correction in soil moisture data assimilation. Water Resour Res 48:W03515

Lacava T, Matgen P, Brocca L, Bittelli M, Pergola N, Moramarco T, Tramutoli V (2012) A first assessment of the SMOS soil moisture product with in situ and modeled data in Italy and Luxembourg. IEEE Trans Geosci Remote Sens 50:1612–1622

Lahoz WA, Khattatov B, Ménard R (eds) (2010a) Data assimilation: making sense of observations. Springer, Berlin

Lahoz WA, Walker S-E, Dammann D (2010b) The NILU SURFEX-EnKF land data assimilation system. NILU technical report TR 2/2010, Jan 2010. Available from http://www.nilu.no

Lakshmi V (2000) A simple surface temperature assimilation scheme for use in land surface models. Water Resour Res 36:3687–3700

Le Moigne P (2009) SURFEX scientific documentation. Note de Centre No. 87, Groupe de Meteorologie a Moyenne Exhell, Centre National de Recherches Météorologiques, Météo-France (SURFEX, V. 5, Issue 1). Available from http://hirlam.org/index.php?option=com_docman&task=doc_downloads&gid=605&Itemid=70

Le Vine DM, Lagerloef GSE, Yueh S, Pellerano F, Dinnat E, Wentz F (2006) Aquarius mission technical overview. IGARSS 2006, pp 1678–1680

Lee H, Seo D, Koren V (2011) Assimilation of streamflow and in situ soil moisture data into operational distributed hydrologic models: effects of uncertainties in the data and initial model soil moisture states. Adv Water Resour 34:1597–1615
Lewis P, Gómez-Dans J, Kaminski T, Settle J, Quaife T, Gobron N, Styles J, Berger M (2012) An earth observation land data assimilation system (EO-LDAS). Remote Sens Environ 120:219–235
Li J, Islam S (1999) On the estimation of soil moisture profile and surface fluxes partitioning from sequential assimilation of surface layer soil moisture. J Hydrol 220:86–103
Li B, Rodell M, Zaitchik BF, Reichle RH, Koster RD, van Dam TM (2012) Assimilation of GRACE terrestrial water storage into a land surface model: evaluation and potential value for drought monitoring in western and central Europe. J Hydrol 446–447:103–115
Liang X, Lettenmaier DP, Wood EF, Burges SJ (1994) A simple hydrologically based model of land surface water and energy fluxes for GSMs. J Geophys Res 99:14415–14428
Liston G, Hiemstra CA (2008) A simple data assimilation system for complex snow distributions (SnowAssim). J Hydrometeorol 9:989–1004
Liu Q, Reichle RH, Bindlish R, Cosh MH, Crow WT, de Jeu R, De Lannoy GJM, Huffman GJ, Jackson TJ (2011) The contributions of precipitation and soil moisture observations to the skill of soil moisture estimates in a land data assimilation system. J Hydrometeorol 12:750–765
Lo M, Famiglietti JS, Yeh PJ, Syed TH (2010) Improving parameter estimation and water table depth simulation in a land surface model using GRACE water storage and estimated base flow data. Water Resour Res 46:W05517
Loew A, Mauser W (2008) Inverse modeling of soil characteristics from surface soil moisture observations: potential and limitations. Hydrol Earth Syst Sci 5:95–145
Loew A, Schwank M, Schlenz F (2009) Assimilation of an L-band microwave soil moisture proxy to compensate for uncertainties in precipitation data. TGRS 47:2606–2616
Loth B, Graf H-F, Oberhuber JM (1993) Snow cover model for global climate simulations. J Geophys Res 98:10451–10464
Lynch-Stieglitz M (1994) The development and validation of a simple snow model for the GISS GCM. J Clim 7:1842–1855
Mackaro S, McNider R, Pour Biazar A (2011) Some physical and computational issues in land surface data assimilation of satellite skin temperatures. Pure Appl Geophys 167:1303–1458
McLaughlin D (2002) An integrated approach to hydrologic data assimilation: interpolation, smoothing, and filtering. Adv Water Resour 25:1275–1286
Madsen H (2003) Parameter estimation in distributed hydrological catchment modelling using automatic calibration with multiple objectives. Adv Water Resour 26:205–216
Maggioni V, Reichle R, Emmanouil A (2011) The effect of satellite rainfall error modeling on soil moisture prediction uncertainty. J Hydrometeorol 12:413–428
Mahfouf J-F (2010) Assimilation of satellite-derived soil moisture from ASCAT in a limited-area NWP model. Q J R Meteorol Soc 136:784–798
Mahfouf J-F, Bliznak V (2011) Combined assimilation of screen-level observations and radar-derived precipitation for soil moisture analysis. Q J R Meteorol Soc 137:709–722
Mahfouf J-F, Noilhan J (1996) Inclusion of gravitational drainage in a land surface scheme based on the force restore method. J Appl Meteorol 35:987–992
Mahfouf J-F, Viterbo P, Douville H, Beljaars A, Saarinen S (2000) A revised land surface analysis scheme in the integrated forecasting system. ECMWF Newsl 88:8–13
Mahfouf J-F, Bergaoui K, Draper C, Bouyssel C, Taillefer F, Taseva L (2009) A comparison of two off-line soil analysis schemes for assimilation of screen-level observations. J Geophys Res 114:D08105
Margulis SA, McLaughlin D, Entekhabi D, Dunne S (2002) Land data assimilation of soil moisture using measurements from the Southern Great Plains 1997 field experiment. Water Resour Res 38:35.1–35.18
Marzahn P, Ludwig R (2009) On the derivation of soil surface roughness from multi parametric PolSAR data and its potential for hydrological modeling. Hydrol Earth Syst Sci 13:381–394
Matgen P, Montanari M, Hostache R, Pfister L, Hoffmann L, Plaza D, Pauwels V, De Lannoy G, De Keyser R, Savenije H (2010) Towards the sequential assimilation of SAR-derived water stages into hydraulic models using the particle filter: proof of concept. Hydrol Earth Syst Sci 14:1773–1785
Mattar C, Wigneron J-P, Sobrino JA, Novello N, Calvet JC, Albergel C, Richaume P, Mialon A, Guyon D, Jimenez-Munoz JC, Kerr Y (2012) A combined optical-microwave method to retrieve soil moisture over vegetated areas. IEEE Trans Geosci Remote Sens 50:1404–1413
Mecklenburg S, Drusch M, Kerr YH, Font J, Martin-Neira M, Delwart S, Buenadicha G, Reul N, Daganzo-Eusebio E, Oliva R, Crapolicchio R (2012) ESA's Soil Moisture and Ocean Salinity Mission: mission performance and operations. IEEE Trans Geosci Remote Sens 50:1354–1366

Ménard R (2010) Bias estimation. In: Lahoz WA, Khattatov B, Ménard R (eds) Data assimilation: making sense of observations. Springer, Berlin, pp 113–135

Meng CL, Li Z-L, Zhan X, Shi JC, Liu CY (2009) Land surface temperature data assimilation and its impact on evapotranspiration estimates from the common land model. Water Resour Res 45:W02421

Merlin O, Rüdiger C, Al Bitar A, Richaume P, Walker JP, Kerr YH (2012) Disaggregation of SMOS soil moisture in Southeastern Australia. IEEE Trans Geosci Remote Sens 50:1556–1571

Mialon A, Wigneron J-P, de Rosnay P, Escorihuela MJ, Kerr YH (2012) Evaluating the L-MEB model from long-term microwave measurements over a rough field, SMOSREX 2006. IEEE Trans Geosci Remote Sens 50:1458–1467

Misra S, Ruf CS (2012) Analysis of radio frequency interference detection algorithms in the angular domain for SMOS. IEEE Trans Geosci Remote Sens 50:1448–1457

Monsivais-Huerteroet A, Graham W, Judge J, Agrawal D (2010) Effect of simultaneous state-parameter estimation and forcing uncertainties on root-zone soil moisture for dynamic vegetation using EnKf. Adv Water Resour 33:468–484

Montanari A, Toth E (2007) Calibration of hydrological models in the spectral domain: an opportunity for scarcely gauged basins? Water Resour Res 43:W05434

Montanari M, Hostache R, Matgen P, Schumann G, Pfister L, Hoffmann L (2009) Calibration and sequential updating of a coupled hydrologic-hydraulic model using remote sensing-derived water stages. Hydrol Earth Syst Sci 13:367–380

Montzka C, Grant J, Moradkhani H, Hendricks-Franssen H-J, Weihermuller L, Drusch M, Vereecken H (2012) Estimation of radiative transfer parameters from SMOS brightness temperatures using data assimilation: implication on soil moisture retrieval. Vadose Zone J (submitted)

Moradkhani H (2008) Hydrologic remote sensing and land surface data assimilation. Sensors 8:2986–3004

Moradkhani H, Hsu K-L, Gupta H, Sorooshian S (2005a) Uncertainty assessment of hydrologic model states and parameters: sequential data assimilation using the particle filter. Water Resour Res 41:W05012. doi:10.1029/2004WR003604

Moradkhani H, Sorooshian S, Gupta HV, Houser PR (2005b) Dual state-parameter estimation of hydrological models using ensemble Kalman filter. Adv Water Resour 28:135–147

Morisette JT, Privette JL, Justice CO (2002) A framework for the validation of MODIS land products. Remote Sens Environ 83:77–96

Nagarajanar K, Judgea J, Graham WD, Monsivais-Huerteroet A (2011) Particle filter-based assimilation algorithms for improved estimation of root-zone soil moisture under dynamic vegetation conditions. Adv Water Resour 34:433–447

Nearing GS, Moran MS, Thorp KR, Collins CDH, Slack DC (2010) Likelihood parameter estimation for calibrating a soil moisture model using radar backscatter. Remote Sens Environ 115:2564–2574

Nearing GS, Crow WT, Thorp KR, Moran MS, Reichle RH, Gupta HV (2012) Assimilating remote sensing observations of leaf area index and soil moisture for wheat yield estimates: an observing system simulation experiment. Water Resour Res 48:W05525

Nichols NK (2010) Mathematical concepts of data assimilation. In: Lahoz WA, Khattatov B, Ménard R (eds) Data assimilation: making sense of observations. Springer, Berlin, pp 13–39

Ni-Meister W (2008) Recent advances on soil moisture data assimilation. Phys Geogr 29:19–37

Ni-Meister W, Houser PR, Walker J (2006) Soil moisture initialization for climate prediction: assimilation of scanning multifrequency microwave radiometer soil moisture data into a land surface model. J Geophys Res 111:1–15. doi:10.1029/2006JD007190

Njoku EG, Chan TK (2006) Vegetation and surface roughness effects on AMSR-E land observations. Remote Sens Environ 100:190–199

Noilhan J, Mahfouf J-F (1996) The ISBA land surface parameterisation scheme. Global Planet Chang 13:145–159

Norman J et al (2008) Simulation of NO and N_2O emissions from a spruce forest during a freeze/thaw event using an N-flux sub-model from the PnET-N-DNDC model integrated to CoupModel. Ecol Model 216:18–30

Oki T, Kanae S (2006) Global hydrological cycles and world water resources. Science 313:1068–1072

Oleson KW et al (2010) Technical description of version 4.0 of the Community Land Model (CLM), NCAR technical note NCAR/TN-478+STR, 257 pp

Oliva R, Daganzo E, Kerr YH, Mecklenburg S, Nieto S, Richaume P, Gruhier C (2012) SMOS radio frequency interference scenario: Status and actions taken to improve the RFI environment in the 1400–1427-MHz passive band. IEEE Trans Geosci Remote Sens 50:1427–1439

Palmer TN, Doblas-Reyes FJ, Weisheimer A, Rodwell MJ (2008) Toward seamless prediction: calibration of climate change projections using seasonal forecasts. Bull Am Meteorol Soc 89:459–470

Pan M, Wood EF (2006) Data assimilation for estimating the terrestrial water budget using a constrained ensemble Kalman filter. J Hydrometeorol 7:534–547

Pan M, Wood EF (2010) Impact of accuracy, spatial availability, and revisit time of satellite-derived surface soil moisture in a multiscale ensemble data assimilation system. J Sel Top Appl Earth Obs Remote Sens 3:49–56

Pan M, Wood EF, Wójcik R, McCabe MF (2008) Estimation of regional terrestrial water cycle using multi-sensor remote sensing observations and data assimilation. Remote Sens Environ 112:1282–1294

Pan M, Wood EF, McLaughlin DB, Entekhabi D, Luo L (2009) A multiscale ensemble filtering system for hydrologic data assimilation: part I, implementation and synthetic experiment. J Hydrometeorol 10:807–819

Paniconi C, Marrocu M, Putti M, Verbunt M (2003) Newtonian nudging for a Richards equation-based distributed hydrological model. Adv Water Resour 26:161–178

Parada LM, Liang X (2008) Impacts of spatial resolutions and data quality on soil moisture data assimilation. J Geophys Res 113:D10101.1–D10101.17

Parajka J, Naeimi V, Blöschl G, Wagner W, Merz R, Scipal K (2006) Assimilating scatterometer soil moisture data into conceptual hydrologic models at the regional scale. Hydrol Earth Syst Sci 10:353–368

Parinussa R, Holmes TRH, Yilmaz MT, Crow WT (2011) The impact of land surface temperature on soil moisture anomaly detection from passive microwave observations. Hydrol Earth Syst Sci 15:3135–3151

Pathmathevan M, Koike T, Lin X, Fujii H (2003) A simplified land data assimilation scheme and its application to soil moisture experiments in 2002 (SMEX02). Water Resour Res 39:SWC6.1–SWC6.20

Pauwels VRN, De Lannoy GJM (2006) Improvement of modeled soil wetness conditions and turbulent fluxes through the assimilation of observed discharge. J Hydrometeorol 7:458–477

Pauwels VRN, De Lannoy GJM (2009) Ensemble-based assimilation of discharge into rainfall-runoff models: a comparison of approaches to mapping observational information to state space. Water Resour Res 45:W08428.1–W08428.17

Pauwels VRN, Hoeben R, Verhoest NEC, De Troch FP (2001) The importance of the spatial patterns of remotely sensed soil moisture in the improvement of discharge predictions for small-scale basins through data assimilation. J Hydrol 251:88–102

Pauwels VRN, Hoeben R, Verhoest NEC, De Troch FP, Troch PA (2002) Improvement of TOPLATS-based discharge predictions through assimilation of ERS-based remotely sensed soil moisture values. Hydrol Process 16:995–1013

Pauwels VRN, Verhoest NEC, De Lannoy GJM, Defourny P, Guissard V, Lucau C (2006) Optimization of a coupled hydrology/crop growth model through the assimilation of observed soil moisture and LAI values using an Ensemble Kalman filter. Water Resour Res 43:W04421.1–W04421.17

Pauwels V, Balenzano A, Satalino G, Skriver H, Verhoest N, Mattia F (2009) Optimization of soil hydraulic model parameters using synthetic aperture radar data: An integrated multidisciplinary approach. IEEE Trans Geosci Remote Sens 47:455–457

Peischl S, Walker JP, Rüdiger C, Ye N, Kerr YH, Kim E, Bandara R, Allahmoradi M (2012a) The AACES field experiments: SMOS calibration and validation across the Murrumbidgee River catchment. Hydrol Earth Syst Sci 16:1697–1708

Peischl S, Walker JP, Ryu D, Kerr YH, Panciera R, Rüdiger C (2012b) Wheat canopy structure and surface roughness effects on multiangle observations at L-band. IEEE Trans Geosci Remote Sens 50:1498–1506

Pipunic RC, Walker J, Western A (2008) Assimilation of remotely sensed data for improved latent and sensible heat flux prediction: a comparative synthetic study. Remote Sens Environ 112:1295–1305

Plaza DA, De Keyser R, De Lannoy GJM, Giustarini L, Matgen P, Pauwels VRN (2012) The importance of parameter resampling for soil moisture data assimilation into hydrologic models using the particle filter. Hydrol Earth Syst Sci 16:375–390

Quets J, De Lannoy G, Pauwels V (2010) Analysis of spectral calibration methods for precipitation-discharge processes. Hydrol Process 24:1048–1062

Rautiainen K, Lemmetyinen J, Pulliainen J, Vehvilainen J, Drusch M, Kontu A, Kainulainen J, Seppanen J (2012) L-band radiometer observations of soil processes in boreal and subarctic environments. IEEE Trans Geosci Remote Sens 50:1483–1497

Reichle RH (2008) Data assimilation methods in the earth sciences. Adv Water Resour 31:1411–1418

Reichle RH, Koster R (2003) Assessing the impact of horizontal error correlations in background fields on soil moisture estimation. J Hydrometeorol 4:1229–1242

Reichle RH, Koster R (2004) Bias reduction in short records of satellite soil moisture. Geophys Res Lett 31:L19501.1–L19501.4

Reichle RH, Koster R (2005) Global assimilation of satellite surface soil moisture retrievals into the NASA Catchment land surface model. Geophys Res Lett 32:L0204.1–L0204.4

Reichle RH, Entekhabi D, McLaughlin DB (2001a) Downscaling of radio brightness measurements for soil moisture estimation: a four dimensional variational data assimilation approach. Water Resour Res 37:2353–2364

Reichle RH, McLaughlin DB, Entekhabi D (2001b) Variational data assimilation of microwave radiobrightness observations for land surface hydrology applications. IEEE Trans Geosci Remote Sens 39:1708–1718

Reichle RH, McLaughlin DB, Entekhabi D (2002a) Hydrologic data assimilation with the ensemble Kalman filter. Mon Weather Rev 120:103–114

Reichle RH, Walker JP, Houser PR, Koster RD (2002b) Extended versus ensemble Kalman filtering for land data assimilation. J Hydrometeorol 3:728–740

Reichle RH, Koster R, Liu P, Mahanama SPP, Njoku EG, Owe M (2007) Comparison and assimilation of global soil moisture retrievals from the Advanced Microwave Scanning Radiometer for the Earth Observing System (AMSR-E) and the Scanning Multichannel Microwave Radiometer (SMMR). J Geophys Res 112:D09108.1–D09108.14

Reichle RH, Crow WT, Keppenne CL (2008) An adaptive ensemble Kalman filter for soil moisture data assimilation. Water Resour Res 44:W03423. doi:10.1029/2007WR006357

Reichle R, Bosilovich MG, Crow WT, Koster RD, Kumar SV, Mahanama SPP, Zaitchik BF (2009) Recent advances in land data assimilation at the NASA Global Modeling and Assimilation Office. In: Park SK, Xu L (eds) Data assimilation for atmospheric, oceanic and hydrologic applications. Springer, New York, pp 407–428

Reichle R, Kumar SV, Mahanama SPP, Koster RD, Liu Q (2010) Assimilation of satellite derived skin temperature observations into land surface models. J Hydrometeorol 11. doi:10.1175/2010JHM1262.1

Reichle RH, Koster RD, De Lannoy GJM, Forman BA, Liu Q, Mahanama SPP, Toure A (2011) Assessment and enhancement of MERRA land surface hydrology estimates. J Clim 24:6322–6338

Reichle R, Crow W, Koster R, Kimball J, De Lannoy G (2012) SMAP Algorithm Theoretical Basis Document: L4 surface and root zone soil moisture product. Tech. Rep. SMAP Project, JPL D-66483, Jet Propulsion Laboratory, Pasadena, CA.

Reichle R, De Lannoy GJM, Forman B, Draper C, Liu Q (2013) Connecting satellite observations with water cycle variables through land data assimilation: examples using the NASA GEOS-5 LDAS. Surv Geophys (in press)

Renzullo LJ, Barrett DJ, Marks AS, Hill MJ, Guerschman JP, Mu Q, Running SW (2008) Multi-sensor model-data fusion for estimation of hydrologic and energy flux parameters. Remote Sens Environ 112:1308–1319

Rodell M, Houser P (2004) Updating a land surface model with MODIS-derived snow cover. J Hydrometeorol 5:1064–1075

Rodgers CD (2000) Inverse methods for atmospheric sounding: theory and practice. World Scientific Publishing Co. Ltd, London

Rowlandson TL, Hornbuckle BK, Bramer LM, Patton JC, Logsdon SD (2012) Comparisons of evening and morning SMOS passes over the Midwest United States. IEEE Trans Geosci Remote Sens 50:1544–1555

Roy A, Royer A, Turcotte R (2010) Improvement of springtime streamflow simulations in a boreal environment by incorporating snow-covered area derived from remote sensing data. J Hydrol 390:35–44

Rüdiger C, Albergel C, Mahfouf J-F (2010) Evaluation of the observation operator Jacobian for leaf area index data assimilation with an extended Kalman filter. J Geophys Res 118:D09111

Sabater JM, Jarlan L, Calvet J-C, Bouyssel F (2007) From near-surface to root-zone soil moisture using different assimilation techniques. J Hydrometeorol 8:194–206

Saha S, Moorthi S, Pan H-L, Wu X et al (2010) The NCEP climate forecast system re-analysis. Bull Am Metereol Soc 91:1015–1057

Sahoo A, De Lannoy GJM, Reichle RH, Houser PR (2013) Assimilation and downscaling of satellite observed soil moisture over the Little River experimental watershed in Georgia, USA. Adv Water Resour 52:19–33. doi:10.1016/j.advwatres.2012.08.007

Sanchez N, Martinez-Fernandez J, Scaini A, Perez-Gutierrez C (2012) Validation of the SMOS L2 soil moisture data in the REMEDHUS network (Spain). IEEE Trans Geosci Remote Sens 50:1602–1611

Santanello J, Peters-Lidard C, Garcia ME, Mocko DM, Tischler MA, Moran MS, Thoma DP (2007) Using remotely-sensed estimates of soil moisture to infer soil texture and hydraulic properties across a semi-arid watershed. Remote Sens Environ 110:79–97

Schlenz F, dall'Amico JT, Loew A, Mauser W (2012) Uncertainty assessment of the SMOS validation in the Upper Danube catchment. IEEE Trans Geosci Remote Sens 50:1517–1529

Schuurmans J, Troch P, Veldhuizen A, Bastiaansen W, Bierkens M (2003) Assimilation of remotely sensed latent heat flux in a distributed hydrological model. Adv Water Resour 26:151–159
Schwank M, Wigneron J-P, Lopez-Baeza E, Volksch I, Matzler C, Kerr YH (2012) L-band radiative properties of vine vegetation at the MELBEX III SMOS cal/val site. IEEE Trans Geosci Remote Sens 50:1587–1601
Scipal K, Holmes T, de Jeu R, Naeimi V, Wagner W (2008) A possible solution for the problem of estimating the error structure of global soil moisture data sets. Geophys Res Lett 35:L24403. doi: 10.1029/2008GL035599
Seneviratne SI, Corti T, Davin EL, Hirschi M, Jaeger EB, Lehner I, Orlowsky B et al (2010) Investigating soil moisture-climate interactions in a changing climate: a review. Earth Sci Rev 99:125–161
Seo D-J, Cajina L, Corby R, Howieson T (2009) Automatic state updating for operational streamflow forecasting via variational data assimilation. J Hydrol 367:255–275
Seuffert G, Wilker H, Viterbo P, Drusch M, Mahfouf J-F (2004) The usage of screen-level parameters and microwave brightness temperature for soil moisture analysis. J Hydrometeorol 5:516–531
Sherwood SC (1999) Convective precursors and predictability in the Tropical Western Pacific. Mon Weather Rev 127:2977–2991
Silvapalan M, Beven KJ, Wood EF (1987) On hydrologic similarity: 2. A scaled model of storm runoff production. Water Resour Res 23:2266–2278
Simmons AJ, Hollingsworth A (2002) Some aspects of the improvement in skill of numerical weather prediction. Q J R Meteorol Soc 128:647–677
Sini F, Boni G, Caparrini F, Entekhabi D (2008) Estimation of large-scale evaporation fields based on assimilation of remotely sensed land temperature. Water Resour Res 44:W06410
Slater AG, Clark M (2006) Snow data assimilation via an ensemble Kalman filter. J Hydrometeorol 7:478–493
Spahni R, Wania R, Neef L et al (2010) Constraining global methane emissions and uptake by ecosystems. Biogeosci Discuss 8:1643–1665
Stieglitz M, Rind D, Famiglietti J, Rosenzweig C (1997) An efficient approach to modeling the topographic control of surface hydrology for regional and global climate modeling. J Clim 10:118–137
Su H, Yang Z, Niu G, Dickinson RE (2008) Enhancing the estimation of continental-scale snow water equivalent by assimilating MODIS snow cover with the ensemble Kalman filter. J Geophys Res 113:D08120
Su H, Yang Z, Dickinson RE, Wilson CR, Niu G-Y (2010) Multisensor snow data assimilation at continental scale: the value of Gravity Recovery and Climate Experiment terrestial water storage information. J Geophys Res 115:D10104.1–D10104.14
Su H, Yang Z-L, Niu G-Y (2011) Parameter estimation in ensemble based snow data assimilation: A synthetic study. Adv Water Resour 34:407–416
Sun S, Jin J, Xue Y (1999) A simple snow-atmosphere-soil transfer (SAST) model. J Geophys Res 104:19587–19597
Sun C, Walker JP, Houser PR (2004) A methodology for snow data assimilation in a land surface model. J Geophys Res 109:D08108.1–D08108.12
Svensson M et al (2008) Bayesian calibration of a model describing carbon, water and heat fluxes for a Swedish boreal forest stand. Ecol Model 213:331–344
Taillefer F (2002) CANARI (code for the analysis necessary for arpege for its rejects and its initialization): Technical documentation, internal CNRM/GMAP report. Available from http://www.cnrm.meteo.fr/gmapdoc/spip.php?article3
Talagrand O (2010a) Variational assimilation. In: Lahoz WA, Khattatov B, Ménard R (eds) Data assimilation: making sense of observations. Springer, Berlin, pp 41–67
Talagrand O (2010b) Evaluation of assimilation algorithms. In: Lahoz WA, Khattatov B, Ménard R (eds) Data assimilation: making sense of observations. Springer, Berlin, pp 217–240
Tapley BD, Bettadpur S, Ries JC, Thompson PF, Watkins MM (2004) GRACE measurements of mass variability in the earth system. Science 305:503–505
Tedesco M, Reichle RH, Loew A, Markus T, Foster JL (2010) Dynamic approaches for snow depth retrieval from spaceborne microwave brightness temperature. IEEE Trans Geosci Remote Sens 48:1955–1967
Trenberth KE, Fasullo J, Kiehl J (2009) Earth's global energy budget. Bull Am Meteorol Soc 90:311–323
USGEO (2010) Achieving and sustaining earth observations: a preliminary plan based on a strategic assessment by the U.S. Group on Earth Observations. Office of Science and Technology Policy, 69 pp. Available online at www.whitehouse.gov/sites/default/files/microsites/ostp/ostp-usgeo-reportearth-obs.pdf
Vachon F, Goïta K, Séve DD, Royer A (2010) Inversion of a snow emission model calibrated with in situ data dor snow water equivalent monitoring. IEEE Trans Geosci Remote Sens 48:59–71

Van den Hurk BJJ, Jia L, Jacobs C, Menenti M, Li Z-L (2002) Assimilation of land surface temperature data from ATSR in an NWP environment—a case study. Int J Remote Sens 23:5193–5209
Van den Hurk B, Ettema J, Viterbo P (2008) Analysis of soil moisture changes in europe during a single growing season in a new ECMWF soil moisture assimilation system. J Hydrometeorol 9:116–131
van Leeuwen PJ (2009) Particle filtering in geophysical systems. Mon Weather Rev 137:4089–4114
Vereecken H, Huisman JA, Bogena H, Vanderborght J, Vrugt JA, Hopmans JW (2008) On the value of soil moisture measurements in vadose zone hydrology: a review. Water Resour Res 44:W00D06
Vrugt JA, Gupta HV, Nualláin BÓ, Bouten W (2006) Real-time data assimilation for operational ensemble streamflow forecasting. J Hydrometeorol 7:548–565
Vrugt J, ter Braak C, Clark M, Hyman J, Robinson B (2008) Treatment of input uncertainty in hydrologic modeling: doing hydrology backward with Markov chain Monte Carlo simulation. Water Resour Res 44:W00B09
Vrugt J, ter Braak C, Diks C, Schoups G (2012) Advancing hydrologic data assimilation using particle markov chain monte carlo simulation: theory, concepts and applications. Adv Water Resour (anniversary issue—35 years). doi:10.1016/j.advwatres.2012.04.002
Walker JP, Houser PR (2004) Requirements of a global near-surface soil moisture satellite mission: accuracy, repeat time, and spatial resolution. Adv Water Resour 27:785–801
Walker JP, Willgoose GR, Kalma JD (2001a) One-dimensional soil moisture profile retrieval by assimilation of near-surface observations: a comparison of retrieval algorithms. Adv Water Resour 24:631–650
Walker JP, Willgoose GR, Kalma JD (2001b) One-dimensional soil moisture profile retrieval by assimilation of near-surface measurements: a simplified soil moisture model and field application. J Hydrometeorol 2:356–373
Walker JP, Willgoose GR, Kalma JD (2002) Three-dimensional soil moisture profile retrieval by assimilation of near-surface measurements: simplified Kalman filter covariance forecasting and field application. Water Resour Res 38:37.1–37.13
Wania R (2007) Modelling northern peatland land surface processes, vegetation dynamics and methane emissions. PhD thesis, University of Bristol, Bristol, 122 pp
Weerts AH, El Serafy GYH (2006) Particle filtering and ensemble Kalman filtering for state updating with hydrological conceptual rainfall-runoff models. Water Resour Res 42:W09403. doi:10.1029/2005 WR004093
Weisheimer A, Doblas-Reyes P, Jung T, Palmer T (2011) On the predictability of the extreme summer 2003 over Europe. Geophys Res Lett 38. doi:10.1029/2010GL046455
Wilker H, Drusch M, Seuffert G, Simmer C (2006) Effects of the near-surface soil moisture profile on the assimilation of L-band microwave brightness temperature. J Hydrometeorol 7:433–442
Wingeron J-P, Olioso A, Calvet J-C, Bertuzzi P (1999) Estimating root zone soil moisture from surface soil moisture data and soil–vegetation–atmosphere transfer modeling. Water Resour Res 35:3735–3745
Xu T, Liang S, Liu S (2011) Estimating turbulent fluxes through assimilation of geostationary operational environmental satellites data using ensemble kalman filter. J Geophys Res 116:621–639
Yilmaz MT, Delsole T, Houser P (2011) Improving land data assimilation performance with a water budget constraint. J Hydrometeorol 12:1040–1055
Zaitchik BF, Rodell M (2009) Forward-looking assimilation of MODIS-derived snow-covered area into a land surface model. J Hydrometeorol 10:130–148
Zaitchik BF, Rodell M, Reichle RH (2008) Assimilation of GRACE terrestrial water storage data into a land surface model: results for the Mississippi river basin. J Hydrometeorol 9:535–548
Zhan X, Houser PR, Walker JP, Crow WT (2006) A method for retrieving high-resolution surface soil moisture from Hydros L-band radiometer and radar observations. IEEE Trans Geosci Remote Sens 44:1534–1544
Zhang S, Shi J, Dou Y (2011) A soil moisture assimilation scheme based on the microwave land emissivity model and the community land model. Int J Remote Sens 33:2770–2797
Zupanski D (1997) A general weak constraint applicable to operational 4DVAR data assimilation systems. Mon Weather Rev 125:2274–2292

DOI 10.1007/978-94-017-8789-5_9
Reprinted from *Surveys in Geophysics* Journal, DOI 10.1007/s10712-013-9270-y

Toward Improved Estimation of the Dynamic Topography and Ocean Circulation in the High Latitude and Arctic Ocean: The Importance of GOCE

J. A. Johannessen · R. P. Raj · J. E. Ø. Nilsen · T. Pripp · P. Knudsen · F. Counillon · D. Stammer · L. Bertino · O. B. Andersen · N. Serra · N. Koldunov

Received: 22 March 2013 / Accepted: 26 November 2013 / Published online: 23 January 2014

Abstract The Arctic plays a fundamental role in the climate system and shows significant sensitivity to anthropogenic climate forcing and the ongoing climate change. Accelerated changes in the Arctic are already observed, including elevated air and ocean temperatures, declines of the summer sea ice extent and sea ice thickness influencing the albedo and CO_2 exchange, melting of the Greenland Ice Sheet and increased thawing of surrounding permafrost regions. In turn, the hydrological cycle in the high latitude and Arctic is expected to undergo changes although to date it is challenging to accurately quantify this. Moreover, changes in the temperature and salinity of surface waters in the Arctic Ocean and Nordic Seas may also influence the flow of dense water through the Denmark Strait, which are found to be a precursor for changes in the Atlantic meridional overturning circulation with a lead time of around 10 years (Hawkins and Sutton in Geophys Res Lett 35:L11603, 2008). Evidently changes in the Arctic and surrounding seas have far reaching influences on regional and global environment and climate variability, thus emphasizing the need for advanced quantitative understanding of the ocean circulation and transport variability in the high latitude and Arctic Ocean. In this respect, this study combines in situ hydrographical data, surface drifter data and direct current meter measurements, with coupled sea ice–ocean models, radar altimeter data and the latest GOCE-based geoid in order to estimate and assess the quality, usefulness and validity of the new GOCE-derived mean dynamic topography for studies of the ocean circulation and transport estimates in the Nordic Seas and Arctic Ocean.

J. A. Johannessen (✉) · R. P. Raj · J. E. Ø. Nilsen · T. Pripp · F. Counillon · L. Bertino
Nansen Environmental and Remote Sensing Center, Bergen, Norway
e-mail: Johnny.Johannessen@nersc.no

P. Knudsen · O. B. Andersen
National Space Institute, Technical University of Denmark, Lyngby, Denmark

D. Stammer · N. Serra · N. Koldunov
Center für Erdsystemforschung und Nachhaltigkeit (CEN), University of Hamburg, Hamburg, Germany

Keywords Oceanography in the Arctic Ocean and surrounding seas · Mean dynamic topography · Surface current · GOCE satellite

1 Introduction

Changes in the dynamic topography and ocean circulation between the northern Atlantic Ocean and the Arctic Ocean result from variations in the atmospheric forcing field and convective overturning combined with changes in freshwater runoff and their pathways, mean sea level, sea ice deformation and water mass transformation. The ocean circulation in this region has been subject to investigations since Helland-Hansen and Nansen (1909). In general, it can be characterized by four regional circulation regimes and cross-regional exchanges and volume transports, namely the Northeast Atlantic, the Labrador Sea and Canadian archipelago, the Nordic and Barents Seas and the Arctic Ocean, as illustrated in Fig. 1.

Accurate knowledge of the ocean transport variability together with understanding of the water mass transformations within and across these regions is highly needed to quantify changes in the overturning circulation with acceptable uncertainty. The Atlantic meridional overturning circulation is, among other factors, influenced by: variations in the upper ocean and sea ice interaction; ice sheet mass changes and their effect on the regional sea-level change; changes in freshwater fluxes and pathways; and variability in the large-scale atmospheric pressure field. For instance, changes in the pathways of the freshwater from the Eurasian runoff forced by shifts in the Arctic Oscillation can lead to increased trapping of freshwater in the Arctic Ocean as presented by Morison et al. (2012) that, in turn, may alter the thermohaline circulation in the sub-Arctic Seas.

Using a new combination of the ice cloud and land elevation satellite (ICESat) laser altimeter and the gravity recovery and climate experiment (GRACE) satellites, along with traditional hydrography, Morison et al. (2012) were able to show that the dominant freshwater changes from 2005 to 2008 were an increase in surface freshwater in the Canada basin balanced by a decrease in the Eurasian basin. These changes were due to a cyclonic (anticlockwise) shift in the ocean pathway of the Eurasian runoff forced by strengthening of the west-to-east Northern Hemisphere atmospheric circulation corresponding to a strengthening of the Arctic Oscillation index. These findings are confirmed in recent results presented by McPhee (2013) and Koldunov et al. (2013). In addition, the regional sea level jointly obtained from tide gauges and ERS-1, 2 and Envisat altimeter satellites together with the gravity field and ocean dynamic topography observations from GRACE and GOCE have also recently allowed new innovative studies of the climate-critical mass changes and freshwater flux variations in the high latitude and Arctic Ocean (e.g., Cheng et al. 2013; Prandi et al. 2012; Henry et al. 2012; Knudsen et al. 2011).

In this paper, a new GOCE-based geoid and mean dynamic topography (MDT) for the high latitude and Arctic Ocean is obtained, assessed and compared to independent steric height observations and state-of-the-art MDTs. Furthermore, comparisons of surface velocity and transport in the Nordic Seas, based on the combination of GOCE gradiometer gravity estimates and in situ hydrographic data, are done with estimates from several forced coupled sea ice–ocean models, ocean surface drifter data and direct measurements. The new findings and results are presented according to the ocean dynamic topography in Sect. 2, ocean surface circulation in Sect. 3 and volume transport in Sect. 4. A summary follows in Sect. 5.

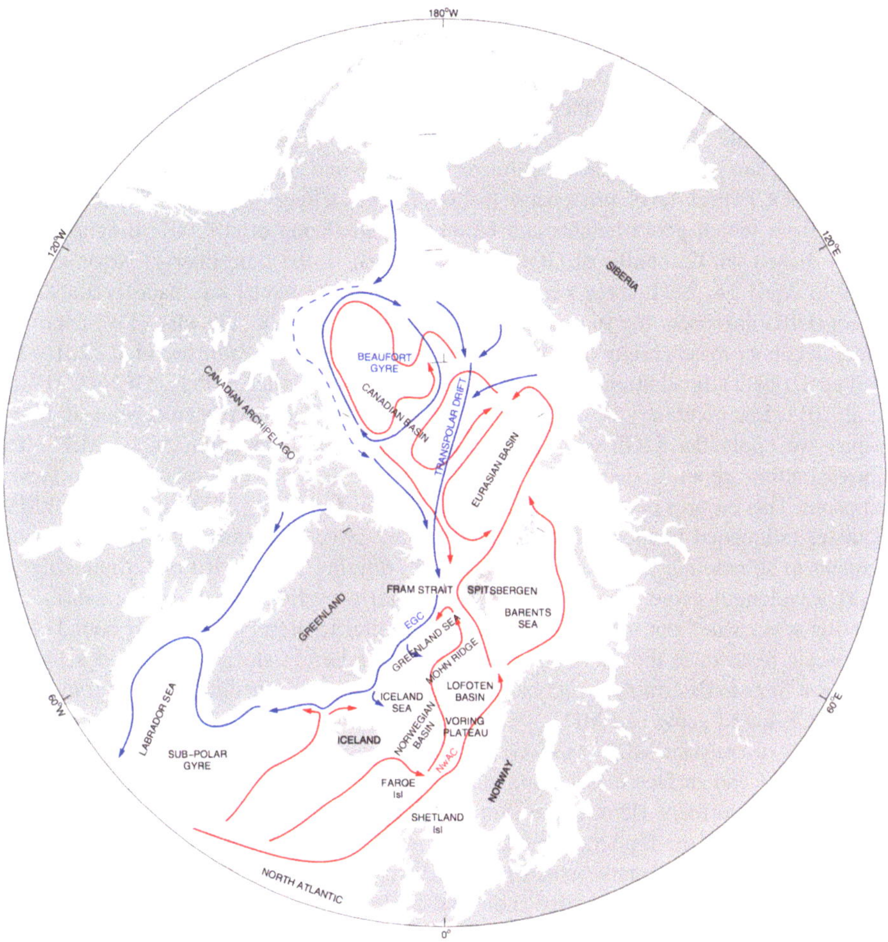

Fig. 1 General circulation of the Arctic Ocean, Nordic Seas, and North Atlantic. Bottom contours are 1000 and 3000 m outlining the shelves and basins. *Red arrows* represent Atlantic Waters, which reside in the surface in the Nordic Seas and submerged in the Arctic Ocean. *Blue arrows* represent Polar Water, residing in the surface. The Norwegian Sea comprises the Norwegian Basin, while Lofoten Basin, while the Nordic Seas are the Norwegian, Iceland and Greenland Seas. Circulation patterns based on AMAP (1998) and Furevik and Nilsen (2005)

2 Ocean Dynamic Topography

Measurements of the sea surface height have been routinely obtained from satellite altimeter missions, such as the TOPEX/POSEIDON (Fu et al. 2001; Shum et al. 2010), in the last 20 years. Today, the annual mean sea surface (MSS) height derived from altimetry is known with millimeter accuracy (e.g., Cazenave et al. 2009) in the open ocean. In addition, knowledge of the marine geoid has drastically improved thanks to satellite gravity measurements from the NASA GRACE (Maximenko et al. 2009) and ESA GOCE (Johannessen et al. 2003; Bingham et al. 2011; Knudsen et al. 2011) missions in the last decade. In turn, the MDT, which is simply the difference between the mean sea surface height (MSS) and the geoid (G) (both referenced to the same ellipsoid as illustrated in

Fig. 2), can now be determined with new and unprecedented accuracy around $\approx$3 cm at 100 km spatial resolution (Bruinsma et al. 2013). In comparison to the use of the reference geoid obtained from the Earth Gravitational Model 2008 (EGM2008), this yields a factor 2 improvement in the MDT at this spatial resolution. However, this accuracy is not necessarily applicable to the Arctic Ocean and the neighboring sub-Arctic seas due to the presence of sea ice, lack of Jason altimeter coverage and shorter dominant spatial scales.

The GOCE high-level processing facility (HPF) delivers the level-2 global gravity model from which geoid heights can be determined (Koop et al. 2007; Bingham et al. 2011). Based on 12 months of GOCE data acquired in the time interval November 01, 2009 to April 14, 2011, three versions of GOCE gravity model are made available: the direct (DIR) approach; the spacewise (SPW) approach; and the timewise (TW) approach. More details of these gravity field models can be obtained from Bruinsma et al. (2010) and Pail et al. (2011). In addition, so-called combination models such as the EIGEN-6C (Förste et al. 2011) that combines the GOCE data with terrestrial data have been developed. In this paper, we apply the EIGEN-6C gravity model for the computation of the MDT. The corresponding geoid is determined in the mean-tide system and relative to a Topex-ellipsoid. This ensures consistency with the Technical University of Denmark (TUD) MSS data set referenced to the time period 1993–2009 (Andersen and Knudsen 2009). Subsequent to subtracting the geoid from the MSS, filtering was carried out eliminating the short wavelength geoid signals, in order to obtain a useful estimate of the MDT. This filtering was carried out using a 80-km Gaussian filter to preserve the upper bound of the mesoscale features in the study area. (Note that Knudsen et al. (2011) applied a 140-km Gaussian filter to determine the global ocean MDT.) In the forthcoming, we refer to this as the GOCE-based geoid and MDT.

Isolines of constant MDT (MSS-G) are usually considered as a stream function for the large-scale ocean surface circulation, which the surface geostrophic currents are directed along. In the Northern Hemisphere (Southern Hemisphere), the flow is clockwise (anticlockwise) around the topographic high. The magnitude of the global spatial MDT variations is around 2–3 meters, which is about two orders of magnitude smaller than the global spatial changes in the marine geoid and the MSS. This makes the computation of the MDT and the handling of errors challenging as it is easy to fail to exploit all of the details in the geoid and the MSS when calculating the MDT because of the need to obtain a smooth solution. Herein, the separation of the MDT from the MSS and the geoid is carried out in the space domain, where the MSS is usually represented using processing tools that are available at the dedicated ESA GOCE User Toolbox (GUT); see Web site http://earth.esa.int/gut/.

The GOCE-based MDT shape and spatial pattern representing the mean from 1993 to 2009 for the North Atlantic, Nordic Seas and the Arctic Ocean is shown in Fig. 3. The total MDT elevation range from the high in the Arctic Ocean to the low in the subpolar gyre in the North Atlantic reaches about 0.9 m. The regional shape of the MDT with the orientation of the dominant slopes in the different sub-domains reveals the presence of the main circulation pathways in: (1) the subpolar gyre south of Greenland; (2) the inflow of Atlantic Water, respectively, between Iceland and the Faroe Islands and between the Faroe and Shetland Islands; (3) the continuous northward flowing Atlantic Water toward the Arctic Ocean; (4) the southward flowing East Greenland Current (EGC); (5) the Beaufort Gyre; and (6) the transpolar drift in the Arctic Ocean.

The MDT in the Arctic Ocean may display some characteristic features that are caused by problems in the data coverage. Both the GOCE data and the altimeter data do not cover the Arctic Ocean entirely, so within 300–400 km from the pole, the data coverage is

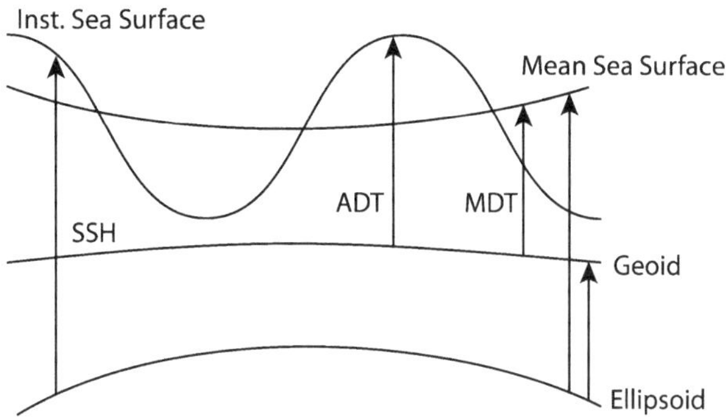

Fig. 2 Schematic illustration of the relationship between the absolute and mean dynamic topography (ADT and MDT), the mean sea surface and the geoid referenced to the same ellipsoid. Note the difference between the instantaneous sea surface and the MDT

insufficient to calculate a reliable MDT. Also, the presence of sea ice may hamper the computation of the MSS and hence the MDT. Though care is taken to avoid erroneous data some of the data that have been used to calculate the MSS may represent the top of the sea ice floes rather than the sea surface. In particular, off the coasts of the Canadian Archipelago and northern Greenland the high values of the GOCE MDT may be caused by the influence of the permanent and thick sea ice cover.

The Arctic Ocean displays an elevation change reaching up to about 0.45 m associated with the high in the Beaufort Gyre, and with the corresponding dominant orientation of the slope mostly aligned from Siberia to the northern shores of Greenland. According to Steele and Ermold (2007), the dynamic height in the Arctic Ocean is predominantly influenced by salinity. In the Nordic Seas, the general shape of the MDT favors the cyclonic circulation pattern displaying steepest MDT slopes of 0.4 m/100 km between the Faroe and Shetland Islands, along the northwest coast of Norway and in the northern part of the EGC. In comparison, the slope across the Gulf Stream reaches 1 m/100 km. This spatial pattern in the MDT agrees well with the spatial pattern in the mean steric height derived from hydrographic data (Nilsen et al. 2008) for the period 1950–2010, respectively, referenced to 500, 1,000 and 1,500 m as shown in Figs. 4 and 5b.

The steric height calculation is done according to Siegismund et al. (2007), where the steric height is referenced to a constant density ρ_0 from salinity of 35 and temperature of 0 °C. More information on the concept and application of the steric height is given by Tomczak and Godfrey (2003). The difference in these height fields primarily reveals the effect of the vertical distribution of temperature and salinity in the upper 1500 m, predominantly influenced by the advection and spreading of the Atlantic Water. Apart from the changes occurring in the Lofoten Basin, the overall structure remains largely unchanged when the density structures from 1,000 to 1,500 m are included. This suggests that the baroclinic circulation in the Nordic Seas is driven by the temperature and salinity structures of the Atlantic Water in the upper 1,000 m.

In the Nordic Seas, the total range in the MDT derived from the combined GOCE and altimetry data is around 0.50–0.55 m as seen in Fig. 5a. In comparison, the range of the mean steric height of 0.30 m (Fig. 5b) suggests that there might be a significant contribution to the MDT pattern from the large-scale atmospheric pressure field and the deep barotropic currents in some of the sub-basins. Siegismund et al. (2007) moreover concluded that the seasonal cycle of the steric height (for the period 1950–1999) is predominantly associated with the temperature variations in agreement with previous studies on global scale (e.g., Gill and Niiler 1973; Stammer 1997; Mork and Skagseth 2005). By subtracting the hydrographic-based steric height associated with the baroclinic structure in

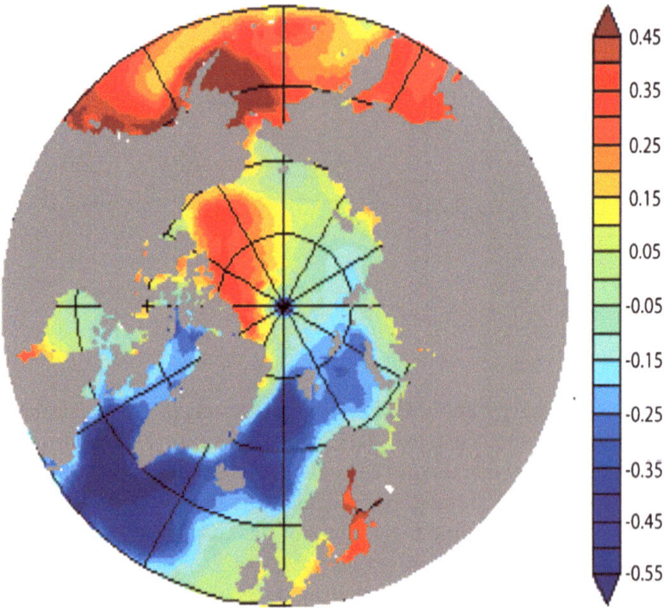

Fig. 3 Mean dynamic topography (MDT) derived from the GOCE gradiometer data (release 3) and altimetry (from 1993 to 2009) with a spatial resolution of about 100 km. Color bar is in units of meter. The structures in the North Pacific are not investigated further in this paper. Note that the GOCE data (release 4) available since March 2013 are more accurate due to more than a doubling in the amount of data

the water masses from the GOCE-based MDT, an estimate of the barotropic contribution to the MDT is derived as shown in Fig. 5c. The barotropic contribution contains distinct elevation changes of about 10 cm having pattern consistent with the known barotropic cyclonic circulations in the Greenland Sea, the Lofoten Basin and in the Norwegian Sea (Nøst and Isachsen 2003). Evidence of this cyclonic barotropic circulation in the Norwegian Sea has also been observed from Argo floats in the intermediate waters below the Norwegian Atlantic Current (Søiland et al. 2008). In summary, the assessment of the GOCE-derived MDT for the Nordic Seas and the Arctic Ocean is promising.

In view of the promising GOCE-based results presented above, they are also providing a new opportunity for inter-comparison and validation of coupled sea ice-ocean models and reanalyses fields. As specified in Table 1, the three models used in this inter-comparison study include the regional setup of the ATL (MITgcm) model (Serra et al. 2010); the MICOM model (Sandø et al. 2012); and the HYCOM model (Bleck 2002; Sakov et al. 2012). The models are either forced by the 6 hourly NCEP reanalysis field (ATL and MICOM) or the ERA Interim field (HYCOM).

Ignoring the offset in the mean MDT, the three coupled sea ice–ocean models in general reproduce comparable overall spatial structure of the MDT in the Arctic Ocean, the Nordic Seas and the North Atlantic, notably the high in the Beaufort Gyre and the depressions in the Nordic Seas and the subpolar gyre (Fig. 6). The model highs in the Beaufort Gyre are circular and located toward the deep Canadian Basin with decreasing values toward the Eurasian Basin, providing an elevation difference of 0.5–0.6 m. The MICOM-field, however, has a gyre that extends into the Eurasian Basin. In comparison, the GOCE-based elevated feature in the Beaufort Sea is shifted more toward the Canadian Archipelago, while the total elevation

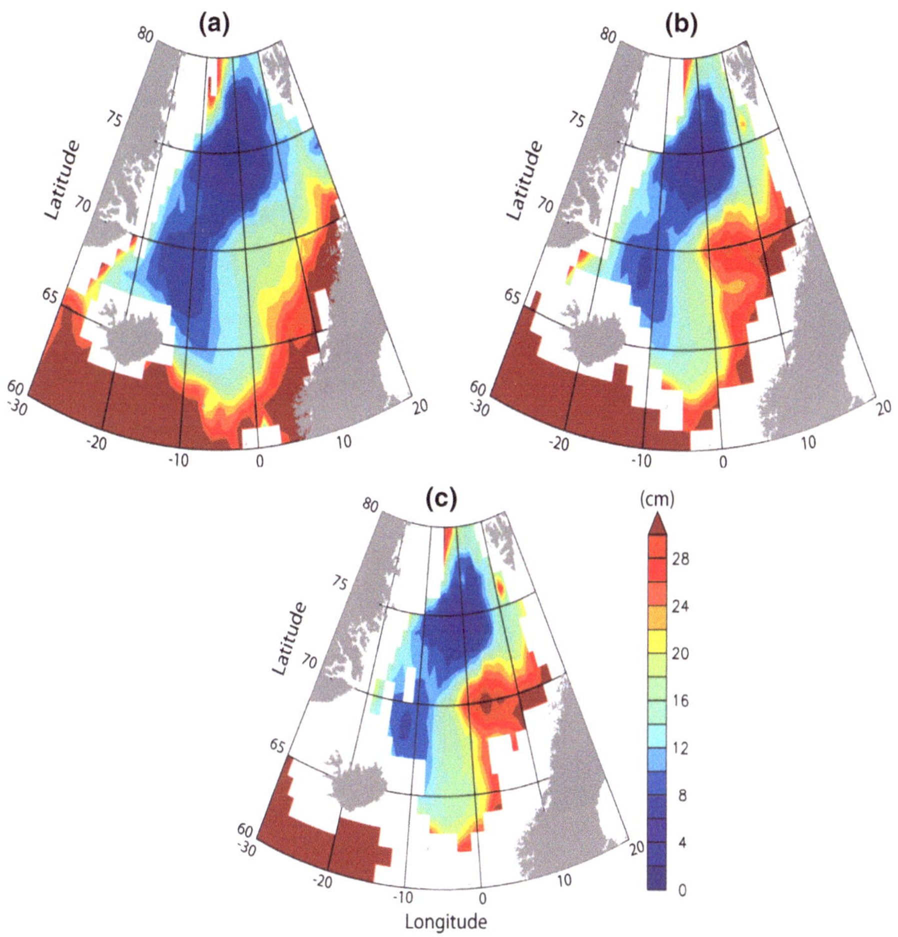

Fig. 4 Observed mean steric height for the period 1950–2008 for the reference depths **a** 500 m, **b** 1,000 m and **c** 1,500 m. The color-scale increment is in cm

difference remains the same. This shift in location is in agreement with the recent findings by Kwok and Morison (2011) and Morison et al. (2012). Overall, the MDT patterns in the model fields for the Arctic Ocean are in reasonably good agreement with the GOCE-based MDT map.

In the central domain of the Norwegian-Greenland Seas, the suppression of the MDT in the three models corresponding to the large-scale cyclonic circulation pattern with the northward flowing Norwegian-North Atlantic Current (NwAC) and the southward flowing EGC is consistent in location. On the other hand the magnitudes and spatial structures of the suppression differ between the models as well as in comparison with the GOCE-based MDT pattern. The largest suppression is found in the ATL model with a deviation from the average of −0.6 m in the northern Greenland Sea being almost twice as large as in the GOCE-based MDT in the same area. Similar tendencies are seen in the subpolar gyre, although the difference in the minima between the ATL model and the GOCE-based MDT now is reduced by a factor of 2.

The most prominent discrepancies are the mismatch in the MDT along the Canadian Archipelago and northern Greenland coast, and the models lack of higher elevations associated with the spread of AW in the Norwegian Sea, notably around the Vøring Plateau. The former

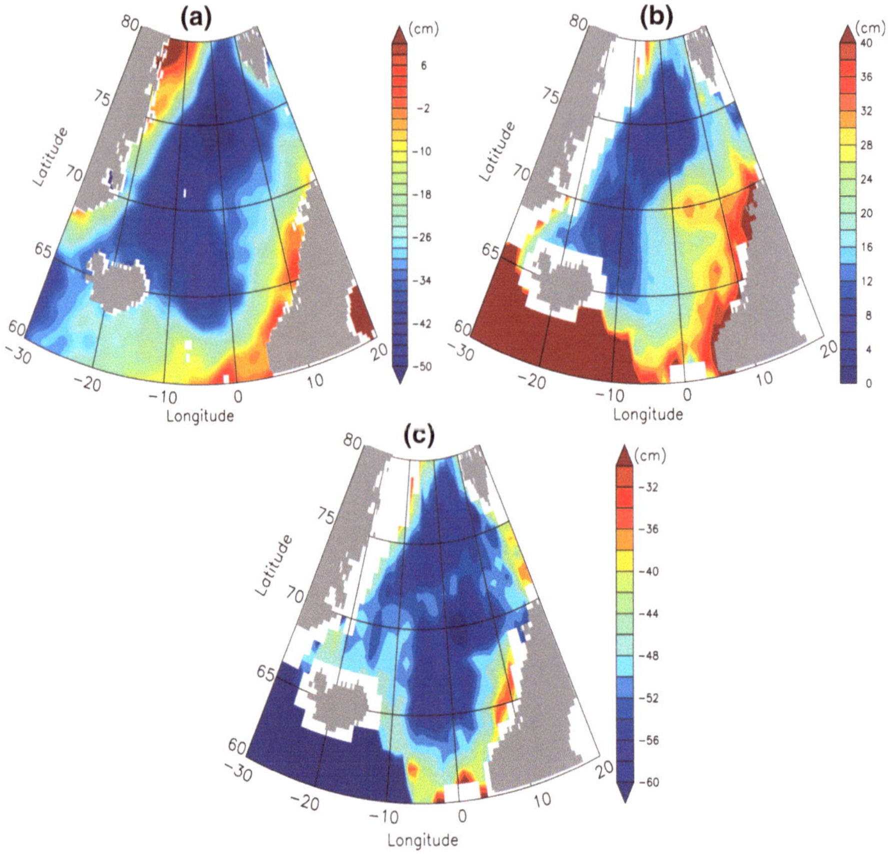

Fig. 5 **a** MDT derived from combined GOCE and altimetry, **b** steric height derived from the in situ hydrographic database where the *white* areas in the 1,500 m reference steric height (see Fig. 4) are filled with steric height values representing every 100 m from 1,400 to 500 m, and **c** difference between (**a**) and (**b**). The color bars represent the height contours in unit of cm. Note the different *color* ranges

might be related to the presence of thicker multiyear sea ice that could influence the estimation of the MSS and thus the GOCE-based MDT. Kwok and Morrison (2011) did not reveal this particular high in the MDT confined to the coastal region from IceSAT data. The latter is related to the topographic steering of the baroclinic western branch of the NwAC (Nilsen and Nilsen 2007), as well as eddy transport of buoyant waters from the slope branch of the NwAC into the Lofoten Basin (Rossby et al. 2009), which are both challenging to model. Furthermore, although totally lacking the broadness of the NwAC, the ATL model is the only model with the doming of the densest waters of the Nordic Seas placed in the correct basin, the Greenland Basin.

These differences in magnitude and spatial structure of the model and GOCE-based MDTs imply different strengths and orientations of the slopes in the MDT. In turn, the mean surface geostrophic currents are expected to have discrepancies that subsequently will lead to differences in the estimation of the associated transport of water masses. This is further assessed in the next sections.

Table 1 Characterization of the three coupled sea ice–ocean models used for inter-comparison to the GOCE-derived MDT and mean surface geostrophic current

Model run	Region	Spatial resolution	Period	Vertical grid, # of layers, forcing
ATL12	Atlantic Ocean north of 33°S including the Nordic Seas and the Arctic Ocean. Uses ETOPO 2-min resolution bathymetry	~8 km	1948–2009 Hindcast	z-coordinates, 50 levels, NCEP—6 h
MICOM	North of 30°S with Nordic Seas and Arctic Ocean included. Uses ETOPO 1 5-min resolution bathymetry	~15 km	1948–2007 Hindcast	Isopycnal, 35 layers, NCEP—6 h
HYCOM	High latitude- Arctic Ocean. Uses GEBCO 1-min resolution bathymetry	~12–16 km	1993–2010 Hindcast	Hybrid coordinates, 28 layers, ERA Interim—6 h

ETOPO 2-min Earth's relief data set with a grid size of 2′ by 2′, from NOAA Geophysical Data Center

NCEP relates to US National Centers for Environmental Prediction

GEBCO 1-min Earth's relief data set with a grid size of 1′ by 1′, from UNESCO and IOC

ERA relates to European Center for Medium-Range Weather Forecasting Reanalyses

3 Surface Circulation

With access to the new GOCE-based MDT with unprecedented accuracy, the uncertainties in mean ocean circulation and transport estimation are expected to improve. The mean surface geostrophic velocities are computed from the MDT, under the assumption of geostrophic balance;

$$u_s = \frac{-g}{f} \cdot \frac{\partial MDT}{\partial y} \tag{1}$$

$$v_s = \frac{g}{f} \cdot \frac{\partial MDT}{\partial x} \tag{2}$$

where u_s and v_s are components of the surface geostrophic velocity, g is the acceleration due to gravity, f is the Coriolis parameter, and x and y are the longitudinal and latitudinal coordinates. The corresponding surface geostrophic current derived from the GOCE MDT for the Nordic Seas over the period 1993–2009 is shown in Fig. 7 and compared to the independently derived CNES_CLS09 MDT Rio et al. 2011 and Maximenko et al. 2009 (which both are using a GRACE-based geoid model together with in situ Argo floats and surface drifter data integrated over the 17-year period from 1992 to 2009), as well as the climatological mean surface velocities (predominantly based on drifters in the Nordic Seas from 1991 to 2010) from the surface drifter data (http://www.aoml.noaa.gov/phod/dac/drifter_climatology.html).

The large-scale cyclonic surface circulation regime is well-reproduced in all three fields. However, while the strongest mean surface currents of the inflowing Atlantic Water to the Norwegian Sea reaching nearly 0.20 m/s are derived from the GOCE MDT, the inflow in the other two surface current fields is clearly weaker with maximum speed not much more than 0.10 m/s. Moreover, it is only the GOCE-based surface geostrophic current that reveals distinct

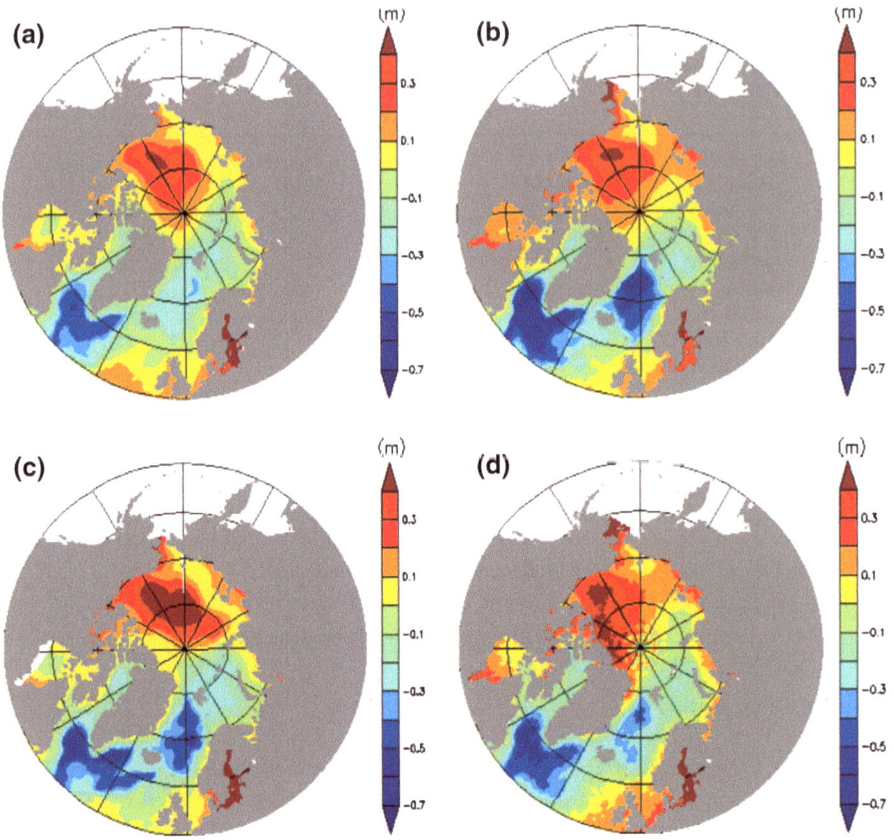

Fig. 6 MDT fields referenced to the full-region average: **a** the HYCOM MDT (free run) from 1993 to 2010, **b** the ATL from 1993 to 2009, **c** the MICOM from 1993 to 2007, and **d** the GOCE-based MDT from 1993 to 2009. The *color bars* are in meters. All the fields are interpolated to a 0.25° resolution grid

expressions of cyclonic circulation in the Greenland Basin, Norwegian Basin and Iceland Sea, as well as the broadening of the NwAC over the Vøring Plateau and in the Lofoten Basin, i.e., signs of a proper western (baroclinic) branch of the northward flowing Atlantic Water. From this inter-comparison and assessment, it is therefore evident that the GOCE-based geoid provides a reliable representation of the MDT and mean ocean surface circulation in the Nordic Seas. Evidently, this is further supported by the mean surface circulation pattern derived from the climatology of the surface drifter data as shown in Fig. 7d.

A comparison of the speed of the GOCE-based mean surface geostrophic currents and corresponding model-based currents for the Nordic Seas is shown in Fig. 8. In general, it must be emphasized that the finer spatial model resolution versus GOCE may favor stronger simulated surface speeds. All models indicate intensified currents at the inflows from the northeast Atlantic Ocean, and in the boundary (slope) currents of the Nordic Seas. The ATL model shows a strengthened component of internal circulation in the Nordic Seas, by very strong currents along all the margins. Regarding the currents over the mid-ocean ridges and other internal topographic features, it is only the MICOM run that shows signs of reproducing the level of intensification shown in the GOCE-based speeds, however only at one location, the Mohn Ridge (as also noticed in Fig. 7d).

For a more detailed study of the seasonal variability induced by the altimetric observations, the surface slopes and meridional velocities across 75°N are presented in Fig. 9 together with the model-derived fields. The seasonal mean meridional velocities are estimated by replacing MDT in Eq. (2) with absolute dynamic topography (ADT). Note that ADT is determined as the sum of MDT and monthly mean sea-level anomaly (SLA) data. The new high-resolution SLA data (obtained from the French CLS-led Sea Level Climate Change Initiative project funded by ESA) are referenced to the time period 1993–2009 and hence consistent with the DTU MSS data used in the calculation of GOCE MDT.

The main expected features of the flow toward and from the Fram Strait is revealed by the mean velocities: the two branch northward flowing West Spitsbergen Current (WSC) around 8° and 15°E; the strong southbound EGC at 10°W; and some minor, possibly cyclonic, circulation features around 0°E, likely related to circulation in the Boreas Basin. Seasonal differences are most pronounced in the WSC. Both branches are strongest in wintertime, with a near doubling of the easternmost branch, which is due to the general (wind driven) intensification of the circulation in the region. This is consistent with velocity retrievals and transport estimates reported by Mork and Skagseth (2005). The western frontal branch stays relatively strong also during the rest of the year, likely due to the summertime spread of buoyant surface water from the coast to the front (as seen further south in the NwAC; Nilsen and Falck 2006), maintaining a steep frontal surface slope.

In comparison, the model-based MDT slopes along 75°N and the corresponding meridional geostrophic velocities across the same latitude consistently reveal that the ATL model has the steepest surface slopes and hence the strongest flow field for both the northward flowing NwAC as well as the southward flowing EGC. Moreover, it is only the ATL model that reproduces the double peak in the WSC current in agreement with the mean and seasonal observation-based findings.

4 Volume Transport

By combining the GOCE-derived MDT and altimetric sea-level anomalies (SLA) with the comprehensive hydrographic database, an estimate of the mean and variable transport of Atlantic Water entering the Nordic seas is obtained for the period 1993–2011 at a spatial resolution of 100 km. Using 44 CTD-sections in the Faroe north section normally taken to represent the Iceland-Faroe Ridge (IFR) inflow (Hansen et al. 2010), 84 CTD-sections for the Faroe–Shetland Channel (FSC) and 76 CTD-sections taken along the Svinøy section (see Fig. 8 for locations), the baroclinic velocity structures in the Atlantic Water defined by salinity values S >35 were estimated across these sections. Combined with the barotropic velocity values, the absolute velocities are then retrieved, and when these are multiplied by the area covered by the Atlantic Water, we obtain estimates of the corresponding volume transports of Atlantic Water across the 3 sections (see Table 2).

From the combination of GOCE, altimetry and hydrography, the mean inflows of Atlantic Water across the IFR and through the FSC are estimated to approximately 3.5 and 4.1 Sv, respectively (1 Sv = 10^6 m^3 s^{-1}). The former is in very good agreement with Hansen et al. (2010), but too low compared to Østerhus et al. (2005), while the latter is too high compared to Østerhus et al. (2005) and too low compared to Sandø et al. (2012). In comparison, the mean transport of the two branches of Atlantic Water crossing the Svinøy section, e.g., the Norwegian Atlantic Slope Current (NwASC) and the Norwegian Atlantic Front Current (NwAFC) is, respectively, 3.9 Sv and 3.0 Sv. The latter value is in acceptable agreement with previous transport estimates for the NwASC reported by Mork

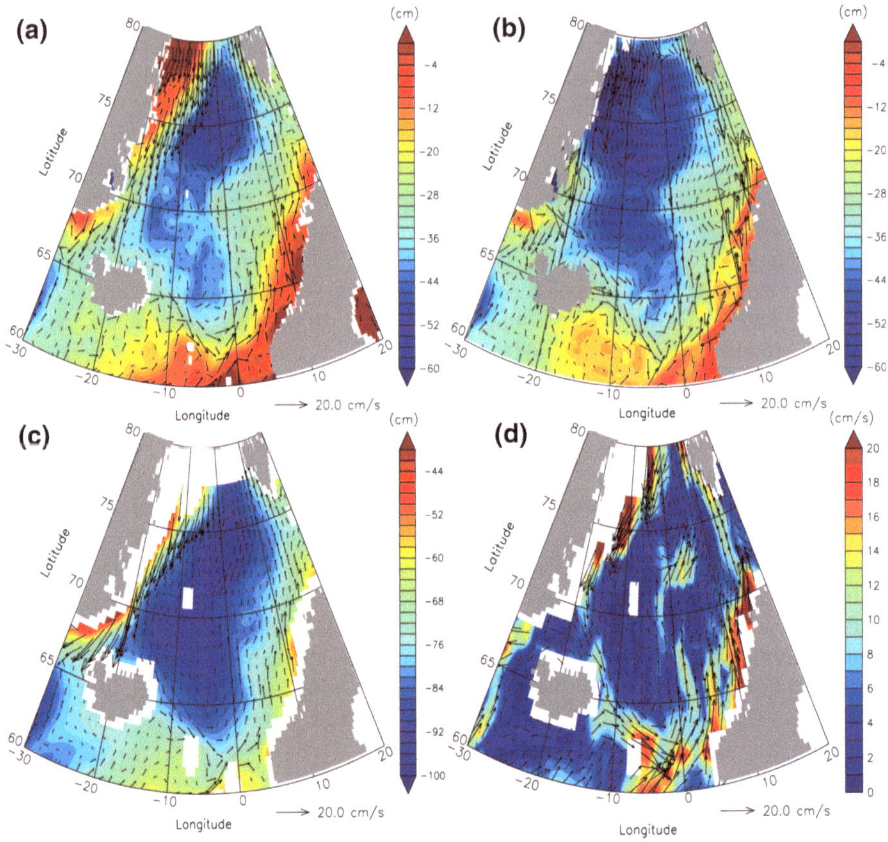

Fig. 7 Mean surface geostrophic velocities shown by vectors superimposed on corresponding mean dynamic topography (MDT) derived from **a** GOCE, **b** CNES_CLS09, **c** Maximenko et al. (2009), and **d** mean surface velocity vectors derived from the climatology of the global surface drifter data. *Color scale* indicates the MDT in cm for (**a**) to (**c**) and speed in cm/s for (**d**). Current-vector scale shown in the *lower right* corner

and Skagseth (2010); Orvik and Skagseth (2003, 2005); Skagseth et al. (2008); and Orvik et al. (2001) as documented in Table 2, taking into account the slight differences in the integration periods. On the other hand, the total combined GOCE-based and hydrographic transport estimates across the Svinøy section is about 35 % larger than other reported findings (e.g., 6.9 vs. 5.1 Sv).

In comparison, the mean (1993–2007) transports estimated from the three models across these sections show quite different values as noticed in Table 2. One explanation for this is partly related to the definition and choice of layers for the transport estimations. For instance, Sandø et al. (2012) defines the Atlantic Water (AW) as water in model layers above the pycnocline (sigma_2 < 36.9 kg m^{-3}), which is representative of the interface between inflowing and outflowing waters throughout the integration. In contrast, Berx et al. (2013) simply uses T >3 °C as definition for the AW in their calculation of the transport of AW across the IFR section. The best agreement between the model and the combined GOCE-based and hydrographic data is clearly obtained for the ATL simulation with transport estimates across the IFR and FSC of 3.5 and 4.2 Sv, respectively.

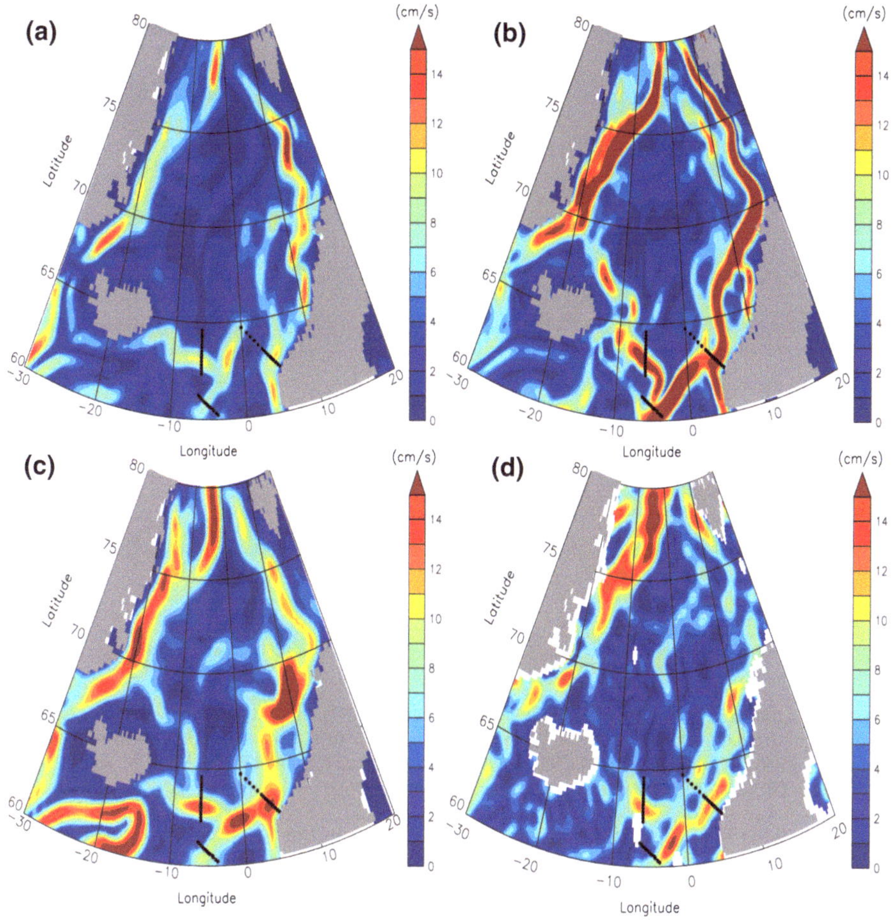

Fig. 8 Inter-comparison of models and GOCE-based mean absolute surface geostrophic velocity from **a** the HYCOM model from 1993 to 2010, **b** the ATL model from 1993 to 2009, **c** the MICOM model from 1993 to 2007 and **d** GOCE. The *color bars* are in cm/s. The three *black dotted lines* mark the position of the Faroe north section, the Faroe–Shetland Channel section and the Svinøy section

For the Svinøy section, the comparison is, in general, less satisfactory. The HYCOM model clearly underestimates the observed transport of 5.1 Sv reported by Mork and Skagseth (2010) as well as the GOCE-based estimate of 6.9 Sv. This is mainly due to a mis-location of the NwAFC in the HYCOM simulation as seen in Fig. 8a. In contrast, the MICOM and ATL models, having comparable mean transport estimates in the range of 8.2–8.5 Sv overestimate both the GOCE-based estimate and the transport reported by Mork and Skagseth (2010). Overall, this large spread in mean transport estimates implies significant differences in the mean northward advection of heat and salt to the Nordic Seas and Arctic Ocean. This, in turn, affects both the evaporation–precipitation fluxes and convective overturning in the Norwegian and Greenland Seas. Further studies are needed to investigate the accuracies of these transport estimates.

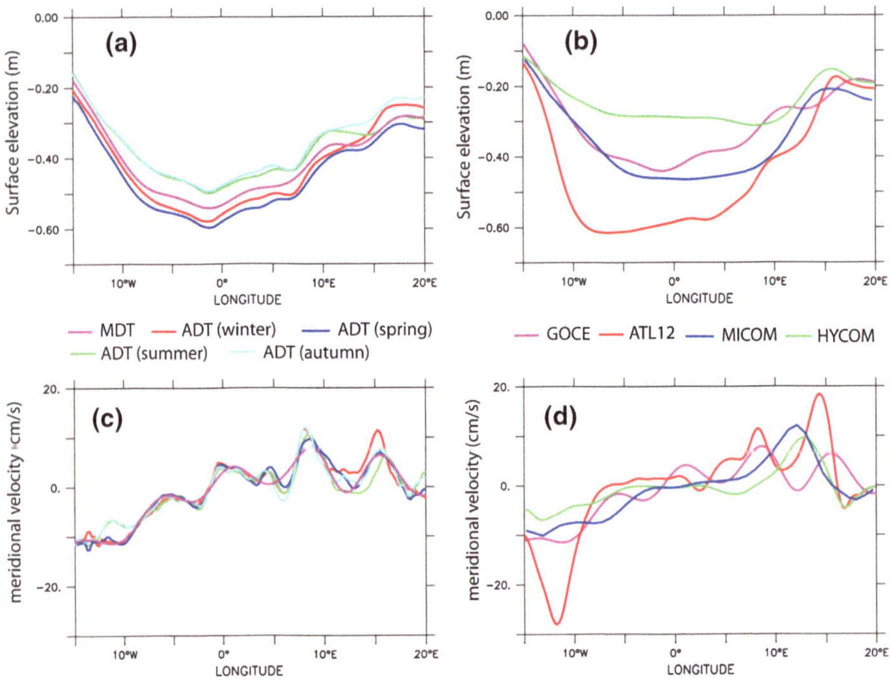

Fig. 9 Climatologies of (**a**, **b**) dynamic topography across 75°N and (**c**, **d**) corresponding meridional absolute geostrophic velocities: (**a**, **c**) seasonal climatologies from combined GOCE-based MDT and altimetry and (**b**, **d**) comparison of GOCE-based MDT with MDT from ATL12, MICOM and HYCOM. Note that in (**b**), each MDT is referenced to its full-region average as defined in Fig. 6

Taking benefit of the temporal variability observed in the SLA and hydrographic data, the mean and seasonal cycle in the transport of the inflowing Atlantic Water for the period 1993–2009 can also be estimated and inter-compared as shown in Fig. 10.

On average, the NwASC contains approximately 57 % (or 3.9 Sv) of the total mean volume transport across the Svinøy section of about 6.9 Sv. The mean seasonal variability reveals a pattern with largest transports (9.3 Sv) in winter being 70 % larger than the summer transport minimum (5.4 Sv). Moreover, the mean seasonal NwASC transport always exceeds the mean seasonal NwAFC transport, while the latter displays a narrower range of seasonal variability in the volume transport. This suggests that the seasonal changes of the transport across the Svinøy section are predominantly controlled by seasonal changes in the transport of the NwASC.

The partitioning of these total transport estimates (both in the mean and seasonal signals) into the respective barotropic and baroclinic components is shown in Fig. 10b, c and reveals distinct differences. While the transport in the NwASC is dominated by the barotropic flow as expected along the shelf break at the Svinøy section, the transport of the NwAFC, in contrast, is clearly larger in the baroclinic component with the exception of the autumn period.

These GOCE-based estimates together with high-quality in situ hydrographic data are providing new and promising abilities to examine the seasonal transport variability (total as well as barotropic and baroclinic components) across key-selected sections. As such, it is also providing an important tool for validations of model circulation and transports between the northeast Atlantic Ocean and the Nordic Seas and Arctic Ocean.

Table 2 Comparison of volume transport estimates from combined GOCE, altimetry and in situ hydrography to previous studies as well as estimates from simulation models for the Island-Faroe Ridge (IFR), Faroe–Shetland Channel (FSC), NwAFC, NwASC in the Svinøy Section and the total Svinøy Section

Source	Data	Period	IFR [Sv]	FSC [Sv]	Svinøy [Sv]		
					NwAFC	NwASC	Total
The current study	GOCE + Altim. + hydr.	1993–2011	3.5	4.1	3.0	3.9	6.9
Mork and Skagseth (2010)	Altim. + hydr.	1993–2009			1.7	3.4	5.1
Skagseth et al. (2008)	Current meter	1995–2006				4.3	
Orvik and Skagseth (2005)	Curr. meters	1995–1999				4.2	
Orvik and Skagseth (2003)	Curr. meters	1998–2000				4.4	
Orvik et al. (2001)	Curr. meters + ADCP + hydr.	1995–1999			3.4	4.2	7.6
Berx et al. (2013)	Altm. + ADCP + hydro	1995–2009	3.5				
Østerhus et al. (2005)	Bottom ADCP + hydr.	1999–2001	3.8	3.8			
Hansen et al. (2010)	Bottom ADCP + hydr.	1997–2008	3.5				
Sandø et al. (2012)	MICOM model	1994–2007	4.7*	4.7			
The current study	HYCOM model	1993–2007	1.8	1.5	2.0	0.6	2.6
The current study	MICOM model	1993–2007	3.5	6.9	3.5	5.0	8.5
The current study	ATL model	1993–2007	3.5	4.2	3.5	4.7	8.2

* Only from 1997 to 2007

5 Summary

In combination with in situ hydrographical data, surface drifters and current meter measurements, coupled sea ice—ocean models and the latest GOCE-derived geoid and MDT (Pail et al. 2011), the paper has investigated the quality, usefulness and validity of the new GOCE data for studies of the ocean circulation and transports in the Nordic Seas and Arctic Ocean. Using the GOCE data from release number 3 (based on 12 months of GOCE data in the time interval 1 November 2009 to 2014 April 2011), the gravity model from the direct approach yields the computation of the GOCE- based geoid, and jointly with the DTU10 MSS data (based on the integration over the period 1993–2011, Knudsen et al. 2011), the MDT (MSS-G) representing the same 18-year integration period has been calculated. In summary, the following key findings and results are highlighted:

1. New knowledge of the shape and spatial pattern of the MDT is derived at a spatial resolution of around 100 km and with an accuracy of around 4–5 cm which is superior to previous existing MDTs for this region.
2. Combined with the steric height estimated from hydrographic data, the pure barotropic contribution to the MDT shows distinct features in consistence with known deep barotropic circulations in the Norwegian and Greenland Seas.
3. The new GOCE-based MDT and surface geostrophic currents compare favorably with existing independent surface velocity calculations derived from combined altimeter data, in situ observations and gravity field models.
4. The transport estimates, both in the mean and seasonal signals, are also favoring the combined use of the GOCE-based surface geostrophic current and hydrographic data.

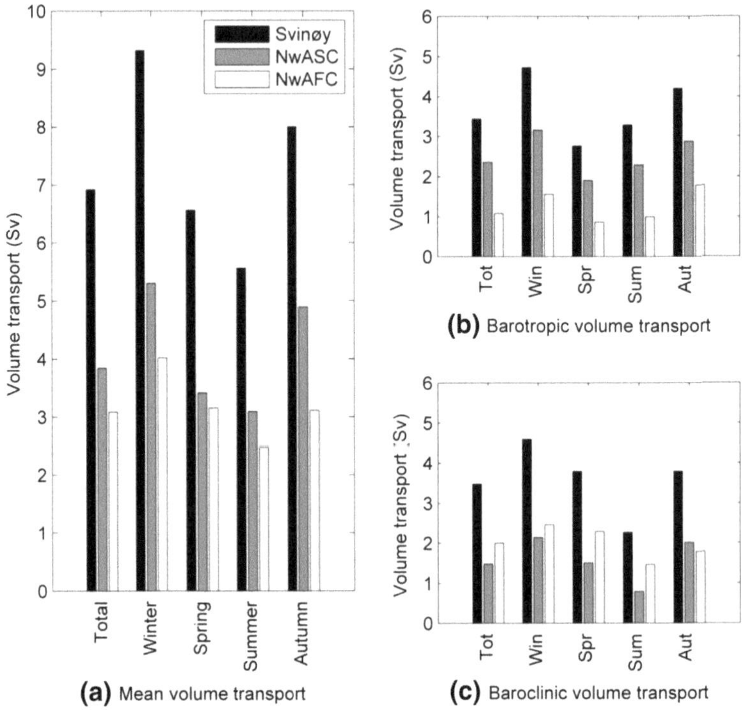

Fig. 10 Mean annual and mean seasonal total volume transport estimates (**a**), the respective barotropic components (**b**) and baroclinic components (**c**) for the Svinøy section including the NwASC and the NwAFC for the period 1993–2010 based on combined use of GOCE, altimetry and in situ hydrography data. The *grayscale* legend is shown in (**a**)

5. New understanding of the relationship between the MDT, the mean surface geostrophic current and the magnitude of the mean ocean volume transport has been derived for the seasonal variability with regard to the inflow of Atlantic Water to the Norwegian Sea at the Svinøy section.
6. The NwASC contains approximately 60 % of the total volume flux across the Svinøy section with a distinct transport maximum in winter (Dec–Jan) and a minimum in summer (Jun–Aug). This transport is moreover dominated by the barotropic component.
7. These data and findings are also excellent for assessment and validation of model-based retrieval of the MDT, the surface geostrophic current and the volume transport across selected sections and straits.

Overall, the findings add new insight into the ocean circulation and transport between the northeast Atlantic Ocean and the Arctic Ocean. They are also considered to be highly valuable for further studies of the regional sea-level change in the Nordic Seas and the Arctic Ocean, notable via the contribution of steric height and changes in the volume transport. Consistent use of the GOCE data for assimilation as suggested by Haines et al. (2011) might also become feasible in near future.

Moreover, as gravity measurements provide an integrated view of the mass variations, their interpretation in terms of mass transport is inherently multidisciplinary. Satellite gravimetry (such as combined GRACE and GOCE) is thus a vital component of a multisensor Earth-observing system, which complements and relates observations of different Earth system constituents in a common and consistent global framework (Panet et al. 2012). Being closely related to changes in sea level, ocean transports, glaciers and ice caps, future mass change observations from satellites (at a 100 km scale not resolved by GRACE today) have the potential to significantly advance the ability to monitor seasonal-to annual-to decadal variability in ocean mass transport.

Acknowledgments The work presented in this paper has partly been supported by European Union 7th Framework Program through the MONARCH-A Collaborative Project, FP7-Space-2009-1 contract no. 242446. In addition, it has been supported by the Research Council of Norway funded projects Number 200408 (SATICE) and number 212020 (GOCE MDT) and the (ESA/NRS) PRODEX project IGOCE, contract number 90377. The study was also partly funded by the Centre for Climate Dynamics at the Bjerknes Centre.

References

AMAP (1998) AMAP assessment report: Arctic pollution issues. Arctic Monitoring and Assessment Programme (AMAP), Oslo, Norway, 859 pp

Andersen OB, Knudsen P (2009) The DNSC08 mean sea surface and mean dynamic topography. J Geophys Res 114:C11. doi:10.1029/2008JC005179

Berx B, Hansen B, Østerhus S, Larsen KM, Sherwin T, Jochumsen K (2013) Combining in situ measurements and altimetry to estimate volume, heat and salt transport variability through the Faroe Shetland Channel. Ocean Sci 9. doi:10.5194/os-9-639-2013

Bingham RJ, Knudsen P, Andersen O, Pail R (2011) An initial estimate of the North Atlantic steady-state geostrophic circulation from GOCE. Geophys Res Lett 38:L01606. doi:10.1029/2010GL045633

Bleck Rainer (2002) An oceanic general circulation model framed in hybrid isopycnic-Cartesian coordinates. Ocean Model 4(1):55–88

Bruinsma SL, Marty JC, Balmino G, Biancale R, Förste C, Abrikosov O and Neumayer H (2010) GOCE gravity field recovery by means of the direct numerical method, presented at the ESA Living Planet Symposium, 27th June–2nd July 2010, Bergen, Norway; See also: earth.esa.int/GOCE

Bruinsma SL, Förste C, Abrikosov O, Marty J-C, Rio M-H, Mulet S, Bonvalot S (2013) The new ESA satellite-only gravity field model via the direct approach. Geophy Res Lett 40:1–6. doi:10.1002/grl.50716

Cazenave et al (2009) Sea level budget over 2003–2008: a re-evaluation from GRACE space gravimetry, satellite altimetry and Argo. Global Planet Change 65(1–2):83–88

Cheng YO, Andersen and Knudsen P (2013) Evaluation of gridded and along-track altimetric data in the Arctic Ocean for climate research, submitted to Marine Geodesy

Förste C, Bruinsma S, Shako R, Marty JC, Flechtner F, Abrikosov O, Dahle C, Lemoine, JM, Neumayer KH, Biancale R, Barthelmes F, König R, Balmino G (2011) EIGEN-6—A new combined global gravity field model including GOCE data from the collaboration of GFZPotsdam and GRGS-Toulouse; Geophysical Research Abstracts, vol. 13, EGU2011-3242-2, EGU General Assembly

Fu L–L, Cheng B, Qiu B (2001) 25-day period large-scale oscillations in the Argentine Basin revealed by the TOPEX/POSEIDON altimeter. J Phys Oceanogr 31:506–517

Furevik T, Nilsen JEØ (2005) Large-scale atmospheric circulation variability and its impacts on the Nordic Seas ocean climate—a review. In: The Nordic Seas: an integrated perspective. AGU Geophysical Monograph Series, vol 158. pp 105–136

Gill AE, Niiler PP (1973) The theory of seasonal variability in the ocean. Deep Sea Res 20:141–177

Haines K, Johannessen JA, Knudsen P, Lea D, Rio MH, Bertino L, Davidson F, Hernandez F (2011) An ocean modelling and assimilation guide to using GOCE geoid products. Ocean Sci 7(1):151–164

Hansen B, Hatun H, Kristiansen R, Olsen SM, Østerhus S (2010) Stability and forcing of the Iceland-Faroe inflow of water, heat, and salt to the Arctic. Ocean Sci 6:1013–1026

Hawkins E, Sutton R (2008) Geophys Res Lett 35:L11603. doi:10.1029/2008GL034059

Helland-Hansen B, Nansen F (1909) The Norwegian Sea: its physical oceanography based upon the Norwegian Researches 1900–1904, Report on Norwegian Fishery and Marine Investigation, vol. II. The Royal Department of Trade, Navigation and Industries, Mallingske, Kristiania, pp 390

Henry O, Prandi P, Llovel W, Cazenave A, Jevrejeva S, Stammer D, Meyssignac B, Koldunov N (2012) Tide gauge-based sea level variations since 1950 along the Norwegian and Russian coasts of the Arctic Ocean: contribution of the steric and mass components. J Geophys Res 117(C6):C06 023. doi:10.1029/2011JC007706

Johannessen JA, Balmino G, Le Provost C, Rummel R, Sabadini R, Sünkel H, Tscherning CC, Visser P, Woodworth P, Hughes CW, LeGrand P, Sneeuw N, Perosanz F, Aguirre-Martinez M, Rebhan H, Drinkwater M (2003) The European gravity field and steady-state ocean circulation explorer satellite mission: impact in Geophysics. Surv Geophy 24:339–386

Knudsen P, Bingham R, Andersen O, Rio Marie-Helene (2011) A global mean dynamic topography and ocean circulation estimation using a preliminary GOCE gravity model. J Geodesy. doi:10.1007/s00190-011-0485-8

Koldunov NV, Serra N, Kohl A, Stammer D, Henry O, Prandi P, Cazenave A, Knudsen P, Andersen OB, Gao Y, Johannessen JA (2013) Arctic Ocean Sea Surface Height variability during the last 40 years, *to be submitted to JGR*

Koop R, Gruber T, Rummel R (2007) The status of the GOCE highlevel processing facility (HPF). In: Proceedings of the 3rd GOCE User Workshop, pp 199–204, European Space Research Institute, European Space Agency, Frascati, Italy

Kwok R, Morison J (2011) Dynamic topography of the ice-covered Arctic Ocean from ICESat. Geophys Res Lett 38(2):L02 501. doi:10.1029/2010GL046063

Maximenko N, Niiler P, Rio M-H, Melnichenko O, Centurioni L, Chambers D, Zlotnicki V, Galperin B (2009) Mean dynamic topography of the ocean derived from satellite and drifting buoy data using three different techniques. J Atmos Ocean Tech 26(9):1910–1919

Mcphee MG (2013) Intensification of geostrophic currents in the Canada Basin, Arctic Ocean. J Climate 26. doi:10.1175/JCLI-D-12-00289.1

Morison J, Kwok R, Peralta-Ferriz C, Alkire M, Rigor I, Andersen R, Steele M (2012) Changing Arctic Ocean freshwater pathways. Nature 481(7379):66–70. doi:10.1038/nature10705

Mork KA, Skagseth Ø (2005) Annual sea surface height variability in the Nordic Seas, in The Nordic Seas: An Integrated Perspective, *Geophys Monogr Ser*, vol. 158, edited by H. Drange et al. pp. 51–64, AGU, Washington, DC

Mork KA, Skagseth Ø (2010) A quantitative description of the Norwegian Atlantic current by combining altimetry and hydrography. Ocean Sci 6:901–911. doi:10.5194/os-6-901-2010

Nilsen JEØ, Hatun H, Mork KA and Valdimarsson H (2008) The NISE Data Set. *Technical Report* 08-01, Faroese Fisheries Laboratory, Box 3051, Torshavn, Faroe Islands

Nilsen JEØ, Falck E (2006) Variation of mixed layer properties in the Norwegian Sea for the period 1948–1999. Prog Oceanogr 70:58–90. doi:10.1016/j.pocean.2006.03.014

Nilsen JEØ, Nilsen F (2007) The Atlantic water flow along the Vøring plateau: detecting frontal structures in oceanic station time series. Deep Sea Res Part 1 54(3):297–319. doi:10.1016/j.dSV.2006.12.012

Nøst OA, Isachsen PE (2003) The large-scale time-mean ocean circulation in the Nordic seas and Arctic Ocean estimated from simplified dynamics. J Mar Res 61:175–210

Orvik KA, Skagseth Ø (2003) Monitoring the Norwegian Atlantic slope current using a single moored current meter. Cont Shelf Res 23:159–176

Orvik KA, Skagseth Ø (2005) Heat flux variations in the eastern Norwegian Atlantic current toward the Arctic from moored instruments, 1995–2005. Geophys Res Lett 32:L14610. doi:10.1029/2005GL023487

Orvik KA, Skagseth Ø, Mork M (2001) Atlantic inflow to the Nordic Seas: current structure and volume fluxes from moored current meters, VM-ADCP and SeaSoar-CTD observations, 1995–1999. Deep-Sea Res I:48. doi:10.1016/S0967-0637(00)00038-8

Østerhus S, Turrrell WR, Jónsson S, Hansen B (2005) Measured volume, heat, and salt fluxes from the Atlantic to the Arctic Mediterranean. Geophys Res Lett 32:L07603. doi:10.1029/2004GL022188

Pail R, Bruinsma S, Migliaccio F, Foerste C, Goiginger H, Schuh W-D, Hoeck E, Reguzzoni M, Brockmann JM, Abrikosov O, Veicherts M, Fecher T, Mayrhofer R, Krasbutter I, Sanso F, Tscherning CC (2011) First GOCE gravity field models derived by three different approaches. J Geodesy 85(11):819–843

Panet I, Flury J, Biancale R, Gruber T, Johannessen JA, van den Broeke MR, van Dam P, Gegout T, Hughes CW, Ramillien G, Sasgen I, Seoane L, Thomas M (2012) Earth system mass transport mission (e.motion): a concept for future earth gravity field measurements from space. Surv Geophy. doi:10.1007/s1072-012-9209-8

Prandi P, Ablain M, Cazenave A, Picot N (2012) Sea level variability in the Arctic Ocean observed by satellite altimetry. Ocean Sci Discuss 9(4):2375–2401. doi:10.5194/osd-9-2375-2012

Rio MH, Guinehut S, Larnicol G (2011) New CNES-CLS09 global mean dynamic topography computed from the combination of GRACE data, altimetry, and in situ measurements. J Geophys Res 116:C07018. doi:10.1029/2010JC006505

Rossby T, Ozhigin V, Ivshin V, Bacon S (2009) An isopycnal view of the Nordic Seas hydrography with focus on properties of the Lofoten Basin. Deep Sea Res Part 1 56:1955–1971. doi:10.1016/j.dsr.2009.07.005

Sakov P, Counillon F, Bertino L, Lisæter KA, Oke PR, Korablev A (2012) TOPAZ4: an ocean-sea ice data assimilation system for the North Atlantic and Arctic. Ocean Sci 8:633–656. doi:10.5194/os-8-633-2012

Sandø AB, Nilsen JEØ, Eldevik T, Bentsen M (2012) Mechanisms for variable North Atlantic–Nordic seas exchanges. J Geophys Res 117:C12006. doi:10.1029/2012JC008177

Serra NRH, Käse A, Stammer Köhl D, Quadfasel D (2010) On the low-frequency phase relation between the Denmark Strait and the Faroe-Bank Channel overflows. Tellus 62:530–550. doi:10.1111/j.1600-0870.2010.00445.x

Shum CK, Hans-Peter Plag, Jens Schröter, Victor Zlotnicki, Peter Bender, Alexander Braun, Anny Cazenave, Don Chamber, Jianbin Duan, William Emery, Georgia Fotopoulos, Viktor Gouretski, Richard Gross, Thomas Gruber, Junyi Guo, Guoqi Han, Chris Hughes, Masayoshi Ishii, Steven Jayne, Johnny A. Johannesen, Per Knudsen, Chung-Yen Kuo, Eric Leuliette, Sydney Levitus, Nikolai Maximenko, Laury Miller, James Morison, Harunur Rashid, John Ries, Markus Rothacher, Reiner Rummel, Kazuo Shibuya, Michael Sideris, Y. Tony Song, Detlef Stammer, Maik Thomas, Josh Willis, Philip Woodworth (2010) Geodetic observations of the ocean surface topography, geoid, currents and changes in ocean mass and volume, Plenary Session Paper, OceanObs09, Venice Italy, 21–25 Sept. 2009, *ESA Publicatiion WPP 306,* doi: 10.5270/OceanObs09

Siegismund F, Johannessen JA, Drange H, Mork KA, Korablev A (2007) Steric height variability in the Nordic Seas. J Geophys Res 112:C12010. doi:10.1029/2007/JC004221

Skagseth Ø, Furevik T, Ingvaldsen R, Loeng H, Mork KA, Orvik KA, Ozhigin V (2008) Volume and Heat Transports to the Arctic Ocean via the Norwegian and Barents Seas. In: Dickson (ed) Arctic-Subarctic Ocean Fluxes (ASOF): Defining the Role of the Northern Seas in Climate. Springer, Berlin, pp 45–64

Søiland H, Prater MD, Rossby T (2008) Rigid topographic control of currents in the Nordic Seas. Geophys Res Lett 35:L18607. doi:10.1029/2008GL034846

Stammer D (1997) Steric and wind-induced changes in TOPEX/POSEIDON large-scale sea surface topography observations. J Geophys Res. doi:102,20,987-21,009

Steele M, Ermold W (2007) Steric sea level change in the northern seas. J Climate 20. doi:10.1175/JCLI4022.1

Tomczak M, Godfrey JS (2003) Regional Oceanograhy: An Introduction, 2nd edn. Daya Publishing House, New Delhi

DOI 10.1007/978-94-017-8789-5_10
Reprinted from *Surveys in Geophysics* Journal, DOI 10.1007/s10712-013-9244-0

Sea Surface Salinity Observations from Space with the SMOS Satellite: A New Means to Monitor the Marine Branch of the Water Cycle

Nicolas Reul · Severine Fournier · Jaqueline Boutin · Olga Hernandez · Christophe Maes · Bertrand Chapron · Gaël Alory · Yves Quilfen · Joseph Tenerelli · Simmon Morisset · Yann Kerr · Susanne Mecklenburg · Steven Delwart

Received: 21 January 2013 / Accepted: 18 June 2013 / Published online: 8 August 2013

Abstract While it is well known that the ocean is one of the most important component of the climate system, with a heat capacity 1,100 times greater than the atmosphere, the ocean is also the primary reservoir for freshwater transport to the atmosphere and largest component of the global water cycle. Two new satellite sensors, the ESA Soil Moisture and Ocean Salinity (SMOS) and the NASA Aquarius SAC-D missions, are now providing the first space-borne measurements of the sea surface salinity (SSS). In this paper, we present examples demonstrating how SMOS-derived SSS data are being used to better characterize

N. Reul · S. Fournier · B. Chapron · Y. Quilfen
Laboratoire d'Océanographie Spatiale, Institut Français de recherche et d'Exploitation de la Mer, Centre Bretagne BP 70, 29280 Plouzané, France

N. Reul (✉)
Centre Méditerranée, Zone Portuaire de Brégaillon, BP 330, 83507 La Seyne-sur-Mer Cedex, France
e-mail: nreul@ifremer.fr

J. Boutin · O. Hernandez · S. Morisset
Laboratoire d'Océanographie et du Climat: Expérimentation et Approches Numériques, UMR 7159 CNRS/UPMC/IRD/MNHN, Paris, France

C. Maes · G. Alory
LEGOS, OMP, CNAP, Université de Toulouse, Toulouse, France

J. Tenerelli
CLS-Radar Division, Bâtiment Le Ponant, Avenue La Pérouse, Technopôle Brest-Iroise, 29280 Plouzané, France

Y. Kerr
CESBIO, Toulouse, France

S. Mecklenburg · S. Delwart
ESA-ESRIN, Frascatti, Italy

key land–ocean and atmosphere–ocean interaction processes that occur within the marine hydrological cycle. In particular, SMOS with its ocean mapping capability provides observations across the world's largest tropical ocean fresh pool regions, and we discuss from intraseasonal to interannual precipitation impacts as well as large-scale river runoff from the Amazon–Orinoco and Congo rivers and its offshore advection. Synergistic multi-satellite analyses of these new surface salinity data sets combined with sea surface temperature, dynamical height and currents from altimetry, surface wind, ocean color, rainfall estimates, and in situ observations are shown to yield new freshwater budget insight. Finally, SSS observations from the SMOS and Aquarius/SAC-D sensors are combined to examine the response of the upper ocean to tropical cyclone passage including the potential role that a freshwater-induced upper ocean barrier layer may play in modulating surface cooling and enthalpy flux in tropical cyclone track regions.

Keywords Sea surface salinity · SMOS satellite · Passive microwave remote sensing · Oceanic freshwater cycle

1 Introduction

Salinity is known to play an important role in the dynamics of the ocean's thermohaline overturning circulation and in large-scale atmosphere–ocean climate signals such as the El Nino Southern Oscillation (ENSO), and is the key freshwater tracer within the oceanic component of the global hydrologic cycle, a branch that comprises most of the global precipitation and evaporation as well as the river runoff (Schmitt 2008). Multi-decadal sea surface salinity (SSS) trends have been documented in tropical and high latitudes and associated with signatures of evaporation or precipitation variation that are consistent with global warming scenarios (e.g., Dickson et al. 2002; Gordon and Guilivi 2008; Morrow et al. 2008; Cravatte et al. 2009; Yu 2011; Durack et al. 2012; Terray et al. 2011). These studies highlight the need for well-sampled SSS time series both for monitoring the change and to improve the basic understanding of the respective roles of the atmosphere and ocean dynamics, thermodynamics, air–sea interaction, and land–ocean interaction in the global water cycle context.

Our basic knowledge of the global SSS distribution is derived from the compilations of all available oceanographic data collected over time (e.g., Boyer and Levitus 2002). The SSS in situ observing system has expanded significantly during the last decade due mostly to the full deployment of the Argo profiling float array and now provides a monthly SSS estimate on a grid of roughly 300–400 km^2. Notwithstanding these recent gains, this sampling density is still too sparse to resolve climatologically important intraseasonal, seasonal, and interannual to decadal signals at the 300-km spatial scale within which SSS is known to vary significantly (Lagerloef et al. 2010). The recent launches of the ESA/Soil Moisture and Ocean Salinity (SMOS, see Kerr et al. 2010; Font et al. 2010) and NASA/ Aquarius SAC-D (Lagerloef et al. 2008; Lagerloef 2012) mission satellites represent contributions toward filling this gap using passive microwave remote sensing.

Salinity remote sensing is based on the measurement of sea surface microwave emission at the lower end of the microwave spectrum and from a surface skin layer having a thickness of $O(1\ \text{cm})$. This emission depends partly on the dielectric constant of sea water, which in turn can be related to salinity and temperature. Thus, given sea surface temperature (SST), theory predicts some ability to invert SSS information. In practice, however, numerous additional external factors (extra-terrestrial sources, atmosphere, ionosphere, and surface roughness) also contribute to the satellite-observed emission, and these must be corrected to allow accurate ocean salinity estimates. The SMOS and Aquarius sensors are both ocean microwave radiometers operating at a frequency of ~1.4 GHz (L-band, wavelength of 21 cm), a band chosen for the relatively strong sensitivity to change in salinity and because this is a transmission-free, or protected, frequency. An additional and important benefit for this choice is minimization of atmospheric signal contributions.

Based on the observed SSS variability and need to better resolve it, the satellite missions aim to produce salinity estimates with an accuracy of 0.1–0.2 over the so-called Global Ocean Data Assimilation Experiment scales of 100 km, 1 month or 200 km, and 10 days. This is a challenging objective for several reasons. First, the sensitivity of L-band brightness temperatures to variations in SSS is on average 0.5 degK per salinity scale. This sensitivity is very weak given that spatial and temporal variability in open ocean SSS does not exceed several units and that the instrument noise is typically 2–5 degK. Note that salinity computations are based on the Practical Salinity Scale PSS-78 and reported with no units (United Nations Educational, Scientific and Cultural Organization 1985). Second, there are many geophysical sources of brightness at L-band that corrupt the salinity signal, and correction models for these factors have uncertain accuracy. Moreover, the technical approach developed in order to achieve adequate radiometric accuracy and spatiotemporal resolution for SMOS is polarimetric interferometric radiometry, the first such space-borne system. The complex SMOS image reconstruction data processing includes contamination by different errors and induces residual inaccuracies in SSS estimates. Finally, there is significant radio frequency interference emanating from sources along the many coastlines that contaminate data collected over many ocean regions. Nevertheless, much work at ESA SMOS level 2 expert centers and the CNES/IFREMER Centre Aval de Traitement des Données SMOS (CATDS) has addressed these issues, leading to the first global satellite SSS estimates (Font et al. 2013; Reul et al. 2012; Boutin et al. 2012a).

Two examples of monthly composite SMOS SSS maps are shown in Fig. 1. They show salient basin scale features, including the elevated salinity in the Atlantic relative to the other basins, and the general correspondence of lower SSS with known river runoff and tropical precipitation regions. SMOS data validation efforts using in situ observations reveal an overall SSS accuracy on the order of 0.3 (Boutin et al. 2012a; Reul et al. 2012; Banks et al. 2012; Font et al. 2013), but with degraded quality at high latitudes partly because of reduced sensitivity in colder waters. While further improvements are in progress, many interesting features of the global SSS could be already evidenced.

This paper reviews preliminary results addressing several key applications of these new satellite SSS data. Given the reduced SMOS sensitivity in cold waters, the focus is on tropical ocean data where SMOS measurements have proven to be the most accurate. We also attempt to highlight combined use of other satellite and in situ observations (altimetry, SST, ocean color, river discharge, evaporation, and precipitation). It is shown that these new data are proving useful in the monitoring of intraseasonal to interannual variability across major tropical freshwater pools of the world ocean. SMOS-detected SSS freshening events within intense precipitation zones (e.g., the Inter Tropical Convergence Zone) are also shown to provide promising new information related to the ocean surface response to rainfall. Finally, SMOS SSS data are used to address interactions between wind-driven phenomena, such as upwelling and tropical cyclones (TCs), and some of the world's largest fresh pools. The data sets used in these cases are described in Sect. 2. SMOS monitoring capabilities for the major tropical river plumes are given in Sect. 3. In Sects. 4 and 5, we illustrate rain impacts detected in SMOS SSS data; then, their application improved the understanding of freshwater pools interaction with the atmosphere. Conclusions and perspectives are given in Sect. 6.

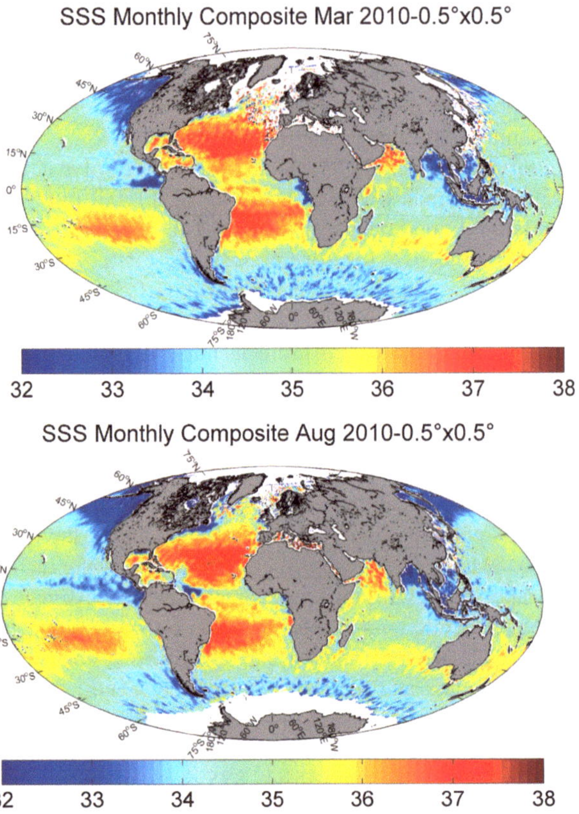

Fig. 1 Monthly composites of the sea surface salinity at a spatial resolution of 0.5° × 0.5° deduced from SMOS data (CATDS v2) for the months of March (*Upper*) and August (*Lower*) 2010

2 Data

A range of satellite and in situ data sets are used in the present study with focus on the years 2010–2012 following the SMOS launch date. The data products are described below.

2.1 SMOS SSS Data

SMOS (Soil Moisture and Ocean Salinity) is the European Space Agency (ESA)'s water mission (Kerr et al. 2010; Mecklenburg et al. 2012), an Earth Explorer Opportunity Mission approved under the Living Planet Program. SMOS was launched in November 2009, and the technical approach developed to achieve adequate radiometric accuracy, as well as spatial and temporal resolution compromising between land and ocean science requirements, is polarimetric interferometric radiometry (Ruf et al. 1988; Font et al. 2010) at L-band (frequency of ~1.4 GHz). ESA produces so-called level 2 SSS, or L2 products, which correspond to instantaneous SSS retrievals under the satellite swath.

In the present study, level 2 SMOS SSS are from the first SMOS/ESA annual reprocessing campaign in which ESA level 1 v5.04 and level 2 v5.50 processors have been used. In these versions, significant improvements with respect to the flaws discovered in the first products (e.g., Reul et al. 2012) have been implemented (see a complete description in the Algorithm Theoretical Basis Document (ATBD) available at http://www.argans.co.uk/smos/docs/deliverables/). Nevertheless, accuracy of these instantaneous SSS retrievals is rather low ($\sim$0.6–1.7), and space–time averaging of the level 2 products is needed (so-called level 3 SSS) to decrease the noise level in the retrievals.

Here, we used two types of composite SSS level 3 products generated in laboratories participating to the Expertise Center of the Centre Aval de Traitement des Données SMOS (CATDS, http://www.catds.fr), which is the French ground segment for the SMOS data. These products are built either from ESA level 1 products (Reul and Tenerelli 2011) or from ESA level 2 products (Boutin et al. 2012b).

These research products aim at assessing the quality of SMOS operational products (ESA level 2 and CATDS-OP level 3) and at studying new processing to be implemented in the future in operational chains. Main characteristics of these products are detailed in Table 1. CEC-IFREMER products have been used in Sects. 3, and 5, CEC-LOCEAN products in Sect. 4.

Overall accuracy of the 10-day composite products at 25-km resolution is on the order of 0.3 practical salinity scale in the tropical oceans (Reul and Tenerelli 2011).

2.2 Ocean Surface Currents

Here, we used the 1/3° resolution global surface current products from Ocean Surface Current Analyses Real time (OSCAR) (Bonjean and Lagerloef 2002; http://www.oscar.noaa.gov), directly calculated from satellite altimetry and ocean vector winds.

Table 1 Summary of characteristics of CATDS-CEC SSS level 3 products

	CEC-IFREMER	CEC-LOCEAN
SSS retrieval method	SSS retrieved from first Stokes parameter (Reul and Tenerelli 2011)	SSS retrieved from polarized Tbs along dwell lines using an iterative retrieval (see ESA L2OS ATBD)
Region of the instrument field of view (FOV) considered for SSS retrieval	Alias free field of view only	Alias free field of view (AFFOV) and extended AFFOV along dwell lines with at least 130 Tb data samples in AFFOV ($\sim\pm$300 km from the swath center)
Tb filtering method	Determined from interorbit consistency in incidence angles classes and thresholding	Determined from consistency along dwell lines as reported in ESA level 2 products
Galactic model	Geometrical optics model	Kirchoff's approx. scattering at 3 m/s
Roughness/foam models	Empirical adjustment of Tb dependencies to wind speed	Empirical adjustment of parameters in roughness model and foam coverage models (Yin et al. 2012)
Calibration	Single ocean target transformation (OTT) + daily 5° × 5° adjustment wrt World Ocean 2001 SSS climatology	Variable OTT (every 2 weeks synchronized with noise injection radiometer as defined in ESA reprocessing)
Average	Simple average	Average weighted by theoretical error on retrieved SSS and spatial resolution

The OSCAR data processing system calculates sea surface velocities from satellite altimetry (AVISO), vector wind fields (QuikSCAT), as well as from sea surface temperature (Reynolds–Smith) using quasi-steady geostrophic, local wind-driven, and thermal wind dynamics. Near real-time velocities are calculated both on a 1° × 1° and 1/3° × 1/3° grid and on a ~5-day time base over the global ocean. Surface currents are provided on the OSCAR Web site (http://www.oscar.noaa.gov) starting from 1992 along with validations with drifters and moorings. The 1/3° resolution is available for FTP download through ftp://esr.org/pub/datasets/SfcCurrents/ThirdDegree.

2.3 Rain, Evaporation and River Discharge Data

To estimate the rain rate over the oceans, we used three different satellite products.

One is the monthly Tropical Rainfall Measuring Mission (TRMM) Composite Climatology (TCC) of surface precipitation based on 13 years of data from the TRMM. The TCC takes advantage of the information from multiple estimates of precipitation from TRMM to construct mean value maps over the tropics (36°N–36°S) for each month of the year at 0.5° latitude–longitude resolution. The first-time use of both active and passive microwave instruments on board TRMM has made it the foremost satellite for the study of precipitation in the tropics and has led to a better understanding of the underlying physics and distribution of precipitation in this region. The products are available at NASA Goddard Space Flight Center Global Change Master Directory (http://gcmd.nasa.gov).

The second type of satellite rain rate estimates that we used in the present study are the so-called "TRMM and Other Satellites" (3B42) products, obtained through the NASA/Giovanni server (http://reason.gsfc.nasa.gov/OPS/Giovanni). The 3B42 estimates are 3 hourly at a spatial resolution of 0.25° with spatial extent covering a global belt (−180°W–180°E) extending from 50°S to 50°N latitude. The major inputs into the 3B42 algorithm are IR data from geostationary satellites and Passive Microwave data from the TRMM microwave imager (TMI), special sensor microwave imager (SSM/I), Advanced Microwave Sounding Unit (AMSU), and Advanced Microwave Sounding Radiometer-Earth Observing System (AMSR-E).

The Special Sensor Microwave Imager (SSM/I) F16 and F17 orbits cross SMOS orbits within −20 min and +40 min. Hence, numerous SMOS level 2 are collocated with SSMI rain rates (RR) within this range of time. In addition to the TRMM 3B42 products, we therefore used SSM/Is data sets to perform colocations between SMOS SSS and rain estimates. SSM/Is RR version 7 was used and downloaded from http://www.remss.com.

The evaporation (E) data set was taken from the version 3 products of the Objectively Analyzed air-sea Fluxes (OAFlux) project (Yu and Weller 2007).

Finally, the discharge data for the Amazon, Orinoco, and Congo rivers were obtained from the Environmental Research Observatory HYBAM (geodynamical, hydrological, and biogeochemical control of erosion/alteration and material transport in the Amazon basin) Web site (http://www.ore-hybam.org/).

2.4 Ocean Color Products

To study the spatiotemporal coherency between SSS signals from some major tropical river plumes and ocean color properties, we used the level 3 daily, 4-km resolution estimates of the absorption coefficient of colored detrital matter (CDM) at 443 nm. These products processed and distributed by ACRI-ST GlobColour service are supported by the EU FP7 MyOcean2 and the ESA GlobColour Projects, using ESA ENVISAT MERIS data,

NASA MODIS and SeaWiFS data. These products have been averaged at the SMOS L3 product 0.25° resolution, with a 10-day running mean.

2.5 In Situ Data

Salinity measurements from Argo floats are provided by the Coriolis data center (http://www.coriolis.eu.org/). The upper ocean salinity values recorded between 4- and 10-m depth will be referred to as Argo SSS following Boutin et al. (2012b).

Global SSS maps are derived from delayed time quality checked in situ measurements (Argo and ship) by IFREMER/LPO, Laboratoire de Physique des Oceans, using the In Situ Analysis System (ISAS) optimal interpolation (D7CA2S0 re-analysis product) (see a method description on http://wwz.ifremer.fr/lpo/SO-Argo-France/Products/Global-Ocean-T-S/Monthly-fields-2004-2010 and in (Gaillard et al. 2009)). The choice for the time and space scales used in that method results from a compromise between what is known of ocean time and space scales and what can actually be resolved with the Argo array (3°, 10 days); two length scales are considered: the first one is isotropic and equal to 300 km, the second one is set equal to 4 times the average Rossby radius of deformation of the area. As a result, we expect these maps being smoother, especially in tropical areas, than SMOS SSS maps averaged over 0.25° × 0.25° or 1° × 1°.

3 SMOS Monitoring of the Major Tropical Atlantic River Plumes

Rivers are important variables in oceanography as their freshwater affects SSS and the buoyancy of the surface layer, and they represent a source of materials exotic to the ocean and important to biological activity. Obviously, they are key hydrologic components of the freshwater exchanges between land and ocean. Despite this importance, tracing major tropical river water (e.g., Amazon, Congo, and Ganges) over large distances has not been straightforward previously principally because of a lack of SSS observations. Tracing those very large rivers over great distances now become an important endeavor, as sufficient data are available from surface salinity sensors placed aboard satellites.

Occurrence of patches of low surface salinity (<35 practical salinity scales) in the tropical Atlantic Ocean is closely related to the presence of the mouths of the world's largest rivers in terms of freshwater discharge (e.g., Amazon, Congo, and Orinoco) and their subsequent spreading of freshwater by the upper ocean circulation. Another key freshwater source here is the Inter Tropical Convergence Zone (ITCZ), associated with relatively intense precipitation that migrates latitudinally over the tropical Atlantic throughout the year (Binet and Marchal 1993). One of these major low-salinity pools is formed by the Amazon and Orinoco river plumes spreading offshore from the South America northeastern coasts, and influencing a large fraction of the western tropical North Atlantic (Neumann 1969; Lentz 1995; Muller-Karger et al. 1988; Dessier and Donguy 1994). The Gulf of Guinea situated in the northeastern equatorial Atlantic is also an important location for the freshwater budget in the tropical Atlantic. It is a region of intense precipitation with as much as 30 cm of rain falling per month during the rainy season (Yoo and Carton 1988). Furthermore, into this area flows the Congo River, the largest freshwater input to any eastern ocean boundary. These large-scale low-salinity "lenses" at the tropical Atlantic surface can be traced over distances ranging from several hundred up to thousands of kilometers in the upper ocean. They are characterized by very distinct and in general strong seasonally varying spatial extents.

3.1 Amazon and Orinoco River Plume Monitoring

The Amazon is the world's largest river in terms of freshwater discharge (Milliman and Meade 1983; Perry et al. 1996). It drains a large fraction of the South American continent, discharging on average $1.55 \pm 0.13 \times 10^5\ m^3\ s^{-1}$ of freshwater into the equatorial Atlantic Ocean (Perry et al.1996). This is about 15 % of the estimated global river discharge on an annual basis. The Amazon River is by far the largest single source of terrestrial freshwater to the ocean and contributes about 30 % of total river discharge to the Atlantic Ocean (Wisser et al. 2010). The structure of the Amazon plume is strongly influenced by a variety of physical processes, which are present on the northern Brazilian shelf: the North Brazil Current (Flagg et al. 1986; Richardson and McKee 1984), trade winds (Hellerman and Rosenstein 1983) and strong currents associated with the tide (Nittrouer and Demaster 1986). These physical processes play a very significant role in the dispersal and spreading of Amazon discharge (freshwater and suspended sediment) on the northern continental shelf of South America.

Previous studies have shown that Amazon plume water can be traced offshore and northwestward along the north Brazilian coast, covering most of the continental shelf from 11°S to 5°N (Muller-Karger et al. 1988, 1995) into the Caribbean (e.g., Steven and Brooks 1972; Froelich et al. 1978; Hellweger and Gordon 2002; Cherubin and Richardson 2007), and over 1,000 km eastward into the North Atlantic depending on the season. Beyond this region, the Amazon's water has been traced northwestward into the Caribbean Sea and eastward in the North Atlantic (Muller-Karger et al. 1988, 1995; Johns et al. 1990; Hellweger and Gordon 2002). Hydrographic surveys by Lentz and Limeburner (1995) revealed that the Amazon plume over the shelf is typically 3–10 m thick and between 80 and >200 km wide. Beyond the shelf, freshwater within the plume gradually attenuates with depth as it travels away from the source, with a penetration depth of 40–45 m as far as 2,600 km offshore (Hellweger and Gordon 2002; Hu et al. 2004).

Both chlorophyll (Chl) concentration and primary productivity are the greatest in the river plume–ocean transition zone, where the bulk of heavy sediments are deposited (Smith and Demaster 1996). The combination of riverine nutrient input and increased irradiance availability creates a highly productive transition zone, the location of which varies with the discharge from the river. High phytoplankton biomass and productivity of over 25 mg Chl-a $m^{-3}$and 8 g cm^{-2} day^{-1}, respectively, are found in this transition region (Smith and Demaster 1996). Because of this, the North Brazil shelf acts as a significant sink for atmospheric CO_2 (Ternon et al. 2000).

The northwestern tropical Atlantic is also an area where another major river in the world, the Orinoco, enters the ocean. The Orinoco River originates in the southern part of Venezuela and discharges waters from about 31 major and 2000 minor tributaries into the western tropical Atlantic. These waters are most of the time transported into the southeastern Caribbean sea, and during the rainy season, a larger but unquantified fraction of the plume also flows east around Trinidad and Tobago into the Caribbean. The Orinoco is considered to be the third largest river in the world in terms of volumetric discharge (after the Amazon and the Congo), discharging an average of $\sim 3.6 \times 10^4\ m^3\ s^{-1}$ (Meade et al. 1983; Muller-Karger et al. 1989; Vörösmarty et al. 1998). Low discharge occurs during the dry season (January–May) and high discharge during the rainy season (July–October) as a result of the meridional migration of the ITCZ.

The freshwater discharges from the Amazon and Orinoco Rivers spread outward into the western equatorial Atlantic Ocean while continually mixing with surrounding salty ocean surface water. The averaged geographical distribution of the low-salinity signatures of the Amazon and Orinoco River plumes can be revealed with historical in situ surface salinity data. However, only satellite remote sensing data are known to provide means to monitor the wide surface dispersal of these two fresh pools, with ocean color data being the first to illustrate Amazon plume reach to well beyond 1,000 km (Muller-Karger et al. 1988). Since these first observations, the application of ocean color, altimetry, and SST satellite mapping in this region has increased in its sophistication, showing the ability to track surface plume area (e.g., Hu et al. 2004; Molleri et al. 2010), fronts along the shelf to the northwest (Baklouti et al. 2007), and northward propagating eddies or waves shed near the North Brazil Current (NBC) retro reflection region, the so-called NBC rings (Ffield 2005; Goni and Johns 2001; Garzoli et al. 2004). In each case, the satellite data are able to provide time-resolved information on advective processes up to certain limits that include cloud cover, minor SST and ocean color gradients, non-conservative dilution processes for the ocean color to salinity conversions (Salisbury et al. 2011), and baroclinicity and subgrid variability of the altimetry sea surface height anomaly tracking of the NBC rings. As first evidenced by Reul et al. 2009, passive remote sensing data at low microwave frequencies can be successively used to complement these more "classical" satellite observations to better follow the temporal evolution and spatial distribution of surface salinity within and adjacent to the Amazon River plume.

To illustrate this new capability, we first show in Fig. 2 comparisons between collocated SMOS SSS and in situ conductivity–temperature–depth (CTD) measurements acquired during the Geotraces West Atlantic cruise leg 2 across the Amazon River plume in June 2010. This campaign was conducted on RV Pelagia in the frame of the GEOTRACES international program (see http://www.geotraces.org/).

Comparison between satellite and 3-m depth in situ SSS data reveals an overall good agreement with a standard deviation of the difference $SSS_{SMOS}–SSS_{CTD}$ of $\sim$0.45. In particular, the strong gradient and $\sim$3-unit drop observed as the R/V Pelagia leg crossed the Amazon River plume is well detected by the satellite observations.

New SSS products from satellite platforms such as SMOS allow in particular to gain insights into the advection pathways of the freshwater Amazon and Orinoco rivers plume along surface currents. For the first time, SMOS sampling capability thus enables imaging the plume structure almost every 3 days with a spatial resolution of about 40 km.

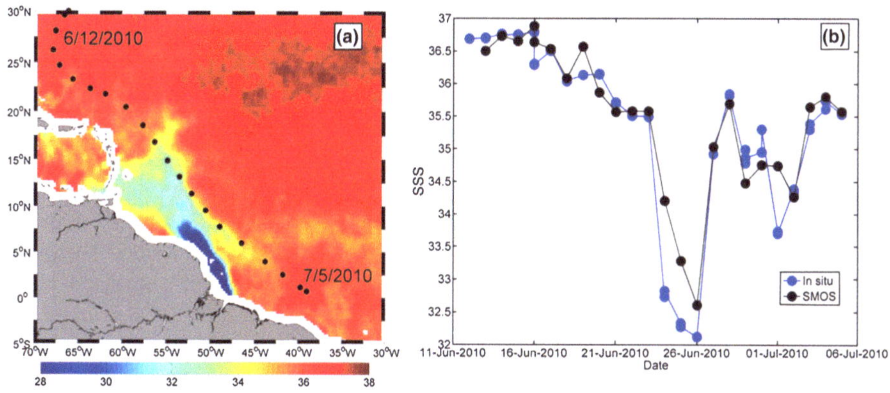

Fig. 2 **a** *Black dots*: location of the CTD stations conducted during the Geotraces West Atlantic cruise leg 2 (RV Pelagia) from 11 June to 5 July superimposed on the SMOS averaged SSS from June 12 to July 5, 2010. **b** Colocated surface salinity between SMOS and in situ data along the leg. SMOS data have been averaged at 50-km resolution with a ±5-day running temporal window

Combining SMOS SSS with altimeter-derived geostrophic currents and wind-driven (Ekman) estimated motions (Lagerloef et al. 1999), the advection of the spatial patterns of low salinity discharged from the major river mouths can now be analyzed systematically with an unprecedented resolution.

As illustrated by the Fig. 3 and by the animation available at http://www.ifremer.fr/naiad/salinityremotesensing.ifremer.fr/altimetry_amazon_atl.gif, a very good visual consistency is found between the geostrophic and Ekman surface current pattern estimates and the SMOS SSS spatiotemporal distribution along the year.

Mignot et al. (2007) show a long-term seasonal to monthly climatology that highlights two freshwater offshore pathways—the north passage to the warm pool and eastward entrainment into the North Equatorial Counter Current (NECC)—but they cannot clearly confirm or track this laterally with time in a given year.

SMOS SSS data combined with altimetry and surface wind information now enable to follow the spatiotemporal evolution of the plume along these two freshwater offshore pathways.

As illustrated in Fig. 3 (top), the surface freshwater dispersal patterns of the Amazon River plume are closely connected to the surface current topology derived from the merged altimeter and wind field product. As also evidenced earlier from several hydrographic surveys (e.g., Hellweger and Gordon 2002), it is clearly apparent in the satellite imagery that the NBC rings are key factors in modulating the freshwater pathways of the Amazon plume from the river mouth at the equator toward higher latitudes up to 20°–22°N.

Eastward entrainment of low-salinity water from the mouth of the Amazon River into the NECC is also evident in the SMOS data for the second half of the year 2010 (see Fig. 3, bottom). During that period, freshwater dispersal structure exhibits a zonal wavy pattern centered around ~ 8°N induced by current instability waves shed near the NBC retroflection region (52°W, 8°N). To analyze the freshwater plume transport and the evolution of salinity along Lagrangian paths following such wavy patterns, hypothetical drifters were dropped around the mouth of the river at the beginning of June and temporally advected with the surface currents deduced from merged altimeter and wind products. The evolution of SSS from SMOS L-band and AMSR-E C-band sensors (see Reul et al. 2009 for details on the AMSR-E SSS product), sea surface temperature analysis products and merged MERIS-MODIS colored dissolved organic matter (CDOM) absorption coefficient was estimated by interpolating the data in space and time along the path of such drifters.

As further illustrated by the example shown in Fig. 4, it takes approximately 6 months to cover a distance of 3,700 km for a freshwater particle (SSS ~ 26–28) in the proximity of the Amazon mouth to relax to an open ocean surface salinity of ~36. At the beginning of the period, the low SSS of water particles is modulated by mixing processes with saltier waters transported westward by the NBC rings shed at the NBC retroflection. The particle-following SSS signal modulation observed here is clearly consistent with the ocean color signal (anti-correlated with SSS), fresher water being systematically associated with colored waters showing high CDOM values, typical of the brackish plume waters. The drifter is then advected eastward along the NECC, remixed with "younger" advected plume waters in August and reached an eastern position slightly north of 8°N–38°W with an SSS of about 32 at the beginning of October. The SSS change along the drifter pathway is progressively and quasi-linearly relaxing to the open ocean values during the next 3-month period.

The link between the SSS and ocean color properties moreover enables investigations of the interactions between bio-optical and bio-chemical properties of the ocean and hydrological fluxes of terrestrial origin. Along with the freshwater, the Amazon provides the

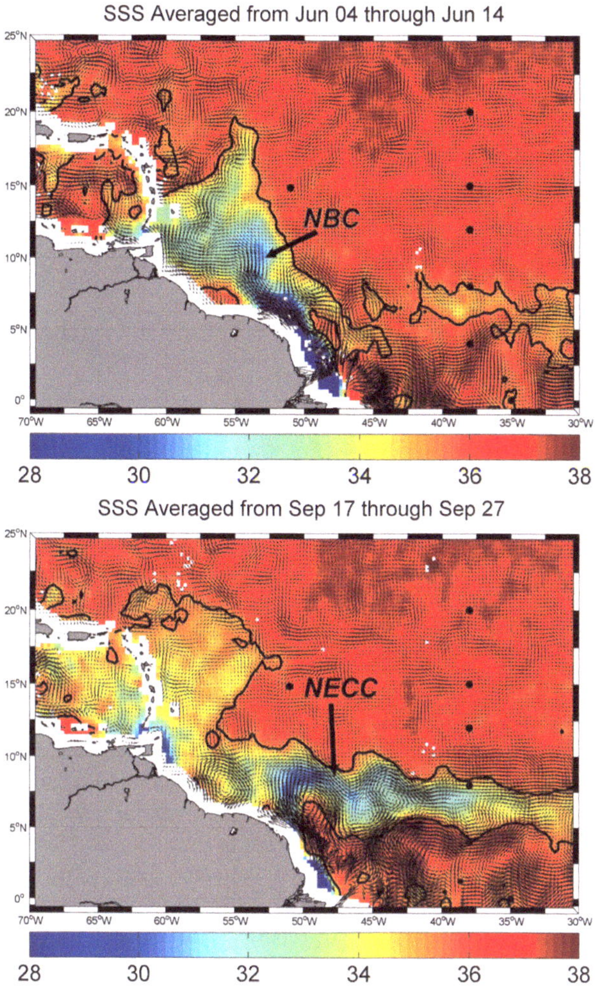

Fig. 3 Major pathways for the freshwater Amazon–Orinoco River plume detected by SMOS in 2010. Surface salinity fields from SMOS are superimposed with coinciding surface OSCAR currents estimated from altimetry and surface wind data. *Top*: the freshwater Amazon River plume is advected northwestward along the Brazilian Shelf by the North Brazilian Current (NBC) during boreal spring. *Bottom*: during boreal summer to fall period, the Amazon plume is carried eastward by the NECC. Note also the signal from the Orinoco River plume extending northeastward along the southern lesser Antilles. In both plots, the *thick black curve* is indicating the 35 SSS contour

largest riverine flux of suspended (1,200 Mt year^{-1}) and dissolved matter (287 Mt year^{-1}), which includes a dissolved organic matter (DOM) flux of 139 Mt year^{-1} (Meybeck and Ragu 1997). These fluxes can have a dramatic effect on regional ecology as they represent potential subsidies of organic carbon, nutrients, and light attenuation into an otherwise oligotrophic environment (Muller-Karger et al. 1995).

In the regions closest to the Amazon plume, light attenuation by suspended detritus acts as the main limitation to phytoplankton growth (DeMaster et al. 1996). Away from this region, as mineral detritus is removed by sinking, absorption attributable to organic

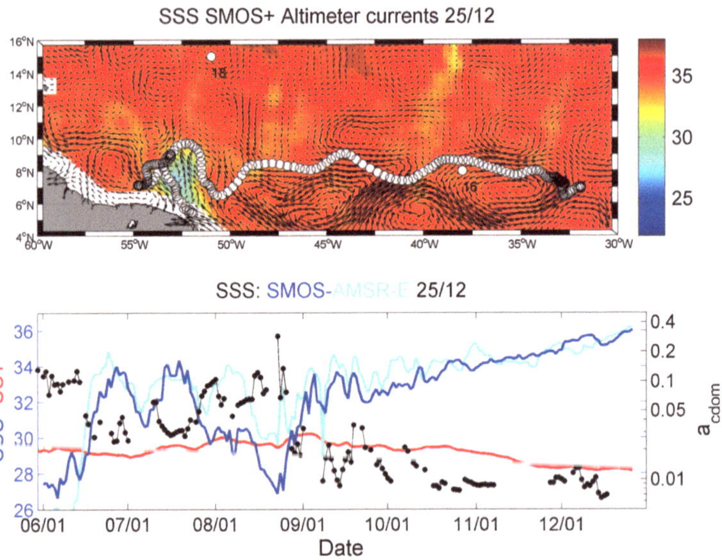

Fig. 4 *Top*: spatiotemporal evolution of the location of an hypothetical drifter (*white dots*) dropped at 52°W 6°N at the beginning of June 2010 and advected with surface currents estimated from altimetry and surface winds (*arrows*). Superimposed are the ±5 days averaged daily SSS fields from SMOS and the surface currents (*black arrows*). *Bottom*: time series of the colocalized SSS from SMOS (*blue*) and from AMSR-E (*cyan*), the analyzed SST (*red*), and the merged daily CDOM (*black*) along the drifter path

substances begins to dominate the attenuation of light in surface waters. Del Vecchio and Subramaniam (2004) studied such conditions in the Amazon plume and characterized the relative contributions of CDOM, particulate organic material, and phytoplankton to the total absorption field. In the coastal ocean adjacent to river sources, CDOM tends to behave as a freshwater tracer, decreasing away from the river source with increasing salinity. Linear correlations between CDOM and salinity in river plume waters are well documented in the ocean color literature with reported relationships robust enough to allow salinity retrievals from CDOM and vice versa (e.g., Ferrari and Dowell 1998; Palacios et al. 2009; D'Sa et al. 2002; Conmy et al. 2009).

Linearity in the CDOM–salinity relationship implies conservative mixing dominated by two distinct endmembers. Departures from linearity can occur when additional water masses are present (Blough and Del Vecchio 2002), or by in situ subsidies of CDOM released via net phytoplankton growth (Yamashita and Tanoue 2004; Twardowski and Donaghay 2001), microbial utilization (e.g., Moran et al. 1999; Obernosterer and Herndl 2000), or photochemical oxidation (e.g., Miller and Zepp 1995).

Based upon preliminary satellite microwave SSS data from AMSR-E sensor and ocean color products, Salisbury et al. (2011) recently demonstrated the spatial coherence between surface salinity and the absorption coefficient of CDOM at 443 nm in the Amazon and Orinoco river plume-influenced waters. Given the new SMOS data, the spatial and temporal coherence between SSS and optical properties of the river plumes, e.g., CDOM, can now be systematically analyzed.

As illustrated in Fig. 5, the amplitude of the annual cycle of the Amazon River discharge peaks in June–July and was apparently more important in 2010 and 2011 compared to the averaged "climatological" cycle since 1968. In comparison, the discharge from Orinoco is much lower and peaks in September. Based upon the Amazon River discharge cycle, four main periods can be distinguished as shown in Fig. 6. From November to April (low flow and ascending periods), the plume is carried northwestward with the NBC, while the summer and fall display a plume mostly carried eastward as the seasonal NECC retroflection strengthens. In comparison, the spatial pattern in the distributions of the CDOM is in general very similar to SSS during the river discharge seasonal cycle. However, the CDOM patterns can deviate from the SSS patterns at large distances from the mouth of the river for some period of the seasonal cycle. This is particularly evident in the region around the northern Antilles and the Caribbean during the high-flow season of 2010 (Fig. 6, third panel from top) whereby high CDOM values are detected north of the low-salinity plume extent (contours at SSS = 35.5 on the right panels), suggesting the presence of dissolved organic matter concentrations that are non-correlated with the Amazon River plume dilution. Altogether, this demonstrates the strength in combining satellite SSS observations with complementary satellite observations in order to better characterize the variability of the pathway of freshwater runoff along with the corresponding mixing processes at seasonal to interannual time scales.

Quasi-linear relationships between SMOS SSS and the MERIS/MODIS CDOM absorption coefficient (acdm) estimated for year 2010 are illustrated in Fig. 7. Acdm values were averaged over SSS bins with 0.5 bin width. As evidenced, while CDOM mixing processes seem to be conservative on average, clear departure from linearity is observed below 30 pss during the descending and low-flow seasons. This fact potentially indicates changes in the endmember values at the mouths of the rivers and tributaries and/or illustrate the occurrence of non-conservative mixing processes as listed above. Thanks to the new satellite observations, departure from conservative mixing and the interannual sources of variability will be certainly more detailed in the next future.

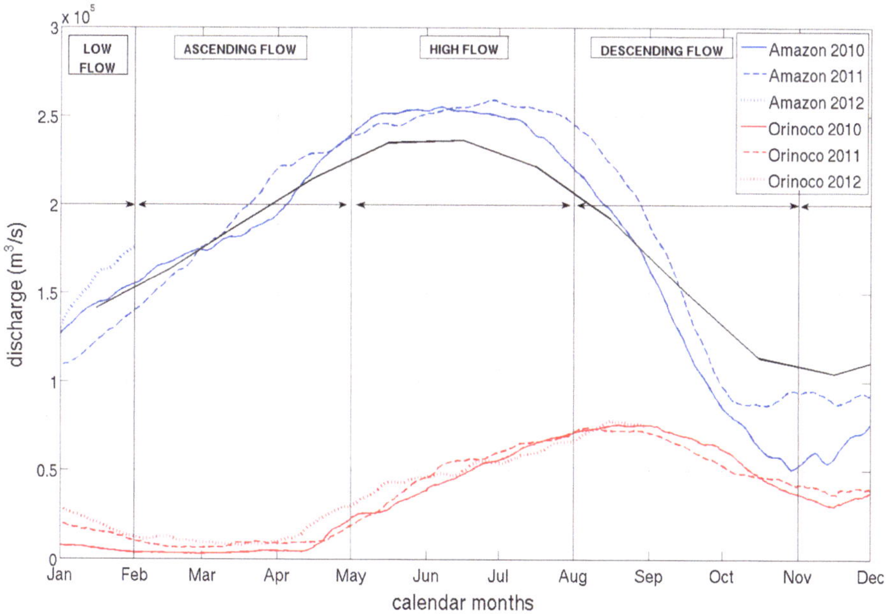

Fig. 5 Amazon (*blue*) and Orinoco (*red*) river discharge cycles measured, respectively, at Obidos and Bolivar gauges, during the period 2010–2012. The *black curve* is showing the Amazon River discharge climatology from 1968 to 2012

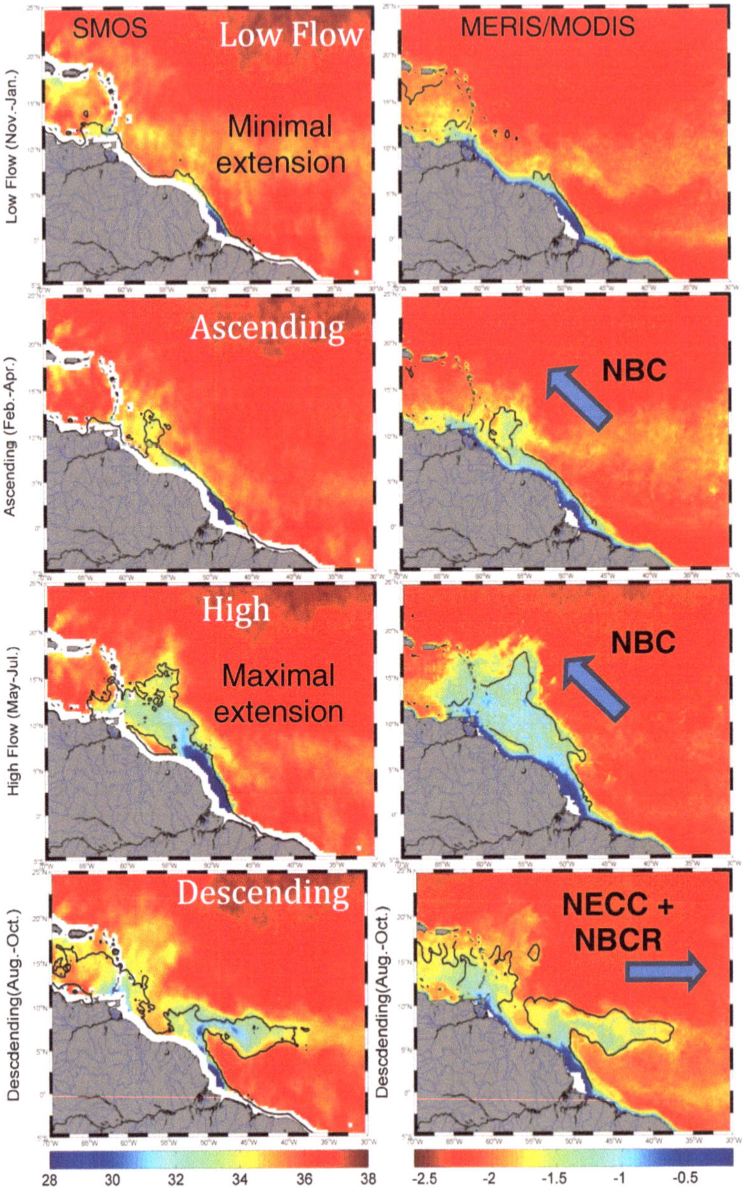

Fig. 6 Seasonal cycle of the freshwater Amazon and Orinoco river plume signals for year 2010. *Left*: SSS from SMOS averaged over the different periods of the discharge cycle. From top to bottom: low flow (November–January); ascending flow (February–April); high flow (May–July); descending flow (August–October). *Right*: corresponding CDOM absorption coefficient averaged from the merged MERIS/MODIS products. The *color bar* is logarithmic in unit of 1/m

3.2 Eastern Tropical Atlantic Freshwater Pools Monitoring

The eastern tropical Atlantic (ETA) Ocean 8°W–12°E, 6°N–20°S is a region of intense upwelling and where the second largest river in the world, the Congo, enters the ocean together with the Niger, Volta and numerous other smaller rivers (Fig. 8). In addition, intense precipitations also decrease SSS in the Guinea current and northeastern Gulf of Guinea (Hisard 1980; Merle 1980). The ETA is therefore characterized with a highly complex hydrographic system, largely influenced by the Congo River, intense precipitation, and strong seasonal coastal and equatorial upwelling in the boreal summer.

Maximum discharge from the Congo River occurs in December and minimum discharge in March through April. The outflow is hardly detectable from SST or sea level data. In chlorophyll, however, the mouth of the Congo River shows a strong signal all year round with large plumes extending offshore. While these ocean color signals highlight real oceanographic features of the plume, frequent cloud cover found in this region during the rainy season strongly inhibits the spatiotemporal evolution of the Congo plume structure to be monitored.

Hitherto the knowledge about the seasonal extension and spreading of the Congo River plume is therefore mainly relying on dedicated in situ surveys (e.g., see Meulenbergh 1968;

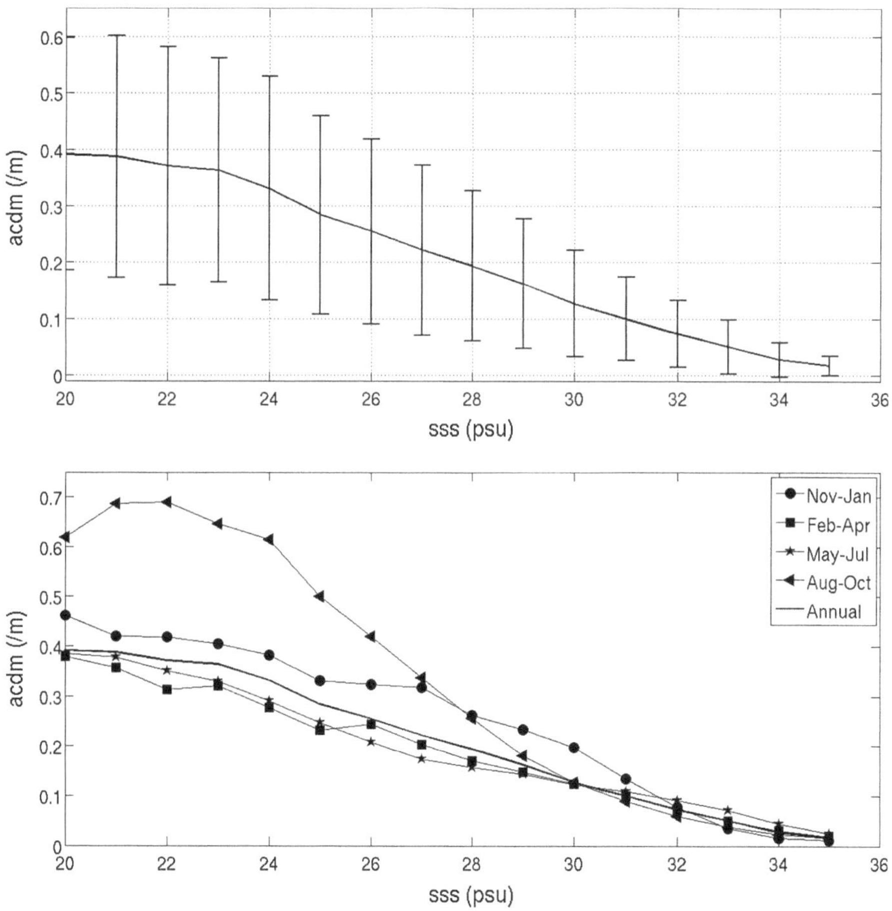

Fig. 7 $a_{CDOM}(490)$ to SMOS SSS dependence in the western tropical North Atlantic averaged over years 2010–2012 for all seasons of the Amazon River Discharge cycle (*Top*) and for each season separately (*bottom*). In the upper panel, the mean $a_{CDOM}(490)$ per 0.5 bins is shown as a *solid black line* ± 1 standard deviation (*vertical bars*)

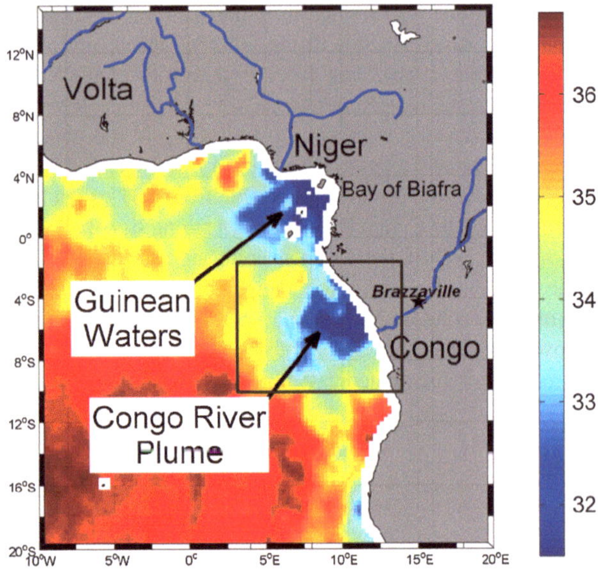

Fig. 8 Map of SMOS SSS in the Gulf of Guinea and Southeast Atlantic Ocean indicating the two largest pools of low-salinity waters in the eastern tropical Atlantic: the Bight of Biafra (Guinean waters) and the Congo River plume. The map was generated by averaging SMOS data over 2010–2012 considering only data acquired during months of April

Koleshnikov 1973;Bornhold 1973; Wauthy 1977;Van Bennekom and Jager 1978; Eisma and Van Bennekom 1978; Van Bennekom and Berger 1984; Piton and Wacongne 1985; Braga et al. 2004; Reverdin et al. 2007; Vangriesheim et al. 2009; Lefèvre 2009). However, the ensemble of in situ SSS data collected during the period 1977–2002 in the ETA is sparse, and only enabled retrievals of low-resolution ($1° \times 1°$) monthly climatology of the SSS field (Reverdin et al. 2007), as displayed in Fig. 9. Note that since 2003, the in situ SSS sampling has however improved with the increasing deployments and operations of Argo floats.

The monthly averaged SMOS SSS maps shown in Fig. 10 were generated by combining SSS data over the SMOS 3-year life period. As evidenced in detail by these maps, consistent with historical in situ observations, the Congo River plume is spreading northwestward along the coast and mixes with southwestward flowing freshwater from the bight of Biafra during February and March (Koleshnikov 1973; Wauthy 1977). In May (Van Bennekom and Jager 1978), June–July (Bornhold 1973; Wauthy 1977), and August (Koleshnikov 1973), the two fresh pools are disconnected with the Congo plume directed in westerly direction, extending up to 800–1,000 km offshore, as far as 8°E. In November, a "jet stream" of low-salinity water is ejected from the estuary with a large velocity and protrudes in WNW direction (Wauthy 1977). The plume extent can also show southward and southwestward legs depending on the prevailing windstress in the Angola Basin (Van Bennekom and Berger 1984; Dessier and Donguy 1994).

The dispersal patterns of the Congo River plume during all seasons can mostly be included inside the rectangle domain shown in Fig. 8. The 10-day running mean time series of the SMOS SSS averaged over that spatial domain is shown in Fig. 11 together with the time series of the river discharge measured at Brazaville gauge station during the period 2010–2012. Maxima in the averaged SSS within that region occur regularly in August at the time of the Congo River minimum discharge. Minima in SSS (detected

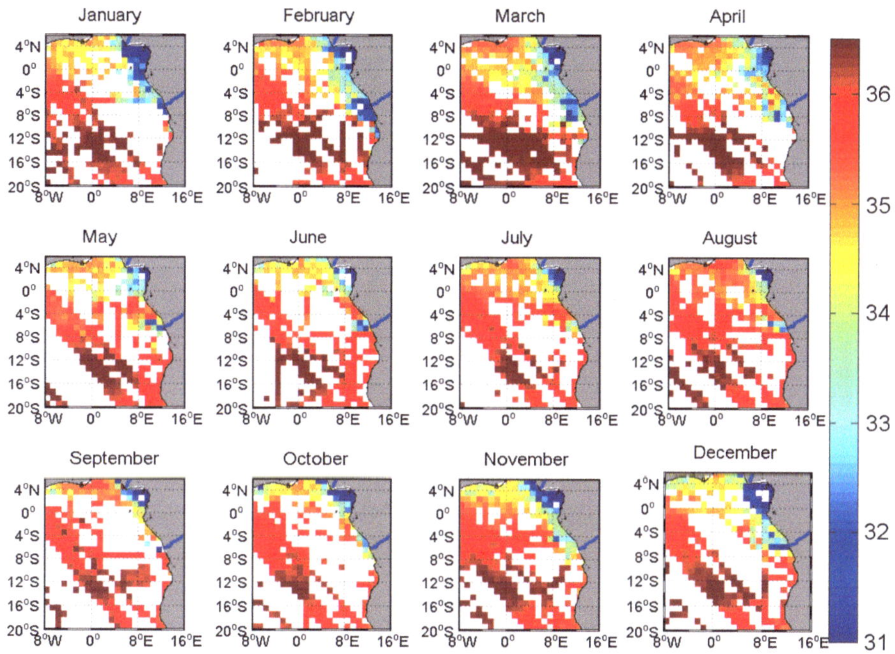

Fig. 9 Maps of the monthly averaged SSS in the ETA derived from the ensemble of in situ measurements collected during the period 1977–2002 and used to build up Reverdin et al. (2007) climatology

around April), however, lag by approximately 4 months the maxima in the river discharge at Brazaville station (found around December–January). These lags probably indicate the time for the freshwater masses to be transported from Brazaville to the river mouth and then to be further advected by surface currents far offshore. However, the interannual variability in the amplitude of the seasonal cycle of SSS and river discharge are not correlated. While the river discharge reached significantly different minimum values of $\sim 3.3 \times 10^4 m^3/s$ and $\sim 2.3 \times 10^4 m^3/s$ in 2010 and 2011, respectively, the maxima in the averaged SSS are constantly found at ~35.5 pss. Similarly, the maximum discharge level of $\sim 5.8 \times 10^4 m^3/s$ measured over the period is found in January 2012, while the minimum in the averaged SSS (~31.9) occurred in April 2011.

While understanding the observed satellite SSS trend in that region is still an undergoing activity, combining satellite information on surface currents, SST, rain rates and SSS together with river discharge levels will certainly help in the near future to better quantify the sources of variability in the local hydrological cycle of the Gulf of Guinea. The terrestrial and atmospheric hydrological fluxes in this region also act as a dominant modulator of the local fishery. The regular SMOS SSS data can therefore help to better understand the mechanisms involved in the biophysical interplay and its relevance for the fishery with potentially significant socioeconomic impact in that region.

In addition, similarly to the Amazon–Orinoco River plumes, conservative mixing laws for bio-optical properties of the major river plume in the ETA region can now be systematically studied using SMOS data as shown in Fig. 12. Examples of the conservative mixing linear laws for the CDOM coefficient deduced only from space-borne measurements are shown for year 2010 around the Congo and Niger rivers.

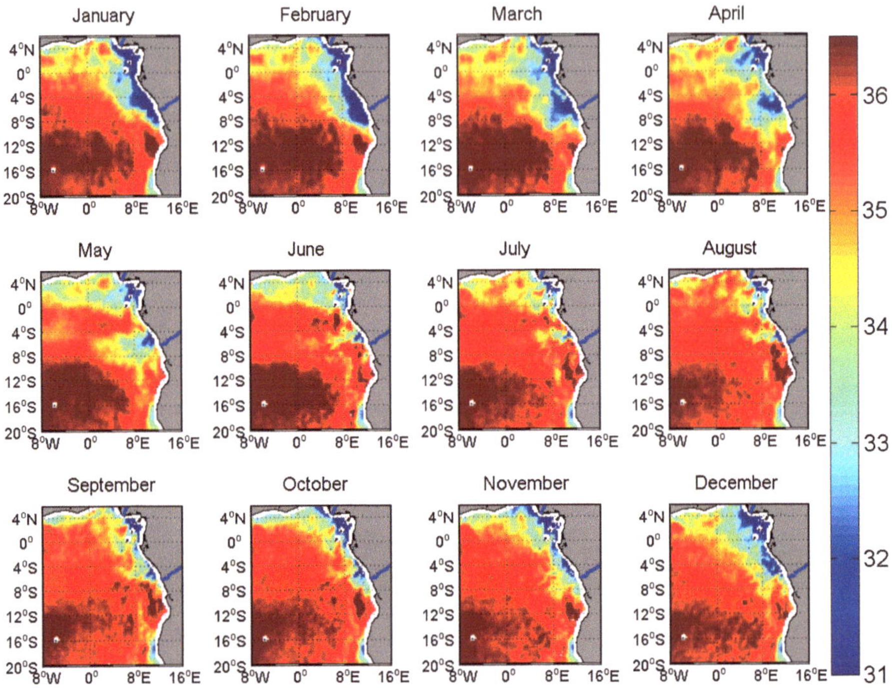

Fig. 10 2010–2012 Monthly averaged seasonal cycle of surface salinity in the eastern tropical Atlantic derived from SMOS observations

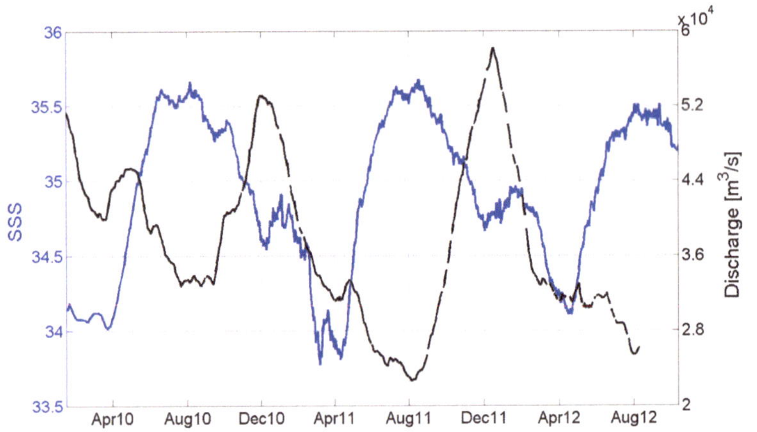

Fig. 11 Times series of (i) the SMOS SSS averaged over the spatial domain (3°–14°E;10°–2°S) illustrated by the *black rectangle* in Fig. 8 (*blue*) and (ii) of the Congo discharge level measured at Brazaville (*black*)

4 Precipitation Signatures in SSS Data from Space

Large vertical gradients can develop in the upper few meters of the ocean after a heavy rainfall, as first evidenced during the Tropical Oceans-Global Atmosphere Coupled Ocean–Atmosphere Response Experiment (TOGA COARE) (Soloviev and Lukas 1996;

Schlössel et al. 1997; Wijesekera et al. 1999). The downward freshwater flux at the sea surface establishes a haline diffusive molecular layer (or freshwater skin of the ocean) (Katsaros and Buettner 1969) that is characterized by a salinity gradient, with salinity differences across this freshwater skin sometimes greater than 4 salinity units. The residual effects of the rain-induced skin layers can even be stronger at the highest rain rates (Schlössel et al. 1997). This freshwater skin stabilizes the near-surface layer (Ostapoff et al. 1973) and tends to dampen free convection in the upper oceanic boundary layer.

These conditions motivate the development of autonomous SSS drifters able to monitor the salinity at less than 50-cm depth. Using such instruments, Reverdin et al. (2012) documented salinity freshening between 15-cm and 50-cm depth in the tropical oceans. Sudden salinity decreases are often associated with local rainfall and vertical salinity gradients that last for a few hours, depending, among other factors, on wind speed conditions. The haline molecular diffusion layer that is established in the upper ocean during rainfall can thus be important for the radiometric observation of the sea surface at low microwave frequencies. At centimeter wavelengths, the dielectric constant is modified by the sea surface salinity (e.g., Klein and Swift 1977; Yueh et al. 2001) and any change of the latter might cause interpretation problems when comparing remotely measured surface salinity at these frequencies to deeper in situ measurements.

Hence, under rainy conditions (or just after a rainfall), the satellite-derived SSS better characterizes the salinity at the ocean–atmosphere interface rather than the 1–10-m deep in situ samples. Whether accumulated precipitation can be estimated from changes in salinity at the ocean surface as observed from space remains, however, an open question, as assumptions have to be made about the penetration depth of the freshwater. In addition, assimilation of the new satellite SSS data into ocean circulation models having limited vertical resolution also challenges our modeling perspectives concerning the dynamics of the first centimeters to first meter of the ocean surface.

In the following section, we discuss signatures of precipitation detected in the new SMOS SSS data. First, the strong SSS spatiotemporal variability associated with rain

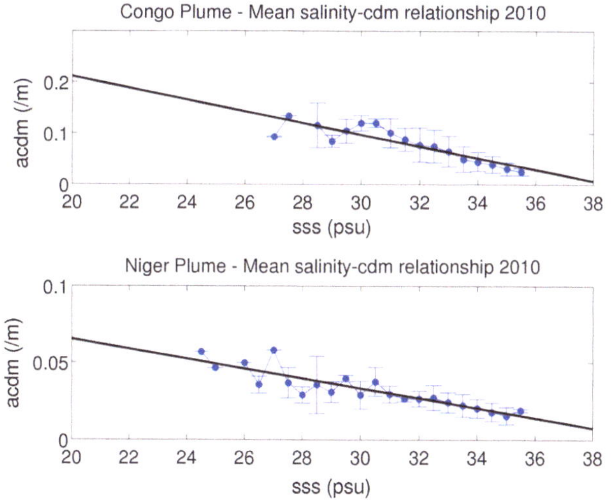

Fig. 12 $a_{CDOM}(490)$ to SMOS SSS dependence in the eastern tropical Atlantic averaged over year 2010 for the Congo(*Top*) and Niger (*Bottom*) River Plumes. The mean $a_{CDOM}(490)$ per 0.5 bins is shown as a *solid black line* ± 1 standard deviation (*vertical bars*)

events as seen both by space-borne and in situ sensors in the Pacific Ocean Inter Tropical Convergence Zone is presented. Second, it is revealed that the SSS from space is systematically showing lower values (negative bias) with respect to the deeper 5–10 m depth of Argo upper salinity in this area. These effects are shown to be statistically correlated with rain. Third, long-lived, large-area, and large-amplitude SMOS SSS anomaly patterns in the tropical Atlantic are shown to follow local anomaly patterns in the evaporation–precipitation (E–P) budget. Finally, some preliminary results concerning the interannual variability of the SMOS SSS signal in the Indian and in the tropical Pacific oceans and connections to key climate indexes will be presented and discussed.

4.1 SSS Temporal Variability Associated with Rain Events

Although satellite observations provide a better sampling of the global ocean than the in situ observing systems, such as the Argo float array, individual SSS measurements are obtained in rainy regions with a strong temporal variability seen on both SMOS and Argo SSS. In Fig. 13, we show such an example of colocated SMOS and Argo profiler measurements in the Inter Tropical Convergence Zone of the Tropical Pacific, indicating a significant surface freshening associated with a rain event. On August 11, 2010, the Argo float WMO id#4900325 detected a freshening of 0.9 between 20- and 5.5-m depth (Fig. 13a). In contrast, the Argo profile derived on 22 August shows that the salinity between 30- and 5-m depth is much more homogeneous with more saline water at 5-m depth compared to the one recorded on 11 August.

The TRMM satellite rain rate (RR) estimates averaged over a 2° × 2° box centered on the Argo float location indicate a significant rain rate of 1–2 mm h^{-1}on 11 August that lasted for at least a day before the Argo profile raised to the surface (Fig. 13c). Contrarily, negligible precipitation occurred on 22 August and during the preceding week. The first SMOS pass collocated with the 11 August Argo profile (Fig. 13a) was acquired also during rainy conditions and showed a low SSS of ~32.8 (0.1 saltier than the Argo SSS taken 6:30 h later, Fig. 13c). The second SMOS pass on the 16 August occurred under non-rainy condition (Fig. 13c) and is 0.5 saltier. Consistent with the 22 August Argo profile (Fig. 13b) observations, the collocated SMOS SSS during these rain-free conditions (Fig. 13c) are also significantly saltier by 0.4–0.6. The large SSS variation (0.7) measured by this Argo float at a 10-day interval and by the collocated SMOS measurements over several SMOS passes clearly demonstrates the influence of the rain timing on the SMOS-Argo SSS differences.

4.2 Systematically Fresher Skin SSS in Rainy Regions

The SMOS SSS map averaged over July–September 2010 is compared to optimally interpolated in situ ISAS map averaged over the same period shown in Fig. 14. At large scale, SSS spatial variability sensed by SMOS is consistent with ISAS. A striking visual feature of the SMOS SSS map compared to the ISAS map in the tropics is the freshest SSS in the North Tropical Pacific, under the location of the ITCZ (particularly west of 120°W).

When SMOS SSS are precisely colocated around Argo SSS in various regions of the global ocean (see Boutin et al. 2012a, b), a more negative bias (~−0.1 than in other regions) and larger standard deviation are systematically observed between 5° and 15°N in the Pacific Ocean with respect to other regions (Table 2).

To investigate whether a systematic negative bias of ~0.1 between the satellite skin depth SSS and the ~5-m depth Argo floats data could be related to rain-induced vertical

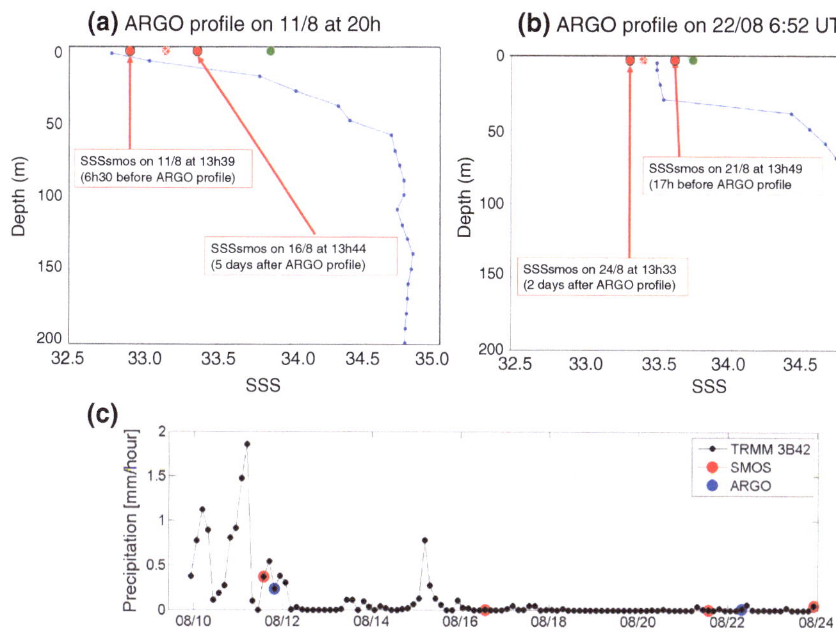

Fig. 13 Two successive Argo profiles taken by float 4900325 (*blue curve*) in the eastern tropical Pacific on **a** 11 August 20:00 UTC (latitude = 12.4°N; longitude = 117.6°W) and **b** 22 August 6:52 UTC (latitude: 12.2°N; longitude: 117.8°W). Mean SMOS SSS collocated within a 5-day window and a radii of 50 km with these profiles are indicated by red dashed point. In each case, two SMOS passes have participated to these collocations: mean SMOS SSS corresponding to each pass is indicated as red filled point. The corresponding ISAS SSS in August is indicated by the green point. The time series of the 3-hourly satellite rain rate from TRMM 3B42 and averaged over (11°–13°N; 116°–118°W) is provided in (**c**). The time at which SMOS and Argo acquired SSS data is indicated by red and *blue dots*, respectively

stratification, a triple collocation between Argo, SMOS level 2 products (at ~40-km resolution, non-averaged in time) and SSMI satellite rain rate (RR) data was conducted. SMOS and SSMI RR data were colocated within a temporal window of −40 min and +80 min, while a ±5-day windows was considered to colocate SMOS and Argo data.

The theoretical error on the SMOS SSS retrieved level 2 data used in this colocation exercise is ~0.5. Without any RR sorting, the statistical distribution of the differences ΔSSS is skewed toward negative values (Fig. 15, Table 3); when only SMOS non-rainy events are considered, the negative skewness disappears, and statistics of the SMOS–Argo differences in the tropical Pacific Ocean become close to the ones in the subtropical Atlantic Ocean (Tables 2, 3). Largest skewness toward negatives differences is obtained when only SMOS SSS close to rain events are considered. For these rainy SMOS cases, we find a negative dependency of the SMOS–Argo SSS differences with respect to SSMIs RR of −0.17 pss/mm^{-1} h, i.e., a freshening of 1.7 for a SSM/I RR of 10 mm h^{-1}(Boutin et al. 2012a, b).

The non-sorting of SMOS measurements close in time with rain events in SMOS–Argo collocated data sets (within 10 days and 100 km) is responsible for (1) a mean −0.1 negative bias over 3 months between 5° and 15°N in the tropical Pacific region with respect to non-rainy conditions and with respect to the subtropical Atlantic region and (2) a negative skewness of the statistical distribution of SMOS minus Argo SSS difference (Fig. 15). Given that the whole set of SMOS–Argo collocations also includes the situations with rainy Argo measurements collocated with non-rainy SMOS measurements, these results indicate a

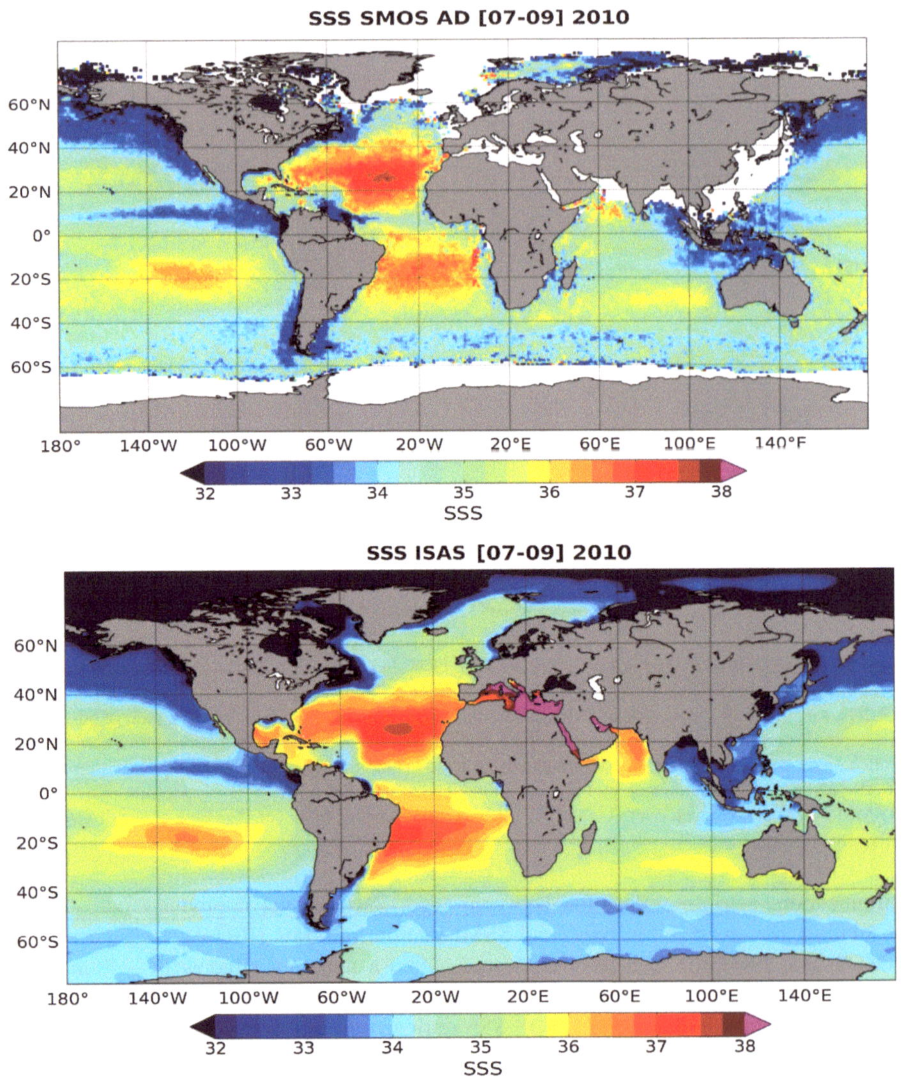

Fig. 14 Maps of SSS averaged from July to September 2010, derived from (*top*) SMOS ascending and descending orbits and ISAS (*bottom*)

systematic freshening of SMOS SSS in rainy conditions and are likely a signature of the vertical salinity stratification between the first centimeter of the sea surface layer sampled by SMOS and the 5-m depth sampled by Argo. For more detail on the vertical SSS stratification induced by rain, the reader is also referred to Boutin et al. (2012b).

4.3 SSS as a Tracer of the Evaporation–Precipitation Budget in the Oceanic Mixed Layer

The SMOS-derived SSS can also be used to investigate the consistency between observed SSS variability and the evaporation minus precipitation budget in the ITCZ of the tropical

Table 2 Comparison of SMOS SSS (10 day, 100 × 100 km^2 average) values, in pss, collocated with a total of N Argo upper depth measurements

	Mean (ΔSSS)	Std (ΔSSS)	*N*
Subtropical Atlantic Ocean (15°–30°N; 45°–30°W)	−0.13	0.28	206
Tropical Pacific Ocean (5°–15°N; 180°–110°W)	−0.23	0.35	692
Southern Indian Ocean (40°–30°S; 70°–90°E)	0.04	0.39	114
Southern Pacific Ocean (50°–40°S; 180°–100°W)	−0.08	0.51	467

$\Delta SSS = SSS_{smos} - SSS_{argo}$ Only SMOS ascending orbits are considered. Std (ΔSSS) primarily reflects the decreasing signal to noise ratio with decreasing SST. Note that subtropical Atlantic Ocean and tropical Pacific Ocean have similar SST

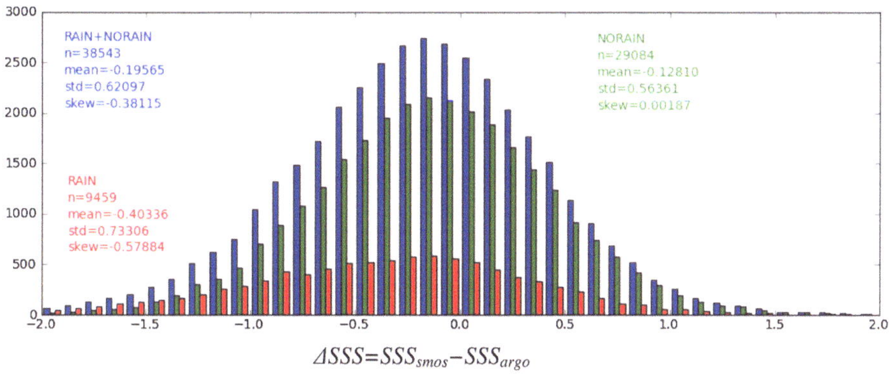

Fig. 15 Statistical distribution of SSS differences $\Delta SSS = SSS_{smos} - SSS_{argo}$ in the tropical Pacific Ocean for various sorting on colocated SSSM/I rain rates. *Blue*: all collocations (without any rain sorting); *green*: for non-rainy cases (SSM/I rain rates less than 0.1 mm h^{-1}); *red*: rainy cases (SSM/I rain rates larger than 0.1 mm h^{-1}). Corresponding statistics are indicated in Table 3

Table 3 Statistics for the SSS differences $\Delta SSS = SSS_{smos} - SSS_{argo}$as a function of rain rate (RR) in the northern tropical Pacific Ocean

	Mean (ΔSSS)	Std (ΔSSS)	Skew (ΔSSS)	*N* (ΔSSS)
Tropical Pacific (5°–15°N; 110°–180°W)				
All colocations	−0.20	0.62	−0.38	38,543
No rain (RR < 0.1 mm h^{-1})	−0.13	0.56	0.01	29,084
Rainy (RR ≥ 0.1 mm h^{-1})	−0.40	0.73	−0.58	9,459

Atlantic based upon the SSS and SST relationship in the ocean mixed layer (OML). The salt conservation budget in the OML with depth *h* can be expressed as follows (Michel et al. 2007; Yu 2010, 2011):

$$\frac{\partial S}{\partial t} = \frac{(E - P - R)S}{h} - \vec{u} \cdot \nabla S - \Gamma(w_e)\frac{w_e(S - S_h)}{h} + k\nabla^2 S \tag{1}$$

where *S* is the surface salinity, *t* is time, *E* and *P* the evaporation and precipitation rates, respectively, *R* the freshwater input by river runoffs, *h* the mixed layer depth, $\vec{u}$ the (vertically averaged) current vector within the OML, and w_e the vertical entrainment rate.

S_h is the salinity just below the OML, k is the horizontal diffusivity coefficient ($k \sim 2{,}000\ \text{ms}^{-2}$). The total entrainment term must be treated differently in case of upward or downward entrainment, so it is multiplied by a step function Γ in Eq. (1). Indeed, when additional water is included into the mixed layer, its properties are affected by mixing with the deeper layer: $\Gamma(w_e) = w_e$ if $w_e > 0$. On the contrary, if water is removed from the mixed layer, the properties of the remaining water are conserved and only its depth h can change: $\Gamma(w_e) = 0$ if $w_e < 0$. The vertical processes are conveniently represented by a single entrainment term, consisting of the vertical Ekman advection and the OML conditions.

The first term in the right-hand side of Eq. (1) is the net freshwater flux. The impact of this flux on the surface water strongly depends on the salinity itself. Moreover, SSS has no direct feedback on the surface flux. These particularities have important consequences on the salt budget and on the duration of SSS anomalies. The second term is the horizontal advection of salinity by surface currents that can be separated into a wind-induced component, the Ekman transport, and the geostrophic current. Ekman transport is due to wind friction on the sea surface, which is rotated by the Coriolis force as it penetrates in depth. The Ekman layer depth is systematically lower than the mixed layer depth, because both increases with the wind stress, although the depth of the mixed layer also deepens in response to other processes. Thus, the Ekman transport occurs entirely in the OML. In addition, the geostrophic current that arises from the balance between the horizontal pressure force and the Coriolis force can usually be considered constant, with the mixed layer resulting from the homogeneous density structure.

The value of the SMOS SSS at a fixed point, $S(t, r)$, is obtained by averaging individual SMOS swath SSS measurements over a considerable time interval ($t - \tau/2$, $t + \tau/2$), say 10 days, which is enough to filter out noise in the SSS. Suppose that the climate mean, or norm, of this SSS (provided by climatology) is $\overline{S(t,r)} = S_o(t,r)$. In the following, we define the SSS anomaly as the departure of the SSS from the norm:

$$\Delta S(t,r) = S(t,r) - S_o(t,r)$$

Following approaches traditionally used for studying large-area SST anomalies (Piterbarg, and Ostrovskii 1997), a formal definition can be introduced for the large-area SSS anomalies. For example, large-area and large-amplitude SSS anomaly comprises the connected components of the set:

$$\{(x,y) : |\Delta S(t,r)| > S_T\}$$

where $r = (x, y)$ and S_T is a threshold that can be taken either as a fixed salinity value, for example, 0.2 pss or as a function of the standard deviation of SSS anomalies, σ_S, for example, 0.5 σ_S. This choice for the threshold depends on the magnitude of the anomaly of interest.

In the tropical Atlantic, Michel et al. (2007) and Yu (2011) have shown that the dominant terms of the mixed layer salinity balance are horizontal advection by Ekman and geostrophic currents and the atmospheric forcing fluxes ($E - P - R$). In that context, the salinity balance equation in the OML can be simplified as follows:

$$\frac{\partial S}{\partial t} \cong \frac{(E - P - R)S}{h} - \vec{u} \cdot \nabla S \tag{2}$$

Using OSCAR surface current products (which comprise contributions of both Ekman and geostrophic currents), the horizontal salt advection term $\vec{u} \cdot \nabla S$ can be deduced from

SMOS observations. The following residual SSS anomaly can then be estimated from SMOS temporal observations of salinity $S(t, r)$ at point r following:

$$\Delta S(t,r) = S(t,r) - S_o(t,r) - \vec{u}(t,r) \cdot \nabla S(t,r) \tag{3}$$

According to the simplified salinity balance (Eq. 2), a priori valid for the tropical Atlantic, the resulting SSS anomaly given by Eq. 3 shall be strongly correlated with the net freshwater flux forcing term. Examples for such SSS anomaly analysis are shown in Fig. 16 for a selected point in the middle of the north tropical Atlantic (16°N–35°W). From TRMM precipitation and OAFLUX daily evaporation fluxes, large-area P and E anomalies were also evaluated:

$$\Delta P(t,r) = P(t,r) - P_o(t,r)$$

$$\Delta E(t,r) = E(t,r) - E_o(t,r)$$

where P_o and E_o are the local climate mean for the precipitation and evaporation.

As illustrated in Fig. 16 (middle right panel), very significant long-lived negative $\Delta S(t, r)$ values are detected in SMOS anomalies at the selected point in the north tropical Atlantic during September/October months (days 250–300) of 2010. Apparently, this happened just after a strong positive anomaly in the precipitation rate as detected from TRMM during the passage of the ITCZ in August (bottom right panel).

The spatiotemporal consistency between the large-area and large-amplitude S, P and E anomalies can be further analyzed over all the tropical Atlantic. This is illustrated in Fig. 17 for two selected months of 2010. The spatial distribution of the large-area and long-lived (monthly averaged) SSS anomalies generally matches well the spatial patterns for the large E–P anomalies. In particular, north–south oscillation in $\Delta S(t, r)$ around the ITCZ (centered on 5°N in March and 8°N in July) follows the ΔE–$\Delta P(t, r)$ far from the Amazon plume area, with negative $\Delta S(r, t)$ corresponding to positive $\Delta P(t, r)$ and positive $\Delta S(r, t)$ found in region of positive $\Delta E(t, r)$. The average relationship between SMOS SSS anomalies and the corresponding anomalies in the net atmospheric freshwater flux in the tropical Atlantic (defined here by 5°S–20°N;75°W–15°E) was further evaluated over year 2010 by binning $\Delta S(t, r)$ values as function of ΔE–$\Delta P(t, r)$ as shown in Fig. 18.

Despite a significant scatter in the data, the results clearly indicate the strong coherency between SMOS SSS anomalies and the evaporation minus precipitation flux signal in the tropical Atlantic. On average, SMOS SSS are thus systematically fresher than the SSS climatology when precipitation rate exceed evaporation rate with respect climatological means, and vice versa. As expected by the skin layer effects (Zhang and Zhang 2012), satellite SSS anomalies are weakly sensitive to excess evaporation showing an almost constant value whatever positive values for ΔE-ΔP. Nevertheless, and as discussed in Sect. 4, the average 0.3 salinity unit excess amplitude found for ΔS in evaporative zones is significantly larger than the expected evaporation-induced effect on the satellite ~0.01 pss. The source for such observed signal amplitude is not yet understood. Other physical processes, not yet well accounted for in the SSS retrieval algorithm, may systematically affect the L-band brightness temperature in strongly evaporative zone (e.g., skin effects in SST, badly accounted for roughness effects at low winds).

Nevertheless, Fig. 18 clearly shows that SSS anomalies become increasingly negative as the precipitation anomalies progressively exceed the evaporation anomalies.

This shows that it is important to monitor SSS from space in the rainy regions as it makes a good oceanic rain gauge for the changing water cycle (Cravatte et al. 2009; Yu 2011; Terray et al. 2011), and therefore help to maintain a continuous observation network

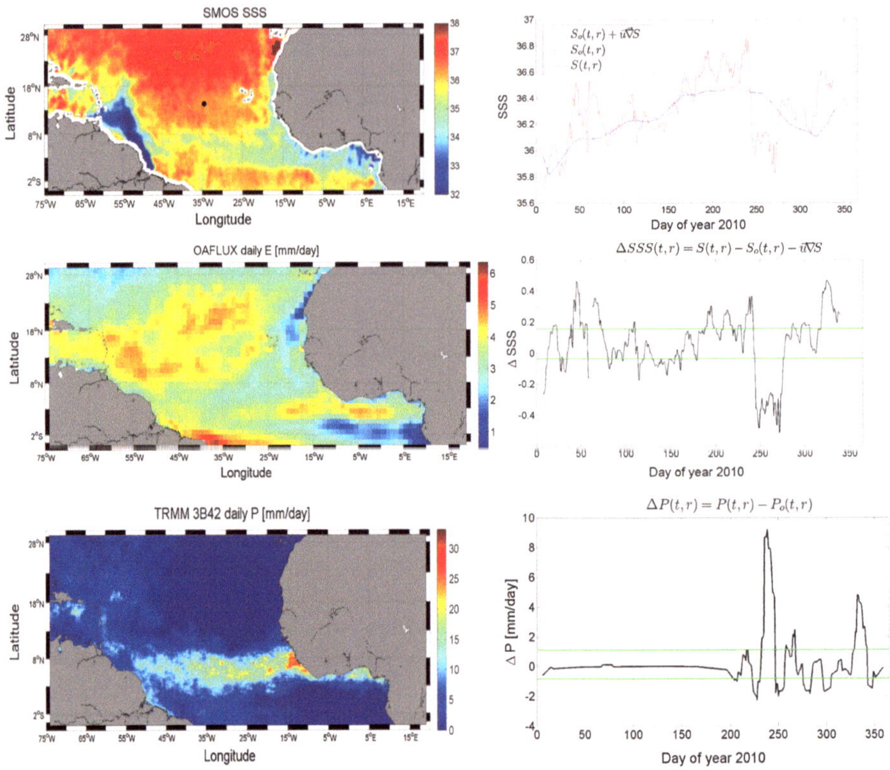

Fig. 16 *Top left*: SMOS 10-day SSS field in June 2010. *Top right*: time series of the surface salinity S(t) at the *black point* shown in the top left figure (35°W; 16°N). *Red*: SMOS SSS, *blue curve*: local mean climatological annual cycle at that point $S_o(t)$. The resulting time series for the SMOS anomaly ΔSSS at that point is shown in the middle panel, right plot. The *green horizontal lines* are indicating ± one standard deviation of the local SSS anomalies, σ_S. In the *middle* and *bottom left panels*, we show the corresponding OAFlux evaporation and TRMM 3B42 precipitation field (mm/day). The time series of the precipitation anomaly at the point is shown in the *bottom right panel*

in these key regions of the marine branch of the global hydrological cycle. In that context, SMOS SSS may therefore be an interesting data set for assimilation into ocean models in the perspective of better constraining oceanic precipitation forcing terms.

4.4 Large-scale SSS Interannual Variability in Tropical Indian and Pacific Oceans

In the Indian and Pacific oceans, the precipitation impact on the large-scale SSS variability can also be observed from SMOS and ISAS monthly maps.

The 2010–2011 period was characterized by a strong La Niña event lasting from July 2010 to March 2011 and by an Indian Ocean Dipole (IOD) index in negative phase in September–November 2010 and in positive phase during about the same months in 2011 (see Fig. 19). Such events are known to generate large-scale SSS signatures in the tropics (e.g., Gouriou and Delcroix 2002; Singh et al. 2011; Grunseich et al. 2011) and are clearly depicted in the SSS signals in both the ISAS and the SMOS monthly difference maps between 2010 and 2011 for both July and November (Fig. 20).

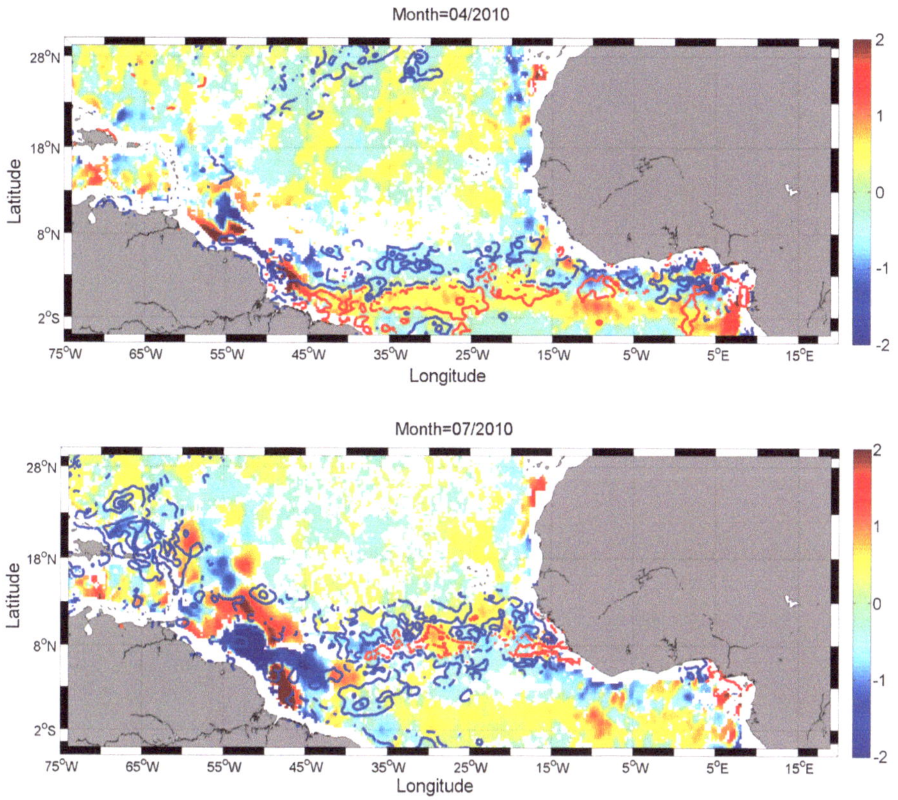

Fig. 17 Maps of the monthly averaged large-amplitude SSS anomalies deduced from SMOS data for 2 selected months of 2010 (*Top*: month of March 2010. *Bottom*: month of July 2010). The threshold value used to derive the anomaly is defined by 1 σ_S, the local standard deviation of SMOS anomaly. Superimposed are the contours of the large positive amplitude precipitation anomalies (*blue*) and positive evaporation anomalies (*red*)

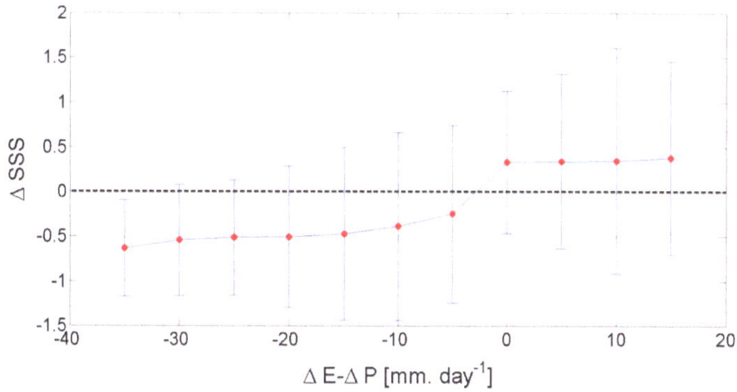

Fig. 18 Average relationship between SMOS SSS anomalies and the net atmospheric freshwater flux anomalies ΔE–ΔP in the tropical Atlantic (defined here by 5°S–20°N;75°W–15°E) over year 2010

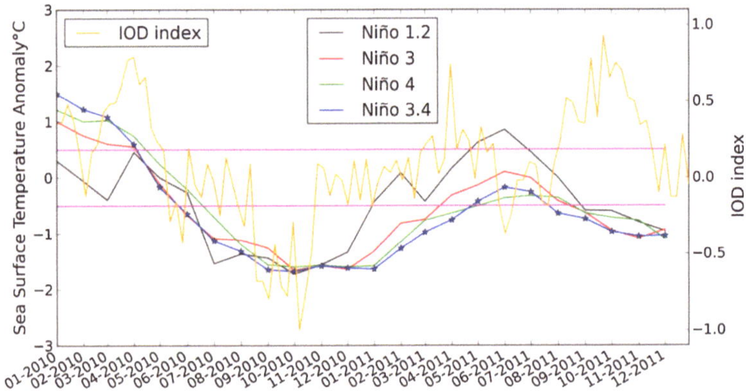

Fig. 19 Time series of SST anomalies in the four Niño regions from http://www.cpc.ncep.noaa.gov/data/indices/sstoi.indices in 2010–2011 and corresponding Indian Ocean Dipole (IOD) Index (SST difference between eastern and western equatorial Indian Ocean) from the Australian bureau of Meteorology (BOM)

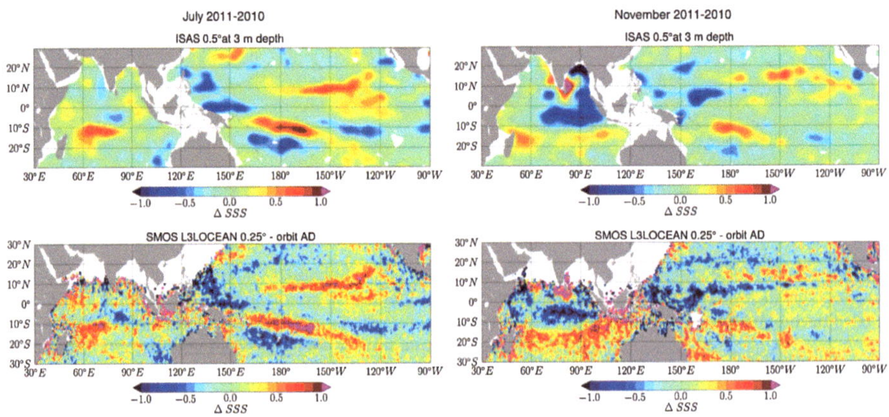

Fig. 20 Differences in the monthly averaged SSS between year 2011 and 2010 for months of July (*left*) and November (*right*). *Top panels* show the $\Delta SSS = SSS_{2011} - SSS_{2010}$ results obtained from in situ OI analysis products ISAS and bottom ones from SMOS data

The differences in rain rate as derived from SSM/I F17 sensor between 2011 and 2010 for several selected months as shown in Fig. 21 further demonstrate that part of the observed SSS interannual variability for July and November is associated with large precipitations anomalies during previous months, associated with displacements of the ITCZ and of the South Pacific Convergence Zone. In the Indian Ocean, SSS differences $\Delta SSS = SSS_{2011} - SSS_{2010}$ observed in November indicate saltier SSS in 2010 than in 2011 in the eastern equatorial Indian Ocean within the band [10°–0°S; 70°–95°E] associated with a smaller rain rate ($RR_{2010} < RR_{2011}$) in the surrounding region during preceding months, as evidenced by the rain rate difference on the October and November maps shown in Fig. 21. Between ~10°S and 20°S, SSS are fresher in 2010 than in 2011; this is associated with higher precipitation in 2010 than in 2011($RR_{2011} < RR_{2010}$) in the eastern basin but not over the whole basin. Patterns of positive SSS anomalies in the eastern equatorial Indian Ocean and negative anomalies in the eastern part of the region south of ~10°S are quite consistent with SSS anomalies already reported during negative IOD coupled with a strong La Niña event (see Fig. 8 of Grunseich et al. 2011).

Although patterns of 2011–2010 SSS differences are similar on SMOS and ISAS monthly maps, the differences are often more contrasted in the SMOS data (e.g., Fig. 20, left part and Fig. 22).

This originates from fresher SSS seen in the SMOS SSS maps than in the ISAS SSS maps (Fig. 22). In addition, the spatial extent of the low SSS region appears wider in the SMOS map, as illustrated in Fig. 22 left, around 8°N. This is possibly due to the in situ

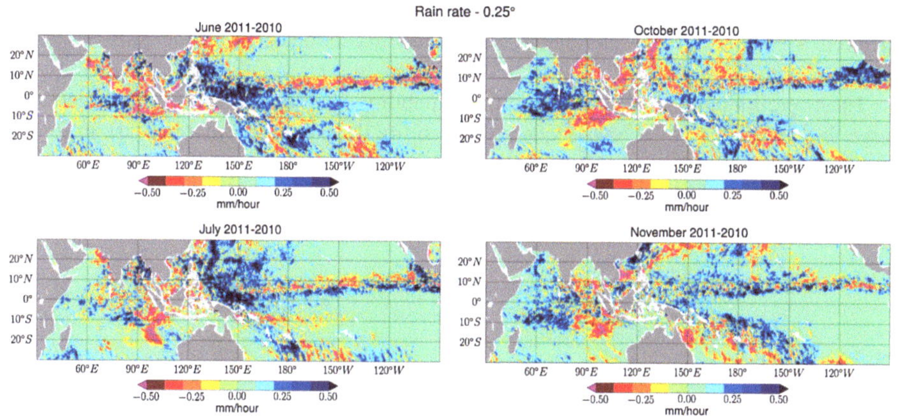

Fig. 21 Rain rate differences $\Delta RR = RR_{2011} - RR_{2010}$ derived from SSM/I F17 between 2011 and 2010 for months of June (*top left*); July (*bottom left*), October (*top right*), and November (*bottom right*)

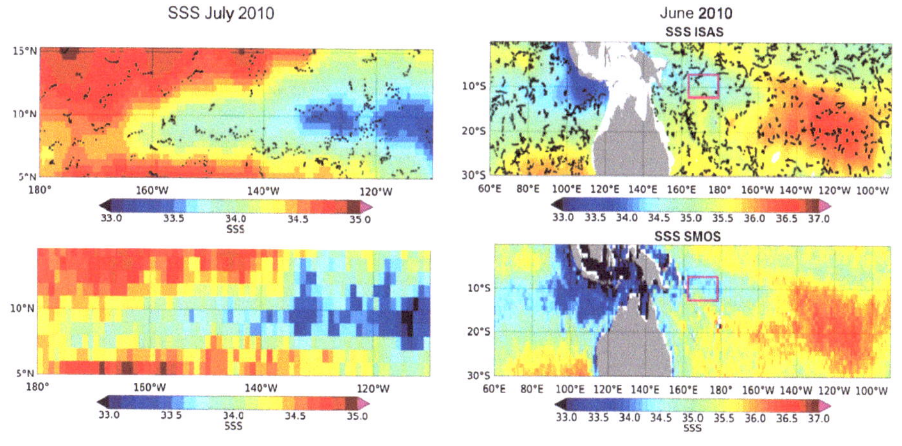

Fig. 22 *Left*: July 2010 SSS maps in the northern tropical Pacific Ocean from ISAS (*top*) and SMOS (*bottom*). *Right*: June 2010 SSS maps in south Pacific–Indian tropics from ISAS (*top*) and SMOS (*bottom*). In both top panels, the *small black dots* represent the locations of the in situ data samples used in the objective analysis. The *purple square* on the *right figure* indicates the region where the drifter discusses in Fig. 23 evolved

measurements undersampling and/or smoothing by the OI applied to the ISAS. In addition, the SMOS freshening could be linked to the different depth of the measurements (SMOS at 1 cm and in situ SSS measured at several meters depths) as described in Sects. 4.1 and 4.2.

Finally, to illustrate the potential impact of the vertical stratification effect on the ΔSSS differences between satellite and in situ, we compare along the drifter trajectory the salinity measured at 45-cm depth by a surface float (Reverdin et al. 2012) in the 2010 rainy western Pacific with monthly SSS maps (Fig. 23). The drifter SSS data clearly indicate a large signature of rainy events, with typical freshening events 1 pss for more than 1 day. The ISAS SSS is on the upper range of the drifter SSS, while monthly SMOS SSS is systematically on the lower range in this rainy region. While more work is certainly needed to determine the physical sources for these observed differences, the vertical SSS stratification associated with rain events, as illustrated by this case, is a likely contributor to the different signatures in the interannual SSS variability as detected by the SMOS satellite SSS data and the Argo data

These preliminary results confirms the capability of L-band radiometry in detecting large SSS signals and their low-frequency variability (here over a 2-year period), in spite of much noisier satellite than in situ measurements. In general, this results from much better satellite-based temporal coverage and with a better spatial resolution, thus offering complementary information to existing in situ measurements.

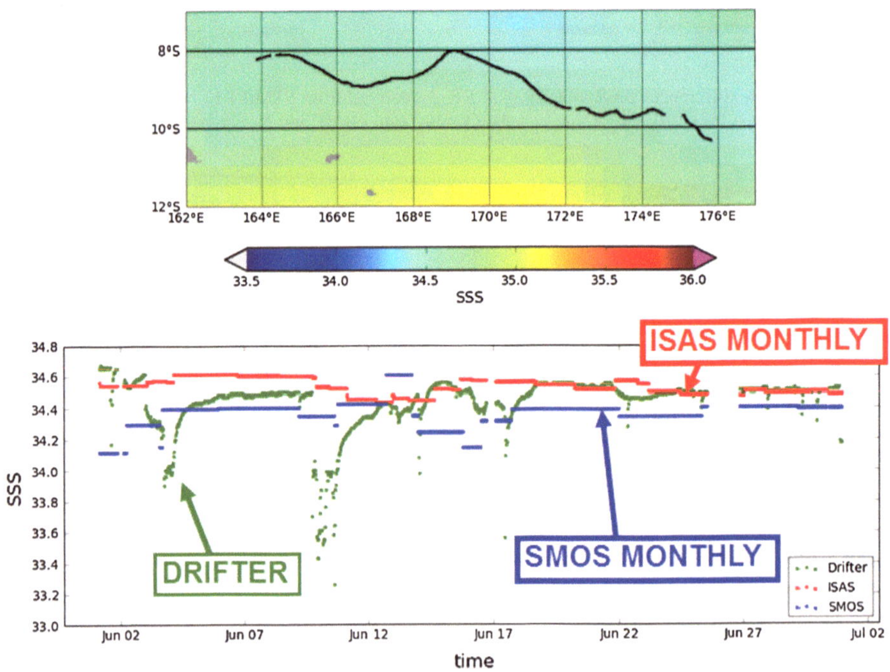

Fig. 23 *Top*: trajectory of a surface velocity program (SVP) float in the western Pacific region measuring conductivity and temperature at 45-cm depth. *Bottom*: SSS along the drifter trajectory measured by the drifter (*green*), derived from SMOS monthly map (*blue*), from ISAS monthly map (*red*)

5 Fresh Pool Interactions with Wind-driven Processes

In this section, two specific SMOS observation cases of wind-driven phenomena are presented. The first example illustrates the erosion of the Far Eastern Pacific Fresh Pool by the gap-wind-driven Panama upwelling processes, whereas the second focuses on the salty wake left behind hurricanes after their passing over the Amazon–Orinoco river plumes.

5.1 An Example of Fresh Pool Erosion by Wind-driven Upwelling

The eastern tropical Pacific Ocean between about 120°W and South America is unique in many respects. Lying in an environment predominantly influenced by the south and northeastern trades and the doldrums, and seasonally affected by the winds from the Caribbean, this region is characterized by complicated and large seasonal variations in the wind field, current pattern, temperature and salinity structure.

The region exhibiting the lowest SSS of the tropical Pacific Ocean, the Eastern Pacific Fresh Pool (EPFP), is found between the warm pool characterized by a mean sea SST greater than 28 °C centered on 15°N along the coast of Central America and the cold and fresh equatorial region, with SSS values lower than 33 pss off the Panama isthmus and lower than 34 pss extending as far as 130°W from the equator to 15°N (Fig. 24).

The EPFP reflects both the conditions of excess precipitation over evaporation beneath the ITCZ and inputs of freshwater from the Andes and Caribbean regions (Benway and Mix 2004). Analysis of a recent gridded in situ SSS product (Delcroix et al. 2011) points out that interannual variations are relatively weak in the EPFP but that seasonal variations are the strongest within the tropical Pacific. Large-scale analysis suggests that the SSS seasonal balance is mostly driven by precipitation in the part of the EPFP covered by the ITCZ, but more complex in the far east as advection and entrainment become important processes (Bingham et al. 2010; Alory et al. 2012).

By focusing on seasonal SSS variations along a well-sampled Voluntary Observing Ship (VOS) line from Panama to Tahiti, Alory et al. (2012) recently showed that this fresh pool dynamically responds to strong regional ocean–atmosphere–land interactions. First, monsoon rains (and associated river runoff) give birth to the fresh pool in the Panama Bight during summer and fall. Second, strong currents driven by topography-induced winds extend the pool westward in winter, while it eventually disappears by mixing with upwelled saltier waters to the east. These dynamic features also generate steep SSS fronts at the edges of the fresh pool (sometimes larger than $\sim$4 pss/° of longitude at the eastern edges).

These SSS fronts and the amplitude of their seasonal cycle are large enough to be detected by the new SMOS satellite mission. Compared to in situ data, SMOS satellite data provide a more homogeneous coverage with finer spatial resolution. Examples of SMOS SSS maps averaged over 10 days and centered at selected dates in December 2010, February and April 2011 are presented in Fig. 24. Remarkably, all the major features observed with in situ VOS data as detailed in Alory et al. (2012) are well reproduced in the SMOS analysis, notably the westward expansion of the fresh pool (SSS $<$ 33 pss) from 85° W in December to 95°W in April, the steep SSS front east of the 32 pss isohaline and SSS minimum of 28 pss in the Panama Bight in December, and the strong SSS increase to around 35 pss in the Panama Bight in April. Moreover, SSS changes occurring between December and April are qualitatively consistent with the expected effects of winter climatological currents, including the Panama Bight upwelling.

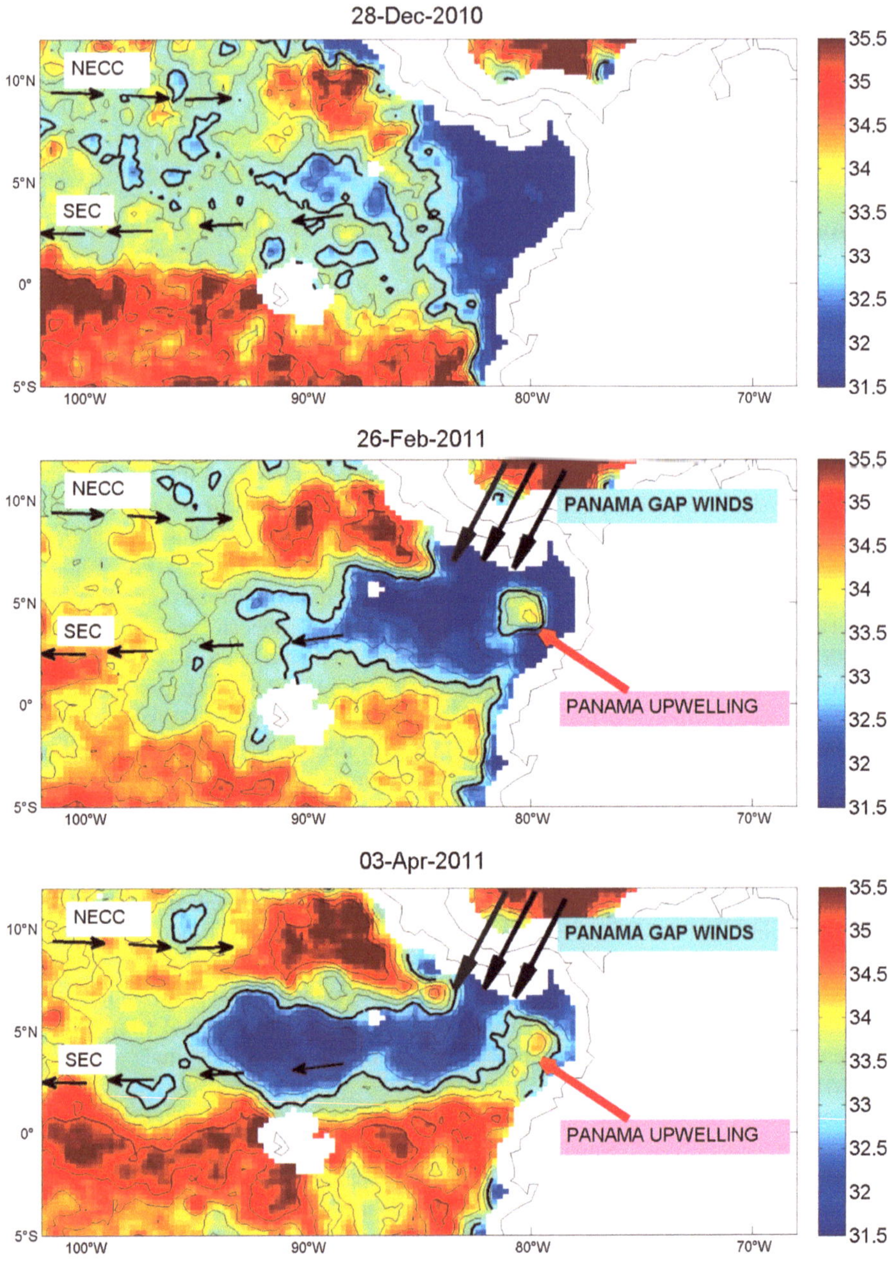

Fig. 24 10-Day averaged SMOS SSS fields centered on the December 28, 2010 (*top*), February 16, 2011 (*middle*), and April 3, 2011 (*bottom*). *Small black arrows* indicate the major surface currents, namely the south equatorial current (SEC) and NECC. *Thick black contour* is indicating the 32 pss isohaline

The freshwater pool disruption as observed by SMOS in the Panama Bight (Fig. 24, middle and bottom panels) is associated with the following processes: during the boreal winter, as the ITCZ moves southward, the northeasterly Panama gap wind creates a southwestward jet-like current in its path with a dipole of Ekman pumping/eddies on its flanks. As a result, upwelling in the Panama Bight brings cold and salty waters to the surface that erode the fresh pool on its eastern side while surface currents stretch the pool westward.

Interestingly, SMOS data are also able to detect other meso-scale features in the region around the fresh pool such as the near-equatorial SSS front or the local SSS maximum in the Costa Rica dome.

Therefore, SMOS SSS data will help in exploring qualitatively the seasonal dynamics of the fresh pools from their birth to their final erosion by wind-driven and turbulent processes (surface current stirring and wind-driven upwelling). Quantifying the relative contribution of the different mechanisms on SSS variations would require a model-based synergetic data analysis scheme to establish the mixed layer salt budget. Also, the regional occurrence of SSS fronts and barrier layers (de Boyer Montégut et al. 2007) suggests, by analogy with the western tropical Pacific, a link between surface and subsurface salinity which could give additional value to the satellite SSS data (Maes 2008; Bosc et al. 2009). As barrier layers can play an active role on the tropical climate (e.g., Maes et al. 2002, 2005), studying their impacts in the region seems worthwhile. This could be done through regional modeling combined with the analysis of subsurface/surface in situ and satellite data. Also, interannual variations of the fresh pool, even if quantitatively smaller than its seasonal variations, need further investigation as ENSO is a strong climate driver in the eastern Pacific. Now that 3 years of SMOS data are available, such type of analysis can be initiated.

5.2 Fresh Pool Interactions with Tropical Cyclones

Because of the buoyant plume of freshwater that forms in the Atlantic due to discharge from the Amazon and Orinoco rivers, the northwestern tropical Atlantic is a region where the salt-driven upper ocean stratification may significantly impact ocean–atmosphere interactions under tropical Cyclones. The spreading of the Amazon–Orinoco River plume exhibits a seasonal cycle coinciding with the Atlantic hurricane season (1 June–30 November) with river influenced minimum salinities observed farthest eastward and northwestward during the height of the hurricane season (mid-August to mid-October). As shown by Ffield (2007), for the 1960–2000 time period, 60 and 68 % of all category 4 and 5 hurricanes, respectively, passed directly over the plume region, revealing that the most destructive hurricanes may be influenced by plume–atmosphere interaction just prior to reaching the Caribbean. Historical in situ data reveal that average ocean surface temperatures first encountered by tropical cyclones moving westward between 12° and 20°N is only 26 °C, but upon reaching the northern reaches of the Amazon–Orinoco River plume (e.g., see Fig. 25), the average SST encountered by tropical cyclones are 2 °C warmer. These warm ocean surface temperatures may play a role in hurricane maintenance and intensification since hurricanes can only form in extensive ocean areas with a surface temperature greater than 25.5 deg C (Dare and McBride 2011). In addition, as shown by Ffield (2007), the buoyant, and therefore stable, 10- to 60-m-thick layer of the plume can mask the presence and influence of other ocean processes and features just below the plume, in particular cool (during hurricane season) surface temperatures carried by NBC rings. After shedding from the NBC retroflection, the 300–500-km-diameter anticyclonic (clockwise) NBC rings pass northwestward through the Amazon–Orinoco River plume toward the Caribbean. The limited observations reveal that at times the cool upper-layer temperatures of the NBC rings are exposed to the atmosphere, while at other times, they are hidden just underneath warm plume water. Strong winds from the 300–1,000-km-diameter cyclonic (counterclockwise) hurricanes might quickly erode a thin plume, exposing several degrees-cooler NBC ring water to the surface, and potentially contributing to limit further development of hurricanes. As shown by Ffields (2007), the warm temperatures associated with the low-salinity Amazon–Orinoco River plume and the relatively cool temperatures associated with NBC rings are in close proximity to the passing hurricanes. As such, they are expected to actively influence the hurricane maintenance and intensification although the interaction is challenging to accurately quantify.

Vizy and Cook (2010) more recently studied the atmospheric response of the summertime large-scale climate to the Amazon/Orinoco plume sea surface temperature anomaly forcing using a regional climate model. They performed simulations in the presence or absence of the Amazon/Orinoco plume SST anomalies. Results from their simulations indicate that the plume does significantly influence the frequency and intensity of summertime storm systems over the Atlantic, consistent with Ffield (2007). The presence of the plume increases the average number of Atlantic basin storms per summer by 60 %. An increase in storm intensity also occurs, with a 61 % increase in the number of storms that reach tropical storm and hurricane strength. Results from their simulations suggest that Atlantic storms also tend to curve northward further west in the Atlantic basin in the presence of the plume SST anomaly. These results support the premise that the warm and low-salinity combined Amazon–Orinoco River plume play an important role in modulating the air–sea interaction during hurricane passages in a manner similar to persistent freshwater barriers layers.

For instance, when there is a freshwater barrier layer, such as in the northwestern tropical Atlantic, mixing is restricted within shallower mixed layer and entrainment of cool thermocline water into the mixed layer is reduced (e.g., Anderson et al. 1996; Vialard and Delecluse 1998a, b; Foltz and McPhaden 2009). As discussed in Price (2009), if the net salinity anomaly (freshwater layer thickness times salinity anomaly in the initial state) is as

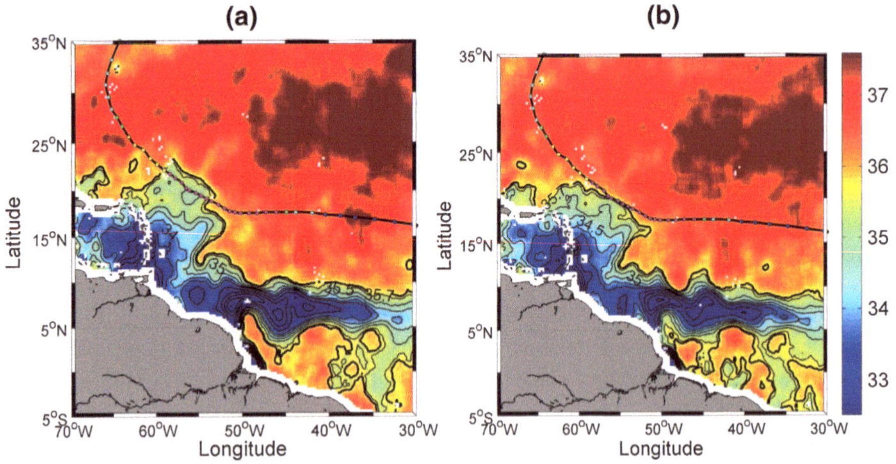

Fig. 25 Two SMOS microwave satellite-derived SSS composite images of the Amazon plume region revealing the SSS conditions **a** before and **b** after the passing of Hurricane Igor, a category 4 hurricane that attained wind speeds of 136 knots in September 2010 during its passage over the plume. *Color-coded circles* mark the successive hurricane eye positions. Seven days of data centered on **a** September 10, 2010 and **b** September 22, 2010 have been averaged to construct the SSS images, which are smoothed by a 1° × 1° block average

large as about 20 m, then the fresh layer will potentially inhibit vertical mixing significantly. As the freshwater surface layer (halocline) of the Amazon and Orinoco river plumes is warmer than the water below (Ffield 2007), salinity stratification acts to reduce the depth of vertical mixing and thus sea surface cooling. The reduced cooling amplitude in the wake of hurricanes passing over the Amazon and Orinoco river plumes, associated with thick barrier layer (BL) effects, might be an important mechanism in favor of hurricane intensification in that region. Similar impact of barrier layers on TC-induced sea surface cooling has been recently evidenced for several case studies such as in the tropical Atlantic (Balaguru et al. 2012), in the Bay of Bengal (Yu and McPhaden 2011; Neetu et al. 2012) and in the tropical Northwest Pacific (Wang et al. 2011).

New insight into the interactions between such extreme atmospheric events and large-scale fresh pools at the ocean surface has been gained from the satellite-based SSS observations as recently reported by Grodsky et al. (2012). They used data from the Aquarius/SAC-D and SMOS satellites to help elucidate the ocean response to hurricane Katia, which crossed the Amazon plume in early fall 2011. As illustrated in their paper, the Katia passage left a 1.5 pss high haline wake covering $>10^5$ km^2 (in its impact on density, the equivalent of a 3.5 °C cooling) due to mixing of the shallow BL.

As illustrated in Fig. 25, very similar observations were also detected from SMOS data alone during the passage of the Category 4 hurricane Igor over the river plume in 2010. The data evidence an erosion of the thin northern reach of the plume fresh surface layer by Igor hurricane-induced mixing, covering an area of $\sim$89,000 km^2 located on the storm right-hand side, where SSS increases by $\sim$1 practical salinity scale while SST cools by 2–3 °C (not shown). On the left side of the storm, much smaller SSS and SST changes are detected after the storm passage. The strong SSS increase in the hurricane wake within the plume is explained by the erosion of the BL. This is supported by Argo profiles collected within the plume (see Grodsky et al. 2012). Mixed layer salinity is lower by 2–4 pss than the water beneath. The shallow haline stratification is destroyed by hurricane-forced entrainment which is stronger on the right side of hurricane eye (Price 2009). It results in a strong SSS signal. Although the hurricane strengthened further along the trajectory, the SSS change is much weaker there corresponding to weak vertical salinity stratification outside the plume.

As further discussed in Grodsky et al. 2012, the fresh (more buoyant) BL limits the turbulent mixing and then the SST cooling in the plume, and thus preserved higher SST and freshwater evaporation than outside. Combined with SST, the new satellite SSS data thus provide a new and better tool to monitor the plume extent and quantify the upper ocean responses to tropical cyclones with important implications for hurricane forecasting.

6 Conclusions and Perspectives

The ocean is the primary return conduit for water transported by the atmosphere. It is the dominant element of the global water cycle, and clearly one of the most important components of the climate system, with more than 1,100 times the heat capacity of the atmosphere. Two new satellite sensors, the ESA SMOS and the NASA Aquarius SAC-D missions, are now providing the first space-borne measurements of the SSS. Synergetic analyses of the new surface salinity data sets together with sea surface temperature, dynamic height and surface geostrophic currents from altimetry, near-surface wind, ocean color, in situ observations, and rainfall estimates will certainly help clarify the freshwater budget in key oceanic tropical areas.

In this paper, we selected illustrative examples to review how the first SSS products derived from the SMOS sensor can readily help to better characterize some of the key processes of the marine branch of the global hydrological cycle. First, we illustrated the new monitoring capabilities for some of the world's largest oceanic freshwater pools generated by the discharge of very large tropical rivers. In particular, we show how SMOS SSS traces the freshwater signals from the Amazon–Orinoco and Congo river plumes. River runoff is an important variable in oceanography as their freshwater affects SSS and the buoyancy of the surface layer, and they represent a source of materials exotic to the ocean and highly important to biological activity. Obviously, they are key hydrologic components of the freshwater exchanges between the atmosphere, land, and ocean. Despite this importance, tracing river freshwater transport over large distances has not been straightforward previously principally because of a lack of SSS data. Tracing those very large rivers over great distances now become an important endeavor, as sufficient data are available from the SMOS and Aquarius sensors that can be further combined with satellite-derived surface geostrophic current data.

Second, we evidenced key oceanic precipitation signatures in the SMOS SSS signal. Satellite radiometry at L-band provides for the first time a global measure of the salinity at the ocean–atmosphere interface (within the upper centimeters). Rain events induce freshening of the ocean surface and are responsible for a high temporal variability in the SSS, consistently detected by both in situ and space-borne sensors. Because of the vertical haline gradient generated by the rain-induced freshening in the upper ocean, fresher surface waters are, however, systematically found from space in rainy area compared with the 1–10-m depth in situ data. These differences challenge calibration/validation activities of the satellite SSS in high precipitation regions. Nevertheless, satellite SSS data certainly provide new information about ocean–atmosphere interfacial freshwater fluxes in these conditions. This was evidenced by comparing spatial patterns and amplitudes of the large-scale SSS anomalies estimated from the SMOS data and the net evaporation minus precipitation fluxes in the tropical Atlantic. Under the Inter Tropical Convergence Zone and sufficiently far away from the river runoff signals, residual SSS anomalies were shown to be highly correlated with the Evaporation minus Precipitaion (E–P) anomalies. In particular, SSS anomalies become increasingly negative as the precipitation anomalies progressively exceed the evaporation anomalies. This demonstrate the importance of monitoring SSS from space in rainy regions, suggesting that the interfacial SSS values might be a good large-scale oceanic rain gauge of the global hydrological water cycle.

The interfacial character of the space-borne measurements also offers new information of interest for ocean circulation models in the perspective of better constraining oceanic precipitation forcing terms.

Finally, the SSS observations from SMOS satellite were used to reveal new aspects of the main tropical fresh pool evolution and interaction with wind-driven atmospheric processes. SMOS imagery thus captures how the large eastern Pacific fresh pool is systematically eroded at the end of the boreal summer on its eastern side by the wind-driven Panama upwelling, which brings cold and salty waters to the surface. Prior to SMOS data availability, the few existing studies of the eastern Pacific describing seasonal variations of SSS did not investigate their cause beyond rainfall (e.g., Fiedler and Talley 2006). Thanks to the new SMOS data, SSS variability associated with wind-driven processes in that region, such as the Panama upwelling signal recently evidenced by Alory et al. (2012), can now be characterized more deeply.

Because of the buoyant character of the freshwater that forms at the ocean surface due to large river discharges or intense local precipitation, the upper ocean stratification in several key tropical oceans regions (e.g., northwestern tropical Atlantic, eastern and western Pacific fresh pools, Bay of Bengal) is mostly controlled by salinity. In such freshwater pool regions, a uniform density mixed layer is found to form the so-called Barrier Layers (BL) at shallower depth than the uniform temperature layer. Because of stable halocline, the BL are acting to inhibit surface cooling and vertical mixing under the action of surface wind stresses. Therefore, there can be some feedback mechanisms between atmospheric, or terrestrial, freshwater fluxes to the ocean and intense atmospheric processes. About 68 % of hurricanes that finally reached category 4 and 5 have thus crossed the Amazon/Orinoco plume (Ffield 2007) where the presence of Barrier Layers can enhance their growth rate by 50 % (Balaguru et al. 2012). Under an intense hurricane, the halocline, which is above the thermocline, is first mixed. This produces a SSS wake that is by a few pss saltier than initial SSS in the plume. By analyzing SMOS SSS data before and after the passage of several intense hurricanes over the Amazon River plume in 2010 and 2011, SSS changes >1 pss over areas exceeding 10^5 km^2 were detected. These abrupt changes have implications for SSS climate, since SSS is more long-lived and not damped like SST. In addition, destruction of the BL is apparently associated with a decreased SST cooling in the plume that, in turn, preserves higher SST and evaporation than outside the BL. This difference in SST cooling is explained by additional work required to mix the BL. Thus, BL leads to a reduction in hurricane-induced surface cooling that favors hurricane intensification, as the resulting elevated SST and high evaporation enhance the hurricane's maximum potential intensity. The geographic location and seasonality of the Amazon/Orinoco plume make hurricane overpasses a frequent occurrence. Indeed, the expansion of the plume in August–September coincides with the peak of the production of Cape Verde hurricanes, which includes many of the most intense (category 4–5) hurricanes. Thus, the results presented here strongly suggest that the role of the salinity stratification in mixed layer dynamics should be taken into account when forecasting tropical cyclone growth over freshwater pools that are generating thick BL (Amazon plume, Bay of Bengal, eastern and western Pacific fresh pools). The availability of satellite SSS from Aquarius and SMOS along with in situ Argo measurements is critical to making such model improvements practical.

Acknowledgments Work presented in this paper was partly done under ESA support in the context of the development of the SMOS level 2/Expert Support Laboratory, ESA Support to Science Element SMOS + SOS and SMOS + STORM projects. CNES also partly funded these activities in the frame of the Centre Aval de Traitement des Données SMOS (CATDS) and of the SMOS/ESA GLOSCAL Cal/Val projects. We are indebted to Micha Rijkenberg (Koninklijk Nederlands Instituut voor Onderzoek der Zee Department of Biological Oceanography (BIO)) for giving us access to the Geotraces Leg 2 CTD data, to F. Gaillard (LPO/IFREMER) for providing ISAS maps version 6 (see http://wwz.ifremer.fr/lpo/SO-Argo/Products/Global-Ocean-T-S). Gilles Reverdin (LOCEAN/CNRS) is responsible for French salinity drifter deployment program; he supervised the quality control of these drifter data set and is warmly acknowledged. These data are available on www.locean-ipsl.upmc.fr/smos/drifters. We thank Nicolas Martin for data processing. We are also indebted to JéromeVialard, Matthieu Lengaigne and Emmanuel Vincent for their very helpful comments about SSS variability in the tropical Indian Ocean and about tropical cyclones interactions with the Amazon plume, Thierry Delcroix about SSS variability in the tropical Pacific Ocean. SSM/I data have been processed by Remote Sensing System (www.ssmi.com). The global ocean heat flux and evaporation products were provided by the WHOI OAFlux project (http://oaflux.whoi.edu) funded by the NOAA Climate Observations and Monitoring (COM) program. We thank the two anonymous referees and Doug Vandemark for their very useful comments and detailed corrections, which were very constructive and which helped us to improve our paper.

References

Alory G, Maes C, Delcroix T, Reul N, Illig S (2012) Seasonal dynamics of sea surface salinity off Panama: the far eastern Pacific fresh pool. J Geophys Res 117:C04028

Anderson SP, Weller RA, Lukas R (1996) Surface buoyancy forcing and the mixed layer of the western Pacific warm pool: observations and 1D model results. J Clim 9:3056–3085

Baklouti M, Devenon JL, Bourret A, Froidefond JM, Ternon JF, Fuda JL (2007) New insights in the French Guiana continental shelf circulation and its relation to the North Brazil Current retroflection. J Geophys Res 112:C02023

Balaguru KP, Saravanan SR, Leung LR, Xu Z, Li M, Hsieh JS (2012) Ocean barrier layers' effect on tropical cyclone intensification. PNAS. doi:10.1073/pnas.1201364109

Banks CJ, Gommenginger CP, Srokosz MA, Snaith HM (2012) Validating SMOS ocean surface salinity in the Atlantic with Argo and operational ocean model data. IEEE Trans Geosci Remote Sens 50:1688–1702

Benway HM, Mix AC (2004) Oxygen isotopes, upper-ocean salinity, and precipitation sources in the eastern tropical Pacific, Earth Planet. Sci Lett 224:493–507

Binet D, Marchal E (1993) The large marine ecosystem of shelf areas in the Gulf of Guinea: long-term variability induced by climatic changes. In: Sherman K et al (eds) Large marine ecosystems: stress, mitigation, and sustainability. American Association for the Advancement of Science, Washington, pp 104–118

Bingham FM, Foltz GR, McPhaden MJ (2010) Seasonal cycles of surface layer salinity in the Pacific Ocean. Ocean Sci 6:775–787

Blough NV, Del Vecchio R (2002) Chromophoric DOM in the coastal environment. In: Hansell DA, Carlson CA (eds) Biogeochemistry of marine dissolved organic matter. Academic, San Diego, pp 509–546

Bonjean F, Lagerloef GSE (2002) Diagnostic model and analysis of the surface currents in the tropical Pacific Ocean. J Phys Oceanogr 32:2938–2954

Bornhold BD (1973) Late quaternary sedimentation in the eastern Angola Basin. Massachusetts Institute of Technology, pp 1–213 (thesis)

Bosc C, Delcroix T, Maes C (2009) Barrier layer variability in the western Pacific warm pool from 2000 to 2007. J Geophys Res 114:C06023

Boutin J, Martin N, Yin X, Reul N, Spurgeon P (2012a) First assessment of SMOS data over open Ocean: part II-sea surface salinity. IEEE Trans Geosci Remote Sens 50(5) Part 1: 1662–1675

Boutin J, Martin N, Reverdin G, Yin X, Gaillard F (2012b) Sea surface freshening inferred from SMOS and Argo salinity: impact of rain. Ocean Sci Discuss 9:3331–3357

Boyer TP, Levitus S (2002) Harmonic analysis of climatological sea surface salinity. J Geophys Res 107(C12):8006

Braga ES, Andrie C, Bourles B, Vangriesheim A, Baurand F, Chuchla R (2004) Congo river signature and deep circulation in the eastern Guinea Basin. Deep Sea Res Part I Oceanogr Res Papers 51(8):1057–1073

Cherubin LM, Richardson PL (2007) Caribbean current variability and the influence of the Amazon and Orinoco freshwater plumes. Deep Sea Res Part I Oceanogr Res Papers 54(9):1451–1473

Conmy RN, Coble PG, Cannizzaro JP, Heil CA (2009) Influence of extreme storm events on West Florida shelf CDOM distributions. J Geophys Res 114:G00F04

Cravatte S, Delcroix T, Zhang D, McPhaden MJ, Leloup J (2009) Observed freshening and warming of the western Pacific warm pool. Clim Dyn 33:565–589

Dare RA, McBride JL (2011) The threshold sea surface temperature condition for tropical cyclogenesis. J Clim 24:4570–4576

De Boyer Montégut C, Mignot J, Lazar A, Cravatte S (2007) Control of salinity on the mixed layer depth in the world ocean: 1. General description. J Geophys Res 112:C06011

Del Vecchio R, Subramaniam A (2004) Influence of the Amazon River on the surface optical properties of the Western Tropical North Atlantic Ocean. J Geophys Res 109:C11001

Delcroix T, Alory G, Cravatte S, Corrège T, McPhaden MJ (2011) A gridded sea surface salinity data set for the tropical Pacific with sample applications (1950–2008). Deep Sea Res Part I 58(1):38–48

DeMaster DJ, Smith W, Nelson DM, Aller JY (1996) Biogeochemical processes in Amazon shelf waters: chemical distributions and uptake rates of silicon, carbon and nitrogen. Cont Shelf Res 16:617–643

Dessier A, Donguy J (1994) The sea-surface salinity in the tropical Atlantic between 10-degrees-S and 30-degrees-N—seasonal and interannual variations (1977–1989). Deep Sea Res Part I Oceanogr Res Papers 41(1):81–100

Dickson RR, Yashayaev I, Meincke J, Turrell W, Dye S, Holfort J (2002) Rapid freshening of the deep North Atlantic Ocean over the past four decades. Nature 416:832–837

D'Sa EJ, Hu C, Muller-Karger FE, Carder KL (2002) Estimation of colored dissolved organic matter and salinity fields in case 2 waters using SeaWiFS: examples from Florida Bay and Florida Shelf. J Earth Syst Sci 111:197–207

Durack PJ, Wijffels SE, Matear RJ (2012) Ocean salinities reveal strong global water cycle intensification during 1950 to 2000. Science 336:455–458

Eisma D, Van Bennekom AJ (1978) The Zaire River and estuary and the Zaire outflow in the Atlantic Ocean. Neth J Sea Res 12(3/4):255–272

Ferrari GM, Dowell MD (1998) CDOM absorption characteristics with relation to fluorescence and salinity in coastal areas of the southern Baltic Sea. Estuar Coast Shelf Sci 47:91–105

Ffield A (2005) North Brazil current rings viewed by TRMM microwave imager SST and the influence of the Amazon Plume. Deep Sea Res I 52:137–160

Ffield A (2007) Amazon and Orinoco River plumes and NBC rings: bystanders or participants in hurricane events? J Clim 20:316–333

Fiedler PC, Talley LD (2006) Hydrography of the eastern tropical Pacific: a review. Prog Oceanogr 69(2–4):143–180

Flagg CN, Gordon RL, McDowell S (1986) Hydrographic and current observations on the continental slope and shelf of the western equatorial Atlantic. J Phys Oceanogr 16(8):1412–1429

Foltz GR, McPhaden JM (2009) Impact of barrier layer thickness on SST in the central tropical North Atlantic. J Clim 22:285–299

Font J, Camps A, Borges A, Martín-Neira M, Boutin J, Reul N, Kerr YH, Hahne A, Mecklenburg S (2010) SMOS: the challenging sea surface salinity measurement from space. Proc IEEE 98(5):649–665

Font J, Boutin J, Reul N, Spurgeon P, Ballabrera-Poy J, Chuprin A, Gabarró C, Gourrion J, Guimbard S, Hénocq C, Lavender S, Martin N, Martínez J, McCulloch M, Meirold-Mautner I, Mugerin C, Petitcolin F, Portabella M, Sabia R, Talone M, Tenerelli J, Turiel A, Vergely JL, Waldteufel P, Yin X, Zine X, Delwart S (2013) SMOS first data analysis for sea surface salinity determination. Int J Remote Sens. doi:10.1080/01431161.2012.716541

Froelich PN, Atwood DK, Giese GS (1978) Influence of Amazon River discharge on surface salinity and dissolved silicate concentration in the Caribbean Sea. Deep Sea Res 25(8):735–744

Gaillard F, Autret E, Thierry V, Galaup P, Coatanoan C, Loubrieu T (2009) Quality control of large Argo datasets. J Atmos Ocean Tech 26:337–351

Garzoli SL, Ffield A, Johns WE, Yao Q (2004) North Brazil current retroflection and transports. J Geophys Res 109:C01013

Goni GJ, Johns W (2001) A census of North Brazil current rings observed from T/P altimetry: 1992–1998. Geophys Res Let 28(1):1–4

Gordon A, Guilivi CF (2008) Sea surface salinity trends over 50 years within the subtropical North Atlantic. Oceanography 20(1):20–29

Gouriou Y, Delcroix T (2002) Seasonal and ENSO variations of sea surface salinity and temperature in the South Pacific Convergence Zone during 1976–2000. J Geophys Res 107:3185

Grodsky SA, Reul N, Lagerloef G, Reverdin G, Carton JA, Chapron B, Quilfen Y, Kudryavtsev VN, Kao HY (2012) Haline hurricane wake in the Amazon/Orinoco plume: a QUARIUS/SACD and SMOS observations. Geophys Res Lett 39:L20603

Grunseich G, Subrahmanyam B, Murty VSN, Giese BS (2011) Sea surface salinity variability during the Indian Ocean Dipole and ENSO events in the tropical Indian Ocean. J Geophys Res 116:C11013

Hellerman S, Rosenstein M (1983) Normal monthly wind stress over the world ocean with error-estimates. J Phys Oceanogr 13(7):1093–1104

Hellweger F, Gordon A (2002) Tracing Amazon River water into the Caribbean Sea. J Marine Res 60:537–549

Hisard P (1980) Observation de réponses de type "El Niño" dans l'Atlantique tropical oriental, Golfe de Guinée. Oceanol Acta 3:69–78

Hu C, Montgomery ET, Schmitt RW, Muller-Karger FE (2004) The dispersal of the Amazon and Orinoco River water in the tropical Atlantic and Caribbean Sea: observation from space and S-PALACE floats. Deep Sea Res Part II 51:1151–1171

Johns WE, Lee TN, Schott FA, Zantopp RJ, Evans RH (1990) The north Brazil current retroflection: seasonal structure and eddy variability. J Geophys Res 95:22103–22120

Katsaros KB, Buettner KJK (1969) Influence of rainfall on temperature and salinity at the Ocean surface. J Appl Meteorol 8:15–18

Kerr Y, Waldteufel P, Wigneron JP, Delwart S, Cabot F, Boutin J, Escorihuela MJ, Font J, Reul N, Gruhier C (2010) The SMOS mission: new tool for monitoring key elements of the global water cycle. Proc IEEE 98(5):666–687

Klein LA, Swift CT (1977) Improved model for dielectric constant of seawater at microwave frequencies. IEEE Trans Antennas Propag 25:104–111

Koleshnikov AG (1973) Equalant I and Equalant II, physical oceanography. UNESCO Paris 1:1–289

Lagerloef GSE (2012) Satellite mission monitors Ocean surface Salinity. Eos Trans AGU 93(25):233. doi:10.1029/2012EO250001
Lagerloef GSE, Mitchum G, Lukas R, Niiler P (1999) Tropical Pacific near surface currents estimated from altimeter, wind and drifter data. J Geophys Res 104:23313–23326
Lagerloef GSE, Colomb FR, Le Vine D, Wentz F, Yueh S, Ruf C, Lilly J, Gunn J, Chao Y, deCharon A, Feldman G, Swift C (2008) The Aquarius/SAC-D mission: designed to meet the salinity remote-sensing challenge. Oceanography 21(1):68–81
Lagerloef GSE, Boutin J, Chao Y, Delcroix T, Font J, Niiler P, Reul N, Riser S, Schmitt R, Stammer D, Wentz F (2010) Resolving the global surface salinity field and variations by blending satellite and in situ observations, Oceanobs'09: sustained Ocean observations and information for society, Venise, Italy, 21–25 September 2009, ESA Publication WPP-306
Lefèvre N (2009) Low C02 concentrations in the gulf of guinea during the upwelling season in 2006. Mar Chem 113(1–2):93
Lentz SJ (1995) Seasonal-variations in the horizontal structure of the Amazon plume inferred from historical hydrographic data. J Geophys Res 100(C2):2391–2400
Lentz SJ, Limeburner R (1995) The Amazon River Plume during AMASSEDS: spatial characteristics and salinity variability. J Geophys Res 100:2355–2375
Maes C (2008) On the ocean salinity stratification observed at the eastern edge of the equatorial Pacific warm pool. J Geophys Res 113:C03027
Maes C, Picaut J, Belamari S (2002) Salinity barrier layer and onset of El Niño in a Pacific coupled model. Geophys Res Lett 29(24):2206
Maes C, Picaut J, Belamari S (2005) Importance of salinity barrier layer for the buildup of El Niño. J Clim 18:104–118
Meade RH, Nordin CF, Hernandez DP, Mejia A, Godoy JMP (1983) Sediment and water discharge in Rio Orinoco, Venezuela and Colombia. In: Proceedings of the 2nd international symposium on river sedimentation. Water Resources and Electric Poers Press, Beijing, China, pp 1134–1144
Mecklenburg S, Drusch M, Kerr YH, Font J, Martin-Neira M, Delwart S, Buenadicha G, Reul N, Daganzo-Eusebio E, Oliva R, Crapolicchio R (2012) ESA's soil moisture and Ocean salinity mission: mission performance and operations. IEEE Trans Geosci Remote Sens 50(5) Part 1:1354–1366
Merle J (1980) Variabilité thermique annuelle et interannuelle de l'océan Atlantique équatorial est. L'hypothèse d'un " El Niño " Atlantique. Oceanol Acta 3:209–220
Meulenbergh J (1968) Diffusion des eaux du fleuve Congo dans les eaux de l'Atlantique Sud. Acad R Sci Outre Mer Bruxelles 16(6):148
Meybeck M, Ragu A (1997) River discharges to the oceans: an assessment of suspended solids, major ions, and nutrients. IAHS Publ 243:1–245
Michel S, Chapron B, Tournadre J, Reul N (2007) Sea surface salinity variability from a simplified mixed layer model of the global ocean. Ocean Sci Discuss 4:41–106
Mignot J, de Boyer Montégut C, Lazar A, Cravatte S (2007) Control of salinity on the mixed layer depth in the world ocean: 2. Tropical areas. J Geophys Res 112:C10010
Miller R, Zepp RG (1995) Photochemical production of dissolved inorganic carbon from terrestrial organic matter: significance to the oceanic organic carbon cycle. Geophys Res Lett 22:417–420
Milliman JD, Meade RH (1983) World-wide delivery of river sediment to the oceans. J Geol 91(1):1–21
Molleri GSF, de Novo EMLM, Kampel M (2010) Space-time variability of the Amazon River plume based on satellite ocean color. Cont Shelf Res 30(3–4):342–352
Moran MA, Sheldon WM, Sheldon JE (1999) Biodegradation of riverine dissolved organic carbon in five estuaries of the southeastern United States. Estuaries 22:55–64
Morrow R, Valladeau G, Sallee J-B (2008) Observed subsurface signature of Southern Ocean sea level rise. Prog Oceanogr 77(4):351–366
Muller-Karger FE, McClain CR, Richardson PL (1988) The dispersal of the Amazon's water. Nature 333:56–59
Muller-Karger FE, McClain CR, Fisher TR, Esaias WE, Varela R (1989) Pigment distribution in the Caribbean Sea: observations from space. Prog Oceanogr 23:23–64
Muller-Karger FE, Richardson PL, McGillicuddy D (1995) On the offshore dispersal of the Amazon's plume in the North Atlantic. Deep Sea Res Part I 42:2127–2137
Neetu S, Lengaigne M, Vincent EM, Vialard J, Madec G, Samson G, Ramesh Kumar MR, Durand F (2012) Influence of upper-ocean stratification on tropical cyclones- induced surface cooling in the Bay of Bengal. J Geophys Res 117:C12020
Neumann G (1969) Seasonal salinity variations in the upper strata of the western tropical Atlantic ocean-I. Sea surface salinities. Deep Sea Res 16:165–177

Nittrouer CA, Demaster DJ (1986) Sedimentary processes on the Amazon continental shelf: past, present and future research. Cont Shelf Res 6:5–30
Obernosterer I, Herndl GJ (2000) Differences in the optical and biological reactivity of the humic and non-humic DOC component in two contrasting coastal marine environments. Limnol Oceanogr 45:1120–1129
Ostapoff F, Tarbeyev Y, Worthem S (1973) Heat flux and precipitation estimates from oceanographic observations. Science 180:960–962
Palacios SL, Peterson TD, Kudela RM (2009) Development of synthetic salinity from remote sensing for the Columbia River Plume. J Geophys Res 114:C00B05
Perry GD, Duffy PB, Miller NL (1996) An extended data set of river discharges for validation of general circulation models. J Geophys Res 101:21339–21349
Piterbarg LI, Ostrovskii AG (1997) Advection and diffusion in random media: implications for sea surface temperature anomalies. Kluwer Academic, New York
Piton B, Wacongne S (1985) Unusual amounts of very saline subsurface water in the eastern Gulf of Guinea in May 1984. Tropical Ocean-Atmosphere Newsletter, pp 325–328
Price JF (2009) Metrics of hurricane-ocean interaction: vertically-integrated or vertically- averaged ocean temperature? Ocean Sci 5:351–368
Reul N, Tenerelli J (2011) SMOS level 3 SSS research products—product validation document—reprocessed year 2010. CECOS/CNES/IFREMER/CATDS Techn Doc. Available from www.catds.fr
Reul N, Saux-Picart S, Chapron B, Vandemark D, Tournadre J, Salisbury J (2009) Demonstration of ocean surface salinity microwave measurements from space using AMSR-E over the Amazon plume. Geophys Res Lett 36:L13607
Reul N, Tenerelli J, Boutin J, Chapron B, Paul F, Brion E, Gaillard F, Archer O (2012) Overview of the first SMOS sea surface salinity products. Part I: quality assessment for the second half of 2010. IEEE Trans Geosci Remote Sens 50(5):1636–1647
Reverdin G, Kestenare E, Frankignoul C, Delcroix T (2007) Sea surface salinity in the Atlantic Ocean (30°S–50°N). Prog Oceanogr 73:311–340
Reverdin G, Morisset S, Boutin J, Martin N (2012) Rain-induced variability of near sea- surface T and S from drifter data. J Geophys Res 117:C02032
Richardson PL, McKee T (1984) Average seasonal variation of the Atlantic equatorial currents from historical ship drifts. J Phys Oceanogr 14(7):1226–1238
Ruf CS, Swift CT, Tanner AB, Le Vine DM (1988) Interferometric synthetic aperture microwave radiometry for the remote sensing of the Earth. IEEE Trans Geosci Remote Sens 26:597–611
Salisbury J, Vandemark D, Campbell J, Hunt C, Wisser D, Reul N, Chapron B (2011) Spatial and temporal coherence between Amazon River discharge, salinity, and light absorption by colored organic carbon in western tropical Atlantic surface waters. J Geophys Res 116:C00H02
Schlössel P, Soloviev A, Emery WJ (1997) Cool and freshwater skin of the ocean during rainfall. Bound Layer Meteorol 82(3):439–474
Schmitt RW(2008) Salinity and the global water cycle. Oceanography 21(1):12–19
Singh A, Delcroix T, Cravatte S (2011) Contrasting the flavors of El Niño-Southern Oscillation using sea surface salinity observations. J Geophys Res 116:C06016
Smith WO, Demaster DJ (1996) Phytoplankton biomass and productivity in the Amazon River plume: correlation with seasonal river discharge. Cont Shelf Res 16(3):291–319
Soloviev A, Lukas R (1996) Observation of spatial variability of diurnal thermocline and rainformed halocline in the Western Pacific warm pool. J Phys Oceanogr 26:2529–2538
Steven DM, Brooks AL (1972) Identification of Amazon River water at Barbados, W. Indies, by salinity and silicate measurements. Mar Biol 14(4):345–348
Ternon JF, Oudot C, Dessier A, Diverres D (2000) A seasonal tropical sink for atmospheric $C0_2$ in the Atlantic Ocean: the role of the Amazon River discharge. Mar Chem 68(3):183–201
Terray L, Corre L, Cravatte S, Delcroix T, Reverdin G, Ribes A (2011) Near-surface salinity as nature's rain gauge to detect human influence on the tropical water cycle. J Clim 25:958–977
Twardowski MS, Donaghay PL (2001) Separating in situ and terrigenous sources of absorption by dissolved material in coastal waters. J Geophys Res 106(C2):2545–2560
UNESCO (1985) The international system of units (SI) in oceanography, UNESCO technical papers no. 45, IAPSO Pub. Sci. No. 32, Paris, France
Van Bennekom AJ, Berger GW (1984) Hydrography and silica Budget of the Angola Basin. Neth J Sea Res 17(2–4):149–200
Van Bennekom AJ, Jager JE (1978) Dissolved aluminium in the Zaire River plume. Neth J Sea Res 12(3/4):358–367

Vangriesheim A, Pierre C, Aminot A, Metzl N, Baurand F, Caprais JC (2009) The influence of Congo river discharges in the surface and deep layers of the gulf of guinea. Deep Sea Res Part II Top Stud Oceanogr 56(23):2183–2196
Vialard J, Delecluse P (1998a) An OGCM study for the TOGA decade. Part I: role of salinity in the physics of the Western Pacific Fresh Pool. J Phys Oceanogr 28:1071–1088
Vialard J, Delecluse P (1998b) An OGCM study for the TOGA decade. Part II: barrier layer formation and variability. J Phys Oceanogr 28:1089–1106
Vizy EK, Cook KH (2010) Influence of the Amazon/Orinoco Plume on the summertime Atlantic climate. J Geophys Res 115:D21112
Vörösmarty CJ, Fekete B, Tucker BA (1998) River discharge database, version 1.1 (RivDIS v1.0 supplement). Institute for the Study of Earth, Oceans, and Space, University of New Hampshire, Durham, NH
Wang X, Han G, Qi Y, Li W (2011) Impact of barrier layer on typhoon-induced sea surface cooling. Dyn Atmos Oceans 52(3):367–385
Wauthy B (1977) Révision de la classification des eaux de surface du golfe de Guinée. Cah Orstom Sér Océanogr 15(3):279–295
Wijesekera HW, Paulson CA, Huyer A (1999) The effect of rainfall on the surface layer during a westerly wind burst in the western equatorial Pacific. J Phys Oceanogr 29:612–632
Wisser D, Fekete BM, Vörösmarty CJ, Schumann AH (2010) Reconstructing 20th century global hydrography: a contribution to the global terrestrial network-hydrology (GTN-H). Hydrol Earth Syst Sci 14:1–24
Yamashita Y, Tanoue E (2004) In situ production of chromophoric dissolved organic matter in coastal environments. Geophys Res Lett 31:L14302
Yin X, Boutin J, Martin N, Spurgeon P (2012) Optimization of L-Band Sea surface emissivity models deduced from SMOS data. IEEE Trans Geosci Remote Sens 50:1414–1426
Yoo JM, Carton JA (1988) Spatial dependence of the relationship between rainfall and outgoing longwave radiation in the tropical Atlantic. J Clim 1:1047–1056
Yu L (2010) On sea surface salinity skin effect induced by evaporation and implications for remote sensing of Ocean salinity. J Phys Oceanogr 40(1):85
Yu L (2011) A global relationship between the ocean water cycle and near-surface salinity. J Geophys Res 116:C10025
Yu L, McPhaden MJ (2011) Ocean pre-conditioning of Cyclone Nargis in the Bay of Bengal: interaction between Rossby waves, surface fresh waters, and sea surface temperatures. J Phys Oceanogr 41:1741–1755
Yu L, Weller RA (2007) Objectively analyzed air-sea heat fluxes (OAFlux) for the global ocean. Bull Am Meteorol Soc 88(4):527–539
Yueh SH, West R, Wilson WJ, Li FK, Njoku EG, Rahmatsamii Y (2001) Error sources and feasibility for microwave remote sensing of ocean surface salinity. IEEE Trans Geosci Remote Sens 39:1049–1060
Zhang Y, Zhang X (2012) Ocean haline skin layer and turbulent surface convections. J Geophys Res 117:C04017

DOI 10.1007/978-94-017-8789-5_11
Reprinted from *Surveys in Geophysics* Journal, DOI 10.1007/s10712-013-9274-7

Role of Ocean in the Variability of Indian Summer Monsoon Rainfall

Porathur V. Joseph

Received: 2 September 2012 / Accepted: 6 April 2013 / Published online: 12 February 2014

Abstract Asian summer monsoon sets in over India after the Intertropical Convergence Zone moves across the equator to the northern hemisphere over the Indian Ocean. Sea surface temperature (SST) anomalies on either side of the equator in Indian and Pacific oceans are found related to the date of monsoon onset over Kerala (India). Droughts in the June to September monsoon rainfall of India are followed by warm SST anomalies over tropical Indian Ocean and cold SST anomalies over west Pacific Ocean. These anomalies persist till the following monsoon which gives normal or excess rainfall (tropospheric biennial oscillation). Thus, we do not get in India many successive drought years as in sub-Saharan Africa, thanks to the ocean. Monsoon rainfall of India has a decadal variability in the form of 30-year epochs of frequent (infrequent) drought monsoons occurring alternately. Decadal oscillations of monsoon rainfall and the well-known decadal oscillation in SST of the Atlantic Ocean (also of the Pacific Ocean) are found to run parallel with about the same period close to 60 years and the same phase. In the active–break cycle of the Asian summer monsoon, the ocean and the atmosphere are found to interact on the time scale of 30–60 days. Net heat flux at the ocean surface, monsoon low-level jetstream (LLJ) and the seasonally persisting shallow mixed layer of the ocean north of the LLJ axis play important roles in this interaction. In an El Niño year, the LLJ extends eastwards up to the date line creating an area of shallow ocean mixed layer there, which is hypothesised to lengthen the active–break (AB) cycle typically from 1 month in a La Niña to 2 months in an El Niño year. Indian monsoon droughts are known to be associated with El Niños, and long break monsoon spells are found to be a major cause of monsoon droughts. In the global warming scenario, the observed rapid warming of the equatorial Indian ocean SST has caused the weakening of both the monsoon Hadley circulation and the monsoon LLJ

P. V. Joseph (✉)
Nansen Environmental Research Centre India, 6A, Oxford Business Center, Sreekandath Road, Ravipuram, Cochin 682016, India
e-mail: joporathur@gmail.com

P. V. Joseph
Cochin University of Science and Technology, Cochin, India

which has been related to the observed rapid decreasing trend in the seasonal number of monsoon depressions.

Keywords Monsoon onset · Active–break cycle · Rapid warming of Indian Ocean · Decadal variability of monsoon · Tropospheric biennial oscillation · Long break monsoon spells

1 Introduction

The continental land mass (Eurasia) north of India gets heated up, and a giant low pressure area is created there in the low levels of the atmosphere in summer as the Sun moves to the northern hemisphere. Six months later with the Sun overhead the southern oceans, radiative heat loss cools this continental land mass when the giant low pressure area gets replaced by a large high pressure area. In response to these large-scale surface pressure changes on the continental scale, we get the summer and winter monsoons of south Asia. That continental scale land–sea thermal contrast (a huge land and sea breeze system) is the main reason that monsoons exist held sway since the time of Halley (1686). That the ocean with its Intertropical Convergence Zone (ITCZ) and the deep convective (cumulonimbus) clouds associated with it plays a major role in generating the deep monsoon wind systems due to the heating of a deep layer of the atmosphere by the latent heat released in the ITCZ clouds is increasingly being considered as the main driving force of the monsoons. The modelling study by Chao and Chen (2001) and the discussions in the following section on monsoon onset over south Asia show that ocean is as important as the land in the Asian summer monsoon. Figures 1, 2 show that, as the ITCZ with its cloud systems moved from south of the equator to north of it in the annual cycle, the winter monsoon wind system changes to the summer monsoon wind system. Figure 2b shows the low-level jetstream (LLJ) at 850 hPa level as a mean of several July and August months.

Summer monsoon rainfall is a precious natural resource for the countries of south Asia, but it is highly variable in space and time. Prediction of monsoon rainfall 7–10 days in advance (medium range) has great applications in agricultural operations. Planners particularly in the government require monthly and seasonal predictions (long range) and also predictions of inter-annual and decadal scale variability of monsoon. Recent research has shown that monsoon variability on these time scales is highly dependent on the ocean below, particularly on sea surface temperature (SST). We have to understand the coupling between the atmosphere and the ocean on these time scales by data analysis, process studies and ocean–atmosphere coupled modelling. In this paper, some aspects of the role of the ocean in the variability of the Indian Summer Monsoon Rainfall (ISMR) of June to September will be presented and discussed.

2 Monsoon Onset Over India

In January, the centre of the warm pool of the tropical oceans lies over southwest Pacific Ocean. In the annual cycle, the central region of the warm pool gets shifted to the north Indian Ocean by May as may be seen from Fig. 3. This large-scale change in the ocean SST is necessary for monsoon onset over south Asia. The warm pool of May attracts

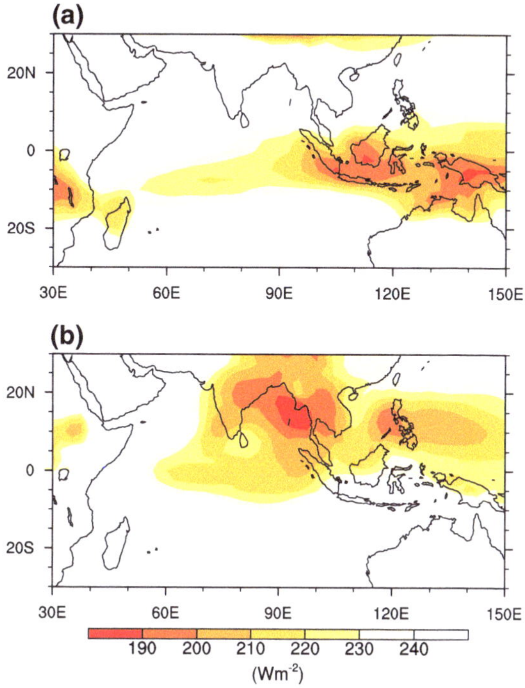

Fig. 1 Outgoing longwave radiation as mean of 1979–2010 **a** January and February and **b** July and August

moisture convergence and over a period of more than a month builds up the vertically integrated moisture content of the atmosphere over south Asia and the oceans around to about 45 kilograms per square metre which is needed for monsoon onset to take place—Pearce and Mohanty (1984) and Joseph et al. (2006). There is a fine structure to the changes in the warm pool over north Indian Ocean. About 40 days (8 pentads) before the monsoon onset over the southern state of India (Kerala), the central Bay of Bengal (BoB) develops high SST and to its south, in the area of large north–south SST gradient, a band of convective clouds form near the equator. These clouds cause westerly low-level winds to its south and this cloud-wind system moves north during the following few pentads cooling the ocean and in many years bringing onset of monsoon rains over southeast Asia. Later a similar warming occurs in central Arabian Sea when a cloud band forms over south Arabian Sea close to the equator and, as it develops and moves north, the cross-equatorial LLJ as documented by Joseph and Raman (1966) and Findlater (1969) forms and intensifies. These changes herald the monsoon onset over Kerala, at the southern tip of India. Figure 4 shows the warming and cooling of the BoB and the Arabian Sea in association with the monsoon onset during a typical year 2003. Pentad zero is the pentad around monsoon onset over Kerala.

The birth of LLJ coincides with the time of monsoon onset over Kerala (Joseph et al. 2006), and it lasts during the four monsoon months June to September with major fluctuations in its strength and spatial location during the active–break (AB) cycle of the monsoon as shown by Joseph and Sijikumar (2004). Summer monsoon onset over south Asia (Kerala) is found to be related to the time of transition across the equator of the ITCZ over the Indian and west Pacific oceans from south of the equator to the north of it, which is related to the large-scale SST anomalies in both these oceans around the equator as

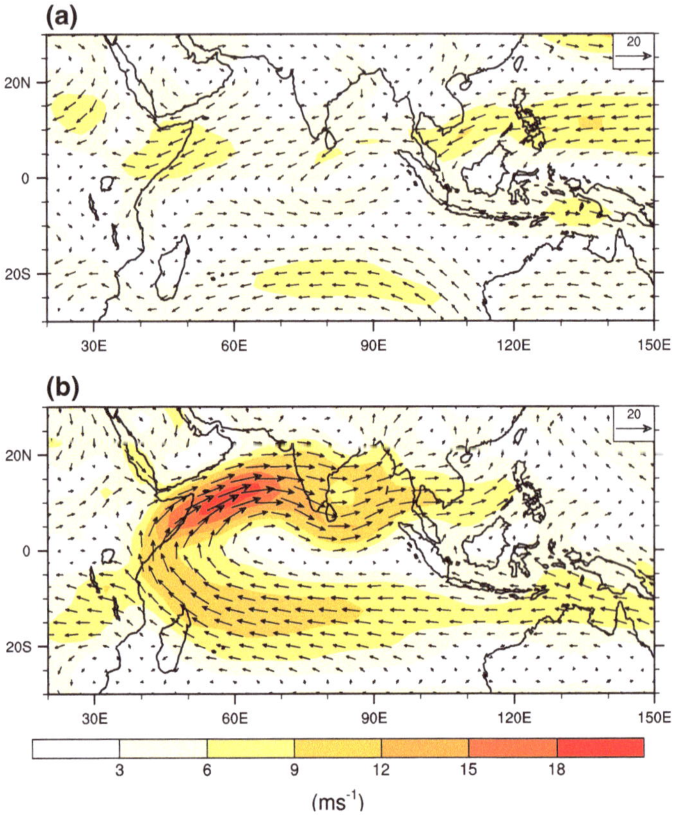

Fig. 2 850 hPa wind as mean of 1979–2010. **a** January and February and **b** July and August

shown by observational studies (Joseph et al. 1994, 2006) and modelling studies (Ju and Slingo 1995; Soman and Slingo 1997; Annamalai et al. 2005). The correlation coefficient between the date of monsoon onset over Kerala and the SST (Fig. 5) shows that when SST anomaly is positive south of the equator and negative north of it, monsoon onset is delayed. This correlation pattern has persistence of several months prior to the monsoon onset.

3 Active–Break Cycle of Monsoon

Monsoon has an active–break (AB) cycle of period varying between 30 and 60 days. This is the most important intra-seasonal variability of the monsoon. There is a large volume of observational and modelling studies that demonstrate the air–sea interactions associated with the monsoon active–break cycle, an excellent review of which may be found in Goswami (2005). During active monsoon spells, there is an east–west band of raining convective clouds passing through India in the latitude belt 10°N–20°N from longitude 70°E extending beyond 120°E and the associated LLJ is located just south of this cloud band from central Arabian sea through peninsular India and the BoB to the west Pacific ocean. LLJ serves as a conduit for the transportation of the large quantity of moisture needed for the monsoon rainfall of Asia and for providing the dynamics for the monsoon

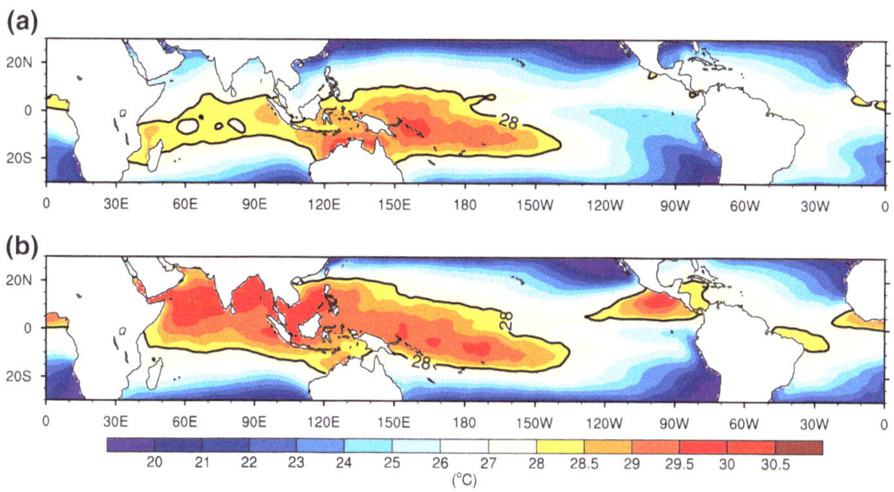

Fig. 3 Warm pool **a** January and **b** May shown by the mean Hadley Centre Sea Ice and Sea Surface Temperature (HadISST) data set of 1961–1990. Note the north–westwards shift of the centre of the warm pool from January to May

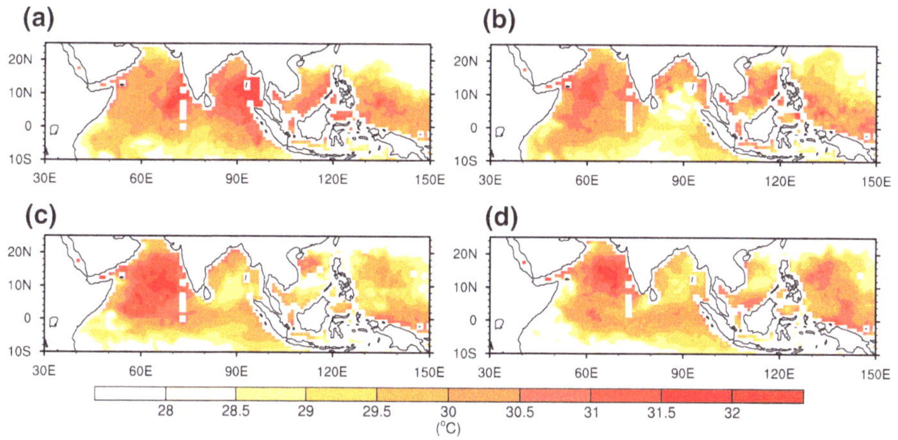

Fig. 4 The mean TRMM Microwave Imager (TMI) SST of pentads. **a** −7, **b** −5, **c** −2 and **d** 0 during the year 2003

rainfall (cyclonic vorticity in the low levels associated with the LLJ together with the friction in the atmospheric boundary layer produce vertical upwards motion of the moist monsoon air) and for the genesis of monsoon depressions in north BoB. When the cloud band of the active monsoon passing through central India and the BoB weakens after a life of 4 or 5 pentads, the LLJ shifts to a position south of India and the raining area gets located around the equator. This is the beginning of the break phase of the monsoon when rainfall becomes deficient over central India and the BoB. During this period of break monsoon, north BoB warms and we get the next active monsoon spell. In a normal monsoon season, there are 3 active–break cycles of average period of about 40 days. LLJ

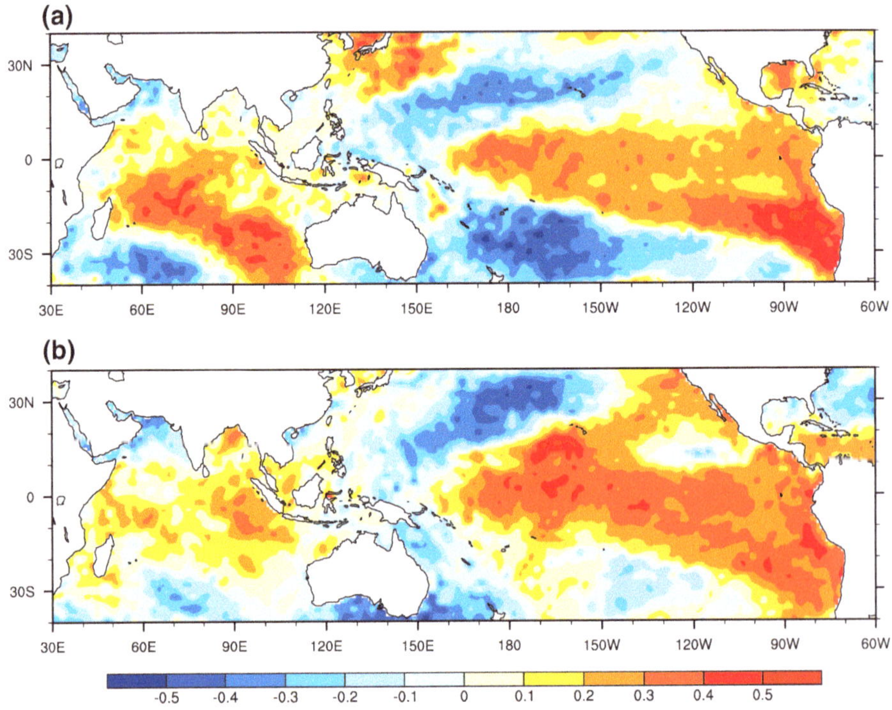

Fig. 5 The correlation between the objective dates of monsoon onset over Kerala as derived by Pai and Rajeevan (2009) and the HadISST of **a** January **b** April using data of the years 1971–2000

and the convective cloud bands (represented by OLR) during active and break monsoon as average of several such days are shown in Figs 6 and 7 of Joseph and Sijikumar (2004).

Joseph and Sabin (2008a) showed that the ocean interacts with the atmosphere in the AB cycle of the Asian summer monsoon. Prior to the beginning of an AB cycle, the SST anomaly over north BoB reaches maximum positive value. At this time, the positive SST anomaly zone extends from the Arabian Sea to about longitude 150°E in the west Pacific Ocean as may be seen in Fig. 6 taken from Joseph and Sabin (2008a) which gives the mean SST anomaly of 11 AB cycles in the 8 pentads (of an average AB cycle of period 40 days). In the SST gradient area to the south of maximum SST anomaly, a convective cloud band forms after about a pentad that in the following 2–3 days generates an LLJ through peninsular India and the active phase of the monsoon begins. The cloud band thus formed (reducing the incident solar radiation) and the strong winds of the LLJ (by causing evaporation at the ocean surface) cools the ocean there, when the convection weakens and the LLJ moves south to an equatorial location in the Indian Ocean which has warmer SST, where a new cloud band forms. This is the break monsoon phase. North BoB has large amplitude SST variations in response to the net heat flux changes in the AB cycle as the ocean mixed layer depth (MLD) there is shallow (typically 20 metres). See also Sengupta et al. (2001). That convection forms in the SST gradient area has been shown empirically by Shankar et al. (2007) and Sabin et al. (2012) and with modelling studies by Lindzen and Nigam (1987) and Back and Bretherton (2009).

The eastwards extent of the LLJ is up to longitude 120°E in La Niña years. In El Niño years, LLJ extends up to 180°E when between longitudes 120°E and 180°E the seasonal

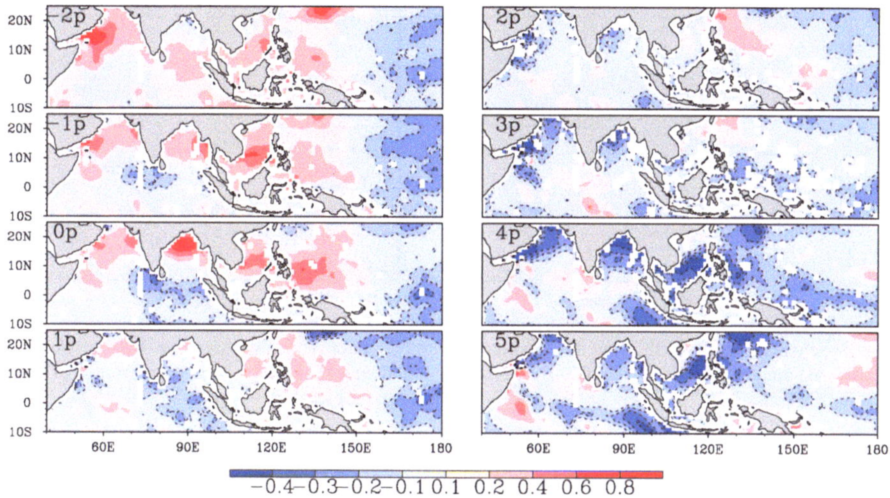

Fig. 6 TMI SST anomalies in degree Celsius of the 11 case composite AB cycle for pentads −2 to +5. At 0 pentad, SST is maximum over north Bay of Bengal and to its east and west. Note the minimum SST over the same area at pentad +4

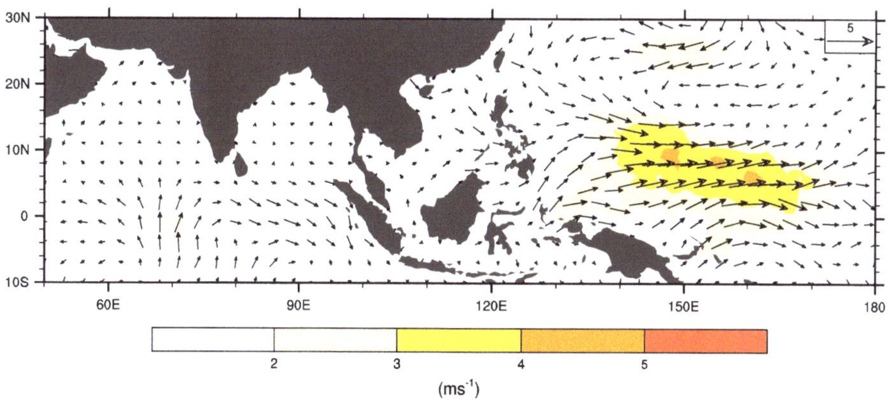

Fig. 7 Surface wind anomaly (Quikscat) July and August as mean of El Niños 2002, 2004 and 2009 minus mean of non El Niño years 2000, 2001, 2003, 2005, 2006, 2007 and 2008

scale cyclonic wind stress curl north of the LLJ axis is hypothesised to generate a large area of shallow MLD there. During the AB cycle, this area warms rapidly and causes the active phase convection to shift there from west of 120°E before moving to the equatorial Indian ocean, thus prolonging the duration of the AB cycle. In the tropics, it is found that areas of warmest SST attract moisture convergence, increase both the convection and the vertically integrated moisture content of the atmosphere there. Available Argo data of Mixed Layer Depth (MLD) and Quikscat surface wind data support the hypothesis made. Figure 7 gives the surface wind anomaly of July and August as a composite of the latest 3 El Niños (2002, 2004 and 2009). The cyclonic wind area between longitudes 120°E and 180°E has shallow MLD during the monsoon season of recent El Niño years as shown by ARGO data.

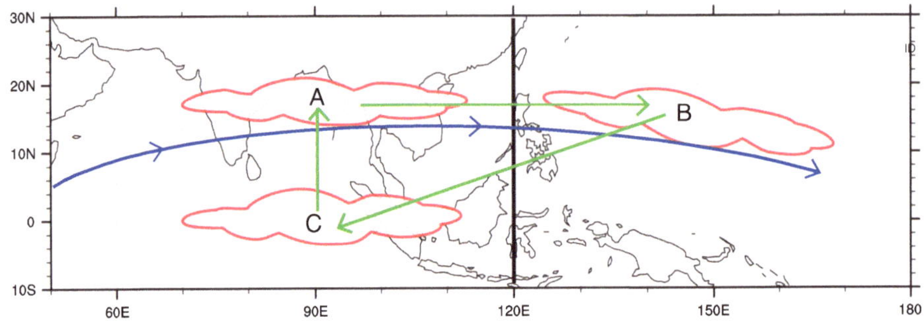

Fig. 8 Schematic diagram showing the location *A*, *B* and *C* of convection in the active–break cycle. In El Niño years, convection shifts from *A* to *B* and then to *C* and back to *A*, in La Niña years from *A* to *C* and back to *A*

Figure 8 gives a schematic of the hypothesis on the lengthening of the AB cycle in El Niño years taken from Joseph and Jayakumar (2012) showing the locations of convection in the AB cycle. In a La Niña case, convection shifts from central India and adjoining oceans (area A) to the equatorial Indian ocean (area C) and then back to A. Convection does not move to the west Pacific (area B) as the SST there with a thick MLD has no intra-seasonal variability, but only a seasonal slow warming and thus has not warmed enough for the convection to shift there. Thus, the AB cycle has short period of about a month. In an El Niño case, convection moves from area A to area B where SST has warmed fast in the intra-seasonal scale as the region is cloud free and with a shallow MLD. After convection and typhoons are active there for about a month, SST cools and the convection shifts to area C. While the convection is active over areas B and C, rainfall over central India is reduced and we get long break monsoon spells as defined by Rajeevan et al. (2006).

Ocean–atmosphere interaction mechanism thus explains the difference in the period of the AB cycle of the monsoon between El Niño and La Niña years. Inter-annual variations

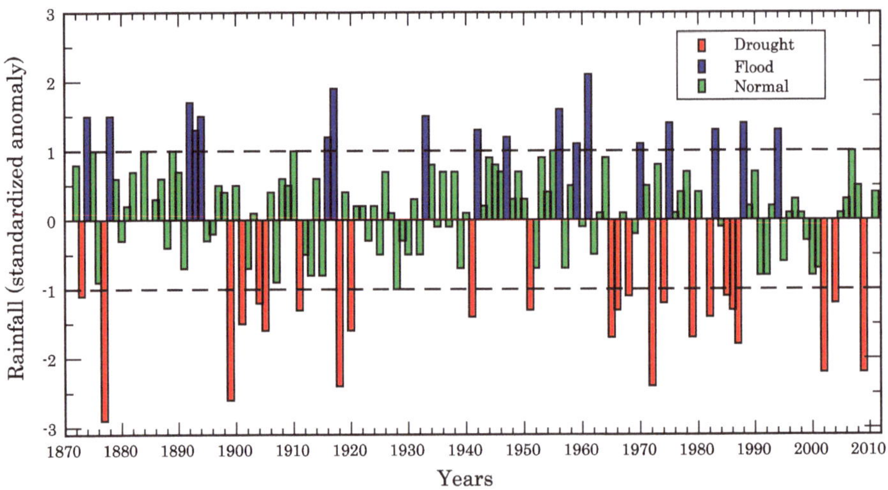

Fig. 9 Standardised anomaly of the Indian Summer Monsoon Rainfall (ISMR) for the years 1871–2010 (data taken from the Indian Institute of Tropical Meteorology (IITM) website www.tropmet.res.in). ISMR is the average June to September rainfall of 306 rainguages well distributed over India. Drought years are marked in *red*

in the thickness of MLD in the western Pacific Ocean play an important role. The amplitude of the SST variations is large on the intra-seasonal time scale where the MLD is shallow. SST variations are mainly in response to the varying net heat flux at the ocean surface—Sengupta et al. (2001). That, in the intra-seasonal time scale, convection moves to oceanic areas of higher SST in the tropics is an accepted hypothesis—see Hendon (2005). Purely atmospheric mechanisms (e.g., MJO) have been proposed to explain the lengthening of the AB cycle in El Niño years—e.g., Joseph et al. (2009). Indian monsoon droughts are known to be associated with El Niños, and it is surmised that it is so mainly because of the long AB cycles associated with El Niño.

4 Inter-annual and Decadal Variability of Monsoon

India gets three fourths of its annual rainfall from its summer monsoon taken as the period 01 June to 30 September. According to the India Meteorological Department, if the Indian Summer Monsoon Rainfall (ISMR) goes below 10 % (about one standard deviation) of its long-term average, it is declared as a drought year. ISMR had very little long-term trend during the period of good rainfall measurements (1871 to date), but it had a prominent decadal variability (see Fig. 9). During the 3 decade long DRY epochs 1901–1930 and 1961–1990, the monsoon scenario was bad, India having droughts on average once in about 3 years. In contrast during the 3 decade long WET epochs 1871–1900 and 1931–1960, the frequency of droughts had been on average once in 10–20 years only. Thus, during the 120 years 1871–1990, we had regular 30-year epochs alternating between DRY and WET. Using the available network having a smaller number raingauge stations over India, it was found that 1841–1870 was a DRY epoch. This regular epochal pattern has undergone a change in recent years. The period 1991–2020 was expected to be a WET epoch. Although the decade 1991–2000 had no monsoon droughts, during the next decade beginning in 2001 we had three droughts (in 2002, 2004 and 2009).

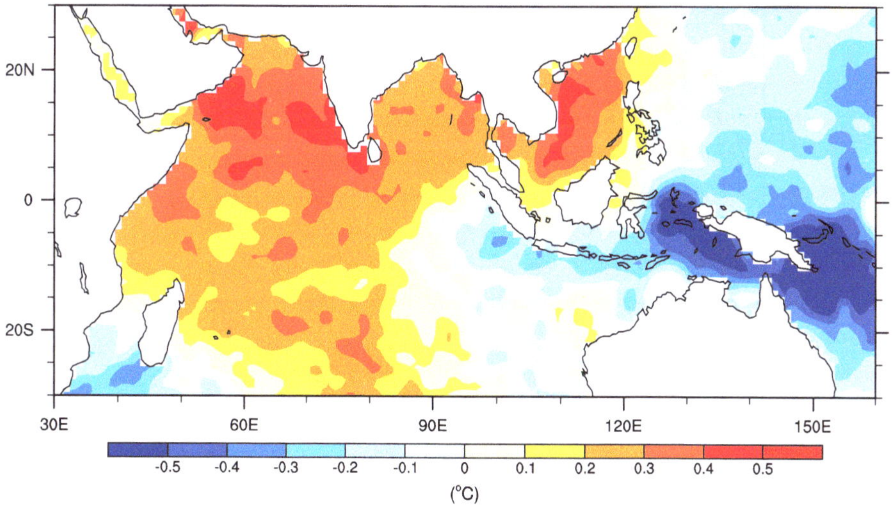

Fig. 10 Mean HadISST anomaly of September to November after five severe drought monsoons of 1965, 1972, 1979, 1982 and 1987

ISMR has large inter-annual variability as may be seen from Fig. 9. Generally, a deficient monsoon year is followed by a normal or excess monsoon year, a sort of biennial oscillation. This phenomenon has been called the tropospheric biennial oscillation (TBO) in which the summer monsoons of India and Australia and the SST of the tropical Indian and west Pacific oceans take part. Meehl (1997) has studied the observational aspects of TBO, and Chang and Li (2000) have given a modelling study. Both these studies have analysed the SST variations on the inter-annual time scale in relation to the heat balance on this scale. The SST anomalies averaged for the five September to November periods following 5 very severe Indian monsoon drought seasons of 1965, 1972, 1979, 1982 and 1987 are shown in Fig. 10. Soon after a season of monsoon drought, the SST of the ocean around India (west Pacific Ocean) warms (cools) and this persists till the following monsoon which has normal or excess ISMR. This is very good for India and makes it different from Africa where drought years have plagued that continent during successive

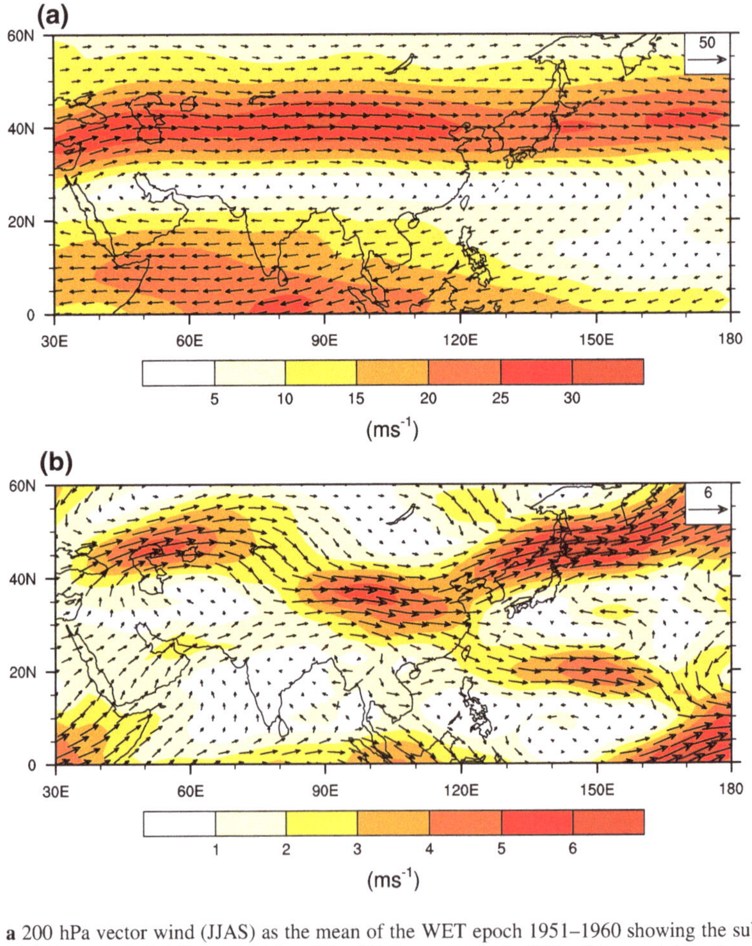

Fig. 11 **a** 200 hPa vector wind (JJAS) as the mean of the WET epoch 1951–1960 showing the subtropical jetstream. **b** Mean wind (JJAS) at 200 hPa of the DRY epoch 1965–1987 of frequent drought monsoons minus the mean wind (JJAS) of the WET epoch 1951–1960 (wind anomaly). Magnitude of the wind is shown by the *shaded* contours. The southwards shift of westerlies over south Asia as a wave number-3 trough in the DRY epoch may be seen

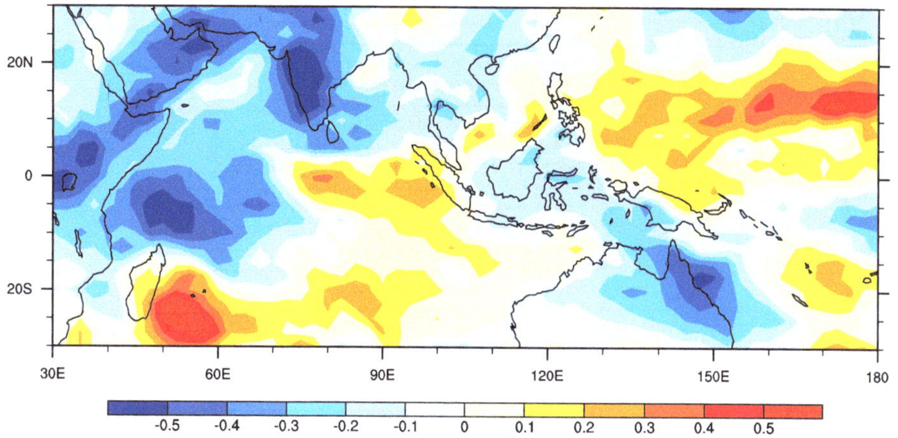

Fig. 12 Correlation between ISMR and mean outgoing longwave radiation (OLR) of July and August using data of 1979–2009 showing the two poles of convection (monsoon heat source). In drought monsoon years, positive convection anomalies (negative OLR anomalies) are over west Pacific Ocean

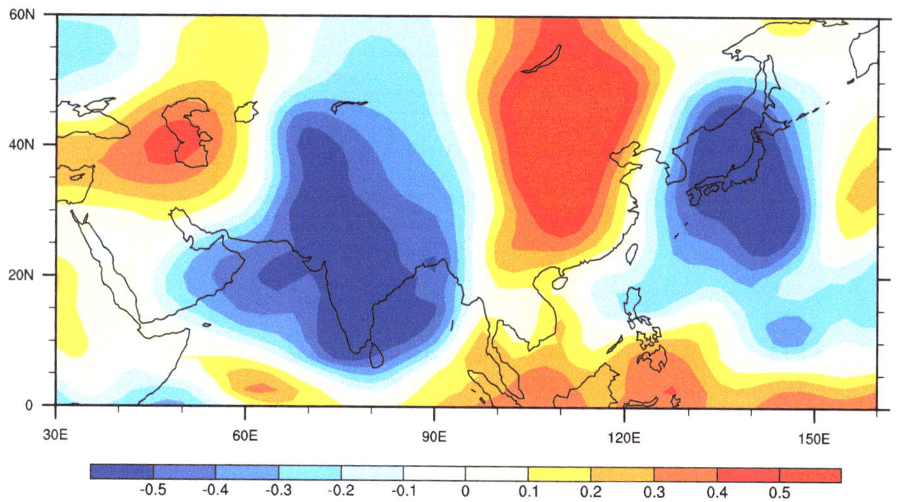

Fig. 13 Correlation between ISMR and 200 hPa meridional wind of June to September showing the Asia–Pacific wave (wave number 6). In drought monsoons, this wave has a trough over northwest India and another trough close to Japanese Islands

years (10–15 consecutive years) as in the Sahelian drought incidences. Such an event is described in the Bible when Egypt was ruled by Pharaohs, one of whom made biblical Joseph governor of Egypt to manage the food problems in a 7-year-long drought episode. The food managers of India have to store food grain to meet the needs of one or at most two consecutive drought years only.

During the DRY epoch 1961–1990 reanalysed wind data showed that subtropical westerlies of the upper troposphere moved to lower latitudes over south Asia as a wave number-3 trough—see Fig. 11. During this epoch SST, monsoon convective heat source

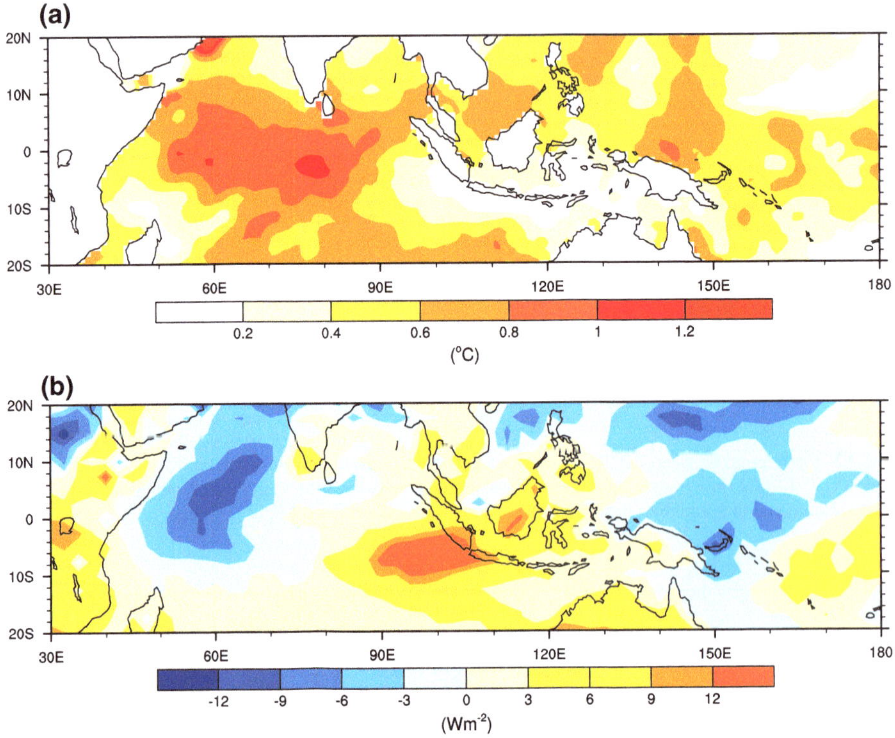

Fig. 14 **a** HadISST of decade (2000–2009) minus (1950–1959) showing the rapid warming of the equatorial Indian Ocean **b** the difference in OLR of (2003–2009) minus (1979–1985) in Watts per square metre

and mid-latitude upper tropospheric westerly winds interacted with each other leading to large inter-annual variability of ISMR. Convective heat source anomalies had large east–west oscillations inter-annually between Indian and west Pacific oceans—see Fig. 12. These heat sources are hypothesised to generate large amplitude wave number-6 Rossby waves in the mid-latitude westerlies which had opposite spatial phase in years of excess and deficient ISMR as shown by Fig. 13 which gives the correlation between ISMR and the meridional wind of 200 hPa level of June to September using data of the period 1961–1990. This large amplitude Rossby wave was named Asia Pacific Wave by Joseph and Srinivasan (1999). The role of this Rossby wave in monsoon variability has been studied by Ding and Wang (2005).

Analysis of SST time series showed that during DRY (WET) epochs, SST anomalies were negative (positive) over areas in Pacific and Atlantic oceans between latitudes 30°N and 60°N and the tropics to high latitude SST gradients over these oceans were large (small). The SST variation in the Pacific Ocean (Atlantic Ocean) is the well-known Pacific decadal oscillation—PDO (Atlantic multidecadal oscillation—AMDO). Decadal oscillations in ISMR and PDO/AMDO had about the same period and temporal phase. Can the large SST gradient phase of PDO/AMDO induce westerly intrusions over south Asia and create large inter-annual variability of ISMR? Modelling studies are needed.

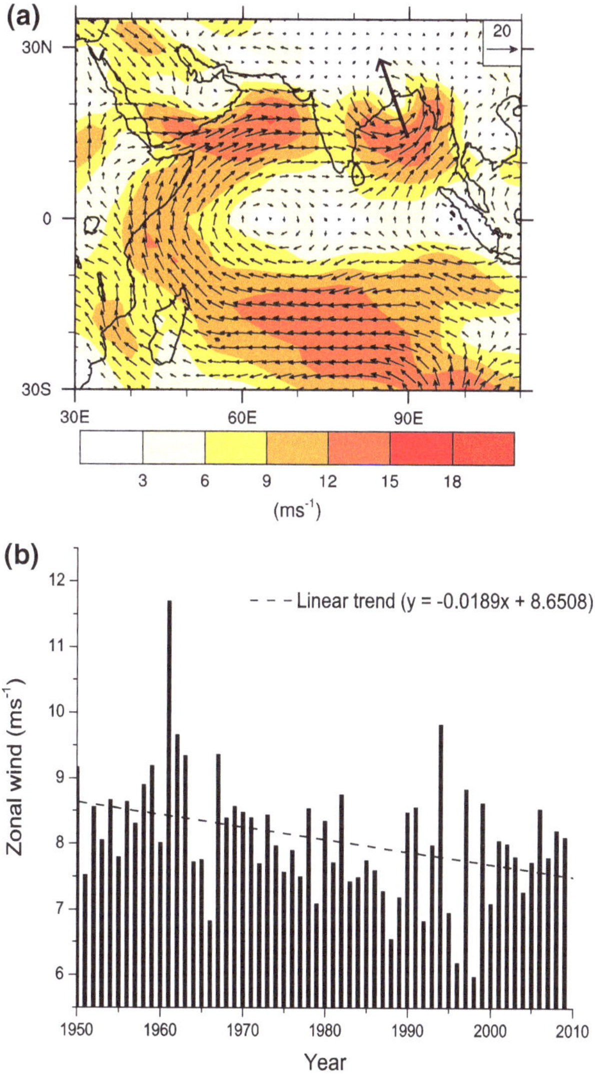

Fig. 15 **a** The 850 hPa wind flow showing the LLJ and a monsoon depression vortex cantered over north BoB **b** The decreasing trend of the mean 850 hPa zonal wind (LLJ) of June to September flowing through peninsular India

5 Rapid Warming of the Indian Ocean

In the global warming scenario, Joseph and Sabin (2008b) reported a rapid increase in the SST of the equatorial Indian Ocean during the monsoon season from 1950 which resulted in an increase in the convection over the area—Fig. 14a, b. This increased equatorial convection resulted in the weakening of the monsoon Hadley circulation and the consequent weakening of the LLJ and the Tropical Easterly Jetstream. The weakening of the LLJ flowing through peninsular India (Fig. 15) could be the possible cause for the observed decreasing trend in the frequency of monsoon depressions that form north of the LLJ in the BoB. As may be seen from Fig. 16, the SST of June to September averaged over an area bounded by latitudes 5°S to 5°N and longitudes 60°E to 90°E showed only a small increasing trend during the period 1891–1960 and the number of monsoon depressions

Fig. 16 **a** The warming trend of the HadISST over the equatorial Indian Ocean of the two periods 1891–1960 and 1951–2010 is shown **b** The decreasing trend of monsoon depression frequency during the two periods 1891–1960 and 1950–2010 is shown

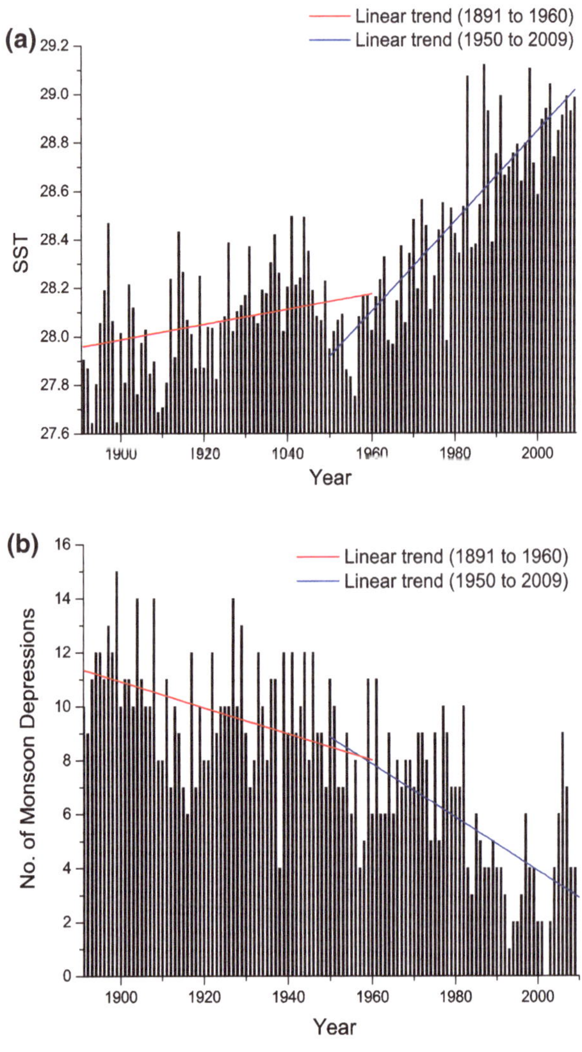

decreased slightly. But, from 1950 to 2010, SST had a rapid increasing trend, and correspondingly, monsoon depressions showed rapid decreasing trend from about 10 depressions in 1950s to about 3 in 2000s. In this connection, reference is drawn to the study by Krishnamohan et al. (2012).

6 Summary and Conclusions

Latent heating of the atmosphere by the deep convection in the ITCZ of the tropical oceans is as important if not more than by the solar heating of the Eurasian land mass in boreal summer for the establishment of the Asian summer monsoon. Monsoon onset over south Asia and its inter-annual variability are closely associated with the changes in SST of the tropical Indian and Pacific Oceans. This paper has highlighted the importance of oceans in the variability of Indian monsoon rainfall on intra-seasonal, inter-annual and inter-decadal time scales. One of the major factors causing drought in ISMR is known to be the El Niño, but it is not clear how monsoon and El Niño are physically related. The shallow MLD area created over the west Pacific Ocean by the El Niño is hypothesised to cause long break monsoon spells (long AB cycle) and thus create large negative anomalies in monsoon rainfall of India. Modelling studies are required to verify this empirical finding. The paper has also given a mechanism for the large inter-annual variability of ISMR during the 30-year-long dry epochs like 1961–1990 occurring during the cold phase of the Pacific decadal oscillation/Atlantic multidecadal oscillation. This mechanism involves the interaction between the monsoon convective heat source and the southwards displaced mid-latitude westerlies (and the subtropical jetstream) over south Asia. Modelling studies are required to test this empirical finding also.

References

Annamalai H, Liu P, Xie SP (2005) Southwest Indian ocean SST variability: its local effect and remote influence on Indian monsoon. J Clim 18:4150–4167

Back LE, Bretherton CS (2009) On the relationship between SST gradients, boundary layer winds and convergence over the tropical oceans. J Clim 22:4182–4196

Chang CP, Li T (2000) A theory for the tropical Biennial oscillation. J Atmos Sci 57:2209–2224

Chao WC, Chen B (2001) The origin of monsoons. J Atmos Sci 58:3497–3507

Ding Q, Wang B (2005) Circumglobal teleconnection in the northern hemisphere summer. J Clim 18:3483–3505

Findlater J (1969) A major low level air current near the Indian Ocean during the northern summer. Q J R Meteorol Soc 95:362–380

Goswami BN (2005) South Asian monsoon. In: Lau WKM, Waliser DE (eds) Intra-seasonal variability in the atmosphere–ocean climate system. Springer, United Kingdom, pp 19–61

Halley E (1686) An historical account of the trade winds and monsoons observable in the seas between and near the tropics with an attempt to assign a physical cause of the said winds. Philos Trans R Soc 16:153–168

Hendon HH (2005) Air-Sea interaction. In: Lau WKM, Waliser DE (eds) Intra-seasonal variability in the atmosphere–ocean climate system. Springer, United Kingdom, pp 223–246

Joseph PV, Jayakumar A (2012) Role of ocean–atmosphere interaction on predictability of the monsoon intraseasonal oscillations. In: Proceedings of IITM Pune Golden Jubilee international conference (OCHAMP) 21–25 February, 2012, IITM website: www.tropmet.res.in

Joseph PV, Raman PL (1966) Existence of low level westerly Jetstream over peninsular India during July. Indian J Meteorol Geophys 17:407–410

Joseph PV, Sabin TP (2008a) An Ocean–atmosphere interaction mechanism for the active break cycle of the Asian summer monsoon. Clim Dyn 30:553–566

Joseph PV, Sabin TP (2008b) Trends in SST and reanalysis 850 and 200 hPa wind data of Asian summer monsoon season during the recent six decades. In: Proceedings of 3rd WCRP international conference on reanalysis—Tokyo, Japan

Joseph PV, Sijikumar S (2004) Intra seasonal variability of the Low Level Jet stream of the Asian summer monsoon. J Clim 17:1449–1458

Joseph PV, Srinivasan J (1999) Rossby waves in May and the Indian summer monsoon rainfall. Tellus 51A:854–864

Joseph PV, Eischeid JK, Pyle RJ (1994) Interannual variability of the onset of the Indian summer monsoon and its association with atmospheric features, El Nino, and sea surface temperature anomalies. J Clim 7:81–105

Joseph PV, Sooraj KP, Rajan CK (2006) The summer monsoon onset process over south Asia and an objective method for the date of monsoon onset over Kerala. Int J Climatol 26:1871–1893

Joseph S, Sahai AK, Goswami BN (2009) Eastward propagating MJO during boreal summer and Indian monsoon droughts. Clim Dyn. doi:10.1007/s00382-008-0412-8

Ju J, Slingo JM (1995) The Asian summer monsoon and ENSO. Q J R Meteorol Soc 122:1133–1168

Krishnamohan KS, Mohanakumar K, Joseph PV (2012) Climate change in tropical cyclones and monsoon depressions of North Indian Ocean. In: (Extended Abstracts) Second WMO international conference on Indian Ocean tropical cyclones and climate change, New Delhi, 2012

Lindzen RS, Nigam S (1987) On the role of sea surface temperature gradients in forcing low level winds and convergence in the tropics. J Atmos Sci 44:2418–2436
Meehl GA (1997) The south Asian monsoon and the Tropospheric Biennial Oscillation. J Clim 10:1921–1943
Pai DS, Rajeevan M (2009) Summer monsoon onset over Kerala: new definition and prediction. J Earth Syst Sci 118:123–135
Pearce RP, Mohanty UC (1984) Onsets of the Asian summer monsoon 1979–1982. J Atmos Sci 41:1620–1639
Rajeevan M, Bhate J, Kale JD, Lal B (2006) High resolution daily gridded rainfall data for the Indian region: analysis of break and active monsoon spells. Curr Sci 91:296–306
Sabin TP, Babu CA, Joseph PV (2012) SST-convection relation over tropical oceans. Int J Climatol. doi:10.1002/joc.3522
Sengupta D, Goswami BN, Senan R (2001) Coherent intraseasonal oscillations of ocean and atmosphere during the Asian summer monsoon. Geophys Res Lett 28(21):4127–4130
Shankar D, Shetye SR, Joseph PV (2007) Link between convection and meridional gradient of sea surface temperature in the Bay of Bengal. Proc Indian Acad Sci (Earth Planet Sci) 116:385–406
Soman MK, Slingo JM (1997) Sensitivity of Asian summer monsoon to aspects of sea surface temperature anomalies in the tropical Pacific Ocean. Q J R Meteorol Soc 123:309–336

DOI 10.1007/978-94-017-8789-5_12
Reprinted from *Surveys in Geophysics* Journal, DOI 10.1007/s10712-013-9245-z

Perspectives in Modelling Climate–Hydrology Interactions

Stefan Hagemann · Tanja Blome · Fahad Saeed · Tobias Stacke

Received: 12 December 2012 / Accepted: 18 June 2013 / Published online: 2 August 2013

Abstract Various land–atmosphere coupling mechanisms exist that may lead to large-scale impacts on climate and hydrology. Some of them are still less understood and not adequately represented in state-of-the-art climate modelling. But, as the current generation of climate models enables consideration and implementation of important coupling processes, the present study provides perspectives for the modelling of relevant climate–hydrology interactions. On a more short-term perspective, these comprise anthropogenic land use and especially irrigation, which has been shown that it may even affect remote regions. On a long-term perspective, the coupling of hydrology to carbon cycle and vegetation becomes important, specifically the dynamics of permafrost and wetlands. Here, we present a review of current knowledge combined with some exemplary studies from a large-scale point of view. Therefore, we focus on climate–hydrology interactions that are relevant on scales utilized in the current or forthcoming global and regional climate modelling exercises.

Keywords Irrigation · Land atmosphere feedbacks · Land-use impacts · Modelling perspectives · Permafrost and wetland dynamics

1 Introduction

The hydrological cycle plays a prominent role within the Earth system and is crucially important to life on Earth including the human society. Thus, the current state of the hydrological cycle and its future development are key issues in environmental research. In

S. Hagemann (✉) · T. Blome · T. Stacke
Max-Planck-Institut für Meteorologie, Bundesstr. 53, 20146 Hamburg, Germany
e-mail: stefan.hagemann@zmaw.de

F. Saeed
Climate Service Center, Helmholtz-Zentrum Geesthacht, Fischertwiete 1, 20095 Hamburg, Germany

studies of global and regional climate change, climate models are the current operational tools. Although the ability of climate models to simulate the various characteristics of the climate or Earth system has considerably improved within the past decades, gaps or large uncertainties in the representation of some specific processes still exist. Consequently, there is a lot of room for improvement. In the following, we will focus on climate–hydrology interactions and provide two major perspectives for their modelling within the framework of Earth system models (ESMs). Here, only those interactions that are relevant on scales utilized in the current or forthcoming global and regional climate modelling exercises are considered.

Harding et al. (2011) give a general overview of current knowledge of the terrestrial global water cycle. Here, they consider aspects of state-of-the-art global hydrology modelling, past and projected future hydrological change in means and extremes, as well as uncertainties in our understanding of the current global water cycle and how it will develop in the future. But in those aspects, the feedback of terrestrial hydrology to the climate is mostly not considered. In order to investigate the interactions between climate and hydrology and how they may behave under climate change conditions, a coupled framework is necessary where both components are adequately represented. Strong interactions between the climate, hydrology and land use occur (Claussen 2004; Falloon and Betts 2010). The snow–climate feedback is well known and described (e.g., Cess et al. 1991). However, feedbacks between CO_2, vegetation, soil moisture, groundwater recharge and climate are less well understood and are not well described in most climate and hydrological models.

Soil moisture controls the partitioning of the available energy into latent and sensible heat flux and conditions the amount of surface runoff. By controlling evapotranspiration, it is linking the energy, water and carbon fluxes (Koster et al. 2004; Dirmeyer et al. 2006; Seneviratne and Stöckli 2008). Seneviratne et al. (2006) stated that a northward shift of climatic regimes in Europe due to climate change will result in a new transitional climate zone between dry and wet climates with strong land–atmosphere coupling in central and eastern Europe. They specifically highlight the importance of soil-moisture–temperature feedbacks (in addition to soil-moisture–precipitation feedbacks) for future climate changes over this region. A comprehensive review of soil moisture feedbacks is given by Seneviratne et al. (2010). Their general principles are known (e.g., Koster et al. 2004, 2006; Teuling et al. 2009), even though there is still room for model improvement.

Soil moisture shows a high variability from daily to interannual timescales. An appropriate knowledge of soil moisture conditions is important for the initialization and quality of seasonal to yearly climate predictions. Fischer et al. (2007) indicated that the record-breaking European heat wave in 2003 was enhanced by the large soil moisture anomalies that were caused by a large precipitation deficit together with early vegetation green-up in the months preceding the extreme summer event. Loew et al. (2009) showed that these soil moisture anomalies were observable using remote sensing sensors. Consequently, the impact of soil moisture memory on the climate is an important scientific topic (e.g., Seneviratne et al. 2006) and is addressed specifically in the BMBF project MiKlip PastLand where its value for seasonal to decadal prediction is investigated.

From the hydrological perspective, two major challenges for modelling climate–hydrology interactions have currently arisen where we will shed some light on in the following. On a more short-term perspective, these comprise anthropogenic land use and especially irrigation. The coupling of hydrology to carbon cycle and vegetation is important on the long-term perspective, specifically the dynamics of permafrost and wetlands.

While the process of anthropogenic emissions due to fossil fuel burning is fairly well established in state-of-the-art climate model simulations, up to now, the possible impact of land-use changes on the climate is mostly neglected in long-term climate simulation. Dale (1997) reviewed the literature dealing with the relationship between land-use change and climate change and concluded that in recent centuries, land-use change has had much greater effects on ecological variables than has climate change. Pielke et al. (2002) documented that land-use change impacts regional and global climate through the surface-energy budget, as well as through the carbon cycle, whereat the surface-energy budget effects may be more important than the carbon-cycle effects. While this is valid for the past climate, results of Cox et al. (2000) indicated that carbon-cycle feedbacks could significantly accelerate climate change over the twenty-first century and pointed out the necessity to consider the potentially large direct human influences on terrestrial carbon uptake through changes in land cover and land management. Changes in the land surface (vegetation, soils, water) resulting from human activities can affect the regional climate through shifts in radiation, cloudiness and surface temperature. Changes in vegetation cover affect surface energy and water balances at the regional scale, so that the impact of land-use change may be very significant for the regional climate over time periods of decades or longer (Denman et al. 2007). The effects of a specific land-use change on the climate depend on the surrounding environment and climate characteristics as a regional modelling study of Gao et al. (2003) over China has shown.

An extreme anthropogenic impact on the local hydrology is the practice of irrigation. Over 18 % of total cultivated land is irrigated (Fischer et al. 2007); additionally, much nonagricultural land has been substantially modified by human activities. Conversion of land to agriculture not only impacts the local evaporation and hydrological response, but may also influence the distribution of rainfall and evaporative demand in the surrounding landscape as well as have remote impacts on the large-scale circulation. The latter will be considered in more detail in Sect. 2. Agriculture and urban development have increased substantially in the past century and will continue to develop in the twenty-first century. Therefore, any assessment of the world's water resources must take into account both the direct and indirect influences of land-use changes and the exploitation of the riverine system.

Earth's climate is determined to a large extent by greenhouse gases (GHG) in the atmosphere, which influence the radiation budget and thus the energy balance of the planet. Thus, fluxes that may change the atmospheric GHG content are of great importance in climate change research. Apart from water vapour and anthropogenic GHG, various components of the global carbon cycle, especially CO_2 and CH_4, play a significant role. In recent years, estimates for the amount of carbon stored in soils have attracted more and more attention, and here especially the consideration of the vast permafrost regions increased numbers drastically (Tarnocai et al. 2009; Zimov et al. 2006; Schuur et al. 2008; McGuire et al. 2009). Permafrost, being defined as ground that is at or below zero degrees Celsius for more than two consecutive years, affects roughly one-quarter of the northern hemisphere (Brown et al. 1997). It is believed to store between 1,400 and 1,800 Pg of C in the upper few metres of the soil (Schuur et al. 2008), which would be twice the amount of the atmosphere's content. The high northern latitudes are one of the critical regions of anthropogenic climate change, where the observed warming is clearly above average due to the so-called Arctic Amplification (Solomon et al. 2007; ACIA 2005). Climate model simulations project this trend to continue (Serreze and Barry 2011). The combination of the high C stocks in subarctic and arctic soils with the pronounced warming in the affected regions could thus lead to a positive feedback through the release of formerly trapped, 'deep-frozen' C into the atmosphere, when near-surface permafrost thaws. For the thawed soils and their biogeochemistry, it is decisive whether dry or wet conditions predominate: Aerobic decomposition is relatively fast and leads to the release of CO_2, while anaerobic decomposition is much slower and leads to the release of CH_4 as the main product of the combustion of organic soil material. Therefore, not only the soil's temperature but also its moisture status and specifically the presence of wetlands are important for the assessment of the biogeochemical response to climatic conditions and thus should be represented in climate or ESMs in a realistic and process-based manner. Thus, the adequate representation of permafrost hydrology is a necessary and challenging task in climate modelling, which will be considered in more detail in Sect. 3.

2 Impact of Anthropogenic Land Use, Especially Irrigation, on Climate

Several studies (e.g., Gordon et al. 2005; Piao et al. 2007; Rost et al. 2008a, b) demonstrated that land cover conversions and water withdrawals have already noticeably changed the partitioning of terrestrial precipitation into evapotranspiration and runoff. Gerten (2013) estimated that these direct human impacts have increased the global river discharge by about 5 %, which is caused by the associated reduction in evapotranspiration. Regionally, the implications of anthropogenic land use may be much larger. Partially, even opposite effects (increased evapotranspiration, reduced runoff) may be induced by land-use change (Destouni et al. 2013) or irrigation (Gerten et al. 2008).

Observations and model studies in tropical forests have shown effects of changing surface energy and water balance on the state of the atmosphere. For example, Marengo and Nobre (2001) found that the removal of vegetation led to decreases in precipitation, evapotranspiration and moisture convergence in central and northern Amazonia. Oyama and Nobre (2004) showed that the removal of vegetation in north-east Brazil would substantially decrease precipitation. Other model studies indicated that increased boreal forest reduces the effects of snow albedo and causes regional warming (Denman et al. 2007). Related to the latter, e.g., Göttel et al. (2008) investigated the influence of changed vegetation fields on the projected regional climate over the Barents Sea region in an off-line coupling experiment with the regional climate model (RCM) REMO and the dynamic vegetation model LPJ-GUESS (Sitch et al. 2003). They projected a forest ratio increase and a shift of the tree line to higher altitudes and latitudes caused by a warmer climate with longer snow-free periods and growing season lengths. The feedback effects to the climate of these changes were one order of magnitude lower than the effects of the greenhouse gas forcing. A further warming in spring could be attributed to the snow-albedo effect, while a cooling in summer was dedicated to changes in roughness length, enhanced transpiration and changes in surface albedo. A more extreme study was conducted by Bathiany et al. (2010) who investigated the effect of large-scale changes in forest cover on global climate. They completely removed tropical forest within the ESM of the Max Planck Institute for Meteorology (MPI-ESM), which resulted in a simulated 0.4 K warming due to an increase in CO_2 concentrations and a decrease in tropical evapotranspiration. A similar experiment for the high northern latitudes led to a global cooling of 0.25 K in case of complete deforestation and an equally large warming in case of afforestation. In both cases, the involved albedo changes (snow-masking effect) are the main drivers of the temperature change.

Land-use changes such as deforestation may have a substantial climate impact in areas located close to strong climatic gradients, such as tropical regions as well as arid and semiarid regions. In this respect, Africa is one of the 'hot spot' areas. Taylor et al. (2002) stated that the climatic impacts of land-use change in the Sahel region are likely to increase rapidly in the coming years. So far, the effect of deforestation and reduced vegetation cover associated with land-use change in Africa has mainly been studied with coarse-grid (300 km resolution or coarser) global climate models in the form of time slice experiments and idealized forcing (see, e.g., Feddema et al. 2005). With these coarse resolution models, effects on the local and regional climate can usually not be resolved. For this purpose, RCMs are an adequate tool, such as done by Paeth et al. (2009) for West Africa who conducted long-term transient climate change experiments with the RCM REMO at 50 km resolution over West Africa where they forced their simulations with increasing greenhouse gas concentrations and land-use changes until 2050. Their results indicate that significant future changes in the near-surface climate may be caused by land-use changes.

A specific form of land use is irrigation, which can considerably affect the regional climate (Boucher et al. 2004; Lobell et al. 2009) and whose feedbacks onto rainfall (ter Maat et al. 2006) may become especially important where irrigation coincides with areas of global hot spots for land surface–atmosphere feedbacks. Koster et al. (2004) identified the Sahel zone as one of the hot spot areas for the feedback of surface soil wetness to subsequent rainfall. In this semiarid region, irrigation is not a major agricultural practice, but an increase in dryland agriculture is possible which is sensitive to rainfall totals. A study of Taylor et al. (2002) showed that future likely changes in land cover could result in a reduction of nearly 10 % in rainfall. Another hot spot of soil moisture–precipitation coupling is located over India (Koster et al. 2004). The Indian subcontinent is one of the most intensely irrigated regions (Fig. 1a) in the world (Sacks et al. 2009), and many studies have shown the role of irrigation in modifying the local climate through feedback mechanisms (e.g., De Rosnay et al. 2003; Douglas et al. 2006, 2009; Lee et al. 2009). Effects of irrigation, embedded in South Asian Summer Monsoon (SASM), affect 22 % of world's population and hence play a crucial role in modifying the water resources, agriculture, economics and human mortality of the region. Therefore, this topic is covered separately in the following Sect. 2.1.

2.1 Impact of Irrigation on the South Asian Summer Monsoon (SASM)

As mentioned above, the effects of irrigation on local climate through feedback mechanisms are well known from earlier studies. Saeed et al. (2011) found that REMO is able to reproduce the general characteristics of the SASM; however, over the land areas of northwest India and Pakistan, a systematic warm temperature bias of more than 5 °C can be noticed (Fig. 2a, c). The too enhanced simulation of the heat low (Fig. 2) is a common systematic error that is present in many regional climate models applied over South Asia (Lucas-Picher et al. 2011). For more than a decade, this heat low is used as an important predictor for SASM rainfall (Singh et al. 1995). The major part of this heat low region falls inside the densely irrigated Indus basin (Fig. 1b), which is the largest contiguous irrigation network in the world, and its surface water is heavily manipulated by building large dams, link canals, watercourses, etc., hence resulting in modification of the amount of water in the soil (Khan et al. 2008). It is estimated that the Indus River drains only one-eighth of the $\sim$400 km^3 water that annually falls on the basin in the form of rain and snow, with the remainder used mostly for irrigation and returned to the atmosphere by evapotranspiration (ET) (Karim and Veizer 2002).

Saeed et al. (2009) applied the regional climate model REMO (Jacob et al. 2007) over South Asia at a resolution of ½ degree ($\sim$55 km) domain, forced with lateral boundary

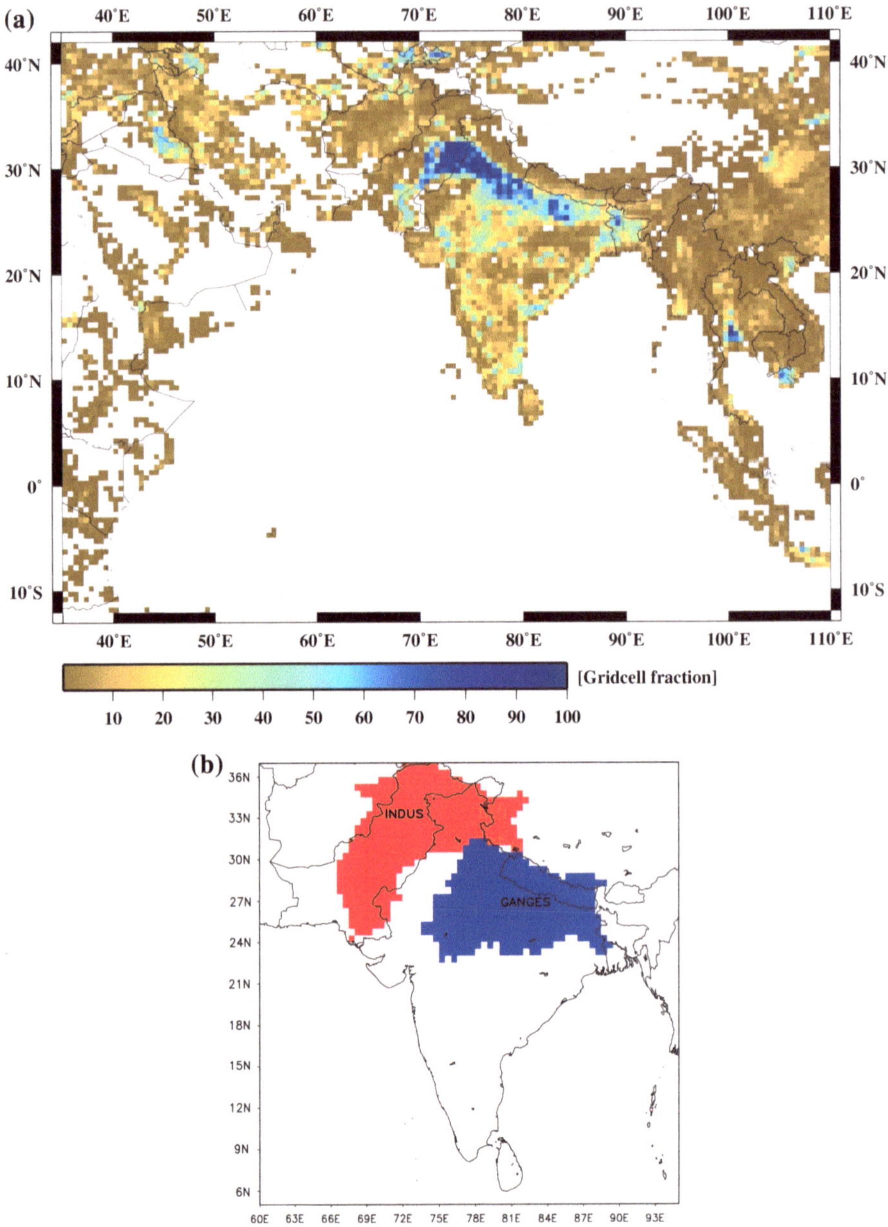

Fig. 1 **a** Fraction of area equipped for irrigation in South Asia based on the reference year 2000 (Siebert et al. 2005). **b** Locations of Indus and Ganges catchments at 0.5° resolution

conditions obtained from the European Centre for Medium Range Weather Forecasts reanalysis (ERA40) (Uppala et al. 2005). In order to take into account the effect of irrigation, a map of areas equipped for irrigation (Fig. 1a; Siebert et al. 2005) was used in the simulation and the results for four SASM summer months June, July, August and September (JJAS) were presented. For the potentially irrigated fraction of a grid box, the

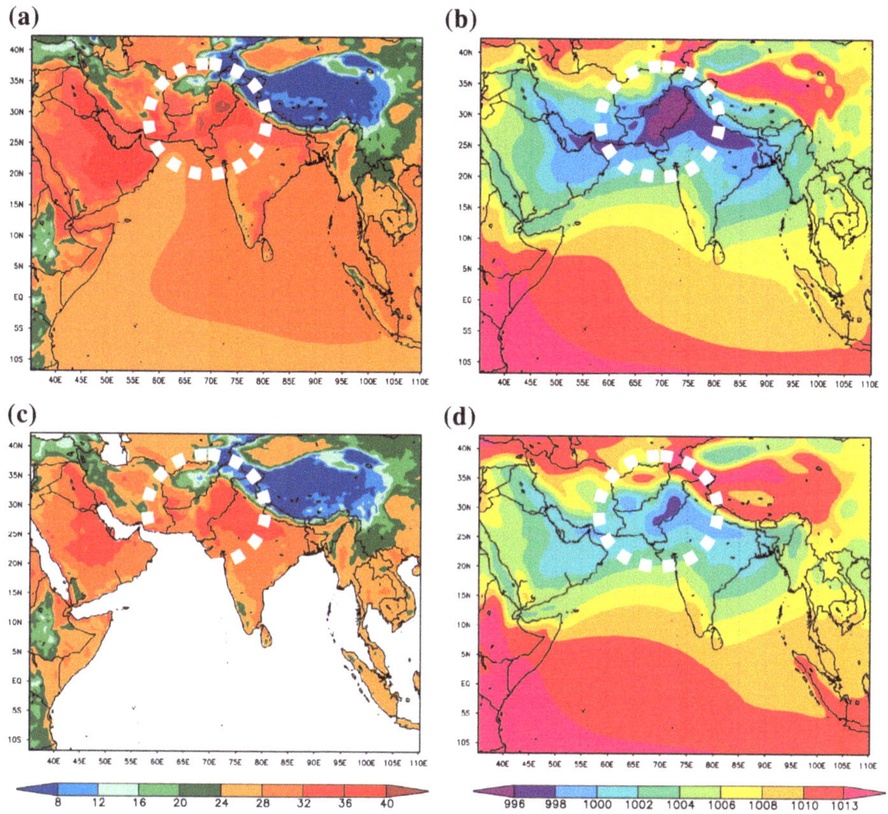

Fig. 2 Simulated (*upper panels*) and observed (*lower panels*) summer climatologies (JJAS) for the period 1961–2000: 2 m temperature [°C] (*left*, REMO and data of Willmott and Matsuura 2009) and mean sea-level pressure [hPa] (*right*, REMO and ERA40). The *white circles* indicate the heat low region

soil wetness was increased to a critical value in each time step, so that potential ET can occur. In this way, it is assumed that irrigation is conducted to fulfil optimal conditions for the vegetation/crops, allowing them to transpire at a potential rate. Note that the main results will not change if water for irrigation is limited by available water from the rivers (Saeed et al. 2012).

In the REMO simulation without irrigation, the overestimation of the heat low (too high temperature, too low pressure; Fig. 2) resulted in increased differential heating between ocean and land, and therefore the overestimation of winds entering into the plains of the Indian subcontinent from the Arabian Sea. This causes a situation unfavourable for westward propagating currents from Bay of Bengal to intrude deep into western India and Pakistan. Therefore, less moisture is advected causing an underestimation of precipitation over this area as well. When irrigation is accounted for, a more realistic behaviour of the simulated climate is yielded. Figures 3 and 4 compare the changes in the REMO simulation with irrigation (Fig. 3b, d) to the reference simulation without irrigation (Fig. 3a, c). An improvement of simulated temperature and MSLP can be seen over the whole region, but statistically significant and most pronounced changes are present over Indus, Ganges (Fig. 1b) and southern India (Fig. 4a, b). For these regions, the standard REMO version simulated the largest systematic biases (Fig. 2); hence, the irrigation led to a better

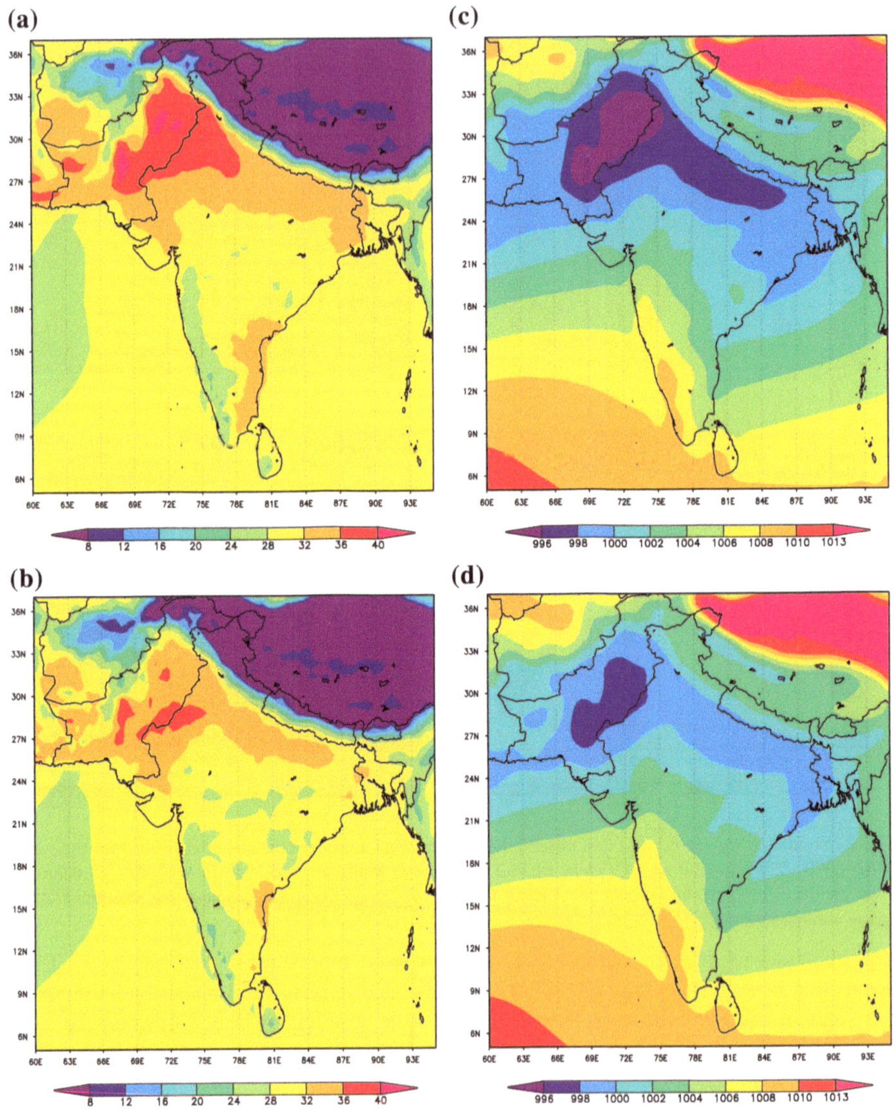

Fig. 3 Simulated REMO summer climatologies (JJAS) for the period 1989–1992: 2 m temperature [°C], **a** reference, **b** REMO with irrigation; and mean sea-level pressure [hPa], **c** reference, **d** REMO with irrigation

representation of these variables. Figure 4c indicates a significant increase in ET over the whole subcontinent region, again with largest increase over Indus and southern India.

The reduced differential heating of the land relative to the ocean leads to a reduction in the too strong westerly winds from the Arabian Sea into the Indian plains (Fig. 4d). This creates conditions favourable for monsoon depressions originating from the Bay of Bengal to intrude deep into the land up to western India and Pakistan. Saeed et al. (2009) could illustrate this behaviour for the development and movement of several monsoon depressions that brought rainfall to the western part of the Indian peninsula and that have been

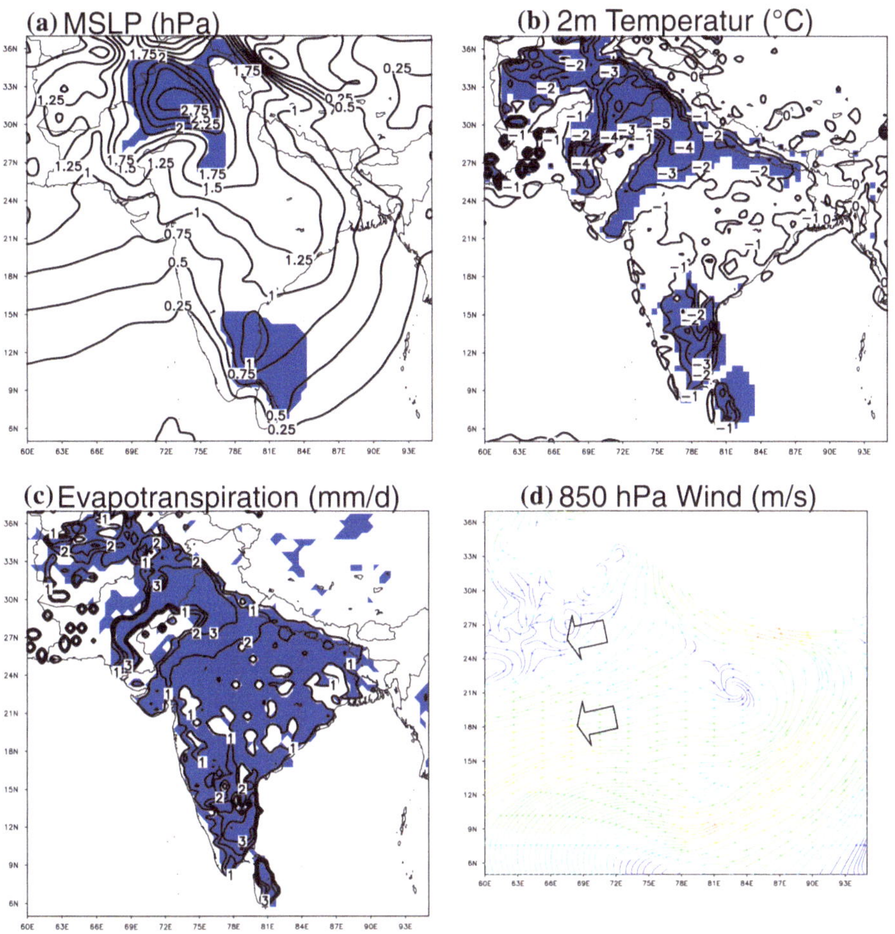

Fig. 4 Differences in several variables from the REMO simulations with and without irrigation averaged for the summer (JJAS) 1989–1992: **a** mean sea-level pressure (hPa), **b** 2 m temperature (°C), **c** evapotranspiration (mm/day) and **d** 850 hPa winds. The *shaded blue areas* indicate significant differences at the 90 % level from a two-tailed *t* test

discussed in earlier published literature. Figure 5 shows one of these cases that occurred from 19 to 21 September 1991 (Mahajan et al. 1995). All these cases have in common that, in the REMO simulation without irrigation, a depression forms in the Bay of Bengal, but it stops at the east coast of India and dissolves. Only in the simulation with irrigation, the depression is able to travel deep into land towards western India and Pakistan and to transport moisture and precipitation into these regions, such as it has been observed. Together with the increased local recycling of moisture due to the increased ET, this leads to an increase in precipitation over central/western India and Pakistan, which reduces the dry precipitation bias in this area.

For the same two model set-ups of REMO, i.e. with and without the representation of irrigation, climate change simulations have been conducted over the South Asian model domain following the A1B emission scenario (Gerten et al. 2011). Figure 6 shows that for the projected 2-m temperature changes (2085–2099 minus 1985–1999), REMO without

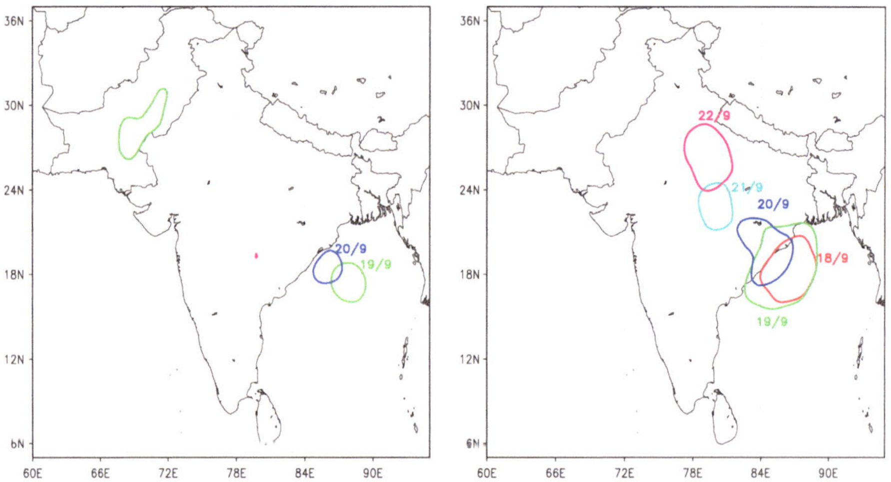

Fig. 5 Simulated development and movement of a monsoon depression that has been observed within the period from 18 to 22 September 1991: The *panels* show the MSLP (mean sea level pressure) contour lines at 999 hPa around the centre of the depression as simulated by REMO without (*left*) and with (*right*) irrigation

irrigation projects an increase of more than 4 °C in general and more than 6 °C over the central Indian region. In contrast, the REMO simulation with irrigation projects much less warming, with a temperature increase ranging from 2 to 4 °C. This highlights the role of irrigation in attenuating the climate change signal over the South Asian region. Thus, it can be concluded that the irrigation performed over the twentieth century may have already masked recent climate change signals over this region.

2.2 Conclusions and Perspectives for the Impact of Irrigation on Climate

The results presented in Sect. 2.1 signify the role of irrigation in effecting the local temperature, which in turn effects large-scale circulations and precipitation of the SASM.

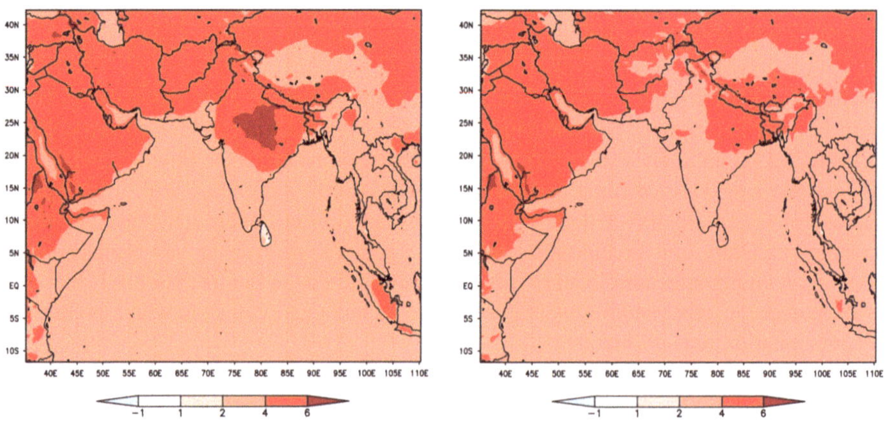

Fig. 6 Projected 2 m temperature changes (2085–2099 minus 1985–1999) in °C for REMO without (*left panel*) and with irrigation (*right panel*)

They also show the potential of irrigation for mitigating climate change effects in the SASM region. The present neglect of irrigation was the main cause of the systematic REMO model error over the heat low region in NW India and Pakistan that led to a too enhanced formation (too warm, too deep) of this heat low. The representation of irrigation has caused the removal of this bias. Similar biases in other RCMs (Lucas-Picher et al. 2011) suggest that they are also related to the missing irrigation process. Consequently, the representation of water used for irrigation in climate models is necessary for the realistic simulation of SASM circulation and associated rainfall. This, together with taking into account land-use change, has also been emphasized by Gordon et al. (2005) for the global scale.

The impact of resolution on the irrigation effects upon the SASM has not been explicitly considered up to now. In a recent study of Tuinenburg et al. (2013), consistent results to those presented in Sect. 2.1 were found across an ensemble of three RCMs and one GCM. Here, the application of irrigation on a large scale led to changes in the large-scale circulation, in which moisture shifted away from the Ganges plain towards the Indus basin and Pakistan. This has confirmed the results found by Puma and Cook (2010) and Asharaf et al. (2012). But generally, a higher resolution leads to an improved simulation of the SASM. Kumar et al. (2013) summarized that most of the GCM studies focusing on the Indian monsoon region concluded that GCMs have difficulties in simulating the mean monsoon climate over India. Due to their coarse horizontal resolution, GCMs have limitations in simulating the complex orographic precipitation over India. Also, several RCM studies have been carried out to simulate the summer monsoon over South Asia, whereat all have reported an improvement in the simulation of SASM spatial and temporal distribution compared to coarser global models (Kumar et al. 2013).

Eighty percentage of Indus basin river flows are attributed to the melt of snow and glacier. Considering the large impact of irrigation on SASM behaviour, one can assume that under global warming the changes in the timings of water inflows would shift towards earlier months, hence causing changes in cropping patterns and subsequently irrigation. As irrigation is impacting the climate change signal over the SASM, changes in irrigation patterns over the Indus basin will also affect the SASM circulation and associated rainfall under climate change. Therefore, not only irrigation itself but also changes in irrigation patterns need to be regarded for climate change studies over the SASM region.

While irrigation seems to have a positive mitigating impact on the SASM climate, the picture looks different for areas where the human consumption of water leads to drying and shrinking of surface waters (Asokan et al. 2010). Here, the associated decrease in evaporation from these surface waters counteracts the direct irrigation effect of increasing evapotranspiration in irrigated land areas. A very prominent example is the Aral Sea, which was the fourth largest lake on the globe until 1960, with a surface area of about 68,000 km^2. But, large irrigation activities in many parts of Middle Asia were mainly responsible for the catastrophic desiccation of the Aral Sea within the last five decades (see Breckle and Geldyeva (2012) and references therein). How irrigation has affected the current climate or may affect the future climate under global warming conditions in other regions is an important subject for future studies. In this respect, a first ESM study was provided by Guimberteau et al. (2012). Irrigation also causes groundwater depletion over many areas of the globe (Döll et al. 2012). How this may affect climate and water resources is a prospect for future studies as the current knowledge of the impacts of changing groundwater on climate is limited.

3 Permafrost

Permafrost and wetlands are two focal points in the coupling of hydrology to biogeochemical processes under climate change conditions. A large part (~24 %) of the northern hemisphere terrestrial land surface is underlain by permafrost (French 1990) that is mainly situated in high latitudes (Fig. 7). Here, climate warming is more pronounced than elsewhere and is very likely to continue to do in the future according to Solomon et al. (2007). Permafrost soils build a globally relevant carbon reservoir as they store large amounts of deep-frozen organic material with high carbon contents. If permafrost thaws under global warming conditions, the stored carbon can be decomposed and released to the atmosphere as additional greenhouse gas, which will lead to a positive feedback. Consequently, relevant scientific questions are the following: How fast, how deep and to what temperature are permafrost soils going to thaw in the future?

Fig. 7 Distribution of permafrost areas in the Arctic according to the International Permafrost Association (1998)

In Sect. 3.1, relevant hydrological processes are described that occur in permafrost areas and that should preferably represented in models simulating interactions of permafrost hydrology with vegetation, climate and the carbon cycle. The current state of the representation of permafrost processes in ESMs is tackled in Sect. 3.2, while Sect. 3.3 deals with the specific topic of wetlands.

3.1 Basic Hydrological Processes in Permafrost Areas

Apart from climatic cold conditions, the occurrence of permafrost is largely controlled by physiographic features such as aspect, slope and elevation. Other factors such as soil types, soil moisture, vegetation cover and disturbances (e.g., wildfire) can also influence the distribution of permafrost (Haugen et al. 1982; Yoshikawa et al. 2002). The most basic process in permafrost areas is the seasonal melting and freezing of soil water in the presence of continuously frozen ground below a certain depth. The depth to which the soil is thawed is called the active layer. Regions that are affected by permafrost or extensive seasonal ground freezing show a specific behaviour of important hydrological variables: (1) Soil moisture is often rather high in near-surface layers, despite low precipitation rates in many regions; (2) river discharge observations display very low wintertime values; and (3) surface runoff shows a steep spring peak after snowmelt that can deliver a substantial part of the annual total runoff (Swenson et al. 2012). Several reasons are responsible for these features, which will be described below.

Firstly, the phase change exerts a drop in liquid water content, and the freezing front can be seen as a water sink within the soil. This leads to the development of a gradient in liquid water content and thus induces water movement towards the freezing front. This process is called cryosuction. It leads to unique characteristics of soil moisture in regions with permafrost and extensive seasonal ground freezing, namely to the increase in total soil moisture in the upper layers as well as to the development of large ice bodies in the ground (see also description of ice wedges below).

Secondly, the permeability for liquid water flow is reduced in frozen soils. This might be the main and most obvious effect frozen soil exerts on hydrology (Niu and Yang 2006). According to Staehli et al. (1999), there are two possible pathways for the flow of liquid water when soil temperature is below 0 °C. Transport channels for slow water flows are provided by thin films of adsorptive and capillary water, which are still existing in liquid phase and whose amount depends mainly on soil texture type. Alternatively, fast water movement is possible through air-filled macropores. Soils contain such pores through structural variations like cracks, holes and channels, e.g., from dead roots and soil inhabitants like worms.

Nearly impermeable soil layers can develop due to the freezing of the soil during winter and spring seasons (Koren et al. 1999) as ice bodies in the ground impede liquid water movement through blocking of the pore space (Swenson et al. 2012). Moreover, a strongly frozen soil will contain only very limited amounts of unfrozen water so that the ability of the subsurface material to conduct liquid water, i.e. the hydraulic conductivity, is decreased, yet not totally set to zero.

Frozen ground and snow cover also influence rainfall–runoff partitioning, the timing of spring runoff and the amount of soil moisture that subsequently is available for evapotranspiration in spring and summer (Koren et al. 1999). For the infiltration of surface water into the soil, the above-explained principles lead to the same general behaviour as for the hydraulic conductivity, as the infiltration process is lastly determined by the soil's ability to conduct water away from the surface. Nevertheless, infiltration can vary even more than the conductivity. When considering two areas with the same climatic conditions, several other locally variable factors influence infiltration:

- Soil texture.
- Orography and slope.
- Snow: thicker snow cover leads to enhanced insulation and thus weaker freezing of the soil, and vice versa.
- Vegetation: similar effects as snow, via weaker or stronger insulating properties of the vegetation cover, and via its roots' effects on the soil structure.

These influencing factors can lead to patterned areas with surface-water-impeding zones in some parts, and with permeable zones, where conditions for infiltration are much more favourable, in others. In consequence, the impact of frozen soil on infiltration and on hydraulic conductivity is strongly scale dependent (Niu and Yang 2006; Koren et al. 1999). Thus, it is important to notice that the falling below the 0 °C threshold does not lead to a complete blocking of infiltration and percolation. These characteristics complicate the implementation of cold regions' soil processes in land surface schemes for climate and Earth system modelling.

The response of the soil to freezing leads to specific variations in the annual cycle of soil hydrology. The snowmelt, which is usually constrained to a very short period of sometimes less than two weeks, delivers a large water input to the land surface, which at this time of the year is still frozen. Infiltration capacity is thus low, and much of the snowmelt water is channelled into surface runoff. The thawing of the active layer begins immediately upon the completion of snowmelt (Boike et al. 1998), dependent upon a number of factors including soil material, duration of snow cover, soil moisture and ice content, and convection of heat by groundwater (Woo 1986). The beginning of the thawing coincides with high surface moisture values, and ice melting in near-surface layers occurs on the top of still frozen, and thus less permeable, deeper layers. Consequently, subsurface water flows are weak, and high soil moisture values develop within the still thin thawed upper layers. Refreezing of infiltrated snowmelt water also contributes to this (Swenson et al. 2012). Over the course of the warm summer season, the thawing and deepening of the active layer increase the water-holding capacity of the soil, resulting in a decreasing surface water contribution during precipitation events and a steadily increasing baseflow contribution (Hinzman and Kane 1991). The latter is a lateral slow subsurface runoff that can develop as the permafrost table forms a barrier to the deeper soil, where again water cannot easily percolate. Due to the enlarged water storage and increased baseflow, upper soil layers can also become drier in this part of the year. The autumn precipitation often coincides with the start of the freezing season; thus, again high surface runoff rates are produced, yet much lower than in spring. During winter, the decreased hydraulic conductivity in frozen soils leads to the observed very low winter baseflow. Permafrost degradation due to a warming trend will likely lead to a decreasing seasonal variability of water flows (Frampton et al. 2011), whereat results of Frampton et al. (2013) show that total runoff will first increase and then decrease as the permafrost degradation progresses further to total thaw.

Apart from the above-mentioned effects of the soil processes on hydrological quantities, perennially frozen ground shows some unique features that are examples for processes that act on both long and short timescales and that are often highly nonlinear. Massive ice wedges are one of these features, which occur in permafrost-dominated landscapes (French 1990). Water enters the soils through frost cracks and, through volume expansion during freeze-up, further increases the cavities in the ground. Cryosuction leads to movement of unfrozen, supercooled water towards the freezing front, and, over time, the ice body can grow to reach several metres in height and thickness. This process happens very slow and lasts many years to decades (French 1990). Ice wedges (and other ice bodies) are the reason for the often oversaturated ice contents in permafrost soils and can be seen as long-term storages of both energy and water in the climate system.

However, the presence of ice wedges may also lead to abrupt changes that occur on the landscape scale when large ice bodies in the soil, after a period of relatively slow and constant warming, collapse in sudden events, e.g., due to intense rain events that bring high heat inputs into the ground. This might lead to considerable change in the landscape in the form of severe soil subsidence, opening of channels and coastal erosion. Belonging to these phenomena are the so-called thermokarst lakes that develop when formerly stable permafrost thaws at the top due to perturbation (e.g., a fire event), and soil subsidence and melting ground ice lead to the formation of a lake (see, e.g., French 2007). This describes a particular process of wetland formation (see also Sect. 3.3). On the other hand, these thermokarst lakes may also drain by catastrophic outflow following lake tapping due to the expansion of adjacent basins or truncation by coastal retreat (see, e.g., Mackay 1988; Walker 1978; Romanovskii et al. 2000). Cycles of slow build-up of ice masses in the ground and relatively short-term collapses in conjunction with the implied morphological changes have happened ever since. Yet, since the atmosphere is warming, and since the atmospheric moisture transport from mid- to high northern latitudes as well as precipitation and circulation patterns is believed to change with anthropogenic climate change, these events might become more abundant in the future. This again has implications for the carbon cycle, as erosion events always bring formerly bound carbon back into the cycle.

3.2 Representation of Permafrost Processes in ESMs

The climate modelling community has a long history in systematic model intercomparison through the climate model intercomparison projects (CMIPs; Meehl et al. 2000). Results from CMIPs provide a good overview of the respective state of ESM model accuracy and performance. Koven et al. (2012) analysed the performance of ESMs from the most recent CMIP5 exercise over permafrost areas. They found that the CMIP5 models have a wide range of behaviours under the current climate, with many failing to agree with fundamental aspects of the observed soil thermal regime at high latitudes. This is partially related to the fact that most of these models do not include permafrost-specific processes, not even the most basic process of freezing and melting of soil water. Moreover, the land surface parameterizations used in GCMs usually do not adequately resolve the soil conditions (Walsh et al. 2005), which often rely on either point measurements or information derived from satellite data.

Although a good understanding of many permafrost-related hydrological processes exists at the point and hillslope scales, this knowledge had not been adequately or systematically incorporated even into process-based mesoscale hydrological models (Vörösmarty et al. 1993) for a long time. Models on point/hillslope scales were generally constrained to one-dimensional domains of vertical extent only (Riseborough et al. 2008), which usually could not be upscaled to larger scales due the complexity of physical interactions in permafrost regions. Also, Bolton (2006) has identified a lack of process-based hydrology models that adequately simulate the soil moisture dynamics at the watershed scale and also include a realistic land–atmosphere exchange in permafrost-dominated regions. But, such models are required to bridge the gap between the point/hillslope scale understanding and the scale of RCMs and GCMs by capturing the hydrological behaviour and variation in individual watersheds. Recent developments started to fill this gap (see, e.g., Schramm et al. 2007; Bense et al. 2009; Frampton et al. 2011), thereby responding to the need for an increased realism of numerical permafrost models highlighted by Riseborough et al. (2008) and Woo et al. (2008).

As the development of mesoscale permafrost-related hydrology models has only gained momentum within the last years, it is not surprising that the situation is comparable or even worse for climate models. Until recently, the representation of frozen ground physics, as well as of above-mentioned characteristics (Sect. 3.1) like reduced permeability of frozen soil and impeded infiltration of spring melt water, was, if at all, represented in a rather simplified way. An intercomparison study of different land surface schemes especially with respect to cold regions' climate and hydrology revealed large differences between the models, even in case the implementation of frozen ground physics was constructed in a similar way (Luo et al. 2003). Due to missing processes and related deficiencies of their land surface schemes, climate models often show substantial biases in hydrological variables over high northern latitudes (Luo et al. 2003; Swenson et al. 2012). Therefore, large efforts are ongoing to extend ESMs in this respect, in order to improve simulated soil moisture profiles and associated ice contents, river discharge, surface and subsurface runoff. The ESM improvement over permafrost areas is, e.g., one of the research objectives of the European Union Project PAGE21 (http://www.page21.org).

Given the substantial range in the level of complexity and advancement of permafrost-related processes implemented in the ESMs, the large variety of results from the CMIP5 models is not surprising (see Fig. 8; Koven et al. 2012). The most comprehensive ESM land surface schemes include freezing and melting of soil water, the dependency of soil thermal properties on water and ice content, multilayer snow schemes with snow on the top of the soil instead of blending upper soil layers and snow and the representation of soil organic matter (e.g., the Community Land Model (CLM) of Lawrence et al. 2011). In contrast, many models incorporate only few of these processes. Soil hydrology is assessed using multilayer schemes that compute vertical flows using Richards law (Richards 1931) or some of its derivations (e.g., Oleson et al. 2004), thereby replacing more and more the formerly used bucket schemes. The reduced permeability can thus be considered via coupling of soil thermodynamics and hydrology, i.e. hydraulic properties are functions of liquid soil water content only, instead of the total water content. This is refined in some models through the implementation of a freezing point depression. Reduction in infiltration at the surface is partially assessed, ranging from very simple approaches like total blocking soils in case of freezing to the consideration of subgrid scale variability, based on power law relationships between infiltration and the degree of soil freezing. Examples for global ESM land surface schemes that are in various states of ongoing development are CLASS (Verseghy 1991), ORCHIDEE (Gouttevin et al. 2012), JSBACH (Ekici et al. 2013) and JULES (Best et al. 2011). The latter three also participate in the Page21 (http://www.page21.eu) model improvement activities. One of the planned Page21 improvements is the development of a global scheme for the formation and drainage of thermokarst lakes that has not been implemented in any ESM up to now.

It is important to note that also for models that represent the same processes the results may diverge markedly. This can be attributed to differences in parameterization schemes and the choice of parameters, e.g., soil column depth and thickness of its layers, as well as to the choice of input data, e.g., soil porosity or heat capacity of the soil's dry material. In addition, initial conditions play a role due to the long spin-up times of model soils. These may comprise several years for liquid and frozen soil water content, several years (de Ridder 2008) up to two decades for soil temperature and several centuries to millennia for soil carbon storages (Wutzler and Reichstein 2007; Hashimoto et al. 2011). In this respect, Christensen (1999) pointed out the importance of an adequate initialization of soil

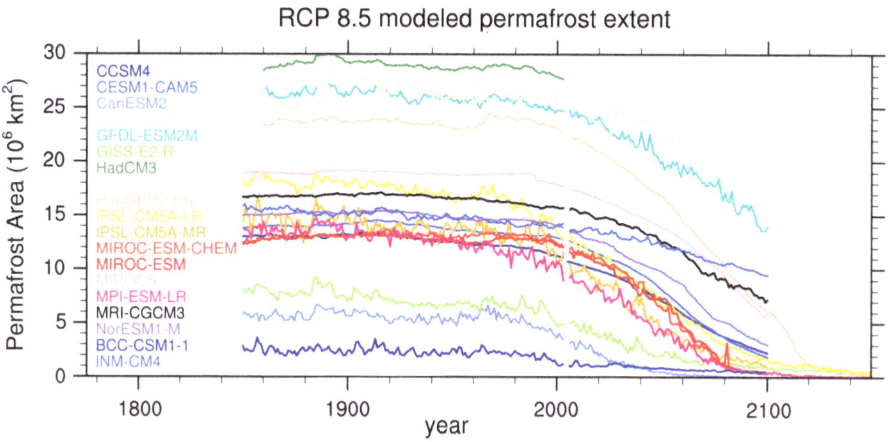

Fig. 8 Simulated total permafrost area for historical twentieth-century climate and future climate following the RCP8.5 scenario for various CMIP5 models. Figure is taken from Koven et al. (2012)

temperature and soil moisture in climate modelling experiments. An inadequate initialization of these fields may lead to transient signals that have to be suppressed as much as possible in modern numerical climate experiments as climate sensitivity experiments operate with quite small signals.

3.3 Wetlands

Thawing permafrost will also contribute to the formation (French 2007) or disappearance (Smith et al. 2005) of wetlands that currently cover about 6–8 % of the land surface. Note that available global wetland observations span a range of potential wetland coverages, partially due to their different wetland definitions which they are based on (see Stacke and Hagemann 2012). Despite this, they agree on many large-scale patterns that can be seen in Fig. 9 showing the ensemble mean coverage of wetlands based on four different data sets. Due to their function as water storage, the majority of research studies found that wetlands regulate river discharge, mitigate flood events and show increased evapotranspiration compared to other land cover types (Bullock and Acreman 2003). However, some exceptions to this general behaviour have been reported (van der Velde et al. 2013) where evapotranspiration is less efficient for wetlands than for other land cover types. The extension of wetlands determines the area where anoxic decomposition instead of oxic decomposition may take place. While CO_2 is released under oxic conditions, the anoxic decomposition yields methane that is a far more active greenhouse gas than CO_2. Here, the water level is an important factor for the wetland's biogeochemistry which results in carbon sequestration or decomposition (e.g., O'Connor et al. 2010, and references therein). Generally, an increase in wetland area will lead to an enhanced methane production. On the other hand, a decrease will reduce moisture fluxes to the atmosphere and may lead to a reduction in precipitation. Thus, their future development is of major interest in climate change studies.

While most studies identify wetlands as net carbon sinks for today's climate conditions (Bohn et al. 2007; Gorham 1991; Friborg et al. 2003), a number of studies concluded that some wetlands might turn into carbon sources in a warmer climate (St-Hilaire et al. 2010; Gorham 1991) due to higher productivity of methane-releasing microbes. Several recent

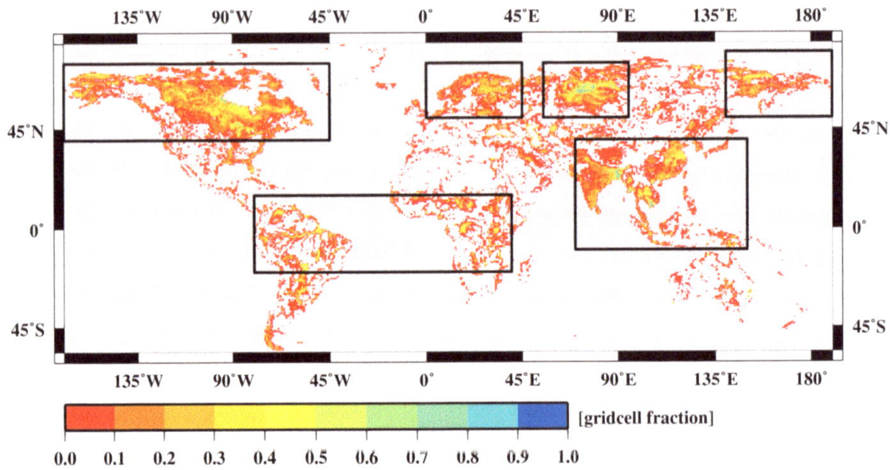

Fig. 9 Observed wetland fraction at 0.5° resolution obtained from the ensemble mean of four global wetland data sets (see Stacke and Hagemann 2012)

studies suspected wetlands to play an important role during the periods of climate change (e.g., Ringeval et al. 2011; Gedney et al. 2004; Levin et al. 2000). However, the representation of the wetland's spatial extent and its temporal variations is still a weak point in today's ESMs and needs to be improved by a better simulation of their hydrological cycle (O'Connor et al. 2010; Ringeval et al. 2010).

However, even without consideration of the carbon cycle, the wetland hydrology in itself is an important key factor in the climate system. Wetlands are often related to regions with open surface water and saturated soil. Such regions have to be considered in ESMs because of their potential feedbacks to the atmosphere (Coe and Bonan 1997). The effect of open water surfaces on the energy and water balance was investigated by several modelling studies, e.g., Bonan (1995) and Mishra et al. (2010), who reported a significant impact of wetlands on the local climate. Generally, they found a cooling of the surface in wetland-dominated regions due to increased evapotranspiration, as well as an increase in the latent heat flux and a decrease in the sensible heat flux. Eventually, this could result in increased precipitation rates as shown by Coe and Bonan (1997) and Krinner et al. (2012). Furthermore, wetlands interact in several ways with the hydrological cycle of their surrounding area. Most studies report wetlands to regulate river flow, mitigate flood events and recharge groundwater (Bullock and Acreman 2003). These observations are consistent with a modelling study by Mishra et al. (2010) who found decreased surface runoff in wetland-dominated regions. However, the range of possible hydrological impacts of wetlands is rather large and depends strongly on additional conditions such as topography and soil properties. This is emphasized by several studies that describe different wetland impacts such as an increased effect on flood peaks and no or a discharging impact on groundwater (Bullock and Acreman 2003). All of these processes are of great interest for impact studies that investigate how climate change might affect the water storage capacities in a region or the characteristics of river flooding.

The modelling of the hydrological cycle in wetlands and their extent dynamics has motivated a large number of modelling studies. Generally, most models follow one of two main approaches for the hydrological representation of wetlands. One approach is concerned with the redistribution of soil moisture in the model grid cell. A widely used example is TOPMODEL (Beven and Kirkby 1979). In this approach a topographical index is computed that depends on the drainage of a given area routed through a point and its slope. This index is then applied to determine the position of the local water table at that point relative to the mean water table of the whole grid cell. The grid cell fraction where the subgrid soil moisture exceeds the soil moisture storage capacity of the grid cell is then regarded as a wetland. The TOPMODEL approach was used and improved in several studies (e.g., Barling et al. 1994; Gedney et al. 2004; Bohn et al. 2007; Kleinen et al. 2012) and is able to compute changes in wetland extent as well. While this approach is an elegant solution, it has one major problem. As the wetland fraction depends on the redistribution of the mean grid cell soil moisture, it follows that there is an upper boundary for the maximum water depth and wetland fraction. For the extreme case of a grid cell with zero slope, no wetland can emerge because the mean soil moisture can obviously not exceed the maximum soil moisture capacity. However, observations indicate that flat regions appear to be more suitable for wetland formation.

The second approach is the explicit modelling of surface water. In this case depressions in the topography are identified and filled with water that results from a positive water balance. On the one hand, this can be done on a continental scale (e.g., Coe 1997, 1998, 2000), but then the quality of the wetland representation is strongly limited by resolution of the model. Alternatively, regional models allow for a higher resolution but then depend strongly on detailed soil property information (e.g., Bowling and Lettenmaier 2010; Yu et al. 2006) or are calibrated for specific catchments (e.g., Bohn et al. 2007). Decharme et al. (2008, 2011) developed a global inundation model, but its focus is concentrated on the representation of floodplains.

In contrast to these sophisticated approaches, Stacke and Hagemann (2012) developed a somewhat simpler hydrological scheme that represents the global distribution and extent variability of very different types of wetlands. The scheme was designed for the application in complex ESMs on global scale with medium to coarse resolutions (50 km or coarser), as the representation of surface water dynamics is—albeit important—not strongly developed in such models. The global-scale hydrological scheme of Stacke and Hagemann (2012) has been implemented in the Max Planck Institute for Meteorology Hydrology Model (MPI-HM). It solves the water balance of wetlands and estimates their extent dynamically. The extent depends on the balance of water flows in the wetlands and the slope distribution within the grid cells. In contrast to most models, this scheme is not directly calibrated against wetland extent observations. Using MPI-HM, the spatial distribution of simulated wetlands agreed well with different global observations for present climate (Fig. 10). The best results were achieved for the northern hemisphere where not only the wetland distribution pattern but also their extent was simulated reasonably well. However, the wetland fraction in the tropical parts of South America and Central Africa was strongly overestimated, which seems to be related to an underestimation of potential evapotranspiration over wet tropical areas by the Penman–Monteith method used in MPI-HM. The simulated extent dynamics correlated well with monthly inundation variations obtained from satellites for most locations. Also, the simulated river discharge was affected by wetlands, resulting in a delay and mitigation of peak flows. Compared to simulations without wetlands, locally increased evaporation and decreased river flow into the oceans were generated due to the implemented wetland processes.

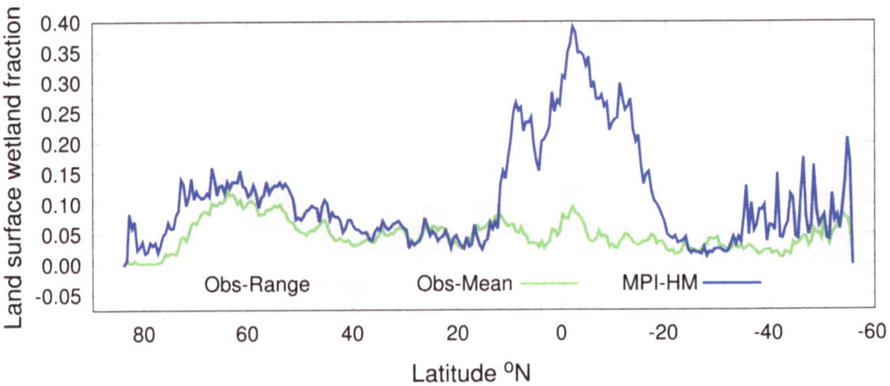

Fig. 10 Zonal means of land surface wetland fraction. The *grey* area indicates the range of the observation data sets, the *green curve* shows their mean extent (cf. Fig. 9), and the *blue curve* shows wetland fractions as simulated by the MPI-HM

4 Concluding Remark

In the present review, we have highlighted some noticeable deficiencies in climate modelling with respect to the hydrological cycle, which provide perspectives for the modelling of relevant interactions between climate and terrestrial hydrology. These interactions are often imposed by different land–atmosphere coupling mechanisms. Over many regions, the wet state of the soil (soil moisture, wetlands, irrigation) determines feedback characteristics. These feedbacks not only impact the local scale but also often act on the large scale. In this respect, human land use may affect remote regions as has been shown for irrigation over the South Asian monsoon region. As the characteristic of associated feedbacks varies for different regions and may change under future climate conditions, they have to be regarded in respective modelling studies. The coupling to biogeochemistry, i.e. carbon cycle and vegetation, is important to quantify feedbacks related to wetlands and permafrost. The representation of their complex dynamics within ESMs is a challenging task, but it is nevertheless necessary to investigate ongoing and future climate changes over the high-latitude regions.

Acknowledgments The authors acknowledge the financial support of T. Blome by the European Union FP7-ENV project PAGE21 under contract number GA282700. S. Hagemann was supported by funding from the European Union within the EMBRACE project (grant no. 282672). We also acknowledge previous work that was supported by funding from the European Union within the WATCH project (contract No. 036946).

References

ACIA (2005) Arctic climate impact assessment. Cambridge University Press, Cambridge, p 1042. http://www.acia.uaf.edu

Asharaf S, Dobler A, Ahrens B (2012) Soil moisture-precipitation feedback processes in the Indian summer monsoon season. J Hydrometeorol 13:1461–1474. doi:10.1175/JHM-D-12-06.1

Asokan SM, Jarsjö J, Destouni G (2010) Vapor flux by evapotranspiration: effects of changes in climate, land-use and water-use. J Geophys Res 115:D24102. doi:10.1029/2010JD014417

Barling R, Moore I, Grayson R (1994) A quasi-dynamic wetness index for characterizing the spatial distribution of zones of surface saturation and soil water content. Water Resour Res 30:1029–1044. doi:10.1029/93WR03346

Bathiany S, Claussen M, Brovkin V, Raddatz T, Gayler V (2010) Combined biogeophysical and biogeochemical effects of large-scale land-cover change in the MPI Earth system model. Biogeosciences 7:1383–1399. doi:10.5194/bg-7-1383-2010

Bense VF, Ferguson G, Kooi H (2009) Evolution of shallow groundwater flow systems in areas of degrading permafrost. Geophys Res Letters 36:L22401. doi:10.1029/2009GL039225

Best MJ, Pryor M, Clark DB, Rooney GG, Essery RLH, Ménard CB, Edwards JM, Hendry MA, Porson A, Gedney N, Mercado LM, Sitch S, Blyth E, Boucher O, Cox PM, Grimmond CSB, Harding RJ (2011) The joint UK land environment simulator (JULES), model description—part 1: energy and water fluxes. Geosci Model Dev 4:677–699. doi:10.5194/gmd-4-677-2011

Beven K, Kirkby M (1979) A physically based, variable contributing area model of basin hydrology. Hydrol Sci Bull 24:43–69

Bohn T, Lettenmaier D, Sathulur K, Bowling L, Podest E, McDonald K (2007) Methane emissions from western Siberian wetlands: heterogeneity and sensitivity to climate change. Environ Res Lett 2:045015. doi:10.1088/1748-9326/2/4/045015

Boike J, Roth K, Overduin PP (1998) Thermal and hydrologic dynamics of the active layer at a continuous permafrost site (Taymyr Peninsula, Siberia). Water Resour Res 34(3):355–363

Bolton WR (2006) Dynamic modelling of the hydrologic processes in areas of discontinuous permafrost. Ph.D. dissertation, University of Alaska, Fairbanks, US, p 163

Bonan G (1995) Sensitivity of a GCM simulation to inclusion of inland water surfaces. J Clim 8:2691–2704

Boucher O, Myhre G, Myhre A (2004) Direct human influence of irrigation on atmospheric water vapor and climate. Clim Dyn 22:597–603. doi:10.1007/s00382-004-0402-4

Bowling L, Lettenmaier DP (2010) Modeling the effects of lakes and wetlands on the water balance of arctic environments. J Hydrometeorol 11:276–295. doi:10.1175/2009JHM1084.1

Breckle S-W, Geldyeva GV (2012) Dynamics of the Aral Sea in geological and historical times. In: Breckle S-W et al (eds) Aralkum—a man-made desert: the desiccated floor of the Aral Sea (Central Asia), ecological studies, vol 218. Springer, Berlin, pp 13–35. doi:10.1007/978-3-642-21117-1_2

Brown J, Ferrians OJ Jr., Heginbottom JA, Melnikov ES (eds.) (1997) Circum-Arctic map of permafrost and ground-ice conditions. U.S. Geological Survey in Cooperation with the Circum-Pacific Council for Energy and Mineral Resources. Circum-Pacific Map Series CP-45, scale 1:10,000,000, Washington, DC

Bullock A, Acreman M (2003) The role of wetlands in the hydrological cycle. Hydrol Earth Syst Sci 7:358–389. doi:10.5194/hess-7-358-2003

Cess RD, Potter GL, Zhang MH, Blanchet JP, Chalita S, Colman R, Dazlich DA, Genio AD, Dymnikov V, Galin V, Jerrett D, Keup E, Lacis AA, LE Treut H, Liang XZ, Mahfouf JF, McAvaney BJ, Meleshko VP, Mitchell JF, Morcrette JJ, Norris PM, Randall DA, Rikus L, Roeckner E, Royer JF, Schlese U, Sheinin DA, Slingo JM, Sokolov AS, Taylor KE, Washington WM, Wetherald RT, Yagai I (1991) Interpretation of snow-climate feedback as produced by 17 general circulation models. Science 253:888–892. doi:10.1126/science.253.5022.888

Christensen JH (1999) Relaxation of soil variables in a regional climate model. Tellus 51A:474–685

Claussen M (2004) The global climate. In: Kabat P et al. (eds) Vegetation, water, humans and the climate. Global change—the IGBP series 24. Springer, Heidelberg, pp 33–57

Coe M (1997) Simulating continental surface waters: an application to Holocene Northern Africa. J Clim 10:1680–1689

Coe M (1998) A linked global model of terrestrial hydrologic processes: simulation of modern rivers, lakes, and wetlands. J Geophys Res (D Atmos) 103:8885–8899. doi:10.1029/98JD00347

Coe M (2000) Modeling terrestrial hydrological systems at the continental scale: testing the accuracy of an atmospheric GCM. J Clim 13:686–704. doi:10.1175/1520-0442(2000)013<0686:MTHSAT>2.0.CO;2

Coe M, Bonan G (1997) Feedbacks between climate and surface water in northern Africa during the middle Holocene. J Geophys Res (D Atmos) 102:11087–11101. doi:10.1029/97JD00343

Cox PM, Betts RA, Jones CD, Spall SA, Totterdell IJ (2000) Acceleration of global warming due to carbon-cycle feedbacks in a coupled climate model. Nature 408:184–187

Dale VH (1997) The relationship between land-use change and climate change. Ecol Appl 7:753–769

De Rosnay P, Polcher J, Laval K, Sabre M (2003) Integrated parameterization of irrigation in the land surface model ORCHIDEE: validation over Indian Peninsula. Geophys Res Lett 30:1986. doi:10.1029/2003GL018024

Decharme B, Douville H, Prigent C, Papa F, Aires F (2008) A new river flooding scheme for global climate applications: offline evaluation over South America. J Geophys Res (D Atmos) 113:D11110. doi:10.1029/2007JD009376

Decharme B, Alkama R, Papa F, Faroux S, Douville H, Prigent C (2011) Global off-line evaluation of the ISBA-TRIP flood model. Clim Dyn 38:1–24. doi:10.1007/s00382-011-1054-9

Denman KL, Brasseur G, Chidthaisong A, Ciais P, Cox PM, Dickinson RE, Hauglustaine D, Heinze C, Holland E, Jacob D, Lohmann U, Ramachandran S, da Silva Dias PL, Wofsy SC, Zhang X (2007) Couplings between changes in the climate system and biogeochemistry. In: Solomon S, Qin D, Manning M, Chen Z, Marquis M, Averyt KB, Tignor M, Miller HL (eds) Climate change 2007: the physical science basis. Contribution of working group I to the fourth assessment report of the intergovernmental panel on climate change. Cambridge University Press, Cambridge

Destouni G, Jaramillo F, Prieto C (2013) Hydroclimatic shifts driven by human water use for food and energy production. Nat Clim Change 3:213–217

Dirmeyer P, Koster R, Guo ZAD (2006) Do global models properly represent the feedback between land and atmosphere? J Hydrometeorol 7:1177–1198

Döll P, Hoffmann-Dobrev H, Portmann FT, Siebert S, Eicker A, Rodell M, Strassberg G, Scanlon BR (2012) Impact of water withdrawals from groundwater and surface water on continental water storage variations. J Geodyn 59(60):143–156

Douglas EM, Niyogi D, Frolking S, Yeluripati JB, Pielke RA Sr, Niyogi N, Vörösmarty CJ, Mohanty UC (2006) Changes in moisture and energy fluxes due to agricultural land use and irrigation in the Indian Monsoon Belt. Geophys Res Lett 33:L14403. doi:10.1029/2006GL026550

Douglas EM, Beltrán-Przekurat A, Niyogi D, Pielke RA Sr, Vörösmarty CJ (2009) The impact of agricultural intensification and irrigation on land–atmosphere interactions and Indian monsoon precipitation—A mesoscale modeling perspective. Global Planet Change 67:117–128. doi:10.1016/j.gloplacha.2008.12.007

Ekici A, Beer C, Hauck C, Hagemann S (2013) Improved soil physics for simulating high latitude permafrost regions by the JSBACH terrestrial ecosystem model. Geosci Model Dev Discuss 6:2655–2698. doi:10.5194/gmdd-6-2655-2013

Falloon P, Betts RA (2010) Climate impacts on European agriculture and water management in the context of adaptation and mitigation—The importance of an integrated approach. Sci Total Environ 408:5667–5687

Feddema JJ, Oleson KW, Bonan GB, Mearns LO, Buja LE, Meehl GA, Washington WM (2005) The importance of land-cover change in simulating future climates. Science 310:1674–1678

Fischer EM, Seneviratne SI, Vidale PL, Lüthi D, Schär C (2007) Soil moisture - atmosphere interactions during the 2003 European summer heat wave. J Clim 20:5081–5099

Frampton A, Painter SL, Lyon SW, Destouni G (2011) Nonisothermal, three-phase simulations of near-surface flows in a model permafrost system under seasonal variability and climate change. J Hydrol 403:352–359. doi:10.1016/j.jhydrol.2011.04.010

Frampton A, Painter SL, Destouni G (2013) Permafrost degradation and subsurface-flow changes caused by surface warming trends. Hydrogeol J 21:271–280

French HM (1990) Editorial. Permafr Periglac Process 1:1. doi:10.1002/ppp.3430010102

French HM (2007) The periglacial environment, 3rd edn. Wiley, West Sussex, p 457

Friborg T, Soegaard H, Christensen T, Lloyd C, Panikov N (2003) Siberian wetlands: where a sink is a source, Geophys Res Lett 30:CLM 5–1–CLM 5–4. doi:10.1029/2003GL017797

Gao X, Luo Y, Lin W, Zhao Z, Giorgi F (2003) Simulation of effects of land use change on climate in China by a regional climate model. Adv Atmos Sci 20(4):583–592

Gedney N, Cox P, Huntingford C (2004) Climate feedback from wetland methane emissions. Geophys Res Lett 31:L20503. doi:10.1029/2004GL020919

Gerten D (2013) A vital link: water and vegetation in the Anthropocene. Hydrol Earth Syst Sci Discuss 10:4439–4462. doi:10.5194/hessd-10-4439-2013

Gerten D, Rost S, von Bloh W, Lucht W (2008) Causes of change in 20th century global river discharge. Geophys Res Lett 35:L20405. doi:10.1029/2008GL035258

Gerten D, Hagemann S, Biemans H, Saeed F, Konzmann M (2011) Climate change and irrigation: feedbacks and impacts. WATCH technical report 47

Gordon LJ, Steffen W, Jonsson BF, Folke C, Falkenmark M, Johannessen A (2005) Human modification of global water vapour flows from the land surface. Proc Natl Acad Sci USA 102:7612–7617

Gorham E (1991) Northern peatlands: role in the carbon cycle and probable responses to climatic warming. Ecol Appl 1:182–195. doi:10.2307/1941811

Göttel H, Alexander J, Keup-Thiel E, Rechid D, Hagemann S, Blome T, Wolf A, Jacob D (2008) Influence of changed vegetations fields on regional climate simulations in the Barents Sea Region. Clim Change 87:35–50. doi:10.1007/s10584-007-9341-5

Gouttevin I, Krinner G, Ciais P, Polcher J, Legout C (2012) Multi-scale validation of a new soil freezing scheme for a land-surface model with physically-based hydrology. Cryosphere 6:407–430. doi:10.5194/tc-6-407-2012

Guimberteau M, Laval K, Perrier A, Polcher J (2012) Global effect of irrigation and its impact on the onset of the Indian summer monsoon. Clim Dyn 39:1329–1348
Harding RJ, Best M, Blyth E, Hagemann S, Kabat P, Tallaksen LM, Warnaars T, Wiberg D, Weedon GP, van Lanen H, Ludwig F, Haddeland I (2011) Current knowledge of the terrestrial global water cycle. J Hydrometeorol 12:1149–1156. doi:10.1175/JHM-D-11-024.1
Hashimoto S, Wattenbach M, Smith P (2011) A new scheme for initializing process-based ecosystem models by scaling soil carbon pools. Ecolog Model 222:3598–3602
Haugen RK, Slaughter CW, Howe KE, Dingman SL (1982) Hydrology and climatology of the Caribou-Poker Creeks Research Watershed, Alaska, CRREL report 82-26, p 42
Hinzman LD, Kane DI (1991) Snow hydrology of a headwater arctic basin, 2. Conceptual analysis and computer modeling. Water Resour Res 27(6):95–100
International Permafrost Association (1998) Circumpolar active-layer permafrost system (CAPS), version 1.0
Jacob D, Bärring L, Christensen OB, Christensen JH, Hagemann S, Hirschi M, Kjellström E, Lenderink G, Rockel B, Schär C, Seneviratne SI, Somot S, van Ulden A, van den Hurk B (2007) An inter-comparison of regional climate models for Europe: design of the experiments and model performance. Clim Change 81(Suppl 1):31–52
Karim A, Veizer J (2002) Water balance of the Indus River Basin and moisture source in the Karakoram and western Himalaya: implications from hydrogen and oxygen isotopes in river water. J Geophys Res 107(D18):4362. doi:10.1029/2000JD000253
Khan S, Rana T, Gabriel HF, Ullah M (2008) Hydrogeologic assessment of escalating groundwater exploitation in the Indus Basin, Pakistan. Hydrogeol J 16:1635–1654. doi:10.1007/s10040-008-0336-8
Kleinen T, Brovkin V, Getzieh R (2012) A dynamic model of wetland extent and peat accumulation: results for the Holocene. Biogeosciences 9:235–248. doi:10.5194/bg-9-235-2012
Koren V, Schaake J, Mitchell K, Duan OY, Chen F, Baker JM (1999) A parameterization of snowpack and frozen ground intended for NCEP weather and climate models. J Geophys Res 104:19569–19585
Koster RD, Dirmeyer PA, Guo Z, Bonan G, Chan E, Cox P, Gordon CT, Kanae S, Kowalczyk E, Lawrence D, Liu P, Lu CH, Malyshev S, McAvaney B, Mitchell K, Mocko D, Oki T, Oleson K, Pitman A, Sud YC, Taylor CM, Verseghy D, Vasic R, Xue Y, Yamada T (2004) Regions of strong coupling between soil moisture and precipitation. Science 305:1138–1140
Koster RD, Guo Z, Dirmeyer PA, Bonan G, Chan E, Cox P, Davies H, Gordon CT, Kanae S, Kowalczyk E, Lawrence D, Liu P, Lu CH, Malyshev S, McAvaney B, Mitchell K, Mocko D, Oki T, Oleson KW, Pitman A, Sud YC, Taylor CM, Verseghy D, Vasic R, Xue Y, Yamada T (2006) GLACE: the global land-atmosphere coupling experiment. Part I: overview. J Hydrometeorol 7:590–610
Koven CD, Riley WJ, Stern A (2012) Analysis of permafrost thermal dynamics and response to climate change in the CMIP5 Earth System Models. J Clim. doi:10.1175/JCLI-D-12-00228.1
Krinner G, Lézine A-M, Braconnot P, Sepulchre P, Ramstein G, Grenier C, Gouttevin I (2012) A reassessment of lake and wetland feedbacks on the North African Holocene climate. Geophys Res Lett 39:L07701. doi:10.1029/2012GL050992
Kumar P, Wiltshire A, Mathison C, Asharaf S, Ahrens B, Lucas-Picher P, Christensen JH, Gobiet A, Saeed F, Hagemann S, Jacob D (2013) Downscaled climate change projections with uncertainty assessment over India using a high resolution multi-model approach. Sci Total Environ. doi:10.1016/j.scitotenv.2013.01.051
Lawrence DM, Oleson KW, Flanner MG, Thornton PE, Swenson SC, Lawrence PJ, Zeng X, Yang Z-L, Levis S, Sakaguchi K, Bonan GB, Slater AG (2011) Parameterization improvements and functional and structural advances in version 4 of the community land model. J Adv Model Earth Syst 3:MS000045. doi:10.1029/2011MS000045
Lee E, Chase T, Rajagopalan B, Barry R, Wiggs T, Lawrence P (2009) Effects of irrigation and vegetation activity on early Indian summer monsoon variability. Int J Climatol 29:573–581. doi:10.1002/joc.1721
Levin D, Hopper I, Ernst F, Trivett M, Worthy N (2000) Evidence for a link between climate and northern wetland methane emissions. J Geophys Res (D Atmos) 105:4031–4038. doi:10.1029/1999JD901100
Lobell D, Bala G, Mirin A, Phillips T, Maxwell R, Rotman D (2009) Regional differences in the influence of irrigation on climate. J Clim 22:2248–2255. doi:10.1175/2008JCLI2703.1
Loew A, Holmes T, de Jeu R (2009) The European heat wave 2003: early indicators from multisensoral microwave remote sensing? J Geophys Res 114:D05103
Lucas-Picher P, Christensen JH, Saeed F, Kumar P, Asharaf S, Ahrens B, Wiltshire A, Jacob D, Hagemann S (2011) Can regional climate models represent the Indian monsoon? J Hydrometeorol 12:849–868. doi:10.1175/2011JHM1327.1
Luo LF, Robock A, Vinnikov KY, Schlosser CA, Slater AG, Boone A, Braden H, Cox P, de Rosnay P, Dickinson RE, Dai YJ, Duan QY, Etchevers P, Henderson-Sellers A, Gedney N, Gusev YM, Habets F,

Kim JW, Kowalczyk E, Mitchell K, Nasonova ON, Noilhan J, Pitman AJ, Schaake J, Shmakin AB, Smirnova TG, Wetzel P, Xue YK, Yang ZL, Zeng QC (2003) Effects of frozen soil on soil temperature, spring infiltration, and runoff: results from the PILPS 2(d) experiment at Valdai, Russia. J Hydrometeorol 4:334–351

Mackay JR (1988) Catastrophic lake drainage, Tuktoyaktuk peninsula area, District of Mackenzie. Geological Survey of Canada, paper 88-1D: 83–90

Mahajan PN, Talwalkar DR, Chinthalu GR, Rajamaniet S (1995) Use of INSAT winds for better depiction of monsoon depression over Indian region. Meteorol Appl 2:333–339

Marengo J, Nobre CA (2001) The hydroclimatological framework in Amazonia. In: McClaine M, Victoria R, Richey J (eds) Biogeochemistry of the Amazon basin. Oxford University Press, Oxford, pp 17–42

McGuire AD, Anderson LG, Christensen TR, Dallimore S, Guo L, Hayes DJ, Heimann M, Lorenson TD, Macdonald RW, Roulet N (2009) Sensitivity of the carbon cycle in the Arctic to climate change. Ecol Monogr 79:523–555. doi:10.1890/08-2025.1

Meehl GA, Boer GJ, Covey C, Latif M, Stouffer RJ (2000) The coupled model intercomparison project (CMIP). Bull Am Meteor Soc 81:313–318

Mishra V, Cherkauer K, Bowling L (2010) Parameterization of lakes and wetlands for energy and water balance studies in the great lakes region. J Hydrometeorol 11:1057–1082. doi:10.1175/2010JHM1207.1

Niu G-Y, Yang Z-L (2006) Effects of frozen soil on snowmelt runoff and soil water storage at a continental scale. J Hydrometeorol 7:937–952. doi:10.1175/JHM538.1

O'Connor F, Boucher O, Gedney N, Jones C, Folberth G, Coppell R, Friedlingstein P, Collins W, Chappellaz J, Ridley J, Johnson C (2010) Possible role of wetlands, permafrost, and methane hydrates in the methane cycle under future climate change: a review. Rev Geophys 48:RG4005. doi:10.1029/2010RG000326

Oleson K, Dai Y, Bonan GB, Bosilovich M, Dickinson R, Dirmeyer P, Hoffman F, Houser P, Levis S, Niu G-Y, Thornton P, Vertenstein M, Yang Z-L, Zeng X (2004): Technical description of the community land model (CLM). NCAR technical note NCAR/TN-461 + STR. doi:10.5065/D6N877R0

Oyama MD, Nobre CA (2004) Climatic consequences of a large-scale desertification in northeast Brazil: a GCM simulation study. J Clim 17:3203–3213

Paeth H, Born K, Girmes R, Podzun R, Jacob D (2009) Regional climate change in tropical and northern Africa due to greenhouse forcing and land-use changes. J Clim 22:114–132. doi:10.1175/2008JCLI2390.1

Piao S, Friedlingstein P, Ciais P, de Noblet-Ducoudré N, Labat D, Zaehle S (2007) Changes in climate and land use have a larger direct impact than rising CO2 on global river runoff trends. Proc Natl Acad Sci USA 104:15242–15247

Pielke RA Sr, Marland G, Betts RA, Chase TN, Eastman JL, Niles JO, Niyogi DDS, Running SW (2002) The influence of land-use change and landscape dynamics on the climate system: relevance to climate-change policy beyond the radiative effect of greenhouse gases. Philos Trans R Soc Lond A 360:1705–1719. doi:10.1098/rsta.2002.1027

Puma MJ, Cook BI (2010) Effects of irrigation on global climate during the 20th century. J Geophys Res 115:D16120. doi:10.1029/2010JD014122

Richards LA (1931) Capillary conduction of liquids through porous mediums. Physics 1(5):318–333. doi:10.1063/1.1745010

Ridder De (2008) Soil temperature spin-up in land surface schemes. Theor Appl Climatol. doi:10.1007/s00704-008-0011-y

Ringeval B, De Noblet-Ducoudré N, Ciais P, Bousquet P, Prigent C, Papa F, Rossow W (2010) An attempt to quantify the impact of changes in wetland extent on methane emissions on the seasonal and interannual time scales. Global Biogeochem Cycles 24:GB2003. doi:10.1029/2008GB003354

Ringeval B, Friedlingstein P, Koven C, Ciais P, de Noblet-Ducoudré N, Decharme B, Cadule P (2011) Climate-CH4 feedback from wetlands and its interaction with the climate-CO2 feedback. Biogeosciences 8:2137–2157. doi:10.5194/bg-8-2137-2011

Riseborough D, Shiklomanov N, Etzelmüller B, Gruber S, Marchenko S (2008) Recent advances in permafrost modelling. Permafr Periglac Process 19:137–156. doi:10.1002/ppp.615

Romanovskii NN, Hubberten H-W, Gavrilov AV, Tumskoy VE, Tipenko GS, Grigoriev MN (2000) Thermokarst and land-ocean interactions, Laptev Sea region, Russia. Permafr Periglac Process 11:137–152

Rost S, Gerten D, Bondeau A, Lucht W, Rohwer J, Schaphoff S (2008a) Agricultural green and blue water consumption and its influence on the global water system. Water Resour Res 44:W09405. doi:10.1029/2007WR006331

Rost S, Gerten D, Heyder U (2008b) Human alterations of the terrestrial water cycle through land management. Adv Geosci 18:43–50. doi:10.5194/adgeo-18-43-2008

Sacks WJ, Cook BI, Buenning N, Levis S, Helkowski JH (2009) Effects of global irrigation on the near-surface climate. Clim Dyn 33:159–175. doi:10.1007/s00382-008-0445-z

Saeed F, Hagemann S, Jacob D (2009) Impact of irrigation on the South Asian Summer Monsoon. Geophys Res Letters 36:L20711. doi:10.1029/2009GL040625

Saeed F, Hagemann S, Jacob D (2011) A framework for the evaluation of the South Asian Summer Monsoon in a regional climate model applied to REMO. Int J Climatol. doi:10.1002/joc.2285

Saeed F, Hagemann S, Saeed S, Jacob D (2012) Influence of mid-latitude circulation on upper Indus basin precipitation: the explicit role of irrigation. Clim Dyn. doi:10.1007/s00382-012-1480-3

Schramm I, Boike J, Bolton WR, Hinzman LD (2007) Application of TopoFlow, a spatially distributed hydrological model, to the Imnavait Creek watershed, Alaska. J Geophys Res 112:46. doi:10.1029/2006JG000326

Schuur EAG, Bockheim J, Canadell JG, Euskirchen E, Field CB, Goryachkin SV, Hagemann S, Kuhry P, Lafleur PM, Lee H, Mazhitova G, Nelson FE, Rinke A, Romanovsky VE, Shiklomanov N, Tarnocai C, Venevsky S, Vogel JG, Zimov SA (2008) Vulnerability of permafrost carbon to climate change: implications for the global carbon cycle. Bioscience 58:701–714. doi:10.1641/B580807

Seneviratne SI, Stöckli R (2008) The role of land-atmosphere interactions for climate variability in Europe. In: Brönnimann et al. (ed) Climate variability and extremes during the past 100 years. Advances in Global Change Research 33, Springer, Heidelberg. (Book chapter)

Seneviratne SI, Lüthi D, Litschi M, Schär C (2006) Land-atmosphere coupling and climate change in Europe. Nature 443:205–209

Seneviratne SI, Corti T, Davin E, Hirschi M, Jaeger EB, Lehner I, Orlowsky B, Teuling AJ (2010) Investigating soil moisture-climate interactions in a changing climate: a review. Earth Sci Rev 99:125–161. doi:10.1016/j.earscirev.2010.02.004

Serreze MC, Barry RG (2011) Processes and impacts of Arctic amplification: a research synthesis. Global Planet Change 77:85–96. doi:10.1016/j.gloplacha.2011.03.004

Siebert S, Döll P, Hoogeveen J, Faurés J-M, Frenken K, Feick S (2005) Development and validation of the global map of irrigation areas. Hydrol Earth Syst Sci 9:535–547

Singh D, Bhadram CVV, Mandal GS (1995) New regression model for Indian summer monsoon rainfall. Meteorol Atmos Phys 55:77–86. doi:10.1007/BF01029603

Sitch S, Smith B, Prentice IC, Arneth A, Bondeau A, Cramer W, Kaplan JO, Levis S, Lucht W, Sykes MT, Thonicke K, Venevsky S (2003) Evaluation of ecosystem dynamics, plant geography and terrestrial carbon cycling in the LPJ dynamic global vegetation model. Glob Change Biol 9:161–185

Smith LC, Sheng Y, MacDonald GM, Hinzman LD (2005) Disappearing arctic lakes. Science 308:1429

Solomon S, Qin D, Manning M, Marquis M, Averyt K, Tignor MMB, Miller HL Jr, Chen Z (eds) (2007) Climate change 2007: the physical science basis. Cambridge University Press, Cambridge, p 996

Stacke T, Hagemann S (2012) Development and validation of a global dynamical wetlands extent scheme. Hydrol Earth Syst Sci 16:2915–2933. doi:10.5194/hess-16-2915-2012

Staehli M, Jansson P-E, Lundin LC (1999) Soil moisture redistribution and infiltration in frozen sandy soils. Water Resour Res 35:95–103. doi:10.1029/1998WR900045

St-Hilaire F, Wu J, Roulet N, Frolking S, Lafleur P, Humphreys E, Arora V (2010) McGill wetland model: evaluation of a peatland carbon simulator developed for global assessments. Biogeosciences 7:3517–3530. doi:10.5194/bg-7-3517-2010

Swenson SC, Lawrence DM, Lee H (2012) Improved simulation of the terrestrial hydrological cycle in permafrost regions by the Community Land Model. J Adv Model Earth Syst 4. doi:10.1029/2012MS000165

Tarnocai C, Canadell JG, Schuur EAG, Kuhry P, Mazhitova G, Zimov S (2009) Soil organic carbon pools in the northern circumpolar permafrost region. Global Biogeochem Cycles 23. doi:10.1029/2008GB003327

Taylor CM, Lambin EF, Stephenne N, Harding RJ, Essery RLH (2002) The influence of land use change on climate in the Sahel. J Clim 15:3615–3629

Ter Maat HW, Hutjes RWA, Ohba R, Ueda H, Bisselink B, Bauer T (2006) Meteorological impact assessment of possible large scale irrigation in Southwest Saudi Arabia. Global Planet Change 54:183–201

Teuling AJ, Hirschi M, Ohmura A, Wild M, Reichstein M, Ciais P, Buchmann N, Ammann C, Montagnani L, Richardson AD, Wohlfahrt G, Seneviratne SI (2009) A regional perspective on trends in continental evaporation. Geophys Res Lett 36:L02404. doi:10.1029/2008GL036584

Tuinenburg OA, Hutjes RWA, Stacke T, Wiltshire A, Lucas-Picher P (2013) Effects of irrigation in India on the atmospheric water budget. J Hydrometeorol (submitted)

Uppala SM, Kallberg PW, Simmons AJ, Andrae U, Da Costa Bechtold V, Fiorino M, Gibson JK, Haseler J, Hernandez A, Kelly GA, Li X, Onogi K, Saarinen ACM, Sokka N, Allan RP, Andersson E, Arpe K, Balmaseda MA, Van Beljaars ACM, de Berg L, Bidlot J, Bormann N, Caires S, Dethof A, Dragosavac M, Fisher M, Fuentes M, Hagemann S, Holm E, Hoskins BJ, Isaksen L, Janssen PAEM, McNally AP, Mahfouf JF, Jenne R, Morcrette JJ, Rayner NA, Saunders RW, Simon P, Sterl A, Trenberth KE, Untch A, Vasiljevic D, Viterbo P, Woollen J (2005) The Era-40 re-analysis. Q J R Meteorol Soc 131:2961–3012. doi:10.1256/qj.04.176

van der Velde Y, Lyon SW, Destouni G (2013) Data-driven regionalization of river discharges and emergent land cover-evapotranspiration relationships across Sweden. J Geophys Res 118:1–12. doi:10.1002/jgrd.50224

Verseghy DL (1991) Class—a Canadian land surface scheme for GCMs. I. Soil model. Int J Climatol 11:111–133

Vörösmarty C, Gutowski W, Person M, Chen T-C, Case D (1993) Linked atmosphere-hydrology models at the macroscale. In: Wilkinson W (ed) Macroscale modeling of the hydrosphere, no. 214 in International Association of Hydrological Sciences, pp 3–27

Walker HJ (1978) Lake tapping in the Colville River delta. In: Permafrost, proceedings of the third international conference on permafrost, 10–13 July, Edmonton, Alberta, Canada. National Research Council of Canada, Ottawa, vol 1, pp 233–238

Walsh JE, Anisimov O, Hagen JOM, Jakobsson T, Oerlemans J, Prowse TD, Romanovsky V, Savelieva N, Serreze M, Shiklomanov A, Shiklomanov I, Solomon S, Arendt A, Atkinson D, Demuth MN, Dowdeswell J, Dyurgerov M, Glazovsky A, Koerner RM, Meier M, Reeh N, Sigurosson O, Steffen K, Truffer M (2005) Cryosphere and hydrology. In: Symon C, Arris L, Heal B (eds) Arctic climate impact assessment, chapter 6. Cambridge University Press, Cambridge, pp 184–242

Willmott CJ, Matsuura K (2009) Terrestrial precipitation: 1900–2008 gridded monthly time series. University of Delaware, Newark. http://climate.geog.udel.edu

Woo MK (1986) Permafrost hydrology in North America. Atmos Ocean 24(3):201–234

Woo M, Kane DL, Carey SK, Yang D (2008) Progress in permafrost hydrology in the new millennium. Permafr Periglac Process 19:237–254

Wutzler T, Reichstein M (2007) Soils apart from equilibrium—consequences for soil carbon balance modelling. Biogeosciences 4:125–136

Yoshikawa K, Bolton WR, Romanovsky VE, Fukuda M, Hinzman LD (2002) Impacts of wildfire on the permafrost in the boreal forests of interior Alaska. J Geophys Res 107:8148. doi:10.1029/2001JD000438, [printed 108(D1), 2003]

Yu Z, Pollard D, Cheng L (2006) On continental-scale hydrologic simulations with a coupled hydrologic model. J Hydrol 331:110–124. doi:10.1016/j.jhydrol.2006.05.021

Zimov SA, Davydov SP, Zimova GM, Davydova AI, Schuur EAG, Dutta K, Chapin III FS (2006) Permafrost carbon: Stock and decomposability of a globally significant carbon pool. Geophys Res Lett 33. doi:10.1029/2006GL027484

DOI 10.1007/978-94-017-8789-5_13
Reprinted from *Surveys in Geophysics* Journal, DOI 10.1007/s10712-013-9264-9

Downscaling Satellite Precipitation with Emphasis on Extremes: A Variational ℓ_1-Norm Regularization in the Derivative Domain

E. Foufoula-Georgiou · A. M. Ebtehaj · S. Q. Zhang · A. Y. Hou

Received: 1 April 2013 / Accepted: 5 November 2013 / Published online: 11 December 2013

Abstract The increasing availability of precipitation observations from space, e.g., from the Tropical Rainfall Measuring Mission (TRMM) and the forthcoming Global Precipitation Measuring (GPM) Mission, has fueled renewed interest in developing frameworks for downscaling and multi-sensor data fusion that can handle large data sets in computationally efficient ways while optimally reproducing desired properties of the underlying rainfall fields. Of special interest is the reproduction of extreme precipitation intensities and gradients, as these are directly relevant to hazard prediction. In this paper, we present a new formalism for downscaling satellite precipitation observations, which explicitly allows for the preservation of some key geometrical and statistical properties of spatial precipitation. These include sharp intensity gradients (due to high-intensity regions embedded within lower-intensity areas), coherent spatial structures (due to regions of slowly varying rainfall), and thicker-than-Gaussian tails of precipitation gradients and intensities. Specifically, we pose the downscaling problem as a discrete inverse problem and solve it via a regularized variational approach (variational downscaling) where the regularization term is selected to impose the desired smoothness in the solution while allowing for some steep gradients (called ℓ_1-norm or total variation regularization). We demonstrate the duality between this geometrically inspired solution and its Bayesian statistical interpretation, which is

E. Foufoula-Georgiou (✉)
Saint Anthony Falls Laboratory, Department of Civil Engineering, University of Minnesota, Minneapolis, MN, USA
e-mail: efi@umn.edu

A. M. Ebtehaj
Saint Anthony Falls Laboratory, Department of Civil Engineering, School of Mathematics, University of Minnesota, Minneapolis, MN, USA
e-mail: ebteh001@umn.edu

S. Q. Zhang · A. Y. Hou
NASA Goddard Space Flight Center, Greenbelt, MA, USA
e-mail: sara.q.zhang@nasa.gov

A. Y. Hou
e-mail: arthur.y.hou@nasa.gov

equivalent to assuming a Laplace prior distribution for the precipitation intensities in the derivative (wavelet) space. When the observation operator is not known, we discuss the effect of its misspecification and explore a previously proposed dictionary-based sparse inverse downscaling methodology to indirectly learn the observation operator from a data base of coincidental high- and low-resolution observations. The proposed method and ideas are illustrated in case studies featuring the downscaling of a hurricane precipitation field.

Keywords Sparsity · Inverse problems · ℓ_1-norm regularization · Non-smooth convex optimization · Generalized Gaussian density · Extremes · Hurricanes

1 Introduction

Precipitation is one of the key components of the water cycle and, as such, it has been the subject of intense research in the atmospheric and hydrologic sciences over the past decades. While it still remains the most difficult variable to accurately predict in numerical weather and climate models, its statistical space–time structure at multiple scales has been extensively studied using several approaches (e.g., Lovejoy and Mandelbrot 1985; Lovejoy and Schertzer 1990; Kumar and Foufoula-Georgiou 1993a, b; Deidda 2000; Harris et al. 2001; Venugopal et al. 2006a, b; Badas et al. 2006). These studies have documented a considerable variability spread over a large range of space and timescales and an organization that manifests itself in power law spectra and more complex self-similar structures expressed via nonlinear scaling of higher-order statistical moments (e.g., Lovejoy and Schertzer 1990; Venugopal et al. 2006a). Stochastic models of multi-scale rainfall variability have been proposed based on inverse wavelet transforms (Perica and Foufoula-Georgiou 1996), multiplicative cascades (Deidda 2000), exponential Langevin-type models (Sapozhnikov and Foufoula-Georgiou 2007), among others.

The small-scale variability of precipitation (of the order of a few kms in space and a few minutes in time) is known to have important implications for accurate prediction of hydrologic extremes especially over small basins (e.g., Rebora et al. 2006a, b) and for the prediction of the evolving larger-scale spatial organization of land–atmosphere fluxes in coupled models (Nykanen et al. 2001). This small-scale precipitation variability, however, is not typically available in many regions of the world where coverage with high-resolution ground radars is absent or in mountainous regions where spatial gaps are present due to radar blockage. It is also missing from climate model predictions that are typically run at low resolution over larger areas of the world. As a result, methods for downscaling precipitation to enhance the resolution of incomplete or low-resolution observations from space or numerical weather/climate model outputs continue to present a challenge of both theoretical and practical interests.

To date, multiple passive and active ground-based (i.e., gauges and radars) and spaceborne sensors (i.e., geostationary, polar and quasi-equatorial orbiting satellites) exist that overlappingly measure precipitation with different space–time resolutions and accuracies. Sparsely populated networks of rain gauges provide relatively accurate point measurements of precipitation continuously over time, while ground-based radars detect precipitation in fine enough spatiotemporal scales (e.g., ~6 min at 1 × 1 km) but over limited areal extents. The ground-based radar data are among the most accurate and high-resolution estimates of spatial rainfall. However, this source of information is subject to various shortcomings such as instrumental errors, beam blockage by orographic features, and overshooting range effects (Krajewski and Smith 2002). The only civilian active spaceborne Tropical Rainfall Measuring Mission-Precipitation Radar (PR) sensor (TRMM-PR) provides high-resolution reflectivity of rainfall fields (i.e., $\sim 4 \times 4$ km) over a narrow band in the tropics with relatively low temporal revisiting frequency compared to the other passive spaceborne sensors of lower resolution. The forthcoming Global Precipitation Measuring (GPM) Mission, a constellation of nine satellites, promises to deliver observations of high precision precipitation and cloud dynamics at a global scale (3-h revisiting time) and over varying resolutions and create opportunities for improving climate modeling and hazard prediction at local scales (Flaming 2004).

Precipitation observations from space are especially valuable in regions where no ground observations are available either from rain gauges or from ground radars, such as over the oceans or in underdeveloped regions of the world. It is over these regions, however, that some extreme tropical storms develop for which high-resolution information would provide important means for hazard prediction and warning as well as detailed information on extremes, which could be used in nested models or in a data assimilation setting. These tropical storms have distinct geometrical and statistical structures, as shown below, posing extra demands on the methodologies of precipitation downscaling, data fusion, and data assimilation.

As an illustrative example, Fig. 1 shows a snapshot of the two-dimensional rainfall intensity patterns and the three-dimensional structure of precipitating clouds for typhoon Neoguri, the first typhoon of the 2008 season in the western Pacific Ocean, on April 17, 2008, as observed by the TRMM-PR and the TRMM Microwave Imager (TMI). One notices the geometrically structured precipitation bands embedded within the larger two-dimensional storm system and the localized "towers" of high-intensity rainfall spatially embedded within lower-intensity rainfall background. These localized high-intensity cells and the steep sporadic gradients of precipitation intensity in such a storm are more clearly demonstrated via a one-dimensional cross section as shown in Fig. 2. Specifically, Fig. 2b

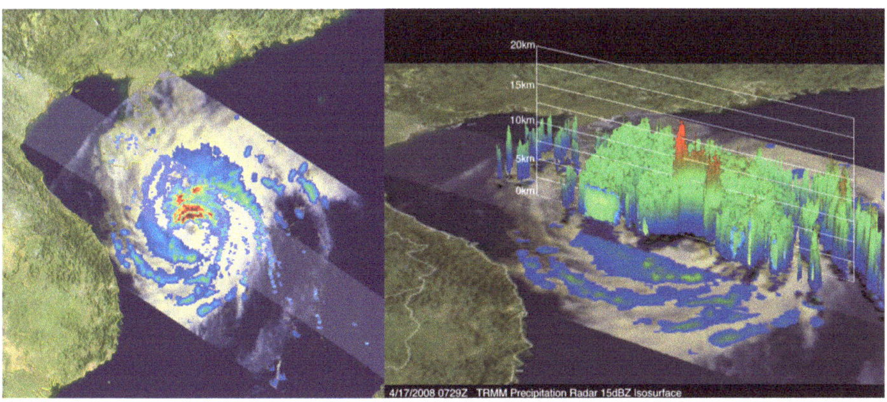

Fig. 1 *Left panel* rainfall pattern of typhoon Neoguri in the western Pacific Ocean, on April 17, 2008. The *dark red* bands indicate regions of the most intense rain. Rainfall rates in the inner swath are from TRMM's-PR, while in the outer swath from the TRMM Microwave Imager (TMI); *Right panel* the three-dimensional structure of precipitating clouds for typhoon Neoguri as observed by the TRMM-PR. This figure illustrates the need for a downscaling scheme that has the ability to reproduce steep rainfall gradients embedded within the storm. *Source*: NASA's Earth Observatory, available online through the TRMM extreme event image archives (http://trmm.gsfc.nasa.gov)

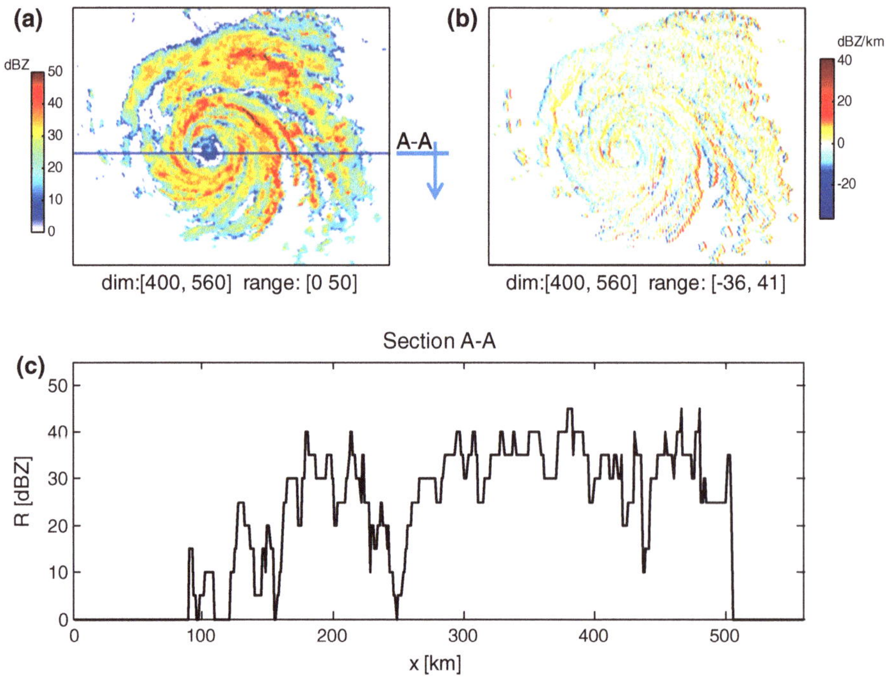

Fig. 2 **a** A high-resolution (HR) snapshot of hurricane Claudette, 07-15-2003, 11:51:00 UTC as monitored by NEXRAD station over Texas at resolution 1 × 1 km and **b** the field of the computed horizontal first-order derivative using the Sobel filter. A horizontal cross section through the storm is shown in (**c**). One observes how the particular geometrical structure of hurricane precipitation projects itself onto an almost piece-wise linear one-dimensional function with sporadic large gradients embedded within regions of almost constant rainfall

demonstrates how the typical circular bands of high rainfall intensity manifest themselves into an almost piece-wise linear structure in the 1D cross section. How is this geometrical structure to be reproduced in downscaling lower resolution and noisy observations of tropical storms, say available at 10-km resolution, down to 1- or 2-km resolution products?

Moving from a geometrical description to a statistical description, we note that coherent precipitation intensity areas (similar intensity in nearby pixels) will result in almost zero values in a derivative space, while the abrupt changes in rainfall intensity (large gradients and discontinuities) will project as high values. In other words, we expect to see a probability distribution in the derivative space that has a large mass close to zero and a few large positive and negative values. Figure 3a shows the histogram of the derivatives of precipitation intensities of hurricane Claudette in the horizontal (zonal) direction (computed via a redundant orthogonal Haar wavelet transform, which is equivalent to using a first-order difference discrete approximation). It is obvious that this histogram is considerably different than a Gaussian probability distribution function (PDF) with a larger mass around zero (capturing the large number of nearby pixels with similar intensity) and much heavier tails than Gaussian (capturing the occasional very steep gradients). How can such a statistical structure be explicitly incorporated in a precipitation downscaling scheme, specifically for hurricanes and tropical storms?

The purpose of this paper is to present a new framework for precipitation downscaling casting the problem as a discrete inverse problem and solving it via a variational

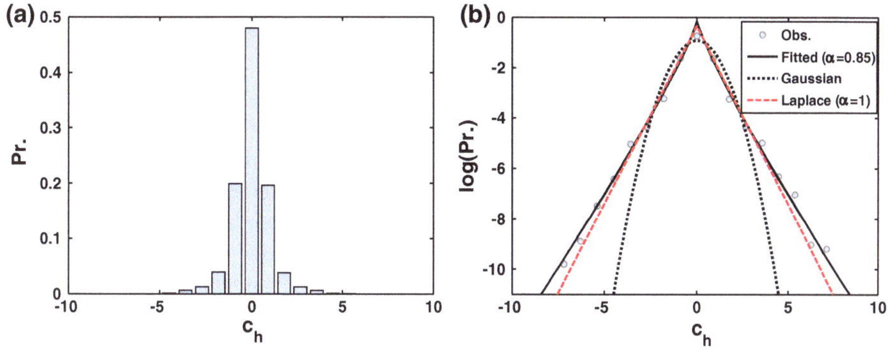

Fig. 3 **a** Histogram of the derivatives in the *horizontal direction* of the hurricane snapshot shown in Fig. 2. The derivative coefficients are obtained by the Sobel operator that produces a second-order discrete approximation of the field derivative. **b** Same histogram plotted on a log-probability scale showing the empirical PDF (*circles*), the fitted generalized Gaussian PDF with parameter $\alpha = 0.85$, the Gaussian PDF ($\alpha = 2.0$), and the Laplace density ($\alpha = 1.0$) for comparison. Note that the assumption of a Laplace density for the rainfall derivatives is theoretically consistent with the proposed ℓ_1-norm variational downscaling framework

regularization approach, which imposes constraints on the specific degree of smoothness (regularity) of the precipitation fields. The proposed regularization is selected to allow the preservation of large gradients while at the same time impose the desired smoothness on the solution. The paper is structured as follows. In Sect. 2, the need for regularization is explained with special emphasis on a total variation regularization scheme (ℓ_1-norm in the derivative space) in order to reproduce steep gradients and to preserve the heavy-tailed structure of rainfall. In this Section, the statistical interpretation of the variational ℓ_1-norm regularization is also explained. In particular, it is elucidated that the downscaled rainfall fields obtained via ℓ_1-norm regularization in the derivative domain is equivalent to the Bayesian maximum a posteriori (MAP) estimate with a Laplace prior distribution in the precipitation derivatives, a special case of the generalized Gaussian distribution $p(x) \propto \exp(-\lambda|x|^{\alpha})$ with $\alpha = 1$ (Ebtehaj and Foufoula-Georgiou 2011). Section 3 presents insights into the problem of an unknown downgrading observation operator or kernel that "converts" the high-resolution rainfall to the lower-resolution observations and discusses an alternative methodology, dictionary-based sparse precipitation downscaling (SPaD), developed in (Ebtehaj et al. 2012). In Sect. 4, we present a detailed implementation of our variational downscaling (VarD) methodology in a tropical (hurricane) storm and compare the results of VarD with those of the SPaD method. Finally, concluding remarks and directions for future research are presented in Sect. 5.

2 Precipitation Downscaling as a Regularized Inverse Problem

2.1 Basic Concepts in the Continuous Space

Consider the true state (or signal) $f(t)$ that is not known but is observed indirectly via a measuring device, which imposes a smoothing on the original state and returns the observation $g(s)$. Let $f(t)$ and $g(s)$ relate via the following linear transformation:

$$\int_0^1 K(s,t)f(t)\mathrm{d}t = g(s)\,0,\ t \leq 1, \tag{1}$$

where $K(s, t)$ is a known kernel, which downgrades the true state by damping its high-resolution components and making it smoother. The problem of recovering $f(t)$ knowing the observation $g(s)$ and the kernel $K(s, t)$ is a well-studied inverse problem, known as the Fredholm integral equation of the first kind. Inverse problems are by their nature ill-posed, in the sense that they do not satisfy at least one of the following three conditions: (1) existence of a solution, (2) uniqueness of the solution, and (3) stability in the solution, i.e., robustness to perturbations in the observation. It can be shown that the above inverse problem is very sensitive to the observation noise, since high frequencies are amplified in the inversion process (so-called inverse noise) and they can easily spoil and blow up the solution (see Hansen 2010). In this sense, even a small but high-frequency random perturbation in $g(s)$ can lead to a very large perturbation in the estimate of $f(t)$. This is relevant to the problem of reconstructing small-scale features in precipitation fields (downscaling) from low-resolution noisy data, when the noise can be of low magnitude but high frequency, e.g., discontinuities in overlapping regions of different sensors or instrument noise.

Therefore, naturally, if we define the distance between the observations and the true state by the following residual Euclidean norm:

$$R(f) = \left\| \int_0^1 K(s,t)\,f(t)\,\mathrm{d}t - g(s) \right\|_2, \tag{2}$$

then minimizing $R(f)$ alone does not guarantee a unique and stable solution of the inverse problem. Rather, additional constraints have to be imposed to enforce some regularity (or smoothness) of the solution and suppress some of the unwanted inverse noise components leading to a unique and more stable solution. Let us denote by $S(f)$ a smoothing norm, which measures the desired regularity of $f(t)$. Then, obtaining a unique and stable solution to the inverse problem amounts to solving a variational minimization problem of the form

$$f(t) = \underset{f}{\operatorname{argmin}}\left\{R(f)^2 + \lambda^2 S(f)\right\}, \tag{3}$$

The value of λ (called the regularization parameter) is chosen as to provide a balance between the weight given to fitting the observations, as measured by the magnitude of the residual term $R(f)$, and the degree of regularity of the solution measured by the smoothing norm $S(f)$. Common choices for $S(f)$ are ℓ_2-norms of the function $f(t)$ or its derivatives, i.e.,

$$S(f) = \|f^{(d)}\|_2^2 = \int_0^1 |f^{(d)}(t)|^2 \mathrm{d}t,\ d = 0, 1, \ldots \tag{4}$$

where $f^{(d)}$ denotes the dth order derivative of f. Another smoothing norm of specific interest in the present study is the ℓ_1-norm of the gradient of f, that is,

$$S_{\mathrm{TV}}(f) = \|f\|_1 = \int_0^1 |f^{(1)}(t)|\,\mathrm{d}t, \tag{5}$$

known as the Total Variation (TV) of the function $f(t)$. Both the $S(f)$ and $S_{\mathrm{TV}}(f)$ norms yield robust solutions with desired regularities but the $S_{\mathrm{TV}}(f)$ penalizes local jumps and isolated

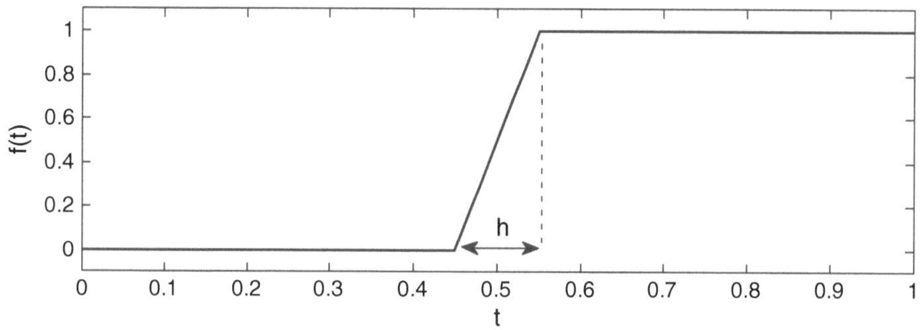

Fig. 4 A piecewise linear function *f(t)* with a slope $f^{(1)} = 1/h$ at the non-horizontal part. As it is easily shown (see text), for this function, the ℓ_1(total variation)-norm $\|f^{(1)}\|_1$ is constant and independent of h while the ℓ_2-norm $\|f^{(1)}\|_2^2 = 1/h$ goes to infinity as h goes to zero (i.e., for a very steep gradient). As a result, the ℓ_2-norm solutions do not allow steep gradients, while the ℓ_1-norm does

singularities in a quite different way than the ℓ_2-norm of *S(f)*. It is important to demonstrate this point as it plays a key role in the proposed downscaling scheme.

Let us consider a piecewise linear function:

$$f(t) = \begin{cases} 0, & 0 \le t < \frac{1}{2}(1-h) \\ \frac{t}{h} - \frac{1-h}{2h}, & \frac{1}{2}(1-h) \le t \le \frac{1}{2}(1+h), \\ 1, & \frac{1}{2}(1+h) < t \le 1 \end{cases} \tag{6}$$

as shown in Fig. 4. It can be shown that the smoothing norms associated with the ℓ_1 and ℓ_2-norms of $f^{(1)}(t)$ satisfy:

$$\|f^{(1)}\|_1 = \int_0^1 |f^{(1)}(t)| \mathrm{d}t = \int_0^h \frac{1}{h} \mathrm{d}t = 1 \tag{7}$$

while

$$\|f^{(1)}\|_2^2 = \int_0^1 |f^{(1)}(t)|^2 \mathrm{d}t = \int_0^h \frac{1}{h^2} \mathrm{d}t = \frac{1}{h}. \tag{8}$$

It is observed that the TV smoothing norm $S_{\mathrm{TV}}(f) = \|f^{(1)}\|_1$ is independent of the slope of the middle part of *f(t)* while the smoothing ℓ_2-norm is inversely proportional to h and, as such, it severely penalizes steep gradients (when h is small). In other words, the ℓ_2-norm of $f^{(1)}$ will not allow any steep gradients and will produce a very smooth solution. Clearly, this is not desirable in solving an inverse problem associated with the reconstruction of small-scale details in precipitation fields, such as in the hurricane storm shown in Fig. 2.

2.2 Discrete Representation

Writing Eq. (1) in a discrete form, the problem of downscaling amounts to estimating a high-resolution (HR) state, denoted in an m-element vector as $\mathbf{x} \in \mathbb{R}^m$, from its low-resolution (LR) counterpart $\mathbf{y} \in \mathbb{R}^n$, where $m \gg n$. It is assumed that this LR counterpart relates to the high-resolution (HR) state via a linear downgrading (e.g., a linear blurring and/or downsampling[1]) operator $\mathbf{H} \in \mathbb{R}^{m \times n}$ as follows:

$$\mathbf{y} = \mathbf{H}\mathbf{x} + \mathbf{v}, \tag{9}$$

where $\mathbf{v} \sim \mathcal{N}(0, \mathbf{R})$ is a zero-mean Gaussian error with covariance $\mathbf{R}$. Due to the fact that the dimension of $\mathbf{y}$ is less than that of $\mathbf{x}$, the operator $\mathbf{H}$ is a rectangular matrix with more columns than rows and thus solving problem (9) for $\mathbf{x}$ is an ill-posed inverse problem (an under-determined system of equations with many solutions). As discussed above, we seek to impose a proper regularization to make the inverse problem well posed.

Following the developments presented above in a continuous setting and replacing $f^{(1)}$ with a discrete approximation derivative operator $\mathbf{L}$, the choice of the smoothing ℓ_2-norm regularization for S($\mathbf{x}$) becomes $\|\mathbf{L}\mathbf{x}\|_2^2$ while for the ℓ_1-norm becomes $\|\mathbf{L}\mathbf{x}\|_1$, where in discrete space $\|\mathbf{x}\|_2^2 = \Sigma_{i=1}^m |x_i|^2$ and $\|\mathbf{x}\|_1 = \Sigma_{i=1}^m |x_i|$.

Thus, the solution (HR state $\mathbf{x}$) can be obtained by solving the following regularized weighted least squares minimization problem:

$$\hat{\mathbf{x}} = \underset{\mathbf{x}}{\operatorname{argmin}} \left\{ \frac{1}{2} \|\mathbf{y} - \mathbf{H}\mathbf{x}\|_{\mathbf{R}^{-1}}^2 + \lambda S(\mathbf{x}) \right\}, \tag{10}$$

It is clear that the smaller the value of λ, the more weight is given to fitting the observations (often resulting in data over-fitting), while a large value of λ puts more weight into preserving the underlying properties of the state of interest $\mathbf{x}$, such as large gradients. The goal is to find a good balance between the two terms. Currently, no closed form method exists for the selection of this regularization parameter and the balance has to be obtained via a problem-specific statistical cross validation (e.g., Hansen 2010). Note that the problem in (10) with $S(\mathbf{x}) = \|\mathbf{L}\mathbf{x}\|_1$ is:

$$\hat{\mathbf{x}} = \underset{\mathbf{x}}{\operatorname{argmin}} \left\{ \frac{1}{2} \|\mathbf{y} - \mathbf{H}\mathbf{x}\|_{\mathbf{R}^{-1}}^2 + \lambda \|\mathbf{L}\mathbf{x}\|_1 \right\}, \tag{11}$$

that is, a non-smooth convex optimization problem as the regularization term is non-differentiable at the origin. As a result, the conventional iterative gradient methods do not work and one has to use greedy methods (Mallat and Zhang 1993) or apply the recently developed non-smooth optimization algorithms such as the iterative shrinkage thresholding method (Tibshirani 1996), the basis pursuit method (Chen et al. 1998, 2001), the constrained quadratic programming (Figueiredo et al. 2007), the proximal gradient-based methods (Beck and Teboulle 2009), or the interior point methods (Kim et al. 2007). In this work, we have adopted the method suggested by Figueiredo et al. (2007).

2.3 Geometrical Versus Statistical Interpretation of the ℓ_1-Norm Regularized Downscaling

As was discussed in the introduction, the motivation for introducing a new downscaling framework lies in the desire to reproduce some geometrical but also some statistical features of precipitation fields. Specifically, the question was posed as to how a downscaling scheme could be constructed that can reproduce both the abrupt localized gradients

[1] Here, by downsampling, we mean to reduce the sampling rate of the rainfall observations by a factor greater than one.

and also the characteristic probability distribution of the precipitation intensity gradients such as that displayed in Fig. 3a.

It can be shown that the solution of (10) obtained via ℓ_2-norm regularization (i.e., $S(\mathbf{x}) = \|\mathbf{Lx}\|_2^2$) is equivalent to the Bayesian maximum a posteriori (MAP) estimator where the transformed variable $\mathbf{Lx}$ is well explained by a Gaussian distribution. On the other hand, considering $S(\mathbf{x}) = \|\mathbf{Lx}\|_1$, the ℓ_1-norm regularized solution of (10), i.e., the solution of Eq. (11), is the MAP estimator where $\mathbf{Lx}$ is well explained by the multivariate Laplace distribution (the generalized Gaussian family with $\alpha = 1$). In other words, the ℓ_1-regularization implicitly assumes that the probability of $\mathbf{Lx}$ goes as $\exp(-\lambda\|\mathbf{Lx}\|_1)$ (Lewicki and Sejnowski 2000; Ebtehaj and Foufoula-Georgiou 2013). We note that for the storm of Fig. 2, the estimated tail parameter α is 0.85 (see Fig. 3), which denotes that the pdf of $\mathbf{Lx}$ goes as $\exp(-\lambda\|\mathbf{Lx}\|_\alpha^\alpha)$, where $\|\mathbf{x}\|_\alpha^\alpha = \Sigma_{i=1}^m |x_i|^\alpha$. This value of α implies that the Laplace distribution ($\alpha = 1$) is only an approximation of the true distribution of the analyzed precipitation (see Fig. 3b for comparison), making thus the proposed ℓ_1-norm regularization solution only an approximate solution in a statistical sense. Finding a solution via regularized inverse estimation that satisfies a prior probability for ($\mathbf{Lx}$) with $\alpha < 1$ requires solving a non-convex optimization, which may suffer from local minima and may be hard to solve for large-scale problems. For this reason, we limit our discussion to the ℓ_1-regularization recognizing the slight sub-optimality of the solution for precipitation applications but also its superiority relative to the Gaussian assumption about the rainfall derivatives.

3 Working with an Unknown Downgrading Operator (H)

In the above formulation of the downscaling problem as an inverse problem, the downgrading operator **H** is assumed to be linear and known a priori. A mathematically convenient form for the downgrading operator is to assume that it can be represented via a linear convolution followed by downsampling. In other words, one may assume that the low-resolution (LR) observation is obtained by applying an overlapping box (weighted) averaging over the HR field and keeping one observation only, typically at the center, per averaging box (downsampling). However, the downgrading operator is not generally known in practice and its characterization might be sensor-dependent. Also often, this operator is highly nonlinear (e.g., the relationship between the radiometer-observed brightness temperature and the precipitation reflectivity observed by the radar) and its linearization may introduce large estimation errors. This nonlinearity may also pose severe challenges from the optimization point of view and may give rise to a hard non-convex problem with many local minima (Bertsekas 1999).

To deal with the problem of an unknown downgrading operator, Ebtehaj et al. (2012) proposed a dictionary-learning-based methodology that allows to implicitly incorporate the downgrading effect via statistical learning without the need to explicitly characterize the downgrading operator. In this methodology, the downgrading operator is being learned via a dictionary of coincidental HR and LR observations (e.g., in practice, TRMM-PR, and ground-based NEXRAD or TMI and NEXRAD). The methodology is explained in detail by Ebtehaj et al. (2012) and is only briefly summarized herein.

In simple terms, the idea is to reconstruct a HR counterpart of the LR rainfall field based on learning from a representative data base of previously observed coincidental LR and HR rainfall fields (e.g., TRMM-PR and NEXRAD observations). As is evident, due to different underlying physics, the shape and patterns of rainfall intensities, viewed in a storm-scale field of view, might be drastically dissimilar. However, the small-scale patterns of rainfall when viewed over smaller windows might be repetitive and "similar" within different regions of the same storm or within different storms. Therefore, the central idea is to (a) collect a representative set of coincidentally observed LR and HR rainfall fields, with some similarities in their underlying physics; (b) zoom down into small-scale patterns (patches) of the given LR rainfall field; (c) for each patch, find few but very similar LR patches in the collected data base; (d) for those similar LR patches, obtain the corresponding HR patches in the data base and then reconstruct the HR counterparts of the LR patch of interest based on an optimality criterion; and (e) repeat this procedure for all possible patches and obtain a HR estimate for the observed LR rainfall field.

To be more specific, let us consider that the representative training set of N coincidental pairs of LR and HR rainfall fields are denoted by $\left\{\mathcal{Z}_l^i\right\}_{i=1}^N$ and $\left\{\mathcal{Z}_h^i\right\}_{i=1}^N$, respectively. As previously explained, for each patch $\mathbf{y}_l$ of the given LR rainfall field, we need to find a few very similar patches in $\left\{\mathcal{Z}_l^i\right\}_{i=1}^N$, where similarity is defined in terms of localized rainfall fluctuations and not in the mean values of the rainfall patches. To this end, all of the LR fields are projected (i.e., $\mathcal{Z}_l^i \to \left(\mathcal{Z}_h^i\right)'$) onto a redundant orthogonal basis (called feature space) to capture the rainfall local fluctuations including horizontal and vertical edges (i.e., zonal and meridional) and curvatures. This was performed by Ebtehaj et al. (2012) via an undecimated orthogonal Haar wavelet, which basically performs a high-pass filtering in each direction using first- and second-order differencing. Then, all of the constituent patches of the transformed LR fields in the data base were extracted, vectorized in a fixed order, and then stored as columns of a matrix $\boldsymbol{\Psi}$, the so-called empirical LR-dictionary.

Clearly, for each coincidental pair $\left(\mathcal{Z}_l^i, \mathcal{Z}_h^i\right)$, a set of "residual fields" can be formed by subtracting the LR fields from their HR counterparts via $\mathcal{R}_h^i = \mathcal{Z}_h^i - \mathbf{Q}\mathcal{Z}_l^i$, where $\mathbf{Q}$ is a readily available interpolation operator (e.g., a nearest-neighbor or bilinear, bicubic interpolator). Notice that, these residual fields contain the rainfall variability and high-frequency (fine spatial-scale) components that are not captured by the LR sensor and need to be recovered. Therefore, all of the constituent patches $\mathbf{r}_h$ of the residual fields can also be collected, vectorized in a fixed order, and then stored in the columns of a matrix Φ, the so-called HR-dictionary. Note that, by the explained construction, the empirical LR and HR dictionaries share the same number of columns while there is a one-to-one correspondence between them. In other words, while the columns of the $\boldsymbol{\Psi}$ contain LR rainfall features, the columns of the Φ contain the corresponding HR residuals, needed for the reconstruction of the HR field.

The premise is that the local variability of any LR patch $\mathbf{y}_l$, denoted by $\mathbf{y}_l'$, in any storm can be well approximated by a linear combination of the elements of the LR dictionary as follows:

$$\mathbf{y}_l' = \boldsymbol{\Psi}\mathbf{c} + \mathbf{v}, \tag{12}$$

where $\mathbf{c}$ is the vector of representation coefficients in the LR dictionary and $\mathbf{v} \sim \mathcal{N}(0, \mathbf{R})$ denotes the estimation error that can be well explained by a Gaussian density.

By analyzing a sample of 100 storms over Texas, it was documented by Ebtehaj et al. (2012) that the vector of representation coefficients $\mathbf{c}$ in the LR dictionary is very sparse. In other words, any desired local rainfall variability in the given LR field can be approximated by a linear combination of only a few columns of the LR empirical dictionary (of the order of 3–5 elements). To impose this sparsity (called "group sparsity") in solving (12) for $\mathbf{c}$, the solution needs to be constrained via an ℓ_1-norm regularization as follows:

$$\hat{\mathbf{c}} = \underset{\mathbf{c}}{\operatorname{argmin}} \left\{ \frac{1}{2} \left\| \mathbf{y}'_l - \mathbf{\Psi c} \right\|^2_{\mathbf{R}^{-1}} + \lambda \left\| \mathbf{c} \right\|_1 \right\}. \tag{13}$$

Using the representation coefficients obtained from (13), one can recover the corresponding residual fields (the details missed by the LR sensor) as follows:

$$\hat{\mathbf{r}} = \mathbf{\Phi} \hat{\mathbf{c}}. \tag{14}$$

Having the estimated residual fields, the HR patch can be obtained as $\hat{\mathbf{x}} = \mathbf{Q}\mathbf{y}_l + \hat{\mathbf{r}}$. Applying the same estimation methodology for all of the patches of the given LR rainfall field, we can recover the entire HR rainfall field (see Ebtehaj et al. 2012). The most important implication of the above framework is that we characterized the pair of $(\mathbf{\Psi}, \Phi)$ empirically without explicit access to the structure of the downgrading operator $\mathbf{H}$, which is the main advantage of this dictionary-based rainfall downscaling method versus the previously explained approach. Since advantage was taken of the rainfall group sparsity (and also implicitly of the sparsity of the precipitation fields themselves), the dictionary-based downscaling methodology was termed SPaD.

4 Results from a Case Study

To demonstrate the proposed downscaling methodology, we have chosen a specific tropical storm, hurricane Claudette, which occurred in July 2003. Claudette began as a tropical wave in the eastern Caribbean on July 8, 2003 and moved quickly westward to the Gulf of Mexico. It remained a tropical storm until just before making landfall in Port O'Conner, Texas, when it quickly strengthened to a category 1 hurricane. Although Claudette produced moderate rainfall across southern Texas, peaking at approximately 6.5 inches (165 mm), it maintained a tropical storm intensity for over 24 h after landfall with winds gusting to 83 mph (134 km/h) at Victoria Regional Airport, Texas. The storm caused excessive beach erosion and damages estimated at 180 million dollars. For this storm, we have available data from a NEXRAD station in Houston, Texas, for which a snapshot at 11:51:00 (UTC) on July 15, 2003 is shown in Fig. 2.

The issues we want to examine here are the following: (1) the ability of the proposed variational downscaling (VarD) scheme to reproduce the steep gradients in precipitation intensities as evidenced by reproducing the tails of the PDF of intensity gradients; (2) the effect of an unknown kernel (smoothing and downsampling operation imposed on the true HR field by a sensor) on the downscaling scheme performance using the proposed methodology; (3) a comparison of the VarD method with a local dictionary-based methodology based on sparse representation (SPaD) as discussed in the previous section; and (4) insights into the ability of the proposed VarD methodology and SPaD to reproduce not only the extreme gradients but also the extreme rainfall intensities, i.e., the tails of the rainfall intensity probability distribution functions (PDFs).

The original HR data at 1 × 1 km (Fig. 5a) were downgraded to 8 × 8 km LR observations via a coarse-graining filter consisting of a simple box averaging of size 8 × 8 followed by downsampling with a factor of 8 (i.e., keeping one observation per box of 8 × 8 km). The resulting LR field is shown in Fig. 5b and is considered to be the field that would be available to us from a satellite sensor. Figure 5c, d shows the results of

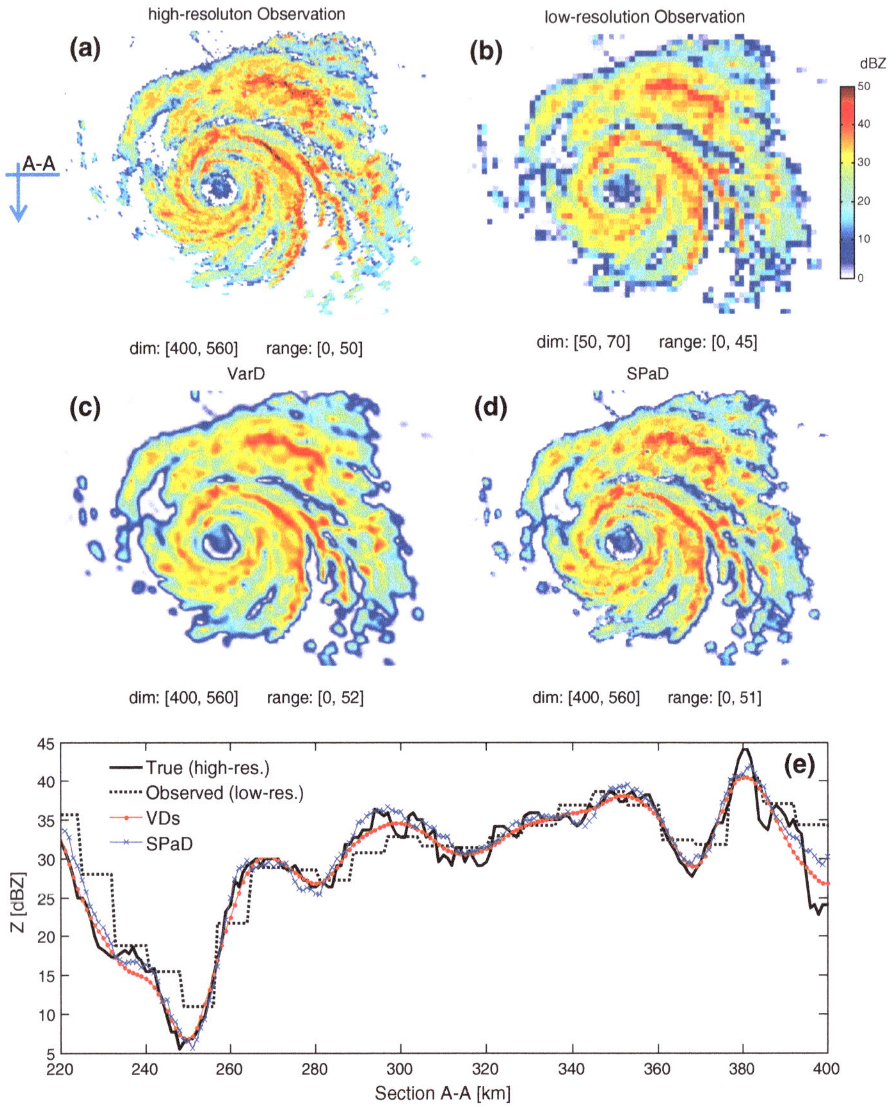

Fig. 5 **a** Original HR base reflectivity snapshot at resolution 1 × 1 km over TX (hurricane Claudette, 08-16-2003, UTC 11:51:00); **b** The synthetic LR observation obtained by coarse graining of the field up to scale 8 × 8 km (smoothing with an average filter of size 8 × 8 followed by downsampling by a factor 8); **c** result of the downscaled field at resolution 1 × 1 km using the variational downscaling (VarD) method; and **d** results of the dictionary-based sparse precipitation downscaling (SPaD) method at resolution 1 × 1 km; **e** intensities averaged over a bandwidth of 8 km centered at a cross section A-A in (**a**), displaying the true HR field, the LR coarse-grained field (observations), and the two downscaled fields

downscaling the 8 × 8 km field to 1 × 1 km resolution using the VarD and SPaD methodologies with $\lambda \approxeq 0.05 \left\| \mathbf{L}^{-\mathrm{T}}\mathbf{H}^{\mathrm{T}}\mathbf{R}^{-1}\mathbf{y} \right\|_{\infty}$ in the original formulation of the problem (11), where $\|\mathbf{x}\|_{\infty} = \max(|x_1|, \ldots, |x_m|)$. Note that in all of our experiments, we empirically found that $0 < \lambda \leq 0.10 \left\| \mathbf{L}^{-\mathrm{T}}\mathbf{H}^{\mathrm{T}}\mathbf{R}^{-1}\mathbf{y} \right\|_{\infty}$ works well for rainfall downscaling in both methods, while it can be shown that the solution of problem (11) is zero for all $\lambda \geq \left\|\mathbf{L}^{-\mathrm{T}}\mathbf{H}^{\mathrm{T}}\mathbf{R}^{-1}\mathbf{y}\right\|_{\infty}$.

As discussed before, VarD assumes the downgrading operator **H** to be known. In our case, we used as **H** the "true" operator, i.e., the same operator we used to coarse grain the HR (1 × 1 km) reflectivity field to the LR (8 × 8 km) one. It is observed that the VarD downscaled field has a smoother appearance than the original field (it does not have the 1 × 1 km pixelized appearance of the original HR field), which is not unexpected given that the ℓ_1-regularization promotes smoothness in the solution while allowing for some steep gradients as demonstrated in the illustrative example of Fig. 4. A one-dimensional cross section shown in Fig. 5e confirms this observation and shows that the downscaled field is much closer to the true field compared to the LR field.

Suppose now that the true filter **H** is not known and only the LR field is given without guidance as to what "filtering" the sensor did to the HR field to return the LR observations. As discussed in the previous section, and in Ebtehaj et al. (2012), we demonstrated that this filter can be "learned" implicitly and locally using coincidental high- and low-resolution images available for a number of similar storms. In that study, a sample of 100 HR summer storms over Texas was used to construct a set of coincidental LR storms (using again a simple box averaging and a downsampling operator). This hundred storm sample was then used to compute the LR and HR dictionaries, which formed the basis of the SPaD method as explained in the previous section. This same dictionary was used herein to recover the 1 × 1 km HR rainfall field of the Claudette storm from 8 × 8 km observations. The results are shown in Fig. 5d.

In general, it is expected that the SPaD method will outperform the VarD method when the operator **H** is not known at all or is locally varying, due, for example, to instrument range effects or cloud interference or different performance of an instrument in low- versus high-resolution rainfall intensities. However, it is noted that, since in our data base the LR and HR fields relate to each other with a simple box averaging operator **H** (by construction), we expect that the dictionary-learning SPaD downscaling will perform comparably to the VarD method. Extra information in SPaD will be gained by the localized nature of the estimation methodology, which might reproduce extra high-frequency (small-scale) features, obtained from the available dictionaries that may not be recovered in the VarD approach.

To more quantitatively compare the two downscaled fields to the true underlying HR field and to each other, we compare in Fig. 6 the PDF of the derivatives in the horizontal direction in terms of their q–q plot (quantiles of the variable of interest vs. standard normal quantiles). We observe that both methods are able to reproduce the heavy tails of the PDF of the precipitation gradients, which are much thicker than those of the Gaussian PDF, and thus, both methods are able to reproduce high gradients in the HR recovered field. VarD is seen to slightly outperform SPaD in reproducing high positive gradients, not surprisingly since, in VarD, the **H** operator was customized to this specific storm, while, in SPaD, information from a suite of other storms was also used.

Turning our attention to the preservation of the statistics of the precipitation field itself, we show in Fig. 7 the comparison of the PDFs of the LR rainfall field with that of the true HR field and the downscaled fields. We recall that although the preservation of the thicker-than-Gaussian (Laplace) tails in the PDF of precipitation intensity gradients is explicitly incorporated in the ℓ_1-norm VarD downscaling methodology, no explicit preservation of the extreme rainfall intensities themselves is accounted for. However, it is clear from Fig. 7 that VarD performs satisfactorily in reproducing extreme rainfall intensities in the

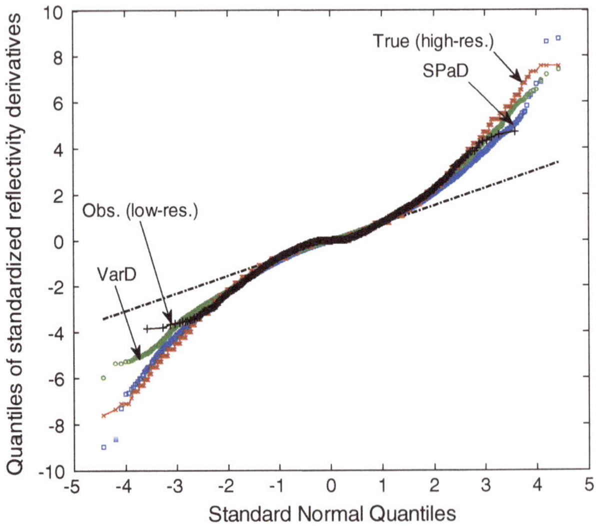

Fig. 6 The quantiles of the standardized horizontal (zonal) derivatives ($\nabla_x Z$) of the true HR field (*red* ×), the LR observation (black +), the VarD downscaled field (*green* o), and the SPaD downscaled field (*blue* □) versus standard normal quantiles. The broken *straight line* represents quantiles of the Gaussian distribution. It is observed that both downscaling methodologies are able to reproduce the extreme reflectivity gradients

downscaled field and is able to enhance substantially the tails of the low-resolution rainfall fields. One of the reasons for reproducing extreme rainfall intensities is that typically extreme gradients are collocated with high rainfall intensities. This was observed and documented by Perica and Foufoula-Georgiou (1996) and is also documented for the Claudette storm in Fig. 8. So, indirectly, VarD is bound to preserve satisfactorily the tails of the PDF of precipitation intensities. From Fig. 7, it is apparent that SPaD outperforms VarD in preserving extreme rainfall. This is attributed to the fact that, in SPaD, the operator is learned directly on the precipitation intensities, and not on the gradients, allowing thus for a more direct reconstruction of extreme intensities, provided that such extremes are available in the data base.

Table 1 presents a comparison of the downscaling methodologies in terms of several quantitative metrics: the mean square error: $\mathrm{MSE} = \|\mathbf{x} - \hat{\mathbf{x}}\|_2^2 / \|\mathbf{x}\|_2^2$, the maximum absolute error: $\mathrm{MAE} = \|\mathbf{x} - \hat{\mathbf{x}}\|_1 / \|\mathbf{x}\|_1$, the peak signal-to-noise ratio: $\mathrm{PSNR} = 20\log_{10}[\max(\hat{\mathbf{x}})/\mathrm{std}(\mathbf{x} - \hat{\mathbf{x}})]$, and the Kullback–Leibler divergence: $\mathrm{KLD}(p_{\mathbf{x}} \| p_{\hat{\mathbf{x}}}) = \Sigma_i \ln[p_{\mathbf{x}}(i)/p_{\hat{\mathbf{x}}}(i)] p_{\mathbf{x}}(i)$ or relative entropy metric, where $p_{\mathbf{x}}(i)$ and $p_{\hat{\mathbf{x}}}(i)$ are the discrete probabilities of the true and estimated rainfall, respectively. The KLD is a non-negative measure that represents a relative degree of closeness of two PDFs in terms of their entropy, while smaller values signify a stronger degree of similarity. It can be seen from Table 1 that both downscaling methods produce HR fields that are closer to the true field compared to the LR field and that the VarD and SPaD methods considerably outperform the "naïve" simple downscaling methods such as the result obtained by the bicubic interpolation scheme. SPaD is seen to outperform VarD in terms of the entropy metric (smaller KLD value) further speaking for the better reproduction of very extreme rainfall intensities.

It is worth presenting here some extra insight into the effect of a misdiagnosed observation filter **H** on the downscaled field. As shown in the illustrative example of Fig. 9, when the observation operator is smoother (a Gaussian filter) as compared to the operator used in the VarD downscaling (a box average filter), the downscaled field exhibits a

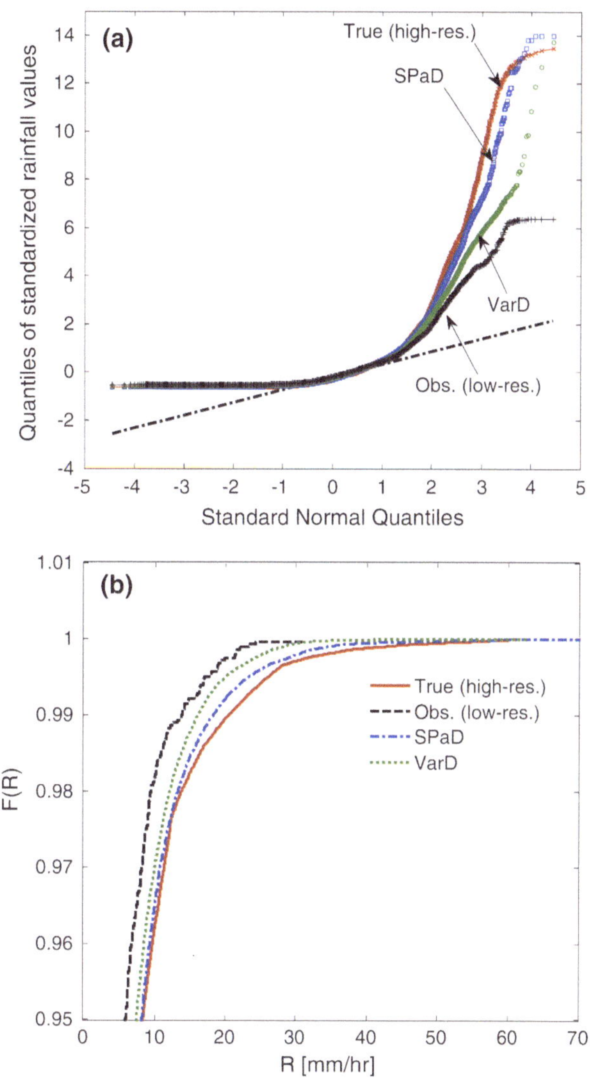

Fig. 7 **a** The quantiles of the standardized recovered rain (i.e., $Z = 300R^{1.4}$) versus standard normal quantiles. The rainfall quantile values are only for the positive rainy part of the fields and are standardized by subtracting the mean and dividing by the standard deviation. **b** Comparison of the cumulative distribution of rainfall intensities focusing on extremes. Both plots show how both the VarD and SPaD downscaling methodologies reproduce extreme rainfall intensities not present in the observed LR fields

blockiness coming from the mismatch between the assumed and true filters. In fact, this blockiness provides a qualitative diagnostic of the filter mismatch, in that it picks up the fact that the underlying true observation filter (the Gaussian in this case) was smoother than the one used for recovery. Apart from the visual inspection of the downscaled field, Fig. 9 (caption) provides the comparison metrics that show the underperformance of this downscaled field relatively to the one obtained using the correct filter (compare values with those in Table 1). The possibility of developing a methodology to learn properties (e.g., smoothness and nonlinearity) of the underlying observation filter in the case that no coincidental LR and HR data sets are available to apply the dictionary-based methodology is appealing and warrants further exploration.

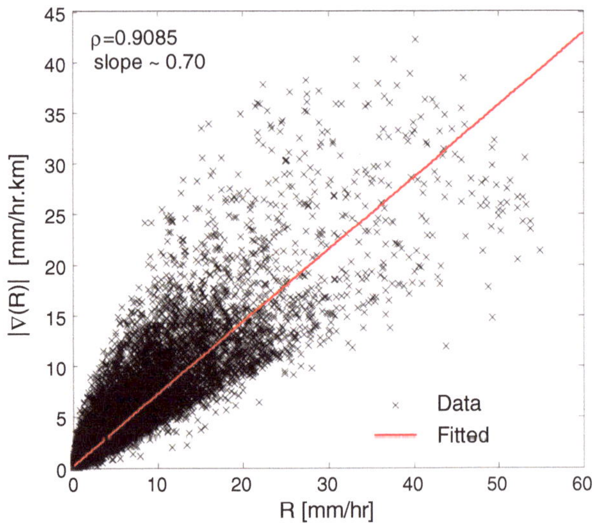

Fig. 8 Magnitude of precipitation gradients $|\nabla(R)| = \sqrt{(\nabla_x R)^2+(\nabla_y R)^2}$ versus precipitation intensity for the nonzero pixels of the Claudette storm at resolution of 1 × 1 km showing that high precipitation gradients are mostly collocated with high precipitation intensities. Only pixels for which the gradient was at least 20 % of the local precipitation intensity were considered

Table 1 Error statistics obtained by comparing the HR precipitation reflectivity image of Hurricane Claudette (true) with the LR one, the downscaled fields via Bicubic interpolation, the VarD, and the SPaD methodologies (see text for definition of these metrics)

	Quality metrics			
	MSE †	MAE	PSNR	KLD
Low. res.	0.305	0.260	17.834	0.089
Bicubic	0.275	0.246	18.742	0.113
VarD	0.194	0.172	22.539	0.065
SPaD	0.209	0.177	22.015	0.044

† *MSE* mean squared error, *MAE* mean absolute error, *PSNR* peak signal-to-noise ratio, *KLD* Kullback–Leibler divergence

5 Concluding Remarks

The problem of downscaling climate variables remains of interest as more spaceborne observations become available and as the need to translate low-resolution (LR) climate predictions to regional and local scales becomes essential for long-term planning purposes. Of special interest are downscaling schemes that can accurately reproduce not only overall statistical properties of rainfall but also specific features of interest, such as extreme rainfall intensities and abrupt gradients. In this paper, such a precipitation downscaling scheme was introduced using a formalism of inverse estimation and solving the (ill-posed) inverse problem by imposing certain constraints that guarantee stability and uniqueness of the solution while also enforcing a certain type of smoothness that allows for some abrupt gradients. Mathematically, this inverse problem is solved via what is called an ℓ_1-norm or total variation regularization. We showed the equivalence of the proposed total variation regularized solution to a statistical maximum a posteriori (MAP) Bayesian solution, which has a Laplace prior distribution in the derivative domain. We demonstrated the

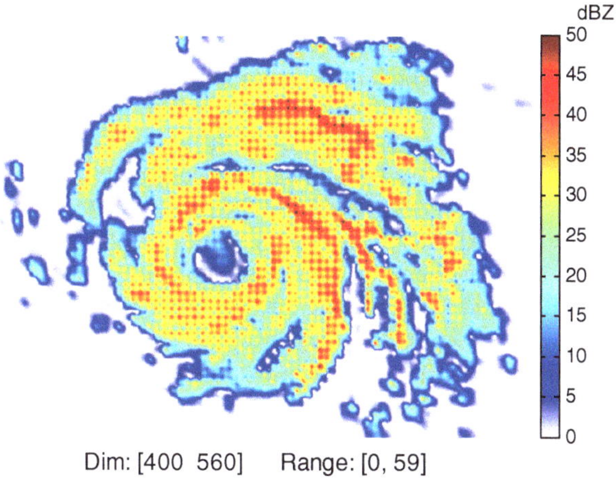

Fig. 9 VarD result for downscaling precipitation reflectivity from scale 8 × 8 to 1 × 1 km with a "wrong" observation operator. In this experiment, the imposed observation operator was a Gaussian filter of size 8 × 8 with standard deviation 2 while in downscaling, we assumed a uniform average filter of the same size. It is clear from the result that the quality of downscaling is blocky and is severely deteriorated because of the misspecification of the observation operator in the downscaling scheme. The selected quantitative measures are as follows: MSE = 0.244; MAE = 0.220; PSNR: 22.0; and KLD = 0.075 (see Table 1)

performance of the proposed downscaling scheme on a tropical storm and concluded that it was able to capture adequately both the extremes of rainfall intensities and gradients.

A practical challenge faced in applying the proposed methodology is that the observation operator (which relates the true unknown HR field to the LR observations) might not be known. In fact, it might be even changing locally due to sensor properties as affected, for example, by range or precipitation intensity and composition. If coincidental high- and low-resolution fields are available in a data base, the data-driven dictionary-based methodology introduced by Ebtehaj et al. (2012) offers promise and, although more computationally intensive, it might offer advantages in capturing more faithfully local details and extremes. However, a lot more work is needed to understand the sensitivity of the dictionary-based methodology to the selection of a data base from environments different than the storm of interest, as well as when the observation filter relates nonlinearly to the underlying field as is the case in problems of retrieval, i.e., estimation of precipitation intensity from radiances recorded by the TRMM microwave imager.

The presence of statistical self-similarity (scaling) in spatial rainfall, manifesting in log–log linearity in the Fourier or wavelet power spectra and also in higher-order statistical moments, has been well documented by now (see discussion in the introduction). This structure, often explained in the context of mono or multifractal formalisms, has guided the development of several stochastic downscaling methodologies (e.g., Rebora et al. 2006a, b; Perica and Foufoula-Georgiou 1996, among many others). The downscaled precipitation fields produced by these models are, by construction, respecting the rainfall scaling laws; however, they are not unique as multiple realizations of plausible high-resolution rainfall fields with the same input parameters can be produced without following a specific optimality criterion. On the other hand, the proposed downscaling methodologies produce unique high-resolution rainfall fields based on the aforementioned optimality criteria that also allow us to partially preserve the underlying non-Gaussian structure of the rainfall fields. An important question that arises then is whether statistical scaling in rainfall fields, although not prescribed in our method, arises as an emergent property. The answer to this question is not obvious. Our preliminary results (not reported herein) demonstrate that statistical scaling indeed arises in both the ℓ_1-norm variational downscaling (VarD) and the SPaD schemes. However, the power law exponents (of the variance of the wavelet coefficients as a function of scale) and the variance of the wavelet coefficients at the smallest scale (similar to the analysis in Perica and Foufoula-Georgiou 1996) seem to be lower than those of the original fields. This might be due to the fact that, although our scheme is able to accurately capture, much better than other statistical schemes, the magnitude of the infrequent localized large gradients in precipitation fields, it might under-produce the variability of the smaller gradients, reducing thus the overall variance. This is an issue that is currently explored both from a theoretical perspective and via simulation, as in most applications one is interested to preserve both the localized extremes but also the overall variance of the smaller magnitude fluctuations.

The work presented herein falls within a larger research direction of using variational regularization approaches or equivalently, Bayesian MAP estimators with heavy-tailed priors in the derivative domain, for estimation problems in hydro-climatology, such as downscaling, multi-sensor data fusion, retrieval, and data assimilation (see Ebtehaj and Foufoula-Georgiou 2013). A relatively small number of abrupt gradients within the field of interest or heavy-tailed PDFs in the derivative domain are associated with the notion of sparsity, that is, the fact that, when the state is projected in a suitable basis, most of the projection coefficients are close to zero and only a few coefficients carry most of the state energy. Estimation problems of sparse states (posed in an inverse estimation setting or in a variational setting of minimizing a functional) require the use of ℓ_1-norm regularization, which results from imposing extra constraints on the solution to enforce sparsity. Motivated by the need to preserve sharp weather fronts in data assimilation of numerical weather prediction models, an ℓ_1-norm regularized variational data assimilation methodology was recently proposed by Freitag et al. (2012) and demonstrated in a simple setting using the advection equation for the state evolution dynamics. In Ebtehaj and Foufoula-Georgiou (2013), data assimilation in the presence of extreme gradients in the state variable was further analyzed using as illustrative example the advection–diffusion equation that forms the basis of many hydro-meteorological problems, such as those dealing with the estimation of surface heat fluxes based on the assimilation of land surface temperature (e.g., see Bateni and Entekhabi 2012). Application of these new non-smooth variational methodologies in real data assimilation problems, and also in combining data assimilation with downscaling of the state, is only in its infancy and is certain to occupy the geophysical community in the years to come.

Acknowledgments This work has been mainly supported by a NASA-GPM award (NNX10AO12G), a NASA Earth and Space Science Fellowship (NNX12AN45H), and a Doctoral Dissertation Fellowship of the University of Minnesota to the second author. The insightful comments of one anonymous referee are also gratefully acknowledged.

References

Badas MG, Deidda R, Piga E (2006) Modulation of homogeneous space-time rainfall cascades to account for orographic influences. Nat Hazard Earth Syst 6(3):427–437. doi:10.5194/nhess-6-427-2006

Bateni SM, Entekhabi D (2012) Surface heat flux estimation with the ensemble Kalman smoother: joint estimation of state and parameters. Water Resour Res 48(3). doi:10.1029/2011WR011542

Beck A, Teboulle M (2009) A fast iterative shrinkage-thresholding algorithm for linear inverse problems. SIAM J Imaging Sci 2(1):183–202. doi:10.1137/080716542

Bertsekas DP (1999) Nonlinear programming, 2nd edn. Athena Scientific, Belmont, MA, p 794
Chen S, Donoho D, Saunders M (2001) Atomic decomposition by basis pursuit. SIAM Rev 43(1):129–159
Chen SS, Donoho DL, Saunders MA (1998) Atomic decomposition by basis pursuit. SIAM J Sci Comput 20:33–61
Deidda R (2000) Rainfall downscaling in a space-time multifractal framework. Water Resour Res 36(7):1779–1794
Ebtehaj AM, Foufoula-Georgiou E (2011) Statistics of precipitation reflectivity images and cascade of Gaussian-scale mixtures in the wavelet domain: a formalism for reproducing extremes and coherent multiscale structures. J Geophys Res 116:D14110. doi:10.1029/2010JD015177
Ebtehaj AM, Foufoula-Georgiou E, Lerman G (2012) Sparse regularization for precipitation downscaling. J Geophys Res 116:D22110. doi:10.1029/2011JD017057
Ebtehaj AM, Foufoula-Georgiou E (2013) Variationl downscaling, fusion and assimilation of hydrometeorological states: a unified framework via regularization. Water Resour Res. doi:10.1002/wrcr.20424
Figueiredo M, Nowak R, Wright S (2007) Gradient projection for sparse reconstruction: application to compressed sensing and other inverse problems. IEEE J Sel Topics Signal Process 1(4):586–597. doi:10.1109/JSTSP.2007.910281
Flaming GM (2004) Measurement of global precipitation. In: Geoscience and remote sensing symposium, 2004. IGARSS'04. Proceedings. 2004 IEEE international, vol 2, p 918–920
Freitag MA, Nichols NK, Budd CJ (2012) Resolution of sharp fronts in the presence of model error in variational data assimilation. Q J Roy Meteor Soc. doi:10.1002/qj.2002
Hansen P (2010) Discrete inverse problems: insight and algorithms, vol. 7. Society for Industrial and Applied Mathematics (SIAM), Philadelphia, PA
Harris D, Foufoula-Georgiou E, Droegemeier KK, Levit JJ (2001) Multiscale statistical properties of a high-resolution precipitation forecast. J Hydrometeor 2(4):406–418
Krajewski WF, Smith JA (2002) Radar hydrology: rainfall estimation. Adv Water Resour 25(8–12):1387–1394. doi:10.1016/S0309-1708(02)00062-3
Kim S-J, Koh K, Lustig M, Boyd S, Gorinevsky D (2007) An interior-point method for large-scale ℓ_1-regularized least squares. IEEE J Sel Topics Signal Process. 1(4):606–617. doi:10.1109/JSTSP.2007.910971
Kumar P, Foufoula-Georgiou E (1993) A multicomponent decomposition of spatial rainfall fields. 2. Self-similarity in fluctuations. Water Resour Res 29(8):2533–2544
Kumar P, Foufoula-Georgiou E (1993) A multicomponent decomposition of spatial rainfall fields. 1. Segregation of large- and small-scale features using wavelet transforms. Water Resour Res 29(8):2515–2532
Lewicki M, Sejnowski T (2000) Learning overcomplete representations. Neural Comput 12(2):337–365
Lovejoy S, Mandelbrot B (1985) Fractal properties of rain, and a fractal model. Tellus A 37(3):209–232
Lovejoy S, Schertzer D (1990) Multifractals, universality classes and satellite and radar. J Geophys Res 95(D3):2021–2034
Mallat S, Zhang Z (1993) Matching pursuits with time-frequency dictionaries. IEEE Trans Signal Proces 41(12):3397–3415. doi:10.1109/78.258082
Mallat S (1989) A theory for multiresolution signal decomposition: the wavelet representation. IEEE Trans Pattern Anal Mach Intell 11(7):674–693. doi:10.1109/34.192463
Nykanen DK, Foufoula-Georgiou E, Lapenta WM (2001) Impact of small-scale rainfall variability on larger-scale spatial organization of land-atmosphere fluxes. J Hydrometeor 2(2):105–121
Perica S, Foufoula-Georgiou E (1996) Model for multiscale disaggregation of spatial rainfall based on coupling meteorological and scaling. J Geophys Res 101(D21):26–347
Rebora N, Ferraris L, Von Hardenberg J, Provenzale A et al (2006) Rainfall downscaling and flood forecasting: a case study in the Mediterranean area. Nat Hazard Earth Syst 6(4):611–619
Rebora N, Ferraris L, Von Hardenberg J, Provenzale A (2006) RainFARM: rainfall downscaling by a filtered autoregressive model. J Hydrometeor 7:724–738
Sapozhnikov VB, Foufoula-Georgiou E (2007) An exponential Langevin-type model for rainfall exhibiting spatial and temporal scaling. Nonlinear Dyn Geosci :87–100
Tibshirani R (1996) Regression shrinkage and selection via the Lasso. J R Stat Soc Ser B Stat Methodol 58(1):267–288
Venugopal V, Roux SG, Foufoula-Georgiou E, Arneodo A (2006) Revisiting multifractality of high-resolution temporal rainfall using a wavelet-based formalism. Water Resour Res 42(6):6. doi:10.1029/2005WR004489
Venugopal V, Roux SG, Foufoula-Georgiou E, Arnéodo A (2006) Scaling behavior of high resolution temporal rainfall: new insights from a wavelet-based cumulant analysis. Phys Lett A 348(3):335–345

DOI 10.1007/978-94-017-8789-5_14
Reprinted from *Surveys in Geophysics* Journal, DOI 10.1007/s10712-013-9263-x

Global Snow Mass Measurements and the Effect of Stratigraphic Detail on Inversion of Microwave Brightness Temperatures

Mark Richardson · Ian Davenport · Robert Gurney

Received: 30 April 2013 / Accepted: 31 October 2013 / Published online: 22 November 2013

Abstract Snow provides large seasonal storage of freshwater, and information about the distribution of snow mass as snow water equivalent (SWE) is important for hydrological planning and detecting climate change impacts. Large regional disagreements remain between estimates from reanalyses, remote sensing and modelling. Assimilating passive microwave information improves SWE estimates in many regions, but the assimilation must account for how microwave scattering depends on snow stratigraphy. Physical snow models can estimate snow stratigraphy, but users must consider the computational expense of model complexity versus acceptable errors. Using data from the National Aeronautics and Space Administration Cold Land Processes Experiment and the Helsinki University of Technology microwave emission model of layered snowpacks, it is shown that simulations of the brightness temperature difference between 19 and 37 GHz vertically polarised microwaves are consistent with advanced microwave scanning radiometer-earth observing system and special sensor microwave imager retrievals once known stratigraphic information is used. Simulated brightness temperature differences for an individual snow profile depend on the provided stratigraphic detail. Relative to a profile defined at the 10-cm resolution of density and temperature measurements, the error introduced by simplification to a single layer of average properties increases approximately linearly with snow mass. If this brightness temperature error is converted into SWE using a traditional retrieval method, then it is equivalent to ±13 mm SWE (7 % of total) at a depth of 100 cm. This error is reduced to ±5.6 mm SWE (3 % of total) for a two-layer model.

Keywords Snow mass · Snow grain size · Remote sensing · Microwave radiometry · Hydrology

M. Richardson · I. Davenport · R. Gurney
Department of Meteorology, University of Reading, Reading, UK

M. Richardson (✉)
ESSC, Harry Pitt Building, 3 Earley Gate, University of Reading, Whiteknights Campus, Reading RG6 6AL, UK
e-mail: m.t.richardson@pgr.reading.ac.uk

1 Introduction

Snow is extremely important hydrologically, with more than one-sixth of the global population situated in areas where snow precipitation is greater than half of annual runoff (Barnett et al. 2005). Snow affects both timing and quantity of runoff as well as the surface energy balance and atmospheric chemistry.

Currently, remote sensing products exist for snow-covered area (SCA), albedo, grain size, surface contaminants, melt and snow water equivalent (SWE). Measurements of snow surface properties such as SCA are regularly used (e.g., Brown and Mote 2009; Dye 2002; Frei et al. 2003) and generally have more well-characterised uncertainties (Hall and Riggs 2007; Rittger et al. 2013) than estimates of bulk properties such as SWE.

Measurement of surface properties has allowed the identification of snow season duration, surface melt (Koskinen et al. 1997) and a determination of snow's contribution to radiative feedback in response to warming (Flanner et al. 2011). In terms of hydrological relevance, Painter et al. (2012) developed a Moderate Resolution Imaging Spectroradiometer (MODIS) algorithm for determining radiative forcing from impurities in near-surface snow. When realistic values of these radiative forcings were included in the analysis of snow in south-western Colorado, it was estimated that the impurities reduced snow cover duration by 21–51 days, increased peak outflow, changed the runoff profile and reduced total seasonal runoff (Skiles et al. 2012).

Despite the successes of surface measurements, there remain large uncertainties in global estimates of SWE, with regional disagreements between products derived from remote sensing, general circulation models (GCMs) and reanalyses.

This study reviews continental-scale SWE products and describes the key techniques and their relative strengths, including the assimilation of remotely sensed passive microwave (PM) observations. A recent product that assimilates PM, Globsnow (Takala et al. 2011), is described in detail as it has been suggested as a suitable product for validation of land surface models (LSMs (Hancock et al. 2013)).

The assimilation of PM observations requires an observation operator, which converts the state vector of snow properties into a vector of observable microwave brightness temperatures. In the case of Globsnow, the snow is described by density, grain size and snow depth of a single layer. The observation operator is the Helsinki University of Technology (HUT) radiative transfer model (Pulliainen et al. 1999), which produces a brightness temperature difference between two PM channels, ΔT_B, for comparison with satellite retrievals.

Although Globsnow assumes a single homogeneous layer, snowpacks typically consist of multiple layers that often feature complex stratigraphy which affects the radiative transfer. The current Globsnow approach neglects this both in the radiative transfer simulation and in calculating the weighting function that determines the size of the PM-driven update to the forecast.

Globsnow's performance might be improved by the relaxation of the one-layer assumption, and here the effect of this relaxation on simulated ΔT_Bs is assessed based on realistic snow profiles obtained from the snowpits of the National Aeronautics and Space Administration Cold Land Processes Experiment (NASA CLPX).

Section 2 reviews the historical methods of snow mass estimation, including separate estimates from snow models, ground stations and PM. Section 3 introduces the principles behind the assimilation of passive microwaves and details Globsnow, identifying its simplified snow stratigraphy as a possible source of error and suggesting that layering might be included in a future scheme.

Section 4 describes the experiments, which attempt to assess the effect of realistic snow layering on radiative transfer. Coincident satellite PM measurements are compared with HUT simulations at CLPX, confirming that HUT simulations are close to observed values. The effect of including or neglecting detailed snow layering is assessed by comparing ΔT_B simulations when snowpits are resampled to different layering profiles. Profiles include between 1 and 5 layers, along with an N-layer case where layers are prescribed based on the snowpit measurement resolution. The N-layer case is taken as truth, and the difference in simulated ΔT_Bs for fewer-layer models relative to this truth allows statistical estimation of the bias and variance introduced through simplification of stratigraphy to fewer layers, which are reported as a function of snow depth and number of layers.

The results are related in Sect. 5 and discussed in Sect. 6 where it is indicated that neglect of statigraphy may mean that Globsnow has unaccounted variance in its assimilation step. The results have relevance to a user who may use these to calculate variance introduced due to simplified stratigraphy, or alternatively may choose an optimal layering structure based on the criteria of computational expense and acceptable levels of variance. However, it is cautioned that these results are only derived for snow typical of that present during CLPX.

2 Current Snow Mass Estimation

2.1 General Circulation Models and Reanalyses

Without the global coverage of space-based remote sensing, alternative methods of snow mass estimation have relied on a combination of models and observations. Coupled GCMs are a modelling approach and have been used to estimate SWE climatologies for current conditions, and the spatial and temporal components of these climatologies have been explored by Clifford (2010) and Roesch (2006) among others.

However, due to the chaotic nature of the system, fully coupled models are only capable of estimating climatology and, in order to produce a time series corresponding to the real-world realisation of weather, regular assimilation of observational data is required.

As such a number of reanalysis products have been produced, coupling LSMs which simulate the snow cover with an atmospheric model. These reanalyses regularly assimilate observations of both the atmosphere and the land surface, although no fully coupled land–atmosphere reanalysis yet assimilates microwave radiances for the purpose of snow mass estimation. Instead, in situ synoptic station measurements of snow depth and estimates of SCA based on satellite data are used.

The full details of these reanalyses and their assimilation schemes are beyond the scope of this paper; the reader is directed to the references in Table 1, which details selected reanalyses and other gridded products which offer snow mass or snow depth.

A number of assessments of reanalysis performance in terms of snow variables have been undertaken. Khan and Holko (2009) noted that reanalyses performed well in much of the Aral Sea Basin, although there were underestimates of snow depth and SWE in mountainous areas. Betts et al. (2009) determined that both the European Centre for Medium Range Weather Forecasts (ECMWF) 40 year and Interim Reanalyses (ERA-40, ERA-Interim) suffer from early snow melt out. Meanwhile, Clifford (2010) reported the spatial and temporal characteristics of different approaches to snow mass estimation in more detail, and that the potential for future improvements remains clear. Improved modelling is one opportunity, with Salzmann and Mearns (2012) comparing SWE

Table 1 Summary of selected gridded time series of snow mass or depth

Name	Type	Start year	End year	Coverage	Resolution	Reference
ERA-40	Reanalysis	1957–1909	2002–2008	Global	T159	Uppala et al. 2005
ERA-Interim	Reanalysis	1989	Present	Global	T255	Dee et al., 2011
MERRA	Reanalysis	1979	Present	Global	0.5° × 0.66°	Rienecker et al. 2010
NCEP-CFSR	Reanalysis	1979	Present	Global	T382	Saha et al. 2010
JRA-25	Reanalysis	1979	2004	Global	T106	Onogi et al., 2007
ECMWF operational	Offline model	1987	Present	Global	T511	Drusch et al. 2004
CNC	Offline model	1979	1997	North America	0.3° × 0.3°	Brown et al. 2003
Kitaev et al.	Observations	1936	1995	Former Soviet Union	3° × 5°	Kitaev et al. 2002
Dyer and Mote	Observations	1960	2000	North America	1° × 1°	Dyer and Mote 2006
SNODAS	Model plus in situ observations	2003–2009-30	Present	Contiguous United States	1 km × 1 km	Carroll et al., 2001
SSMR/SSM/I	Microwave	1978–2011	2007–2005	Global	25 km × 25 km	Armstrong et al. 2005
AMSR-E	Microwave	2002–2006	1978–2011	Global	25 km × 25 km	Tedesco et al. 2004b

modelled by a number of higher resolution regional climate models (RCMs) with snow telemetry observations in the Upper Colorado River Basin. Though the RCMs remained too warm and dry with too little SWE, their simulations better matched observations than the original reanalysis. However, biases remained even with these more computationally expensive models.

2.2 Products Which Prioritise Snow

2.2.1 Surface Observations Only

Outside of reanalyses that attempt to produce complete time series of land and atmosphere properties, a number of snow-specific products have been developed. The simplest approach is to grid weather station snow depth records as performed by Dyer and Mote (2006) in North America and Kitaev et al. (2002) in the Former Soviet Union (FSU). In Kitaev et al.'s work, station number varied from 2 to 25 per $3° \times 5°$ grid square and snow's spatial variability within such areas means that large uncertainties are associated with such sparse measurements. Chang et al. (2005) estimated that across the Northern Great Plains, 10 measurements were required per 1° cell to reduce sampling error of snow depth to ± 5 cm, equivalent to a station density between 6 and 75 times higher than available to Kitaev et al. Furthermore, the sampling distribution of snow stations was found by Brasnett (1999) to be biased to low elevations.

Post hoc assessment of snow mass is possible using river discharge data, although this approach suffers from large uncertainties due to unknowns related to inter-annual terrestrial water storage, periods of river ice and non-snow contributions. This approach has allowed attempts to test seasonal estimates of snow mass in some basins (e.g., Grippa et al. 2005; Rawlins et al. 2007; Yang et al. 2007) and to provide evidence in conjunction with other snow products for intensification of the Arctic hydrological cycle in response to global warming since 1950.

2.2.2 Land Surface Models Assimilating In Situ Observations

Simple areal averaging of snow depth observations cannot account for variation in areas between point measurements, which can be driven by different elevation, meteorological regime or land surface category. LSMs featuring a snow component are in principle able to account for these effects and, furthermore, these models may assimilate measured snow depths when available to improve the analysis.

Brown et al. (2003) used the Canadian Meteorological Centre's analysis scheme developed by Brasnett (1999) to generate a gridded time series of North American snow depth and SWE. A simple snow model was driven by meteorological data from the EC-MWF 15-year Reanalysis (ERA-15), with assimilation of 8,000 snow measurements per day from the USA and Canada. This method relies on relatively intensive daily measurements, for which the authors noted that availability drops off rapidly poleward of 55° N.

A global estimate is published by the ECMWF using a similar approach, and a summary and assessment is provided by Drusch et al. (2004). They note that the observational stations are biased towards lower latitudes and lower elevations and that, without assimilating remotely sensed information on snow-covered area, there are disagreements between the estimated snow-covered areas, and from the Interactive Multisensor Snow and Ice Mapping System (IMS) described in Ramsay (1998).

2.3 Remote Sensing of Snow Mass

The approaches discussed in Sects. 2.1 and 2.2 have been used to estimate snow climatologies and detect climatic changes, but their continued reliance on intensive in situ measurements leaves large uncertainties in some regions. This justifies continued development of remote sensing products, which can provide global coverage for improved estimates of snow mass where station density is insufficient.

Beginning with the work of Frappart et al. (2006) and continuing with Niu et al. (2007) and others, the Gravity Recovery and Climate Experiment (GRACE) gravimetry mission has been used to estimate snow mass based on observing changes in Earth's gravitational field. GRACE responds directly to gravitational changes, suggesting that it should be well suited to retrievals of deep snow or snow in forested areas where traditional remote sensing has to see 'through' the trees. However, further modelling is required to control for other changes in mass of the land surface associated with, for example, other forms of terrestrial water storage. Additionally, GRACE is not suitable for high-resolution measurement with Frappart et al. (2006)'s reported resolution being 660 km. Finally, it is not yet appropriate for assessing long-term changes as the GRACE satellites were only launched in 2002.

As such, efforts for the remote sensing of snow mass have typically focussed on the passive microwave regime, using frequencies near 19 and 37 GHz, for which there has been continuous near-global coverage since the launch of the Scanning Multichannel Microwave Radiometer (SMMR) on Nimbus-7 in late 1978. Many snow products typically utilise the Special Sensor Microwave Imagers (SSM/I) (e.g., Tedesco et al. 2004a) and/or the Advanced Microwave Scanning Radiometer-Earth Observing System (AMSR-E) (Tedesco et al. 2004b).

When observing a typical snowpack, the majority of radiation measured at these wavelengths will have originated from the ground surface, with scattering within the snow the dominant loss mechanism. This scattering is frequency dependent and increases with the quantity of snow, allowing a determination of SWE from the difference between the brightness temperatures in these two channels.

Figure 1 shows simulations of the brightness temperatures over a snowpack at 18.7 and 36.5 GHz horizontal polarisations viewed at 53°. Snow is assumed to be a homogeneous single layer with properties based on those typical of Colorado snowpacks of under 120 cm depth discussed in Davenport et al. (2012). As the amount of snow increases up to 500 mm SWE, the brightness temperature at both frequencies falls, but it falls more quickly at the higher frequency.

By considering the difference in brightness temperatures between the two frequencies, the effect of absolute temperature change is reduced and this led to the simplest approach to SWE retrieval, often called the Chang Algorithm, which was originally developed for SMMR (Chang et al. 1987), a general variant of which is as follows:

$$SWE = A(T_{\mathrm{B}}19H - T_{\mathrm{B}}37H + B) = A(\Delta T_{\mathrm{B,H}} + B) \tag{1}$$

where A and B are constants depending on the exact frequency of the channel and snow properties, $T_{\mathrm{B}}19H$ and $T_{\mathrm{B}}37H$ are the recorded brightness temperatures at the available channels nearest 19 and 37 GHz horizontal polarisation. Figure 1 shows this equation fit to the first 100 mm SWE, and for this snow, the values are $A = 2.54\ \mathrm{mm_{SWE}}\ \mathrm{K}^{-1}$ and $B = 3$ K.

Passive microwave measurements offer the advantage of being largely independent of illumination conditions, precipitation or cloud cover, allowing night time measurements when temperatures are likely to be lower and moisture within the snow is more likely to have refrozen. However, the range of values which can be reliably sensed is limited at the lower end by sensor precision, and, at higher values of SWE, the signal saturates (displayed

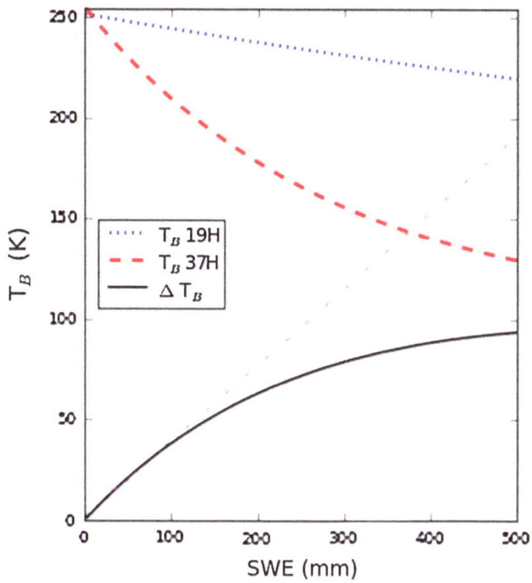

Fig. 1 Simulated brightness temperature for a homogeneous snowpack as a function of snow water equivalent (SWE) at the Advanced Microwave Scanning Radiometer for Earth Observing System (AMSR-E) channels near 19 and 37 GHz horizontal polarisation. Snowpack properties are averages from Cold Land Processes Experiment (CLPX) snowpits where depth was <120 cm. Density of 170 kg m^{-3}, grain size of 0.53 mm, snow temperature −4.6 °C and ground temperature −1.5 °C. The *upper dashed lines* show the brightness temperature at 19H and 37H as labelled in the legend (*centre left*), the *solid line* is the difference in brightness temperature between the two (see Eq. (1)), offset by 3 K to ensure that the value is zero when no snow is present. The *straight dash-dotted line* is the best fit to the first 100 mm of SWE

in Fig. 1 as a flattening of the solid line). The limiting value of this saturation depends on snow properties and was reported by Foster et al. (2005) to be 0.8 m depth.

In reality, the saturation level depends on the snow properties and on the definition of saturation. For the snow properties shown in Fig. 1, saturation could be determined as the point at which the SWE inverted from the brightness temperature difference using a linear fit diverges by more than 10 % away from the true value. In this case, this occurs at a SWE of 143 mm, equivalent to a depth of 79.4 cm which matches well with the value provided by Foster et al. (2005).

An alternative definition of saturation accepts that a more complicated function may be used to map observed brightness temperature difference to SWE. Under this definition, saturation is the point at which the signal is no longer sufficiently sensitive to SWE. This can be determined as the point at which error in inverted SWE, Δ SWE, exceeds some acceptable value Δ SWE_a. The sensitivity of brightness temperature difference $\Delta T_\text{B} = T_\text{B}19H - T_\text{B}37H$ to SWE is $\frac{\partial \Delta T_\text{B}}{\partial \text{SWE}}$ where the partial derivatives indicate that ΔT_B is not only a function of SWE. If the brightness temperature difference observation has a precision $\Delta(\Delta T_\text{B,obs})$, then the SWE at which saturation occurs is defined from:

$$\left(\left.\frac{\partial \Delta T_\text{B}}{\partial \text{SWE}}\right|_\text{SWE}\right)^{-1} \Delta\left(\Delta T_\text{B,obs}\right) > \Delta\ \text{SWE}_\text{a} \tag{2}$$

Therefore, the saturation value depends on the acceptable uncertainty in SWE, the precision of the observing system and the properties and uncertainties associated with the sensitivity of ΔT_B to changes in SWE. This result holds for ideal snow and demonstrates the possibility that the saturation value is not globally fixed. For example, with an uncertainty of 2 K in observed brightness temperature difference and a SWE error threshold of 10 mm for the uniform snowpack in Fig. 1, the saturation SWE rises to 195 mm, equivalent to 108 cm depth.

However, in real situations, this is non-trivial to determine, as simulating the radiative transfer of actual snow profiles leads to additional uncertainty in $\Delta T_B(\text{SWE})$ which must also be considered. Given an optimistic assessment of our ability to simulate radiative transfer in snow and observational uncertainties, this saturation threshold will be assumed to limit the utility of passive microwave measurements to snowpacks of <180 mm SWE or under 1 metre in depth.

Davenport et al. (2012) showed clearly that the functional form of $\Delta T_B(\text{SWE})$ depends on the microstructural properties of the snow. The physical basis of ΔT_B's sensitivity to microstructural properties can be explored by assuming that the snow is a collection of spheres in each other's far fields, for which the single scattering properties can be calculated from Mie theory. In particular, the single scattering albedo is actually a function of the size parameter $x = \frac{2\pi r}{\lambda}$ where r is the radius of the scatterer and λ the wavelength. Single scattering properties for non-spherical grains have also been determined (Teschl et al. 2010), although radiative transfer models (RTMs) generally assume sphericity.

Critically, it is the ratio of scatterer size to wavelength, which determines the single scattering parameters, and so the retrieved signal is strongly affected by the size of the scatterer as well as the wavelength of the light. Figure 2 shows the brightness temperature differences assessed for snow with scatterer diameters ranging from 0.2 to 1.0 mm in 0.2 mm increments, and Table 2 shows how the Chang sensitivity depends strongly on this value. Grains of 0.2 mm diameter are typical of fresh snowfall and 1.0 mm of moderate-sized depth hoar at the bottom of snow layers, although larger and smaller sizes do occur. From Fig. 2 and Table 2, it can be seen that the saturation value of the signal will also depend on the properties of the snow.

The Mie approach provides useful physical insight about scattering of radiation in snow, but any observations of the structure of real snowpacks show that snow is a complex, porous medium and as such these microstructural parameters are accounted for in a number of ways, such as specific surface area (SSA), optical grain size and correlation length. The optical grain size approximation comes from modelling the snow as a collection of spheres in each other's far fields, with the optical grain size defined as the spherical grain size required to reproduce the optical properties of the real snow. This size can vary with wavelength and with grain shape (Macke et al. 1996).

Grenfell and Warren (1999) found that if the optical properties of non-spherical snow were modelled using spheres, then spheres with the same SSA best matched the optical properties of the snow, for which the diameter can be determined from other properties using;

$$D_q = \frac{6M}{\rho S} \tag{3}$$

where M is the total snow mass in a selected volume, ρ the snow density and S the total ice–air interface area.

Correlation length is defined as the gradient of the spatial autocorrelation at a displacement of zero, and like specific surface area is defined independently of grain shape. It can be calculated from mean intercept lengths, by numerical analysis of the autocorrelation

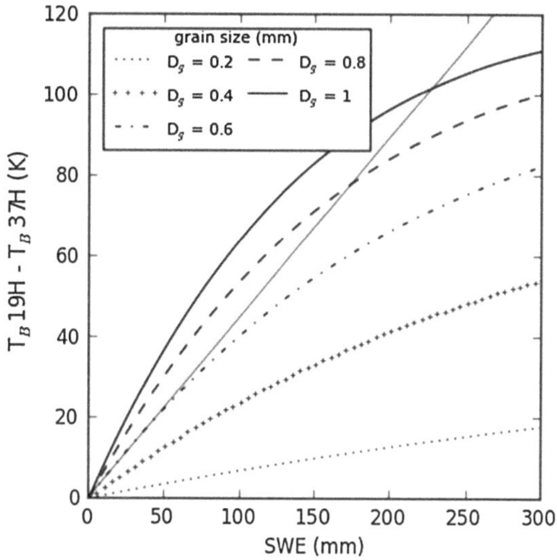

Fig. 2 Brightness temperature difference as a function of snow water equivalent (SWE) for the same snow properties as in Fig. 1, except that grain diameter is varied from 0.2 to 1.0 mm in 0.2 mm increments. The legend (*top left*) indicates which line style refers to each grain size value. The *shaded straight line* is the linear best fit to the first 50 mm SWE worth of the central grain size value

Table 2 Chang sensitivity calculated from the trend in brightness temperature difference for the first 100 mm of snow water equivalent (SWE) for grain diameters of 0.2–1.0 mm

Grain diameter (mm)	Chang sensitivity (mm_{SWE} K^{-1})
0.2	14.64
0.4	4.18
0.6	2.47
0.8	1.85
1.0	1.55

The snow and surface properties used are those from Fig. 1

gradient, or by fitting to an assumed exponential autocorrelation function. There are also functions for converting from correlation length to optical grain size (e.g., Mätzler 2000; Wiesmann et al. 2000; Mätzler 2002).

3 Assimilation of Passive Microwave Observations to Improve Snow Mass Estimation

3.1 Assimilation of Passive Microwave Brightness Temperatures

Microwave-only algorithms retain large uncertainties due to issues with forest coverage and changes in the scattering properties of snow, driven primarily by the snow's microstructure. However, if these effects could be quantified, then an assimilation scheme would be able to extract information from the retrieved brightness temperatures to improve a snow analysis.

Sun et al. (2004) suggested a scheme, which forecasts the snow cover using a LSM before assimilating SWE estimated from PM. They performed an experiment using a synthetic truth generated by the LSM versus two alternative model runs with strongly perturbed initial conditions, one of which assimilated observations from the truth using a Kalman Filter, and one which was left to run as an open loop. They demonstrated that the assimilation scheme returned the analysis state close to the truth within 1 week, and then a later study by Dong et al. (2007) simulated SWE in North America with and without the assimilation of the SMMR SWE product. Assimilation of the SMMR product improved the analysis where SWE < 100 mm, provided the SMMR product was quality controlled.

However, this approach did not account for the changes in snow microstructure, which affect the scattering, as the SMMR-based SWE product is based on a variant of the Chang Algorithm. A more comprehensive approach is detailed in Durand and Margulis (2006), who describe an Ensemble Kalman Filter (EnKF) approach of assimilating microwave brightness temperatures.

The Kalman Filter approach consists of two steps to produce an analysis of the variables of interest, which will be some vector $\mathbf{x}^{\mathrm{a}}$ whose components represent snow properties such as the density and grain size of each snow layer. In the first step, the analysis $\mathbf{x}^{a}_{k-1}$ from the previous time step t_{k-1} is propagated using a model M to produce a forecast $\mathbf{x}^{f}_{\mathrm{k}}$:

$$\mathbf{x}^{\mathrm{f}}(t_k) = M_{k-1}[\mathbf{x}^{\mathrm{a}}(\mathrm{t}_{\mathrm{k}-1})] \tag{4}$$

This forecast is then updated with reference to observations:

$$\mathbf{x}^{\mathrm{a}}(\mathrm{t}_k) = \mathbf{x}^{\mathrm{f}}(t_k) + \mathbf{K}_k\left(\mathbf{y}_k - H_k\left[\mathbf{x}^{\mathrm{f}}(t_k)\right]\right) \tag{5}$$

where $\mathbf{y}_k$ is the observation vector and H_k is an operator, which converts the state vector into an equivalent observation. In the case of snow remote sensing, it is some model of snow's radiative transfer that converts the known snow properties from the state vector into a vector of observable brightness temperatures or some combination thereof. $\mathbf{K}_k$ is the Kalman Gain, which acts as the weighting function and depends on the error covariances of the forecast $\mathbf{P}^{\mathrm{f}}$ and the observations $\mathbf{R}_k$:

$$\mathbf{K}_k = \mathbf{P}^{\mathrm{f}}(t_k)\mathbf{H}_k^T\left[\mathbf{H}_k\mathbf{P}^{\mathrm{f}}(t_k)\mathbf{H}_{\mathrm{k}}^T + \mathbf{R}_{\mathrm{k}}\right]^{-1}. \tag{6}$$

Here $\mathbf{H}_k$ is the linearised approximation of the observation function H_k. It can be seen that as observational error decreases, the Kalman gain increases and greater weight is placed on the observations. The forecast error covariance $\mathbf{P}^{\mathrm{f}}(t_k)$ consists of the model error covariance $\mathbf{Q}_{k-1}$ and the error covariance introduced due to errors in the previous step's analysis, $\mathbf{P}^{\mathrm{a}}(t_{k-1})$:

$$\mathbf{P}^{\mathrm{f}}(t_k) = \mathbf{M}_{k-1}\mathbf{P}^{\mathrm{a}}(t_{k-1})\mathbf{M}_{k-1}^T + \mathbf{Q}_{k-1} \tag{7}$$

where $\mathbf{M}_{k-1}$ is the linearised approximation of the forecast operator M. Estimating this component of the error covariance can be enormously computationally expensive, leading to the attraction of the EnKF where a model ensemble allows the generation of statistics to approximate $\mathbf{M}_{k-1}\mathbf{P}^{\mathrm{a}}(t_{k-1})\mathbf{M}_{k-1}^T$ and therefore allow the calculation of the Kalman Gain.

The Kalman Gain is also required to calculate the error covariance of new analysis $\mathbf{P}^{\mathrm{a}}(t_k)$, which is reduced by the assimilation of observations relative to the forecast:

$$\mathbf{P}^{\mathrm{a}}(t_k) = (\mathbf{I} - \mathbf{K}_k\mathbf{H}_k)\mathbf{P}^{\mathrm{f}}(t_k) \tag{8}$$

Durand and Margulis (2006) tested this approach with a synthetic experiment of snowpack progression in the USA. The system truth was taken to be a single model run with forcing perturbed by doubling the precipitation and adding autocorrelated noise to mimic known issues of gauge undercatch. Synthetic passive microwave observations at SSM/I or AMSR-E frequencies were simulated by the Microwave Emission Model of Layered Snowpacks (MEMLs, Wiesmann and Mätzler (1999)) corrupted with 2 K white noise, and synthetic albedo observations were taken from the truth run with 5 % white noise applied.

Forecasts were generated by an ensemble of 100 LSM replicates with perturbations applied to forcing and model parameters, which allowed the mean forecast state and the forecast error covariance to be determined from the ensemble statistics. Synthetic albedo observations were assimilated daily at 1 pm and passive microwave observations at 1 am to mimic MODIS and AMSR-E overpass times.

The regular assimilation of SSM/I frequencies alone significantly reduced both bias and root mean square error (RMSE) of SWE by approximately 85 % relative to the open-loop simulation. The EnKF approach also allowed an assessment of the contribution of each channel, which indicated that the majority of the SWE improvement occurred due to the assimilation of the 37-GHz channel at both polarisations. The 89-GHz channel appeared to marginally worsen the SWE analysis by nudging it away from truth; however, it significantly improved the grain size analysis, which was vital for the brightness temperature simulations.

Having demonstrated the assimilation approach using a synthetic experiment, the later work of Durand et al. (2008) used data from the University of Tokyo's Ground Based Microwave Radiometer (GBMR-7) and snowpits at NASA's Cold Land Processes Experiment (CLPX) to test the performance of the MEMLS radiative transfer model. Furthermore, they were able to identify accuracy criteria for the snow state variables. They determined that simulated optical grain size should be accurate within ± 0.045 mm and the density of melt-refreeze layers within ± 40 kg m^{-3} in order for predicted brightness temperature errors to be small enough that the assimilation procedure improves the analysis.

Further work has considered the effect of spatial scaling on the analysis, with different spatial resolutions in LSMs and microwaves explored in De Lannoy et al. (2010), while Andreadis et al. (2008) discuss how to account for snow's spatial variability in an assimilation scheme.

3.2 Globsnow

3.2.1 Methodology

The European Space Agency (ESA) Globsnow project's aim is 'production of global long term records of snow parameters intended for climate research purposes on hemispherical scale' (Finnish Meteorological Institute 2012). The Globsnow SWE product is a system where the prior state is estimated from field observations of snow depth, with updates related to the satellite-observed brightness temperature difference ($\Delta T_{B,V}$) at vertical polarisation between channels near 19 GHz (T19 V) and 37 GHz (T37 V).

The use of a brightness temperature difference reduces the sensitivity of the satellite observations to absolute temperatures; if non-snow surfaces are in the field of view and they have the same emissivity at both 19 and 37 GHz, then their effect on the measured brightness temperature difference is dependent only on the area they cover and is independent of their temperature.

Globsnow produces maps of SWE across the Northern Hemisphere on a 25-km Equal Area Scalable Earth (EASE) grid, with areas defined as too watery (>50 % open water) or too mountainous (standard deviation of elevation >200 m) masked out. Largely based on the approach of Pulliainen (2006), its methodology is explained in detail in Takala et al. (2011) and proceeds as follows:

1) The forecast snow depth map is generated by ordinary kriging between in situ snow depth observations provided by the ECMWF and SCCONE (Snow Cover Changes Over Northern Eurasia, Kitaev et al. 2002), and the in situ measurements are given an assumed variance of 150 mm^2 based on the comparison with coincident snow surveys. This forecast map at time t contains the a priori snow depth $\hat{D}_{\text{ref,t}}$ and its variance $\sigma^2_{D,\text{ref,t}}$.

2) At each grid point where a snow depth observation exists, the Helsinki University of Technology radiative transfer model is used to simulate the brightness temperature difference ΔT_B = T19V–T37V. The effect of vegetation is included in the radiative transfer, dependent on forest cover fraction in Eurasia, or at 80 kg m^{-3} ha^{-1} stem volume in North America. A single snow layer of 0.24 g cm^{-3} density is assumed, and snow depth is taken from the in situ observation. Grain size is varied at each location i with the result obtained according to the cost function:

$$\min_{d_{0,i}}\left\{\left[\left(T19V_{\text{mod}}\left(d_{0,i},D_{\text{ref,i}}\right)-T37V_{\text{mod}}\left(d_{0,i},D_{\text{ref,i}}\right)\right)-\left(T19V_{\text{obs}}-T37V_{\text{obs}}\right)\right]^2\right\} \tag{9}$$

where $d_{0,i}$ is the grain size at the ith location, which is allowed to vary and $D_{\text{ref},i}$ is the locally measured snow depth. The final grain size (d_0) and its error variance ($\sigma^2_{\text{d0,t}}$) at each measurement location come from the ensemble of the nearest stations ($N = 6$).

3) A full grain size map with variances is generated by kriging between the point grain size estimates from step 2).

4) At each grid cell, the grain size value and an assumed constant density of 0.24 g cm^{-3} is used as input to the HUT radiative transfer model by varying the snow depth D_t to obtain:

$$\min_{D_t}\left\{\left(\frac{(T19V_{\text{mod}}(D_t)-T37V_{\text{mod}}(D_t))-(T19V_{\text{obs}}-T37V_{\text{obs}})}{\sigma_t}\right)^2+\left(\frac{D_t-\hat{D}_{\text{ref},t}}{\sigma_{D,\text{ref},t}}\right)^2\right\} \tag{10}$$

where the variance at time t, σ_t^2 is obtained from a Taylor expansion of $\mathrm{T_B}(D_t, d_{0,t})$ with respect to grain size, which leads to

$$\sigma_t^2=\left(\frac{\partial T_{\text{B}}(D_t,d_{0,t})}{\partial d_0}\right)^2\sigma^2_{d_0,t} \tag{11}$$

This variance provides the weighting of the microwave contribution, allowing a large correction to the forecast when the SWE sensitivity is high but introducing a large cost to microwave-based adjustments when the signal is saturated with respect to SWE but $\partial T_B/\partial d_0$ grows. This effect is seen in Fig. 2 as the increasing spread in simulated ΔT_{B} for different grain sizes.

3.2.2 Limitations of Globsnow

Globsnow was validated with independent in situ snow depth measurements from campaigns in the Former Soviet Union, Finland and Canada. RMSE values of <40 mm were found where SWE was below 150 mm, although errors increase for thicker snow. Assimilating the passive microwave data was found to improve on the forecast, thus demonstrating the utility of microwave retrievals.

Hancock et al. (2013) considered Globsnow and the Chang-based AMSR-E and SSM/I-only SWE products for the purpose of assessing LSMs. The Chang-based products were found to spike towards the end of the season, which was attributed to melt-refreeze cycles forming ice lenses which increase the effective grain size and cannot be accounted for in the Chang-based approach which assumes a static snow microstructure.

However, a number of questionable assumptions remain in the Globsnow approach. The assumption of constant density is not necessarily valid, as snow settles and increases in density during the season due to metamorphism and overburden (Anderson 1976), and variations in density or in the effect of vegetation are included through varying the grain size parameter which is an unphysical approach. Furthermore, the snow depth forecast is produced purely from interpolated observations, which as previously noted are biased towards low latitudes, low altitudes and clearings in forests. These biases could be accounted for by a LSM, which in addition to producing a snow depth forecast could also produce a forecast of the density and grain size.

Durand et al. (2009) showed that even a relatively simple land surface model coupled to a microwave emission model improved snow depth estimates once microwave brightness temperatures were assimilated. The later work of Brucker et al. (2011) and Toure et al. (2011) coupled the snow model Crocus (Brun et al. 1992) to the MEMLS radiative transfer model and found that point observations of microwave brightness temperatures at both H-pol and V-pol were generally well simulated. Brucker et al. (2011) noted that late season grain growth was not well modelled in Crocus, and Toure et al. (2011) indicated that ice lenses must be accounted for. On a larger geographical scale, Dechant and Moradkhani (2011) reported that assimilating brightness temperatures with the SNOW-17 snow model and a soil moisture model showed potential benefit for operational stream flow forecasting.

Naturally, increasing physical complexity leads to increased computational expense and computational expense is also affected by the number of layers in the snow model. Most land surface models typically limit the number of snow layers, with the ECMWF's Tiled ECMWF Scheme for Surface Exchange over Land (TESSEL) limited to one layer, whereas the Joint UK Land Environment Simulator (JULES) can run up to a user-defined number of layers, with new layers only introduced beyond certain thickness thresholds.

In reality, snowpacks almost always have distinct physical layers and this stratigraphic contrast can have important effects on the radiative transfer. Lemmetyinen et al. (2010) compared passive microwave measurements taken in situ with layered snow information and found that the simulated brightness temperature was affected by whether or not snow layering was included. These approaches used field-observed layer properties, but did not consider how the radiative transfer model would perform if provided with profile information as it would be output by a model.

This is assessed here through the experiments detailed in Sect. 4, where measured snow profiles from CLPX snowpits are resampled to differing layering structures. After a scene simulation experiment to confirm that the HUT radiative transfer model is able to reproduce observations within acceptable uncertainties, the simulated $\Delta T_{B,V}$ values for each snowpit when resampled to different layering structures are compared. This comparison across a large number of snowpits allows estimation of the bias and variance introduced when the layering structure is simplified.

4 Methods

4.1 The Cold Land Processes Experiment (CLPX) Resampled Snowpits

The CLPX dataset provides snow profiles from a large number of snowpits over four intensive observation periods (IOPs). Two of these periods, IOP3 and IOP4, coincide with measurements of microwave brightness temperature from both SSM/I and AMSR-E and as such these periods are assessed here. Snowpit profiles provided snow properties for HUT radiative transfer simulations, which were compared with the satellite retrievals to confirm HUT's applicability.

Figure 3 shows the CLPX study area, largely in Colorado, USA, and the snowpits were within the three mesoscale study areas (MSAs) each of which is approximately 25 by 25 km, the size to which AMSR-E and SSM/I 19 and 37 V brightness temperatures are resampled.

For the first $\Delta T_{B,V}$ simulations, snow profiles were produced with layers of 10 cm thickness, limited by the spatial resolution of the temperature and density measurements. These profiles are referred to here as the N-layer case and are assumed to represent truth. In reality, snow layers are of irregular depth, but LSMs often feature prescribed layer depths and the aim of this experiment is to compare LSM-like outputs.

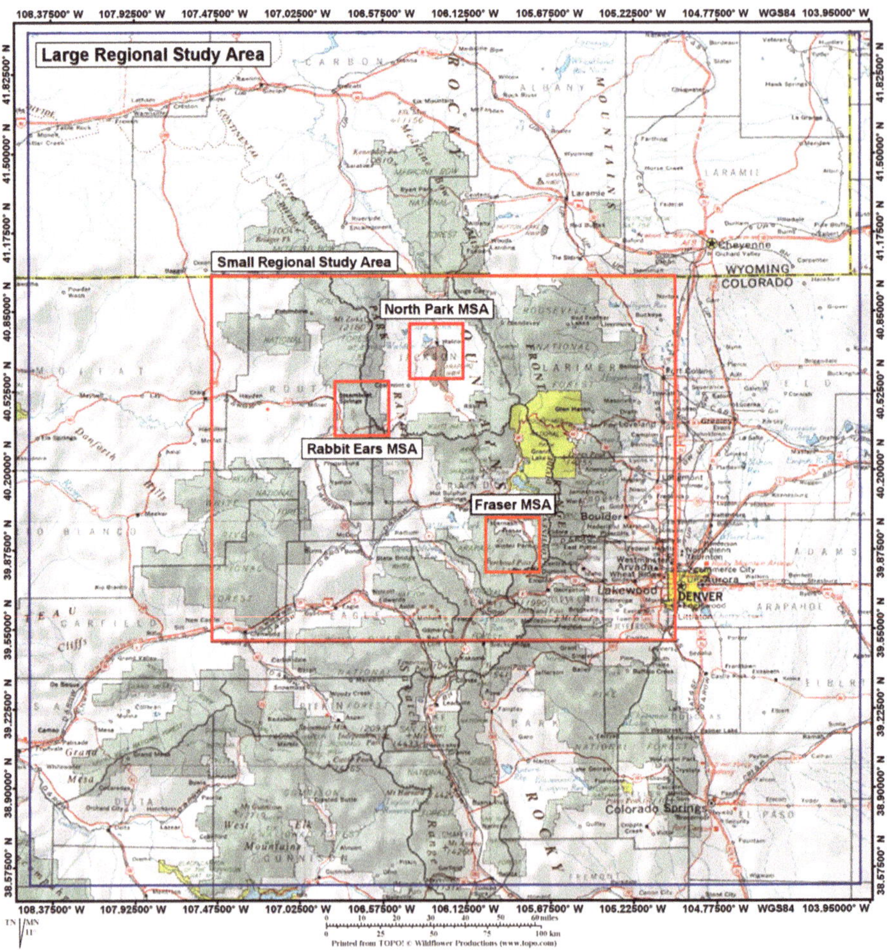

Fig. 3 Map of Cold Land Processes Experiment (CLPX) area showing the mesoscale study areas (MSAs) in which the snowpits were dug. (after http://www.nohrsc.nws.gov/~cline/clpx.html)

Density and temperature were drawn directly from the field measurements, while the mean of the minimum and maximum extent of the average common grain was taken as the grain size, following Davenport et al. (2012). Grain sizes were reported by the observers by stratigraphic layer, rather than at 10-cm intervals, and these were mass weighted onto the 10-cm profile.

As LSMs commonly feature a thin top layer to improve numerical handling of the surface energy exchange, snowpits were resampled from the CLPX observational data with and without a 2-cm surface layer. This did not affect any of the key results, and so arbitrarily the case with a 2-cm surface layer is presented.

The snowpits were then resampled to profiles of 1–5 equally sized layers subject to a minimum layer size of 10 cm, with the mass-weighted grain size, temperature and density from the relevant observational layers applied to each of the resampled layers. Figure 4 illustrates sample layering profiles, where the 2-cm top layer is excluded from the layer count, and for a snowpit of depth d, each of the n main layers is of depth $(d - 2)/n$ cm. The minimum layer size criterion meant that, for example, a snowpit of 12 cm depth would be identical in all layer cases and consist of a single 10-cm layer and the top 2-cm layer, even in the 5- or N-layer cases.

This approach attempts to mimic a depth structure that might be output by an LSM. A key feature is the prescribed layer thicknesses, as each layer depth can be determined uniquely from the total snow depth, thus removing the need for layer thickness components in the snow state vector. However, it is not necessarily representative of an individual LSM snow scheme, as a variety exists, and rather than select some arbitrary combination of layer sizes for each of the 1- to 5-layer schemes, a more simplistic approach was adopted.

Ground surface temperature was taken to be the temperature measured at 0 cm height. Missing data were linearly interpolated, or if they were at the top or bottom of the pack, then the nearest neighbouring value was used. If too many data were missing for this interpolation, then the pit was removed from the analysis.

4.2 Comparison: Layered HUT Scene Simulation Versus Observations

The HUT performance was first assessed by simulating the scene brightness temperatures based on the snowpit information, and comparing these simulations with satellite

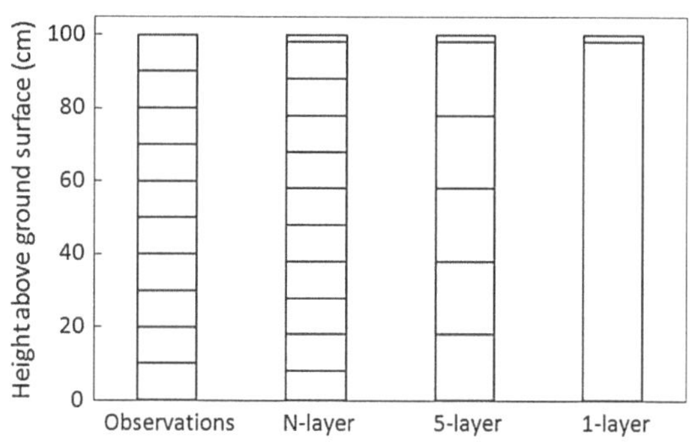

Fig. 4 Example of the how snowpit data were restructured. The *left-hand bar* represents the observed profile where depth and temperature are recorded for each 10 cm of the snow. The N-layer resampling maintains 10 cm layer thicknesses but adds a 2-cm interaction layer at the surface, as is common in a number of land surface models' snow schemes. The other layering schemes apply a 2 cm top layer and then evenly split the remaining snow depth, with density, snow and grain size mass-weighted according to the observations. All layer structures from 1 to 5 inclusive were calculated, but only 1 and 5 are shown here for simplicity

retrievals. The multilayer implementation of the HUT model (Lemmetyinen et al. 2010) simulated the brightness temperatures at 18.7 and 36.5 GHz vertical polarisation for each snowpit with each of the layering structures described in Sect. 4.1. The brightness temperature difference

$$\Delta T_{\mathrm{B,V}} = T_{\mathrm{B}}19V - T_{\mathrm{B}}37V \tag{12}$$

was considered throughout, for consistency with the Globsnow product. This offers the further advantage of being less sensitive to errors in ground or exposed vegetation temperatures.

In addition to the HUT simulations, a Chang algorithm output based on Eq. (1) and using the CLPX coefficients reported in Sect. 2.3 was produced for each snowpit.

In each of the CLPX mesoscale study areas (MSAs), the snowpits were assumed to be representative of the actual snowpack, such that the mean $\Delta T_{\mathrm{B,V}}$ of all of the snowpits within the MSA represents the $\Delta T_{\mathrm{B,V}}$ contribution of the snow within that MSA. The scene brightness temperature has contributions from snow, open ground and vegetation over snow.

$$\Delta T_{\mathrm{B,V,scene}} = A_{\mathrm{open}} \Delta T_{\mathrm{B,V,open}} + (1 - FF) A_{\mathrm{snow}} \Delta T_{\mathrm{B,V,snow}} \tag{13}$$

where A is the fractional area of the pixel that is either open or snow covered, and FF is the forest fraction. With the assumption that $\Delta T_{\mathrm{B,open}} = 0$, the equation becomes

$$\Delta T_{\mathrm{B,V,scene}} = (1 - FF) A_{\mathrm{snow}} \Delta T_{\mathrm{B,V,snow}} \tag{14}$$

Since the distribution of the snowpit properties is assumed to match the distribution of the snow within the scene, then the brightness temperature difference of the snow should be equivalent to the average brightness temperature difference of the N snowpits.

$$\Delta T_{\mathrm{B,V,scene}} = (1 - FF) \frac{A_{\mathrm{snow}}}{N} \sum_{i}^{N} \Delta T_{\mathrm{B,V,i}} \tag{15}$$

where $\Delta T_{\mathrm{B,V,i}}$ is the brightness temperature difference between 18.7 and 36.5 GHz at vertical polarisation for the ith snowpit.

Snow properties were assumed to be static throughout an Intensive Observation Period (IOP) such that all measurements within each IOP could be used in the same analysis.

The fractional area of snow for each IOP and for each MSA was estimated by using the 8-day maximum extent snow cover map from MODIS, taken as the fraction of snow-covered area divided by the total non-cloud-covered area. This offers the advantage of minimising the effect of cloud cover, although can provide inconsistent results if significant snowfall or melt occurs during the 8 days. The MODIS product is at 500 m spatial resolution, so features 2,500 pixels within each 25-km passive microwave grid point.

Forest cover was estimated for each MSA using QuickSCAT data available from Nilsson (2003) and the forest correction factor applied individually for each MSA. More complex forest correction approaches exist, but are not adopted here.

Six sets of simulated scene $\Delta T_{\mathrm{B,V}}$ values were produced, for each of the layering structures (1- to 5-layer plus the N-layer truth), and these were compared with SSM/I and AMSR-E values, where all measurements within a day of each IOP period were recorded. It should be noted that for IOP4, many snowpacks reported temperatures around the melt point, suggesting the presence of liquid water, which acts to reduce the brightness temperature difference through greater absorption and emission at both wavelengths. However, the percentage of snowpits that were noted by fieldworkers as 'wet' in the metadata did not exceed 3 % in either period.

4.3 Comparison: *N*-layer Versus Fewer Layers of Stratigraphic Information

Using the same resampled snowpit data to represent realistic profiles as might be output by an LSM, the $\Delta T_{B,V}$ values simulated by the 1- to 5-layer simulations were assessed relative to the N-layer simulations, which were assumed to be truth.

Uncertainty introduced into the $\Delta T_{B,V}$ by simplification of the model layering was determined from the difference between outputs for each of the 1- to 5-layer models versus the N-layer model. Bias and standard deviation of these residuals is reported in Sect. 5.3 for each of the simpler models as a function of the model layer thickness.

5 Results and Analysis

5.1 Snow Properties at CLPX Sites

Table 3 summarises the main snow properties recorded from the snowpits and MODIS data, including the number of relevant snowpits once those with insufficient data were deleted. Notably, the snow in IOP4 during March was thicker than during IOP3 in February, although the snow cover fraction had declined from universal coverage to around 80 %.

It should be noted that the average depth and SWE is not necessarily a good representation, as the distribution of snowpit values is not symmetric, with a bias towards thin pits in IOP3 and a bimodal distribution in IOP4, with a number of pits showing thin snow (<50 cm) and thicker snow ($\sim$200 cm). The overall distribution is shown in Fig. 5 although thicker snow predominated at Rabbit Ears and Fraser and thinner snow at North Park. Additionally, IOP3 saw generally cooler snow (−3.6 vs. −2.2 °C) and marginally smaller average grain sizes (0.57 vs. 0.60 mm).

5.2 Simulated Scene Brightness Temperatures

Figure 6 shows the simulated scene $\Delta T_{B,V}$ from Eq. (15), using the *N*-layer and 1-layer HUT model compared with AMSR-E and SSM/I retrievals. Here, the scene is represented by the average of all 3 MSAs. The difference between the *N*-layer and 1-layer simulations is minimal (0.01 K in IOP3, 0.40 K in IOP4) compared with the difference between simulations and observations, of approximately 3 K in IOP3 and 2 K in IOP4.

Table 4 shows the brightness temperature difference simulated using different model layering profiles versus the observations. There is a negligible difference in the mean simulated by different layering profiles. The overall Chang estimates are close to observations at IOP3, but are too high during IOP4. The Chang algorithm's poorer performance at individual MSAs (RMSE = 17 K) versus HUT (RMSE = 8 K) is hidden by the averaging over the 3 MSAs. During IOP3, use of the Chang algorithm results in a large $\Delta T_{B,V}$ overestimate at Rabbit Ears MSA, which is counteracted by a large underestimate at North Park MSA. During IOP4, a very large (32 K) overestimate by the Chang algorithm due to saturation in the deep snowpits is partially offset by a 13-K underestimate at North Park.

The largest contributor to the HUT RMSE was due to a large underestimate during IOP4 at North Park, where simulated $\Delta T_{B,V}$ were of order 1 K versus observed values of 14 K.

Table 3 Summary of snow input data for each intensive observation period (IOP) split by mesoscale study area (MSA)

	Dates	MSA	Snowpits	Mean depth (cm)	Mean SWE (mm)	Snow cover (%)
IOP3	20–25th February 2003	North Park	115	14.0	23.1	100
		Rabbit Ears	18	225.6	580	100
		Fraser	48	77.3	189	100
IOP4	26–30th March 2003	North Park	68	5.3	9.6	57
		Rabbit Ears	44	229.0	758	100
		Fraser	48	143.2	381	92

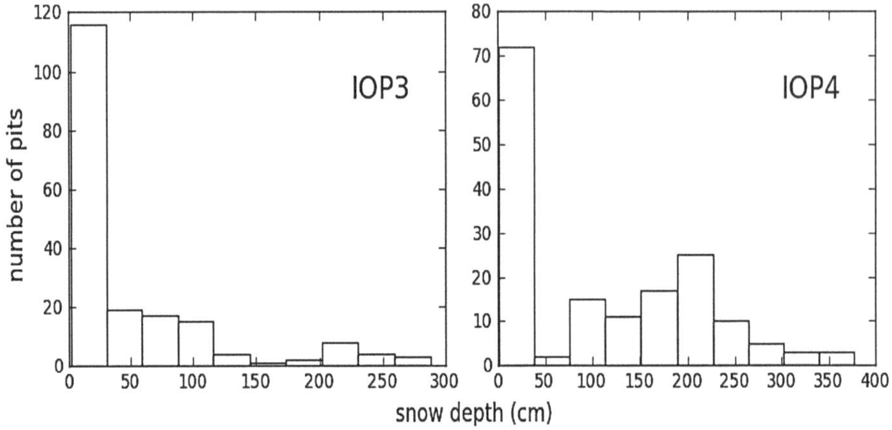

Fig. 5 Count of snowpits by depth for each intensive observation period (IOP)

This is consistent with the results of Davenport et al. (2012) for IOP4 and suggests that the average snowpack was deeper than that sampled by the snowpits. The majority of North Park snowpits occurred near roads and might therefore be biased towards thinner snow, and excluding IOP4 North Park results, the HUT RMSE drops from 8 to 6 K, while Chang increases from 17 to 18 K. The increase in Chang error is due to its North Park error being an underestimate and therefore limited in size to below the 14-K observation.

A number of assumptions contribute to the uncertainty in the $\Delta T_{B,V}$ estimates: static snow properties, dry snowpits, a simplified vegetation model and unbiased sampling of the real snow by the snowpits. Furthermore, other work (Picard et al. 2009) has found that effective microwave grain size does not necessarily match that estimated by field observations. In the light of these assumptions, the HUT simulations are largely consistent with the satellite observations and therefore justify continued assessment using this model.

The small differences in scene simulated $\Delta T_{B,V}$ values when using the *N*-layer or 1-layer HUT realisations does not necessarily indicate that they are equally valid, as the averaging over a large number of simulations will suppress the variance. In the Globsnow assimilation scheme, a single snow profile is taken to represent the snow at each grid point, rather than the average of 59 profiles used here, and for a single simulation the output $\Delta T_{B,V}$ depends on the layering structure provided to the profile.

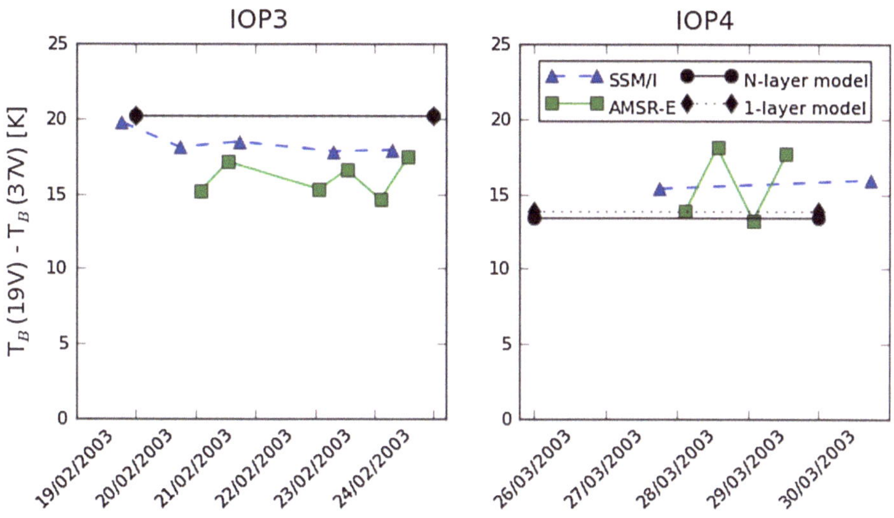

Fig. 6 Brightness temperature difference retrievals for special sensor microwave imager (SSM/I) (*triangles*), advanced microwave scanning radiometer-earth observing system (AMSR-E) (*squares*) and the average simulated snowpit data processed through the Helsinki University of Technology (HUT) microwave emission model. HUT simulations are provided for the N-layer case (*circles*) and for a single-layer case where all properties were averaged to one layer (*diamonds*). Legend in *top right* identifies marker shapes and line styles

Table 4 Average brightness temperature difference for each intensive observation period (IOP) as simulated by inverting the Chang algorithm, using different numbers of layers in the Helsinki University of Technology (HUT) microwave emission model, and the average retrievals for advanced microwave scanning radiometer-earth observing system (AMSR-E) and special sensor microwave imager (SSM/I)

	T_B(19H)−T_B(37H) (K)	
	IOP3	IOP4
Chang	18.58	24.72
1-layer	20.23	13.86
2-layer	20.19	13.18
3-layer	20.24	13.31
4-layer	20.21	13.35
5-layer	20.23	13.32
N-layer	20.24	13.46
AMSR-E	16.03	15.71
SSM/I	18.40	15.64

5.3 Differences Due to Layering Detail

It appears that for the CLPX pits, using the HUT radiative transfer model to generate the scene brightness, temperature difference improves the simulation relative to using the Chang algorithm approach. Furthermore, Lemmetyinen et al. (2010) reported that RMSE and bias were reduced at these microwave channels when HUT accounted for the multiple layering of snow, rather than using bulk averages in a single layer.

It was therefore assumed that the best simulation of $\Delta T_{B,V}$ was provided by the HUT model run with the *N*-layer realisation of the CLPX snowpit properties, and the performance of simplified layer models should be compared to this. Here the same brightness temperature simulations as in Sect. 5.2 were used, based on the CLPX snowpits as these represent realistic snow profiles.

Simulated $\Delta T_{B,V}$ was compared on a pit-by-pit basis, where the simulation for the Chang algorithm and the 1- to 5-layer realisations of HUT were compared to the *N*-layer output. Figure 7 demonstrates the comparison for the Chang algorithm and the 1- and 5-layer HUT output.

For these snow properties, the Chang estimate shows a low bias in simulated ΔT_B across much of the range, but a very large positive bias at high values of SWE. This is due to the CLPX snow properties differing from those assumed by Chang for the low values of SWE and by saturation of the signal at higher SWE values.

For the HUT simulations, bias is much smaller and scatter is reduced relative to the Chang estimate. This scatter is an estimate of the uncertainty introduced by simplifying the model to fewer layers. At the lowest brightness temperature differences, the scatter is zero as the minimum layer depth criterion ensures that for pits of depth <12 cm, the 1-layer and *N*-layer realisations are identical. The scatter is less prominent for IOP3 than for IOP4, possibly due to IOP4 featuring a larger number of thicker snowpacks with potentially more complex stratigraphy.

In an assimilation system, the snow model may output a single profile for each grid point, equivalent to a single point on the graphs in Fig. 7, and the deviation about the 1:1 line indicates that use of a simplified profile will lead to different simulated $\Delta T_{B,V}$ values relative to the best simulation provided by the *N*-layer realisation.

To quantify this deviation, the residuals from the 1:1 fit were considered, i.e. the values

$$\Delta T_{B,i} - \Delta T_{B,N} = (\Delta T_{B,V})_{ilayers} - (\Delta T_{B,V})_{Nlayers} \tag{16}$$

where $\Delta T_{b,i}$ is the brightness temperature difference simulated with *i* layers, *i* is an integer from one to five, and $\Delta T_{B,V,Nlayers}$ is the brightness temperature difference when a maximum of *N*-layers are included in the model profile. As throughout, $\Delta T_{B,V}$ represents the brightness temperature difference described in Eq. (1).

The bias and standard deviation varies with snow thickness and, as such, the residual in Eq. (16) was returned as a function of layer thickness in the 1- to 5-layer models and the results are shown in Fig. 8, where only snow pits of depths up to 100 cm are considered. Beyond this value, signal saturation would reduce the weighting applied to the microwave observational increment in an assimilation scheme, justifying the neglect of thicker pits.

It is apparent that, as layer thickness is increased, the average deviation from the *N*-layer simulation (which contains layers of 10 cm thickness) increases, and there is also an increase in bias, most likely due to layer boundary effects.

As the Globsnow approach is allowed to freely scale grain size at the snow depth observation locations, it is plausible that it accounts for this bias by artificially increasing the grain size depending on snow thickness. Changing the effective grain size is already known to account for variation in vegetation outside the model assumptions.

This would have a secondary level effect on the Globsnow assimilation scheme, by changing the variance of estimated grain size $\sigma_{d_0,t}$ in Eq. (11) if the ensemble of stations used for the averaging have different snow depths (and therefore different grain size biases).

The main concern for the assimilation scheme, however, is the random variance that is introduced, as this means that the simulated $\Delta T_{B,V,mod}$ in the assimilation cost minimisation function (Eq. (10)) should have additional variance associated with the neglect of snow stratigraphy. This variance is not accounted for in Eq. (11), which defines the weight given to the observational increment based on the estimated variance. Instead, it is calculated

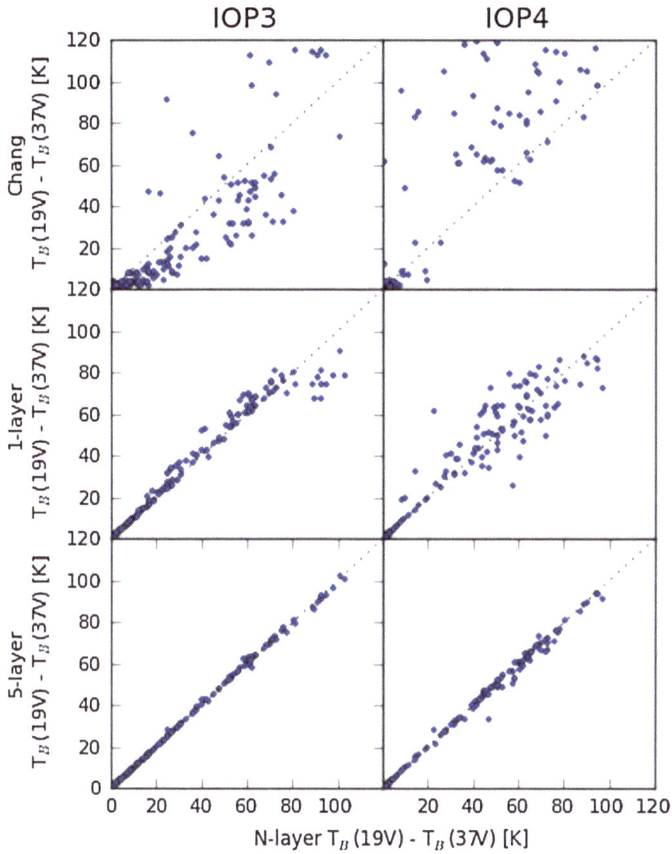

Fig. 7 Brightness temperature differences simulated for each pit during intensive observation period (IOP) 3 (*left*) and IOP4 (*right*). The ordinate in each case is the simulated brightness temperature difference using the *N*-layer model, and the abscissa shows the Chang output (*top*), 1-layer Helsinki University of Technology model (HUT) output (*centre*) and 5-layer HUT (*bottom*) output as labelled. The *dotted line* is the one-to-one correspondence line

from the single-layer model's brightness temperature sensitivity to grain size, and the statistics of the ensemble of nearest station grain sizes.

It was suggested in Sect. 3.2.2 that future implementations of Globsnow might be improved by using a LSM to provide grain size estimates in every grid cell, thus accounting for regional changes in geography and meteorology that are beyond the Globsnow kriging approach, and for the well-noted bias in observation location towards low latitudes, altitudes and canopy cover.

Even if an LSM were to provide the snow state forecasts, the current weighting scheme would not account for the variance introduced by its simplified layering relative to the truth. The LSM could be allowed to increase in complexity and contain more layers, but computational expense would rise both in the forecast step and in solving the update equation as the snow state vector and relevant covariance matrices would grow to contain more layer properties.

A user could apply the approach adopted here to estimate the extra variance introduced to their simulations as a function of the snow depth and their layering structure. Taking the

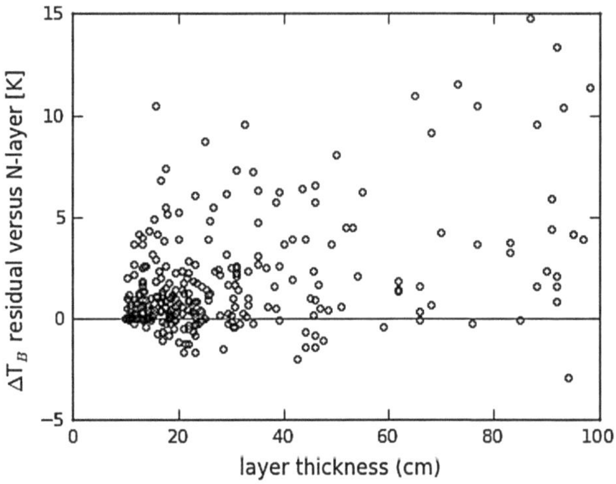

Fig. 8 Residual between brightness temperature difference simulated for models with layer thickness >10 cm versus the 10-cm layer simulation, and plotted as a function of this bulk layer thickness. All data represented as open circles and snowpits of depth >100 cm are excluded

gradient of the standard deviation graph in Fig. 9, the increase in error due to thickening of snow layers beyond 10 cm was found to be 0.053 ± 0.006 K cm^{-1}. The total error from loss of stratigraphic information can therefore be estimated from the snow depth and number of layers and is presented as a function of SWE (calculated by applying the CLPX average density for snowpits <100 cm depth) in Fig. 10 for snowpits whose stratigraphy is averaged to 1, 2 or 3 layers.

Additionally, the estimated error in $\Delta T_{\mathrm{B,V}}$ can be interpreted as an estimate of the SWE error for this regime where the roughly linear relationship between ΔT_{B} and SWE holds, similarly to Eq. (1) but for vertically rather than horizontally polarised microwaves. $\Delta\mathrm{SWE}/\Delta(\Delta T_{\mathrm{B,V}})$ was calculated from linear regression of the known SWE and N-layer calculated $\Delta T_{\mathrm{B,V}}$ values for the pits of depth <100 cm, and found to be 2.45 ± 0.09 mm$_{\mathrm{SWE}}$ K^{-1}, and so error in $\Delta T_{\mathrm{B,V}}$, $\Delta(\Delta T_{\mathrm{B,V}})$ was converted into an approximate SWE error using:

$$\Delta\ \mathrm{SWE} = \frac{\Delta\ \mathrm{SWE}}{\Delta(\Delta T_{\mathrm{B,V}})}\Delta(\Delta T_{\mathrm{B,V}}) \tag{17}$$

For a CLPX snow profile of 100 cm depth (170 mm SWE), simplification of the stratigraphy from the measurement resolution of 10 cm down to a single layer of average properties leads to $\Delta T_{\mathrm{B,V}}$ simulations that contain a 4.8-K error related to the loss of stratigraphic information, equivalent to 13 mm SWE (7 % of total). In a two-layer model, this error would be reduced to 2.1 K (5.6 mm SWE, 3 % of total) and for a three-layer model 1.2 K (3.3 mm SWE, 2 % of total).

The individual user must decide model detail based on the trade-off between precision and computational expense, and it is hoped that this approach will inform such decisions. A user might determine a given threshold for fractional or absolute error in SWE, and from this information could determine the number of layers to use in their model based on the snow depth.

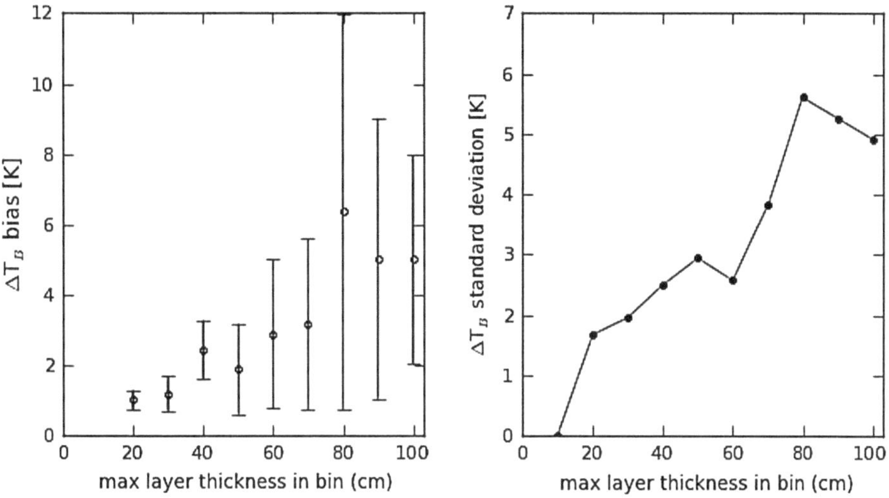

Fig. 9 Bias in simulated brightness temperature difference for snow profiles of thicker layers, relative to an *N*-layer model with 10 cm layer thicknesses, where *error bars* are 2-sigma (*left*). The standard deviation of brightness temperature difference relative to an *N*-layer model is quantified as a function of layer size (*right*), with an approximately linear increase in simulated $\Delta T_{B,V}$ error as snow stratigraphy is simplified into thicker layers

Fig. 10 Absolute error in snow water equivalent (SWE) introduced via simplification of the Cold Land Processes Experiment (CLPX) pit stratigraphy from *N*-layers to fewer layers, as a function of total SWE. *Lines* represent output using a one- (*dotted line*), two- (*dashed line*) or three-layer (*solid line*) snow profile as labelled in legend (*top left*)

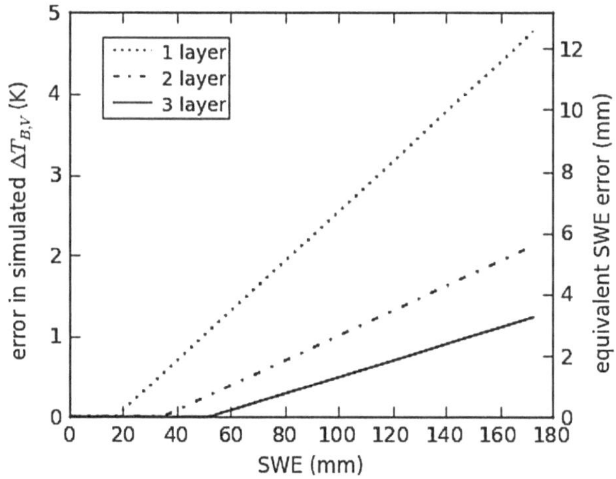

6 Discussion and Conclusions

Snow is extremely important in terms of meteorology, climatology and hydrology. Its surface properties can affect atmospheric chemistry and the energy balance, and an abundance of remote sensing products with well-characterised errors assess a wide variety of these surface properties, from grain size to contaminants to surface melt.

Snow water equivalent is of great interest, and estimates of SWE are most easily defined as a remote sensing, model, reanalysis or station-based product. Each of these approaches provides its own advantages, but each also has drawbacks: reanalyses and station-based products suffer from sparse observations, models have uncertainties due to limitations in the physical modelling of complex processes in snow, and remote sensing products are global, but limited by signal saturation and do not provide a unique SWE solution on inversion due to their high sensitivity to other snow properties.

Data assimilation techniques that use microwave information to update a forecast from other sources have been suggested to improve snow mass estimation. ESA's Globsnow uses modern assimilation techniques to bring together ground observations and remote sensing products and has shown that assimilating microwave measurements does improve SWE estimates. Globsnow isolates and accounts for the snow microstructure's contribution through a grain size parameter, which is obtained by fitting ground measurements to satellite retrievals while assuming a single homogeneous snow layer. The brightness temperature observable chosen by Globsnow is the difference between brightness temperatures at 19 and 37 GHz vertically polarised microwaves, $\Delta T_{B,V}$.

The Globsnow grain size estimate is reliant on point measurements of snow, which may vary greatly over relatively small areas, and it is suggested that physically based snow models could provide an alternate source of information to improve the inversion of the passive microwave signal.

Snow forms in layers and its stratigraphy can be complex, though physical models are capable of reproducing this layering. Lemmetyinen et al. (2010) and Durand et al. (2011) showed that for simulation of brightness temperatures over small areas of snow, this complexity can be an important contribution to the signal. Globsnow ignores this complexity in determining the error covariance for weighting the observational increment in the update step, and this might lead to suboptimal updates.

The HUT radiative transfer model used in Globsnow was able to simulate satellite-observed $\Delta T_{B,V}$ with an RMSE of 8 K, down from the 17 K RMSE associated with estimates made using the Chang algorithm typical of stand-alone microwave SWE products. The HUT RMSE was 6 K excluding one site where it was believed that snowpits were biased towards thin snow.

After confirming that the HUT radiative transfer model used in Globsnow was able to simulate satellite-observed scene brightness temperatures at NASA's CLPX, the HUT-simulated $\Delta T_{B,V}$ values for CLPX snowpits resampled to different layering structures were compared. Simulated $\Delta T_{B,V}$ for snow with the maximum possible level of stratigraphic detail based on the 10-cm resolution of CLPX density and temperature measurements was taken as truth, and deviations from this were treated as due to errors introduced by simplification of the stratigraphy to fewer layers.

Removing layering detail leads to a bias in the simulated $\Delta T_{B,V}$, likely due to the removal of reflection effects at layer boundaries and possibly due to nonlinearities in the $\Delta T_{B,V}$ response to snow grain size and density. Globsnow can freely vary the grain size to account for this, but this is likely to have second-order effects on the assimilation scheme.

Simulated $\Delta T_{B,V}$ values for the same snowpit at different levels of layering detail were found to vary, with the standard deviation increasing approximately linearly with snow depth. For snow of depth 100 cm (172 mm SWE at the CLPX sites), the standard deviation in simulated $\Delta T_{B,V}$ values for a single-layer model versus the N-layer model was estimated at 4.8 K, equivalent to approximately 13 mm SWE (7 % of total). Using two snow layers reduced the $\Delta T_{B,V}$ error to 2.1 K (5.6 mm SWE, 3 % of total).

Globsnow reports RMSE values of 40 mm for SWE <150 mm using a single-layer version of HUT, and the values found here suggest that layering could be a notable component of that RMSE.

Several groups are working on coupling physical snow models with microwave emission models to improve estimates of SWE, and these error assessments are important to help users decide on model complexity, which carries a potentially high computational cost. This paper argues that this is a promising avenue for improving global estimates of SWE, but that not accounting for the effects of detailed stratigraphy can introduce unaccounted-for variance which degrades the performance of an assimilation scheme. The approach adopted here quantifies this variance for the HUT radiative transfer model and presents it in such a way that it could be used to determine the snow layering structure in an LSM for use in an assimilation system.

Acknowledgments MR is funded by National Environment Research Council studentship F3275903, and ID is a member of the National Centre for Earth Observation.

References

Anderson E (1976) A point energy and mass balance model of a snow cover. Office of Hydrology, National Weather Service, Silver Spring

Andreadis K, Liang D, Tsang L, Lettenmaier D, Josberger E (2008) Characterization of errors in a coupled snow hydrology–microwave emission model. J Hydrometeor 9:149–164

Armstrong R, Brodzik M (2000) Validation of passive microwave snow algorithms. Proc IGARSS 2000(4):1561–1563

Armstrong R, Brodzik M, Knowles K, Savoie M (2005) Global monthly EASE-grid snow water equivalent climatology. Boulder, Colorado USA: National Snow and Ice Data Center, URL http://nsidc.org/data/docs/daac/nsidc0271_ease_grid_swe_climatology.gd.html

Barnett TP, Adam J, Lettenmaier DP (2005) Potential impacts of a warming climate on water availability in snow-dominated regions. Nature 438:303–309

Best M, Pryor M, Clark D, Rooney G, Essery R, Ménard C et al (2011) The joint UK (JULES) land environment simulator, model description—Part 1: energy and water fluxes. Geosci Model Dev 4:677–699

Betts A, Köhler M, Zhang Y (2009) Comparison of river basin hydrometeorology in ERA-Interim and ERA-40 reanalyses with observations. J Geophys Res 114

Brasnett B (1999) A global analysis of snow depth for numerical weather prediction. J Appl Meteorol 38:726–740

Brown RD, Mote PW (2009) The response of northern hemisphere snow cover to a changing climate. J Clim 22:2124–2145

Brown R, Brasnett B, Robinson D (2003) Gridded North American monthly snow depth and snow water equivalent for GCM evaluation. Atmos Ocean 41(1):1–14

Brucker L, Royer A, Picard G, Langlois A, Fily M (2011) Hourly simulations of the microwave brightness temperature of seasonal snow in Quebec, Canada, using a coupled snow evolution–emission model. Remote Sens Environ 115:1966–1977

Brun E, David P, Sudul M, Brunot G (1992) A numerical model to simulate snow-cover stratigraphy for operational avalanche forecasting. J Glaciol 38(128):13–22

Budyko M (1958) The heat balance of the Earth's surface. Department of Commerce, Weather Bureau

Carroll T, Cline D, Fall G, Nilsson A, Li L, Rost A (2001) NOHRSC operations and the simulation of snow cover properties for the coterminous U.S. 69th Annual Meeting of the Western Snow Conference. Sun Valley, Idaho USA

Chang A, Hall J, Foster D (1987) Nimbus7 SMMR derived global snow cover parameters. Ann Glaciol 9(9):39–44

Chang A, Kelly R, Josberger E, Armstrong R, Foster J, Mognard N (2005) Analysis of ground-measured and passive-microwave-derived snow depth variations in midwinter across the Northern Great Plains. J Hydrometeor 6:20–33

Clifford D (2010) Global estimates of snow water equivalent from passive microwave instruments: history, challenges and future developments. Int J Remote Sens 31(14):3707–3726

Cline D, Elder K, Davis B, Hardy J, Liston GE, Imel D et al (2002) Overview of the NASA cold land processes field experiment (CLPX-2002). SPIE Proceedings, Hangzhou

Davenport I, Sandells M, Gurney R (2012) The effects of variation in snow properties on passive microwave snow mass estimation. Remote Sens Environ 118:168–175
De Lannoy G, Reichle R, Houster P, Arsenault K, Verhoest N, Pauwels V (2010) Satellite-scale snow water equivalent assimilation into a high-resolution land surface model. J Hydrometeor 11:352–369
Dechant C, Moradkhani H (2011) Radiance data assimilation for operational snow and streamflow forecasting. Adv Water Resour 34(3):351–364
Dee DP, Uppala SM, Simmons AJ, Berrisford P, Poli P, Kobayashi S et al (2011) The ERA-interim reanalysis: configuration and performance of the data assimilation system. Q J Royal Meteorol Soc 137:553–597
Derksen C, Sturm M, Listen G, Holmgren J, Huntington H, Silis A et al (2009) Northwest Territories and Nunavut snow characteristics from a subarctic traverse: implications for passive microwave remote sensing. J Hydrometeorol 10:448–463
Dominé F, Shepson P (2002) Air-snow interactions and atmospheric chemistry. Science 297:1506–1510
Dong J, Walker J, Houser P, Sun C (2007) Scanning multichannel microwave radiometer snow water equivalent assimilation. J Geophys Res 112:D07108
Drusch M, Vasiljevic D, Viterbo P (2004) ECMWF's global snow analysis: assessment and revision based on satellite observations. J Appl Meteorol 43:1282–1294
Durand M, Margulis S (2006) Feasibility test of multifrequency radiometric data assimilation to estimate snow water equivalent. J Hydrometeor 7:443–457
Durand E, Kim E, Margulis S (2008) Quantifying uncertainty in modelling snow microwave radiance for a mountain snowpack at the point-scale, including stratigraphic effects. IEEE Trans Geosci Remote Sens 46:1753–1767
Durand M, Kim E, Margulus S (2009). Radiance assimilation shows promise for snowpack characterization. Geophys Res Lett 29(2)
Durand M, Kim E, Margulis S, Molotch N (2011) A first-order characterization of errors from neglecting stratigraphy in forward and inverse passive microwave modelling of snow. IEEE Geosci Remote Sens Lett 8:730–734
Dutra E, Balsamo G, Viterbo P, Miranda P, Beljaars A, Schär C et al (2010) An improved snow scheme for the ECMWF land surface model: description and offline validation. J Hydrometeorol 11:899–916
Dye D (2002) Variability and trends in the annual snow-cover cycle in Northern Hemisphere land areas, 1972–2000. Hydrol Proc 16(15):3065–3077
Dyer J, Mote T (2006) Spatial variability and trends in observed snow depth over North America. Geophys Res Lett 33(16)
Finnish Meteorological Institute. Globsnow Project Description (2012) http://www.globsnow.info/snow_workshop_2012/presentations/GlobSnow_Fact_Sheet_EuropeanSatelliteSnowMonitoringActivities.pdf. Accessed 8th Aug, 2013
Flanner M, Shell K, Barlage M, Perovich D, Tschudi M (2011) Radiative forcing and albedo feedback from the Northern Hemisphere cryosphere between 1979 and 2008. Nat Geosci 4:151–155
Foster J, Sun C, Walker J, Kelly R, Chang A, Dong J et al (2005) Quantifying the uncertainty in passive microwave snow water equivalent observations. Remote Sens Environ 94:187–203
Frappart F, Ramillien G, Biancamaria S, Mognard N, Cazenave A (2006) Evolution of high-latitude snow mass derived from the GRACE gravimetry mission (2002–2004). Geophys Res Lett 33
Frei A, Miller J, Robinson D (2003) Improved simulations of snow extent in the second phase of the Atmospheric Model Intercomparison Project (AMIP-2). J Geophys Res 108
Grenfell T, Warren S (1999) Representation of a nonspherical ice particle by a collection of independent spheres for scattering and absorption of radiation. J Geophys Res 104(D24):31697–31709
Grippa M, Mognard N, Le Toan T (2005) Comparison between the interannual variability of snow parameters derived from SSM/I and the Ob river discharge. Remote Sens Environ 98:35–44
Hall D, Riggs G (2007) Accuracy assessment of the MODIS snow products. Hydrol Proc 21:1534–1547
Hall DK, Sturm M, Benson C, Chang AT, Foster JL, Garbeil H et al (1991) Passive microwave remote and in situ measurements of arctic [sic] and subarctic snow covers in Alaska. Remote Sens Environ 38(3):161–172
Hallikainen M (1989) Microwave radiometry of snow. Adv Space Res 9(1):267–275
Hancock S, Baxter R, Evans J, Huntley B (2013) Evaluating global snow water equivalent products for testing land surface models. Remote Sens Environ 128:107–117
Haran T (2003) CLPX-Satellite: MODIS radiances, reflectances, snow cover and related grids. MOD10A2. Boulder, Colorado USA: NSIDC: National Snow and Ice Data Center
Khan V, Holko L (2009) Snow cover characteristics in the Aral Sea Basin from different data sources and their relation with river runoff. J Marine Syst 76:254–262
Kitaev L, Kislov A, Krenke A, Razuzaev V, Martuganov R, Konstantinov I (2002) The snow cover characteristics of northern Eurasia and their relationship to climatic parameters. Boreal Environ Res 7:437–445

Koskinen J, Pulliainen J, Hallikainen M (1997) The use of ERS-1 SAR data in snow melt monitoring. IEEE Trans Geosci and Remote Sens 35:601–610

Lemmetyinen J, Pulliainen J, Rees A, Kontu A, Qiu Y, Derksen C (2010) Multiple-layer adaption of HUT snow emission model: comparison with experimental data. IEEE Trans Geosci Remote Sens 48:2781–2794

Liu G (2004) Approximation of single scattering properties of ice and snow particles for high microwave frequencies. J Atmospheric Sci 61:2441–2456

Macke A, Mueller J, Raschke E (1996) Single scattering properties of atmospheric ice crystals. J Atmospheric Sci 53(19):2813–2825

Mätzler C (2000) A simple snowpack/cloud reflectance and transmittance model from microwave to ultraviolet: the ice-lamella pack. J Glaciol 46(152):20–24

Mätzler C (2002) Relation between grain-size and correlation length of snow. J Glaciol 48:166–461

Mie G (1908) Beiträge zur Optik trüber Medien, speziell kolloidaler Metallösungen. Ann Phys 330(3):377–445

Nilsson A (2003) Qscat CLPX data. //ftp.nohrsc.nws.gov/staff/nilsson/qscat/. Accessed 15 April 2013

Niu G-Y, et al. (2007). Retrieving snow mass from GRACE terrestrial water storage change with a land surface model. Geophys Res Lett 34

Onogi K, Tsutsui J, Koide H, Sakamoto M, Kobayashi S, Hatsushika H et al (2007) The JRA-25 reanalysis. J Meteorol Soc Jpn 85(3):369–432

Painter T, Bryant A, Skiles S (2012). Radiative forcing by light absorbing impurities in snow from MODIS surface reflectance data. Geophys Res Lett 39

Picard G, Brucker L, Fily M, Gallée H, Krinner G (2009) Modeling time series of microwave brightness temperature in Antarctica. J Glaciol 55(191):537–551

Pulliainen J (2006) Mapping of snow water equivalent and snow depth in boreal and sub-arctic zones by assimilating space-borne microwave radiometer data and ground-based observations. Remote Sens Environ 101:257–269

Pulliainen J, Grandell J, Hallikainen M (1999) HUT snow emission model and its applicability to snow water equivalent retrieval. IEEE Trans Geosci Rem Sens 37(3):1378–1390

Ramsay B (1998) The interactive multisensor snow and ice mapping system. Hydrol Process 12:1537–1546

Rawlins M, Fanestock M, Frolking S, Vörösmarty CJ (2007) On the evaluation of snow water equivalent estimates over the terrestrial Arctic drainage basin. Hydrol Process 21(12):1616–1623

Rawlins M, Steele M, Holland M, Adam J, Cherry J, Francis J et al (2010) Analysis of the Arctic system for freshwater cycle intensification: observations and expectations. J Clim 23:5715–5737

Rienecker M, Suarez M, Gelaro R, Todling R, Bacmeister J, Liu E et al (2010) MERRA: NASA's modern-era retrospective analysis for research and applications. J Clim 24:3624–3648

Rittger K, Painter T, Dozier J (2013) Assessment of methods for mapping snow cover from MODIS. Adv in Water Resour 51:367–380

Roesch A (2006) Evaluation of surface albedo and snow cover in AR4 coupled climate models. J Geophys Res 111

Saha S, Moorthi S, Pan H-L, Wang J, Nadiga S, Tripp P et al (2010) The NCEP climate forecast system reanalysis. Bull Amer Meteorol Soc 91:1015–1057

Salzmann N, Mearns L (2012) Assessing the performance of multiple regional climate model simulations for seasonal mountain snow in the upper Colorado River Basin. J Hydrometeorol 13:539–556

Skiles S, Painter T, Deems J, Bryant A, Landry C (2012) Dust radiative forcing in snow of the Upper Colorado River Basin: 2. Interannual variability in radiative forcing and snowmelt rates, Water Resour Res 48

Smith C, Guttman L (1953) Measurement of internal boundaries in three-dimensional structures by random sectioning. Trans AIME 5:81–87

Sun C, Walker J, Houser P (2004) A methodology for snow data assimilation in a land surface model. J Geophys Res 109:D08108

Takala M, Luojus K, Pulliainen J, Derksen C, Lemmetyinen J, Kärnä J-P et al (2011) Estimating northern hemisphere snow water equivalent for climate research through assimilation of space-borne radiometer data and ground-based measurements. Remote Sens of Environ 115:3517–3529

Tedesco M, Pulliainen J, Takala M, Hallikainen M, Pampaloni P (2004a) Artificial neural network-based techniques for the retrieval of SWE and snow depth from SSM/I data. Remote Sens of Environ 90:76–85

Tedesco M, Kelly R, Foster J, Chang A (2004b) AMSR-E/Aqua daily L3 global snow water equivalent EASE-Grids V002. updated daily. Boulder, Colorado USA: National Snow and Ice Data Center. URL http://nsidc.org/data/ae_dysno

Teschl E, Randeu W, Teschl R (2010) Microwave scattering from ice crystals: how much parameters can differ from equal volume spheres. Adv in Geosci 25:127–133

Toure A, Goita K, Royer A, Kim E, Durand M, Margulis S et al (2011) A case study of using a multilayered thermodynamical snow model for radiance assimilation. IEEE Trans on Geosci and Remote Sens 49(8):2828–2837

Uppala S, Kållberg P, Simmons A, Andrae U, Da Costa Bechtold V, Fiorino M et al (2005) The ERA-40 re-analysis. Q J of the Royal Meteorol Soc 131(612):2961–3012

Wiesmann A, Mätzler C (1999) Microwave emission model of layered snowpacks. Remote Sens of Environ 70(3):307–316

Wiesmann A, Fierz C, Mätzler C (2000) Simulation of microwave emission from physically modeled snowpacks. Ann of Glaciol 31(1):397–405

Yang D, Zhao Y, Armstrong R, Robinson D, Brodzik, M-J (2007). Streamflow response to seasonal snow cover mass changes over large Siberian watersheds. J Geophys Res Earth Surf 112

DOI 10.1007/978-94-017-8789-5_15
Reprinted from *Surveys in Geophysics* Journal, DOI 10.1007/s10712-013-9262-y

Glaciers in the Earth's Hydrological Cycle: Assessments of Glacier Mass and Runoff Changes on Global and Regional Scales

Valentina Radić · Regine Hock

Received: 4 May 2013 / Accepted: 29 October 2013 / Published online: 11 December 2013

Abstract Changes in mass contained by mountain glaciers and ice caps can modify the Earth's hydrological cycle on multiple scales. On a global scale, the mass loss from glaciers contributes to sea-level rise. On regional and local scales, glacier meltwater is an important contributor to and modulator of river flow. In light of strongly accelerated worldwide glacier retreat, the associated glacier mass losses raise concerns over the sustainability of water supplies in many parts of the world. Here, we review recent attempts to quantify glacier mass changes and their effect on river runoff on regional and global scales. We find that glacier runoff is defined ambiguously in the literature, hampering direct comparison of findings on the importance of glacier contribution to runoff. Despite consensus on the hydrological implications to be expected from projected future warming, there is a pressing need for quantifying the associated regional-scale changes in glacier runoff and responses in different climate regimes.

Keywords Glaciers · Mass balance · Glacier runoff · Sea-level rise · Mass-balance observations · Glacier projections · Modeling

1 Introduction

Mountain glaciers and ice caps, covering 734,400 km^2 on Earth (Gardner et al. 2013), are an integral part of the Earth's hydrological cycle affecting water balances on all spatial

V. Radić (✉)
Department of Earth, Ocean and Atmospheric Sciences, University of British Columbia, Vancouver, BC V6T 1Z4, Canada
e-mail: vradic@eos.ubc.ca

R. Hock
Geophysical Institute, University of Alaska, Fairbanks, AK 99775, USA

R. Hock
Department of Earth Sciences, Uppsala University, 752 36 Uppsala, Sweden

scales. Meier (1984) was the first to recognize that these glaciers outside the two vast ice sheets in Antarctic and Greenland—though only comprising < 1 % of the Earth's total ice volume—are major contributors to global sea-level rise due to worldwide glacier wastage in response to global warming. Various studies have attempted to quantify the mass losses of these ice bodies and their effect on rising sea level indicating that glaciers outside the ice sheets have contributed between one-third and one-half of global sea-level rise during the last decades (Dyurgerov and Meier 2005; Kaser et al. 2006; Cogley 2009a, b; Gardner et al. 2013). The glacier contribution to future sea-level rise is expected to remain significant as the global temperature is expected to further increase (Lemke et al. 2007).

On regional and local scales, glaciers are significant contributors to seasonal riverflow, serving as frozen reservoirs of water that supplement runoff during warm and dry periods of low riverflow. The ongoing glacier retreat has important implications for downstream river flows, regional water supplies, sustainability of aquatic ecosystems, and hydropower generation (e.g., Kaser et al. 2010; Huss 2011; Immerzeel et al. 2010). Glacier runoff is intrinsically linked to the glacier's mass balance, the latter defined as the sum of its total accumulation (mostly due to snowfall, windblown snow, avalanches, and condensation) and ablation (mostly due to melt, calving of icebergs, wind erosion, evaporation, sublimation) over a stated period of time (Cogley et al. 2011). Note that mass loss is defined negatively. Despite the importance of glaciers as modifiers of global and regional water cycles, there are relatively few attempts to assess recent and project future glacier mass changes and quantity their impacts on riverflow on global and regional scales.

Previous review-type publications have focused either on glacier mass changes and their measurement (Braithwaite 2002; Cogley 2011) or on glacier runoff and its characteristics (Jansson et al. 2003; Hock et al. 2005; Hock and Jansson 2005) generally focusing on local catchment or glacier scales. In contrast, here we combine both themes to highlight the links between glaciers and river runoff focusing exclusively on regional and global scales. Our goal is to provide a critical overview of studies that have attempted to quantify recent and future glacier mass changes and to assess the importance of these mass changes in streamflow on larger scales. We only consider glaciers distinct from the two ice sheets in Greenland and Antarctica. First, we will provide an overview on global glacier mass balances including assessment techniques (Sect. 2) and modeling of recent and future changes (Sect. 3). Then, we will discuss the characteristics and definition of glacier runoff (Sect. 4) followed by a discussion of studies exploring the role of glaciers in regional and global hydrology (Sect. 5).

2 Assessing glacier mass balance on regional and global scales

Simulating glacier runoff requires accurate modeling of the components of the glacier mass balance which in turn requires mass-balance measurements for calibration and validation of mass-balance models (Konz and Seibert 2010). Below, we will briefly introduce the techniques for assessing glacier mass balance on global scales before reviewing the results of assessments and projections.

2.1 Assessments by in situ mass-balance measurements

Until recently, all global assessments of the mass balance of glaciers relied on some form of extrapolation of available glacier-wide mass-balance measurements. The most traditional of these techniques, the so-called glaciological method, is based on snow probings and ablation stake measurements (Østrem and Brugman 1991; Kaser et al. 2002; Zemp et al. 2013), and provides a measure of the surface mass balance. The glacier-wide surface mass balance is estimated from extrapolation of the point measurements over the glacier surface. The earliest mass-balance measurements were taken on the Rhône Glacier, Swiss Alps, providing intermittent observations during 1874–1908 (Mercanton 1916). Annual mass-balance measurements have been taken at two stakes on Claridenfirn in Switzerland since 1914 (Müller-Lemans et al. 1994). The longest continuous glacier-wide mass–balance time series exists for Storglaciären, Sweden, reaching back to 1945 (Zemp et al. 2010). In situ measurements of mass balance have been obtained for ~ 340 glaciers worldwide, of which ~ 70 glaciers have continuous annual observations longer than 20 years (Dyurgerov 2010). The records of glacier mass balance are complied and distributed by the World Glacier Monitoring Service (WGMS, Zemp et al. 2009).

Dyurgerov and Meier (1997a, b) provided the first detailed assessment of annual glacier mass balances on global and regional scales followed by updates in Dyurgerov (2002, 2003), Dyurgerov and Meier (2005), and Dyurgerov (2010). Global averages were obtained from area-weighted specific mass balances of smaller subregions whose balances were estimated from the single-glacier observations. A similar approach was taken by Ohmura (2004), while Cogley (2005) used a different approach by applying a spatial interpolation algorithm, fitting a second-degree polynomial to the single-glacier observations to extrapolate the mass-balance observations to all glacierized cells in a $1 \times 1°$ global grid. In contrast to Cogley (2005) who used only glaciological mass-balance measurements, Cogley (2009a, b) also included geodetic observations (Sect. 3.2) from more than 250 glaciers in the interpolation. His estimate was about 30 % more negative than the one derived solely from direct measurements possibly due to a better representation of marine-terminating glaciers (which, in addition to surface melting, lose mass by iceberg calving and submarine melting); however, it was questionable whether the difference represented adequately the global-average ablation by calving and submarine melt (Cogley 2009a, b).

Three global estimates (Ohmura 2004; Dyurgerov and Meier 2005; Cogley 2005) were synthesized into a "consensus estimate" (Kaser et al. 2006) that was used in the Fourth Assessment of the Intergovernmental Panel on Climate Change (IPCC; Lemke et al. 2007). Not surprisingly, since they were based on the same observations, the three assessments agree well with each other. A glacier mass loss rate of 0.50 ± 0.18 mm sea-level equivalent (SLE) year^{-1} was found for the period 1961–2004 and an increased rate of 0.77 ± 0.22 mm SLE year^{-1} for 1991–2004, thus a considerably higher mass loss rate than found for both ice sheets together in both periods (Lemke et al. 2007). Gardner et al. (2013) applied the methods of Cogley (2009a, b) to all glaciers other than the ice sheets and found a mass loss rate of 1.37 ± 0.22 mm SLE year^{-1} (0.92 ± 0.34 mm SLE year^{-1} excluding the glaciers in the Antarctica and Greenlandic periphery) for the period 2003–2009, comparable to the estimate for 2006 by Meier et al. (2007) derived from the interpolation of local glaciological records (1.11 ± 0.26 mm SLE year^{-1}, Table 1).

All these assessments suffer from serious under sampling. Direct observations of glacier mass changes exist on fewer than 1 % of the glaciers worldwide (~ 300 out of ~ 200,000). These are geographically biased with more than 60 % of the records originating from the European Alps, Scandinavia, Western Canada and USA, and parts of the former Soviet Union (Dyurgerov 2010). They are also biased toward smaller, land-terminating glaciers in maritime climates. The assumption that very few benchmark glaciers with observed mass balances over short-term time period (< 10 years) are representative for the region-wide mass balance over ~ 40 years is a major but inevitable

Table 1 Overview of recent studies assessing the global glacier mass budget

Period	Mass budget Gt a^{-1} (mm SLE $year^{-1}$)		Method	Area km^2	Reference
	All glaciers	Excl. A + G			
1961–2004	−182 ± 78 (0.50 ± 0.22)	−155 ± 67 (0.43 ± 0.19)	Glaciological	785,000 *546,000*	Kaser et al.(2006)
2006	−402 ± 95 (1.11 ± 0.26)		Glaciological	763,000	Meier et al. (2007)
1961–2004	(0.79 ± 0.34)	(0.51 ± 0.29)	Modeling	704,000 *518,000*	Hock et al. (2009)
2001–2005		−740 ± 62 (1.12 ± 0.14)	Glaciological and geodetic		Cogley(2009a, b)
1961–2004		−153 (0.42)	Modeling	*535,000*	Hirabayashi et al. (2010)
1902–2009	(1.07 ± 0.05)[a]		Modeling	*590,900*[a]	Marzeion et al. (2012)
1948–2005		(0.45 ± 0.02)[b]	Modeling	*517,000*	Hirabayashi et al. (2013)
1971–2010	−247 ± 190[c] (0.68 ± 0.53)		Modeling based on AAR observations	734,430[c]	Mernild et al. (2013)
10/2003–10/2009		−170 ± 32 (0.47 ± 0.09)	GRACE (method by Jacob et al. 2012)	734,430 *511,800*	Gardner et al. (2013)
10/2003–10/2009		−166 ± 37 (0.46 ± 0.10)	GRACE (method by Wouters et al. 2008)	734,430 *511,800*	Gardner et al. (2013)
10/2003–10/2009	−495 ± 78 (1.37 ± 0.22)	−335 ± 124 (0.92 ± 0.34)	Glaciological (method by Cogley 2009a, b)	734,430 *511,800*	Gardner et al. (2013)
10/2003–10/2009	−259 ± 28 (0.71 + -0.08)	−215 ± 26 (0.59 ± 0.07)	Combination of GRACE, ICESat, glaciological	734,430 *511,800*	Gardner et al. (2013)

Numbers in italic refer to those excluding the glaciers in the Antarctic (A) and Greenland (G) periphery

[a] Here calculated as the average rate over the period with total glacier mass loss of 114±5 mm SLE (Marzeion et al. 2012). Total area of 590,900 km^2 (excluding glaciers peripheral to Antarctic ice sheet) is the modeled area at 2009

[b] Here calculated as the average rate over the period with total mass loss of 25.9±1.4 mm SLE (Hirabayashi et al. 2013)

[c] Here calculated from the average global mass balance of -336±260 kg m^{-2} $year^{-1}$ (Mernild et al. 2013), using glacierized area of 734,430 km^2 (Gardner et al. 2013) and ocean surface area of 362×10^6 km^2

limitation in these assessments. Another source of uncertainty is poor knowledge of glacier inventory data at that time, i.e., data on glacier location and surface area. Nevertheless, these assessments provide continuous annual or pentadal time series of mass balance reaching back into the mid-twentieth century.

2.2 Assessments by geodetic method

With the geodetic method, the glacier mass balance is estimated by repeated mapping, either by ground-based surveys or remote sensing (laser, radar altimetry, stereoscopic imagery). The change in glacier volume (obtained from the difference in glacier surface elevations over the glacier area) multiplied by the average density of the removed or added material gives the change in glacier mass. In contrast to in situ measurements, geodetic observations generally have good regional but poor temporal coverage, since surveys are often separated by multi-annual to multi-decadal gaps. On the other hand, the geodetic method can observe the mass changes of tidewater glaciers, which are not included in traditional glaciological measurements.

Initially, geodetic surveys have been mostly used to assess the mass changes of individual glaciers (Cogley 2009a, b); however, increasing availability of remote sensing data (in particular satellite laser altimetry after the launch of ICESat in 2002) triggered a number of studies covering entire glacier regions. Geodetic estimates of mass changes on regional scale exist for Alaska (Arendt et al. 2002; airborne laser altimetry, Berthier et al. 2010; satellite remote sensing), Arctic Canada (Abdalati et al. 2004; airborne laser altimetry; Gardner et al. 2011; satellite remote sensing), British Columbia (Schiefer et al. 2007; satellite radar altimetry), Svalbard (Nuth et al. 2010; Moholdt et al. 2010; satellite laser altimetry), Iceland (Bjornsson et al. 2013; airborne and satellite remote sensing), Russian High Arctic (Moholdt et al. 2012; satellite laser altimetry), Austrian Alps (Lambrecht and Kuhn 2007; DEM from aerial photographs), Swiss Alps (Paul and Haeberli 2008; satellite radar altimetry), parts of central Asia (Gardner et al. 2013; satellite laser altimetry), Patagonia (Rignot et al. 2003; Willis et al. 2012; satellite remote sensing), the peripheral glaciers in Greenland (Bolch et al. 2013; Gardner et al. 2013; satellite laser altimetry), and the glaciers on the islands surrounding the Antarctic mainland (Gardner et al. 2013; satellite laser altimetry). Most of these estimates are derived for relatively short recent (after 2000) time periods.

2.3 Assessments using satellite gravimetry

Gravimetric measurements have become a popular tool to estimate glacier mass changes since the launch of the satellites of the Gravity Recovery and Climate Experiment (GRACE) in March 2002. GRACE consists of a pair of satellites orbiting together and measuring variations in the terrestrial gravity field, therefore detecting mass movements. The twin satellites orbit the Earth 15 times a day, recording minute variations in the Earth's gravitational pull. When passing over a region of larger gravity, the first satellite is pulled ahead of the trailing satellite, thus increasing the distance between the satellites. Mass changes are derived from the constantly changing distance between the twin satellites combined with precise positioning measurements (Tapley et al. 2004).

GRACE observes mass changes with high temporal resolution (e.g., sub-monthly), but the spatial resolution is relatively poor (roughly 100 $\times$ 100 km). In contrast to the methods above, no density assumptions are needed because mass change is measured directly. However, since the satellites detect the total mass changes over a large area, and are unable to resolve individual components of the mass changes, the signal needs to be decomposed in order to identify the signal due to glacier mass changes. The decomposition is relatively complex and, because it relies on the accuracy of models used to simulate Earth system processes (isostatic rebound, tectonics, hydrology, atmosphere), it may introduce large uncertainties into the derived mass balances (e.g., Jacob et al. 2012).

GRACE-derived regional-scale mass balances have been reported for the Canadian Arctic (Gardner et al. 2011), Alaska (Tamisiea et al. 2005; Chen et al. 2006; Luthcke et al. 2008; 2013; Wu et al. 2010), Patagonia (Chen et al. 2007; Ivins et al. 2011), and High Mountain Asia (Matsuo and Heki 2010). Jacob et al. (2012) were the first to compute GRACE-derived mass-balance estimates for all glacierized regions outside Greenland and Antarctica, followed by Gardner et al. (2013) who updated their estimate and generated a new one based on the methods of Wouters et al. (2008). These two analyses report a total mass budget for these regions of -170 ± 32 Gt year^{-1} and -166 ± 37 Gt year^{-1}, respectively, for the period 2003–2009. Jacob et al. (2012) note that their results are roughly 30 % smaller than the most recent available estimate at that time, obtained from the interpolation of glaciologically derived in situ observations by Dyurgerov (2010).

2.4 Assessments by other approaches

2.4.1 AAR method

Bahr et al. (2009) derived global glacier mass changes using an approach based on the observations of the accumulation area ratio (AAR), i.e., the ratio of the accumulation area to the total glacier area. AAR is closely related to the mass balance of a glacier in the case when calving and submarine melt are negligible (Dyurgerov and Meier 2005). AARs can be relatively easily approximated from aerial and satellite observations of the end-of-summer snowline. For a glacier in balance with the climate, the AAR is equal to its equilibrium value, AAR_0, whose average value from a sample of ~ 100 glaciers has been found to be 58 % (Dyurgerov et al. 2009).

Glaciers with AAR < AAR_0 will retreat to higher elevations, typically over several decades or longer, and the AAR may return to the equilibrium value. Using AAR observations of ~ 80 glaciers collected during 1997–2006, Bahr et al. (2009) computed a mean AAR of 44 ± 2 %, with AAR < AAR_0 for most glaciers in the dataset. Mernild et al. (2013) revised the methodology, expanded and updated their data, and found an average AAR of 34 ± 3 %, for the period 2001–2010. Using the empirical relationship between the ratio AAR/AAR_0 and annual glacier mass balance, Mernild et al. (2013) reconstructed pentadal global glacier mass balances for 1971–2010, showing a good agreement with estimates from Cogley (2009b). However, they also found much larger uncertainties in the global estimate than in the original study by Bahr et al. (2009).

This AAR-based approach has also been used to provide estimates of future glacier area and volume changes assuming that the future climate resembles the one of the recent few decades. Bahr et al. (2009) estimated that, even without additional atmospheric warming, the volume of glaciers must shrink by 27 ± 5 % to return to a balanced mass budget. Assuming that the total volume of the Earth's glaciers and ice caps is 650 mm SLE (Dyurgerov and Meier 2005), the fractional losses would raise global mean sea level by 184 ± 33 mm. With the updated AAR dataset and updated estimate of total glacier volume (430 mm SLE by Huss and Farinotti 2012) and accounting for the larger errors due to regional and global undersampling, Mernild et al. (2013) revised this estimate to 163 ± 69 mm. We note that, because of its simplicity, the AAR-based approach may only give an indication of future mass changes, and projections should preferably be addressed by models describing the physical processes involved and using transient climate scenarios.

2.4.2 Multi-method approach

Gardner et al. (2013) synthesized a consensus global mass-balance estimate for the period October 2003–October 2009 by standardizing existing and generating new, regional estimates for 19 individual glacierized regions (Fig. 1) while investigating the large discrepancies between the estimates obtained from GRACE and those from interpolating local glaciological records (Table 1). The analysis is based on a new globally complete glacier inventory (Randolph Glacier Inventory, RGI, Arendt et al. 2012). ICESat and GRACE estimates agreed well in large glacierized regions, where results from spatial interpolation of local records tended to give considerably more negative mass budgets. Their analyses suggest that available local glaciological records are negatively biased in larger regions, indicating that previous assessments based on spatial interpolation (Sect. 2.1) may have overestimated mass losses. GRACE results tend to have large uncertainties in regions with little ice cover. Therefore, averages of available ICESat and GRACE estimates were generally used for the larger glacierized areas while results from spatial interpolation of local measurements updated from Cogley (2009a) were adopted for the smaller ($< 5{,}000$ km^2) regions where the density of in situ measurements tends to be high.

Results show that all glaciers other than the ice sheets lost 259 ± 28 Gt year^{-1} accounting for 29 ± 13 % of the observed sea-level rise of 2.50 ± 0.54 mm SLE year^{-1} during 2003–2009, thus matching approximately the combined contribution of the two large ice sheets (Shepard et al. 2012). Glacier mass was lost in all 19 regions during this period with the largest losses from Arctic Canada, Alaska, and peripheral Greenland.

3 Modeling glacier mass balance on regional and global scales

State-of-the-art simulations and projections of global mass changes of glaciers and ice caps have relied on low-complexity models of surface mass balance and glacier dynamics. These modeling studies have commonly assumed that the main drivers of glacier mass balance are air temperature and precipitation, while glacier dynamics, involved in changes of glacier area and thickness, are assumed to be successfully simulated by scaling methods (Bahr et al. 1997). In the following sections, we will briefly discuss a selection of the modeling studies (listed in Tables 1 and 2), narrowing our review to the most recent studies (last few years) and to those that used some type of meteorological/climate data. Methodological approaches fall broadly into two categories: (1) models based on mass-balance sensitivities to temperature and precipitation changes (e.g., Hock et al. 2009; Slangen et al. 2012), and (2) direct modeling of transient surface mass balance (e.g., Raper and Braithwaite 2006; Radić and Hock 2011; Marzeion et al. 2012). Most of the latter studies have used an ensemble of global climate model (GCMs) to provide climate forcing for their models. The use of a multimodel ensemble is a common way to provide a range of projections and uncertainties in any climate change impact studies. Also, studies that evaluated GCM simulations of mean climate on global and regional scales have shown that the multimodel ensemble average is superior to any individual model (e.g., Gleckler et al. 2008; Pierce et al. 2009).

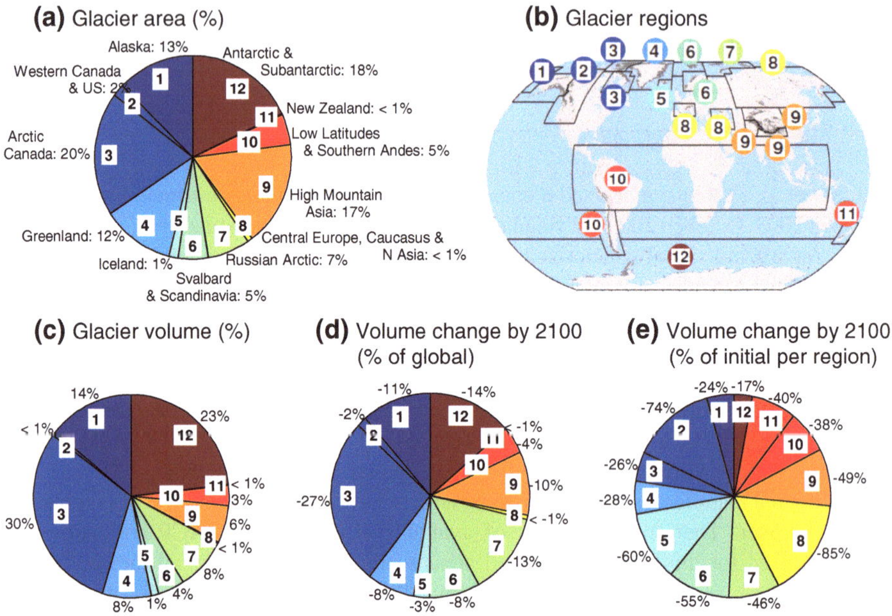

Fig. 1 **a** Regional glacier area in % relative to the global glacier area of 744,430 km^2 (Gardner et al. 2013). **b** 12 glacierized regions containing all the Earth's mountain glaciers and ice caps (Randolph Glacier Inventory; Arendt et al. 2012). Some of the 19 primary regions from Randolph Glacier Inventory (*black polygons*) are combined here. **c** Regional glacier volume in % relative to the global glacier volume of 522 mm SLE (Radić et al. 2013). **d** Regional contribution of glacier volume loss (%) to global volume loss of 155 mm SLE projected for 2006–2100 as a mean of 14 GCMs with RCP4.5 (Radić et al. 2013). **e** Total regional volume change of glaciers over 2006–2100, expressed in % of initial regional glacier volume, as a mean projection from 14 GCMs with RCP4.5 (Radić et al. 2013). Here, each region's pie size is proportional to its regional volume change (therefore the sum over the pie sizes is not 100 %). Numbers in the pies correspond to the regions in (*b*)

To account for glacier area changes, most models apply volume–area (or volume–length) scaling, which states that the volume of a mountain glacier is proportional to its area (or length) raised to a power (Bahr et al. 1997). Considering the lack of data required for higher-order glacier dynamics models, these methods are shown to be a good first-order approximation of glacier dynamics for the assessments of global-wide and region-wide glacier mass changes (Radić et al. 2007, 2008; Adhikari and Marshall 2012). Though simple, the approach allows modeling of the tendency of mountain glaciers to reach a new equilibrium in a warming climate since the specific mass balance (i.e., mass change per unit area) tends to become less negative as the glacier retreats from low-lying, high-ablation altitudes.

3.1 Models based on mass-balance sensitivity

Mass-balance sensitivities refer to the changes in mass balance that result from instantaneous changes in temperature and precipitation, and are generally estimated from mass-balance modeling of glaciers with mass-balance observations (e.g., Braithwaite and Zhang 2000; de Woul and Hock 2005). In combination with the data on temperature (ΔT) and precipitation change (ΔP), glacier mass change (ΔM) over a time window (Δt) can be obtained from

Table 2 Overview of studies projecting global glacier mass changes for the twenty-first century

Period	Projected SLE (mm)		Climate Scenario	Reference
	All glaciers	Excl. A + G		
2001–2100		46, 51	SRES A1B, 2 GCMs	Raper and Braithwaite (2006)
2001–2100	124 ± 37	99 ± 33	SRES A1B, 10 GCMs	Radić and Hock (2011)
2006–2100	148 ± 35		RCP2.6, 13 GCMs	Marzeion et al. (2012)
2006–2100	166 ± 42		RCP4.5, 15 GCMs	Marzeion et al. (2012)
2006–2100	175 ± 40		RCP6.0, 11 GCMs	Marzeion et al. (2012)
2006–2100	217 ± 47		RCP8.5, 15 GCMs	Marzeion et al. (2012)
2000–2099	159 ± 52	116	SRES A1B, 12 GCMs	Slangen et al. (2012)
2001–2100	150 ± 37	114 ± 30	SRES A1B, 10 GCMs	Radić et al. (2013)
2006–2100	155 ± 41	122 ± 36	RCP4.5, 14 GCMs	Radić et al. (2013)
2006–2100	216 ± 44	167 ± 38	RCP8.5, 14 GCMs	Radić et al. (2013)
2006–2099		73 ± 14[a]	RCP8.5, 10 GCMs	Hirabayashi et al. (2013)
2012–2099	102 ± 28	64	SRES A1B, 8 GCMs	Giesen and Oerlemans (2013)

[a] Here calculated by subtracting the reported projections for the period 1948-2005 (25.9±1.4 mm SLE) from the period 1948-2099 (99.0±14.9 mm SLE) in Hirabayashi et al. (2013).

$$\frac{\Delta M}{\Delta t} = S\left(\frac{d\dot{b}}{dT}\Delta T + \frac{d\dot{b}}{dP}\Delta P\right)$$

where $\frac{d\dot{b}}{dT}$ and $\frac{d\dot{b}}{dP}$ are mass-balance sensitivities to temperature and precipitation change, respectively, and S is glacier surface area (e.g., Hock et al. 2009). The specific mass-balance rate, $\dot{b}$, is typically in m water equivalent (w.e.) year^{-1}.

Several studies have found that glaciers in wetter or maritime climates tend to be more sensitive to temperature and precipitation changes than subpolar glaciers or glaciers in continental climates (e.g., Oerlemans and Fortuin 1992; Braithwaite and Zhang 1999; de Woul and Hock 2005). Quantifying the relations between mass-balance sensitivities and climate variables enables extrapolation of the sensitivities to glaciers without mass-balance observations. For the purpose of projecting global glacier mass changes, this approach was first applied in Gregory and Oerlemans (1998), further developed in Van de Wal and Wild (2001), and recently used for regional projections of twenty-first century sea-level change based on IPCC AR4 SRES scenarios (Slangen et al. 2012). In Slangen et al. (2012), the mass balance sensitivity ($\frac{d\dot{b}}{dT}$) is differentiated between summer and non-summer months, accounting for seasonality in glacier mass balance. Future scenarios of temperature and precipitation changes are taken from an ensemble of 12 global climate models (GCMs), and the results for the A1B emission scenario show a glacier contribution to the twenty-first century sea-level rise of 130–250 mm.

Hock et al. (2009) used a mass-balance sensitivity approach to reconstruct the average global glacier mass balance spatially resolved on a 0.5° global glacier grid (Cogley 2003) for the period 1961–2004 using gridded reanalysis temperature and precipitation trends. At the time of publication, this study was the only alternative approach to spatial interpolation of local mass-balance observations (Sect. 2.1). Their global estimate of 0.79 ± 0.34 mm SLE year^{-1} was larger than the consensus estimate of 0.50 ± 0.22 mm SLE year^{-1} by Kaser et al. (2006) for the same period, mainly due to large modeled mass loss of glaciers peripheral to the Antarctic ice sheet (28 % of the global estimate), where large temperature sensitivities, temperature trends, and glacier area combined to yield large mass losses. In contrast, the Gardner et al. (2013) ICESat analyses found very little mass loss for the Antarctic periphery during 2003–2009.

3.2 Models of surface mass balance

This approach directly models the evolution of surface mass balance in time by simulating surface melting and accumulation using climate data. Melt is most commonly modeled by so-called degree-day models, mainly because of their simplicity and the fact that the positive degree days are shown to be good indicators for glacier melt (Ohmura 2001; Hock 2003). Raper and Braithwaite (2006) were the first to perform global-scale projections of glacier mass balance based on a degree-day model. Resulting mass-balance gradients were regressed against annual precipitation and summer temperature from gridded climatology, and the relation applied to all 1° × 1° grid cells with glaciers (Cogley 2003). Based on the initial, calibrated equilibrium line altitudes (ELAs), upscaled glacier size distributions for each glacier grid cell, and derived vertical extent for each glacier, the model was run by perturbing the ELAs according to summer temperature anomalies. The resulting changes in total area and area–altitude distribution were computed annually with a simple glacier geometry model assuming a generic area–altitude distribution triangular in shape between its minimum and maximum altitude. Driven by climate data from two GCMs with A1B emission scenario, the projected sea-level rise for all glaciers, but excluding the glaciers peripheral to the Antarctic and Greenland ice sheet, was 46 and 51 mm for 2001–2100 (Table 2).

Hirabayashi et al. (2010) used a degree-day model specifically designed to feed into a global hydrological model. Consistent with the resolution of the latter model, the mass-balance model was run with daily time steps and on a 0.5 × 0.5° grid, treating each grid cell's glacier area as one large glacier, but allowing for sub-grid elevation bands. The model was initially used for the reconstruction of mass balance for the period 1948–2004, where gridded datasets of daily precipitation and near-surface temperature (Hirabayashi et al. 2005, 2008) were used as forcing. The modeled parameters were tuned to maximize the match between modeled and observed mass balance from 110 glaciers (Dyurgerov and Meier 2005); thus, the modeled global mass balance of 0.42 ± 0.15 mm SLE almost replicated the consensus estimate from Kaser et al. (2006). Recently, the model has been refined and run with the new Randolph glacier inventory (Arendt et al. 2012) to project glacier mass changes in response to the more extreme climate scenario (RCP8.5) from 10 GCMs prepared for the IPCC AR5 (Hirabayashi et al. 2013). They projected global glacier mass loss, excluding glaciers peripheral to the ice sheets, to be 73 ± 14 mm SLE for the period 2006–2099 (Table 2).

Radić and Hock (2011) developed a global-scale mass-balance model for the elevation-dependent mass balance of each individual glacier in the world glacier inventory by Cogley (2009a). The inventory comprised ~ 120,000 glaciers, covering 40 % of the total global glacier area. A degree-day model was calibrated using in situ mass-balance observation from 36 glaciers. The parameter values for all other glaciers were derived from established relationships with climate variables. Projections were made in response to downscaled monthly temperature and precipitation scenarios of ten GCMs from IPCC AR4 based on the A1B emission scenario. For the regions with incomplete glacier inventories, the projected volume changes were upscaled with a scaling relationship between regional ice volume change and regional glacierized area. The multi-model mean suggested sea-level rise of 112 ± 37 mm for the period 2001–2100. In a follow-up study, Radić et al. (2013)

updated the projections by using the new Randolph Glacier Inventory (Arendt et al. 2012). They modeled volume change for each glacier in response to transient spatially differentiated temperature and precipitation projections from 14 GCMs with two emission scenarios (RCP4.5 and RCP8.5) prepared for the IPCC AR5. Radić et al. (2013) arrived at much higher values than Hirabayashi et al. (2013) for the period 2006–2100: 155 ± 41 mm SLE (RCP4.5) and 216 ± 44 mm SLE (RCP8.5), and projected the largest regional mass losses from the Canadian and Russian Arctic, Alaska, and glaciers peripheral to the Antarctic and Greenland ice sheets. Although small contributors to global volume loss, glaciers in Central Europe, low-latitude South America, Caucasus, North Asia, and Western Canada and USA were projected to lose more than 80 % of their volume by 2100 (Fig. 1. Note that the region names are adopted from the Randolph Glacier Inventory).

Marzeion et al. (2012) applied a similar approach to model global mass balances to reconstruct the mass changes in the past and project future glacier mass evolution. Following Radić and Hock (2011), they modeled the surface mass balance of each individual glacier in the Randolph Glacier Inventory and coupled it with volume–area and volume–length scaling to account for glacier dynamics. The model was validated by a cross validation scheme using observed in situ and geodetic mass balances. When forced with observed monthly precipitation and temperature data, the world's glaciers are reconstructed to have lost mass corresponding to 114 ± 5 mm SLE between 1902 and 2009. Using projected temperature and precipitation anomalies for 2006–2100 from 15 GCMs prepared for IPCC AR5, the glaciers are projected to lose 148 ± 35 mm SLE (scenario RCP2.6), 166 ± 42 mm SLE (scenario RCP4.5), 175 ± 40 mm SLE (scenario RCP6.0), and 217 ± 47 mm SLE (scenario RCP8.5). Based on the extended RCP scenarios, glaciers are projected to approach a new equilibrium toward the end of the twenty-third century, after having lost 248 ± 66 mm SLE (scenario RCP2.6), 313 ± 50 mm SLE (scenario RCP4.5), or 424 ± 46 mm SLE (scenario RCP8.5).

Giesen and Oerlemans (2013) provided an alternative to the degree-day modeling approaches and projected global glacier mass changes using a simplified surface energy balance model. The model separates the melt energy into contributions from net solar radiation (computed by multiplying the incoming solar radiation at the top of the atmosphere by atmospheric transmissivity and subtracting the part of the incoming radiation that is reflected by the surface) and all other fluxes expressed as a function of air temperature. The model was calibrated on 89 glaciers with mass-balance observations, whose mass changes were then projected in response to A1B emission scenario from 8 GCMs from IPCC AR4. Volume–area scaling was applied to account for changes in glacier hypsometry. The simulated volume changes from 89 glaciers were then statistically upscaled to all glaciers in Randolph Glacier Inventory larger than 0.1 km^2, resulting in 102 ± 28 mm SLE for the period 2012–2099.

3.3 Model limitations

The models above are subject to large simplifications necessary for operation on global scales. Transferability of model parameters in time and space is questionable (e.g., Carenzo et al. 2009; MacDougall and Flowers, 2011). In addition, some studies have pointed out that variations in solar radiation have a significant effect on glacier mass changes (e.g., Ohmura et al. 2007; Huss et al. 2009). To address these concerns, a better approach than the generally applied degree-day approach would be to apply a physically based mass-balance model, accounting for all energy and mass fluxes at the glacier scale (Hock 2005). These high-complexity models have been applied successfully on many individual glaciers worldwide (e.g., Klok and Oerlemans 2002; Reijmer and Hock 2008; Mölg et al. 2009;

Anderson et al. 2010). However, the models require detailed meteorological input data, obtained at a glacier surface, which often are not available. Alternatively, these data can be obtained by dynamical downscaling of climate reanalysis products, i.e., by running mesoscale atmospheric models at high spatial resolution (less than 1 km in horizontal) over a region of interest. This approach has only recently been attempted in studies of glacier melt over a few summer seasons in Kilimanjaro and Karakoram (Mölg and Kaser 2011; Collier et al. 2013). Despite promising results, the applicability of this approach in order to simulate long-term surface mass balance on regional scale still needs to be investigated. In addition, the validation of surface mass-balance models should ideally be performed on sub-annual temporal scales, e.g., comparing modeled versus observed winter and summer mass balances, rather than only annual net mass balances. However, very few glaciers with annual mass-balance observations have the seasonally resolved components.

The representation of glacier dynamics using volume–area scaling remains a first-order approximation that is necessitated by the lack of input and validation data needed for physically based ice dynamics models. However, as shown by Lüthi (2009), volume–area scaling has some serious shortcomings in modeling glacier volume evolution. Glacier flow models of high complexity have been successfully applied on individual mountain glaciers (e.g., Picasso et al. 2004; Deponti et al. 2006; Jouvet et al. 2009). However, it is challenging to simulate the flow of a full suite of glaciers within a region of complex topography (Jarosch et al. 2012). Such ice-flow models require detailed information of the underlying bedrock topography, which has been observed for fewer than 1 % of glaciers in the world (Huss and Farinotti 2012). In the absence of abundant measured data on glacier thickness and volume, various alternative approaches to derive ice thicknesses have recently been developed (e.g., Clarke et al. 2012; Huss and Farinotti 2012; Linsbauer et al. 2012; McNabb et al. 2012). In particular, promising is the first globally complete dataset of glacier bed topographies derived from inverse modeling by Huss and Farinotti (2012), which will open new avenues for modeling glacier dynamics on the global scale.

To our knowledge, none of the current global-scale modeling studies of glacier volume changes incorporates frontal ablation, i.e., mass loss by iceberg calving or submarine melt of marine-terminating glaciers. Studies on marine-terminating ice caps have shown that calving may account for roughly 30 % to the total ablation (e.g., Dowdeswell et al. 2002, 2008), a significant contribution if widely applicable. Burgess et al. (2013) found that regional-scale losses by frontal ablation in Alaska are equivalent to 36 % of the total annual net mass loss of the region. Gardner et al. (2013) estimated that the present-day percentage of glacierized area (excluding the ice sheets) draining into the ocean is $\sim$ 35 %. Hence, the projections of volume loss, in which only the loss due to the surface mass balance is modeled, represent a lower bound. However, estimates of frontal ablation are scarce and lacking on a global scale. Nevertheless, it may be expected that the fraction of total mass change due to frontal ablation will decrease as warming and terminus retreat proceed (McNabb et al. 2012; Colgan et al. 2012).

4 Glacier runoff

4.1 Effects of glaciers on streamflow

Glaciers significantly modify streamflow both in quantity and timing, even with low percentages of catchment ice cover (e.g., Meier and Tangborn 1961; Fountain and Tangborn 1985; Chen and Ohmura 1990; Hopkinson and Young 1998; see Hock et al.

2005 for review). Characteristics of glacier discharge include pronounced melt-induced diurnal fluctuations with daily peaks reaching several fold the daily minimum flows during precipitation-free days. Glacier runoff shows distinct seasonal variations with very low winter flows and a larger and seasonally delayed summer peak compared to non-glacierized basins. Hence, glaciers can sustain streamflow during dry summer months and compensate for otherwise reduced flows. Year-to-year variability is dampened by the presence of glaciers in a catchment with a minimum reached at 10–40 % of glacierization (Lang, 1986). This so-called glacier compensation effect occurs when glacier runoff offsets precipitation variations. Glaciers may also cause sudden floods, often referred to as Joekulhlaups, posing a potential hazard for downstream populations. Common causes include subglacial volcanic eruptions or sudden drainage of moraine- or ice-dammed glacial lakes (e.g., Lliboutry et al. 1977; Bjornsson 2002).

Annual runoff from a glacierized basin is a function of glacier mass balance, with years of negative balance producing more runoff than years of positive balance. As climate changes and causes specific glacier mass balances to become progressively more negative, total glacier runoff will initially increase followed by a reduction in runoff totals as the glaciers retreat (Janson et al. 2003). With high percentage of ice cover, the initial increase in runoff can be substantial, considerably exceeding the runoff changes to be expected from any other component of the water budget. Adalgeirsdottir et al. (2006) modeled an increase in annual runoff from ice caps in Iceland of up to 60 % until about 2100 followed by a rapid decrease thereafter. However, in the long term, the loss of ice will lead to lower watershed yields of water. Observations from gauge records in glacierized basins show both increases in runoff, for example, along the coast in southern Alaska (Neal et al. 2002) or northwestern British Columbia (Fleming and Clarke 2003) and negative trends in summer streamflow, for example in the southern Canadian Cordillera (Stahl and Moore 2006). The replacement of ice by temperate forest and alpine vegetation will further decrease water yields.

In addition to contributing directly to runoff through ice wastage, glacier coverage within a watershed decreases direct evaporation and plant transpiration, the combination of which can result in substantially higher water yields for watersheds with glaciers compared to unglacierized watersheds (Hood and Scott 2008). In addition, the proportion of streamflow derived from glacial runoff has profound effects on physical (Kyle and Brabets 2001), biogeochemical (Hodson et al. 2008; Hood and Berner 2009; Bhatia et al. 2013), and biological (Milner et al. 2000; Robinson et al. 2001) properties of streams. As a result, changes in watershed glacial coverage also have the potential to alter riverine material fluxes. For example, area-weighted watershed fluxes of soluble reactive phosphorus decrease sharply with decreasing watershed glacial coverage (Hood and Scott 2008). Recent evidence also suggests dissolved organic material contained in glacial runoff has a microbial source and is highly labile to marine heterotrophs (Hodson et al. 2008; Hood et al. 2009).

4.2 What is glacier runoff?

There is substantial ambiguity in the literature with respect to the way the importance of glacier contribution to total runoff is quantified. Different concepts have been used (Table 3), and the importance will depend on how glacier runoff is defined. First, in its most general sense, glacier runoff is defined as the runoff from the glacierized area, and hence it includes all runoff exiting a glacier usually in one or several streams at the glacier terminus (Concept 1 in Table 3; Fig. 2). According to this definition, it is the residual in

Table 3 Different concepts found in the literature to assess the importance of glaciers in total river runoff. Q_g is glacier runoff, P_l is liquid precipitation, E is evaporation, M is melt, R is refreezing melt or rainwater, and C is snow accumulation

Concept	Equation	Description
1[a]	$Q_g = M_{ice,firn,snow} - R + P_l - E$	All runoff from glacierized area
2	$Q_g = M_{ice,firn,snow} - R$	Glacier meltwater runoff
3	$Q_g = M_{ice/firn}$	Ice/firn melt (melt from snow-free surface of the glacier)
4	$Q_g = C - M_{ice,firn,snow} - R$ (for $Q_g > 0$)	Runoff from glacier net mass loss
5	$Q_g = C - M_{ice,firn,snow} - R + P_l - E$	As 4, but including other water balance components (water balance approach)
6	$Q_g = C$	Runoff assuming balanced glacier mass budget if the budget is negative

All quantities are integrated over the glacierized area

[a] Lateral or subglacial inflows/outflows are neglected here.

the water balance equation over the area of the glacier and corresponds to the quantity that is directly measured by gauging the proglacial stream(s) at the glacier terminus. Thus, glacier runoff includes the portion of all water inputs to the glacier through melt, rain, or other inflow at the surface, laterally or subglacially that exit the glacier at the terminus.

Second, the term has also been used to describe only the component of runoff that comes from the melting of the glacier itself, i.e., from glacier ice, snow, and firn (i.e., snow that has survived at least one melt season but has not been transformed to glacier ice yet), hence excluding any rainwater or other inflow to the glacier system (Concept 2). This component is more accurately referred to as glacier meltwater runoff (Cogley et al. 2011).

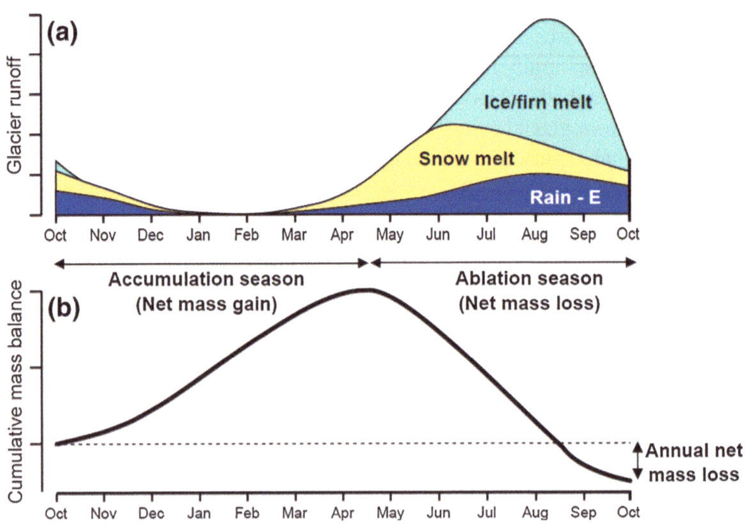

Fig. 2 **a** Schematic seasonal variation of total glacier runoff and its components following Concept 1 (Table 3), E is evaporation. **b** cumulative glacier mass balance in specific units (m w.e. year^{-1}) showing a year with negative annual balance. According to Concept 4 (Table 3), annual glacier runoff corresponds to the annual mass loss

It is important to note that this definition is not equivalent to the glacier's mass balance (budget) since it does not include the accumulation term but only the fraction of meltwater that does not refreeze, and exits the glacier.

Third, sometimes glacier runoff is understood as the meltwater runoff originating solely from ice/firn melt, i.e., melt of snow on the glacier is accounted for separately (Koboltschnik et al. 2007; Weber et al. 2010). This definition is consistent with the view that all other components (snow melt, rain etc.) would exist for the glacierized area even if the glacier did not exist. Hence, only the excess water due to the presence of the glacier is considered (Concept 3). This component is difficult to measure directly since it requires detailed measurements of melt at the surface in concert with the observations of snow line retreat and therefore is better quantified by mass-balance modeling which can separate the components of mass change.

Glacier runoff following Concepts 1–3 will affect river runoff in a glacierized catchment no matter whether or not the glacier over the time span considered had a positive, a negative, or a balanced mass budget. In contrast, glacier runoff is sometimes defined as the runoff component that is due to glacier net mass loss, hence referring only to the water originating from the glacier volume (storage) change (Concept 4, Huss 2011). Lambrecht and Mayer (2009) refer to this component as "excess discharge" since it constitutes additional water due to the reduced storage volume of glaciers that is not available in unglacierized catchments. Accordingly, in contrast to Concepts 1–3, glacier runoff is zero when the glacier's mass budget is balanced or positive, no matter how much meltwater is leaving the glacier. Hence, a glacier only affects runoff if there is a net mass loss during the considered time period. In this case, the glacier runoff is equivalent to the glacier's (negative) mass budget, which can be measured directly using the methods described in Sect. 2. Some studies have extended this definition to include the balance of liquid precipitation and evaporation (P_l-E; Concept 5; Dyurgerov 2010). Finally, Kaser et al. (2010) consider glacier mass loss assuming a balanced annual mass budget, i.e., water from net mass loss is not considered. In this case, annual glacier runoff effectively corresponds to annual snow accumulation (Concept 6).

In summary, definitions vary with respect to the inclusion of water not generated from melt and whether snow accumulation is included. Snowmelt runoff from the glacier can be substantial (Fig. 2) and is included in some, but excluded in other studies. It is obvious that the absolute amounts of glacier runoff and the degree to which glacier runoff affects total runoff of a glacierized catchment depend on the concept used in defining glacier runoff. It is paramount that any investigations aimed at assessing the importance of glacier runoff in total runoff clearly define the quantity used.

5 Assessing global-scale impacts of glaciers on the hydrological cycle

Analyses based on the observations or modeling in individual glacierized river basins have highlighted the role of glaciers in the hydrological cycle and indicated significant hydrological changes in response to climate change, including changes in total water amounts and seasonality (e.g., Braun et al. 2000; Casassa et al. 2006; Rees and Collins 2006; Hagg et al. 2006; Horton et al. 2006; Yao et al. 2007; Huss et al. 2008; Immerzeel et al. 2013; Koboltschnik et al. 2008; Stahl and Moore 2006; Kobierska et al. 2013). However, few studies have investigated the hydrological effects of glaciers on regional or global scales.

Dyurgerov (2010) updated an earlier study by Dyurgerov and Carter (2004) and investigated the role of glaciers in freshwater inflow to the Arctic Ocean by comparing the estimates of river discharge from gauging stations to estimates of meltwater fluxes and annual mass changes of all glaciers draining to the Arctic Ocean including the Greenland ice sheet. Annual glacier runoff (Concept 5, Table 3) was found to have increased substantially from 1961 to 1992 to 1993–2006 (from 146 ± 338 to 202 ± 48 Gt a^{-1}) while glacier mass loss more than doubled. The increase in glacier runoff was the same order of magnitude as the observed increase in river runoff (Bring and Destouni 2011), suggesting an important role of glacier melt in Arctic freshwater budgets.

Neal et al. (2010) adopted a water balance approach to estimate the contribution of glacier runoff to freshwater discharge into the Gulf of Alaska, a 420,230 km^2 watershed covered 18 % by glaciers. Glacier runoff (Concept 1) contributed 47 % of the total runoff (870 km^3 a^{-1}), with 10 % originating from glacier net mass loss alone (Concept 4, Table 3).

Dyurgerov (2010) analyzed all available mass-balance profiles, which describe the distribution of mass change with altitude, and found an increase in both accumulation and ablation in the observed period (1961–2006), with major increases since the late 1980s, and a steepening of the mass-balance gradient. The latter was attributed to an increase in meltwater production at low elevations combined with more snow accumulation at higher elevations and interpreted as evidence of an intensified hydrological cycle in times of global warming.

Huss (2011) assessed the contribution of glaciers to runoff from large-scale drainage basins in Europe with areas up to 800,000 km^2 over the period 1908–2008 based on modeled monthly mass budget estimates for all glaciers in the European Alps. The glacier runoff defined as the water due to glacier mass change (Concept 4, Table 3) was computed for each month and compared to monthly river runoff measured at gauges along the entire river lengths. Although glacierization of the investigated basins did not exceed 1 % of the total area, the maximum monthly glacier contributions during summer ranged from 4 to 25 % between catchments, indicating that seasonal glacier contributions can be significant even in basins with little ice cover. Comeau et al. (2009) analyzed annual runoff in a large catchment in Western Canada and found that reductions in glacier volume due to receding glaciers (Concept 4, Table 3) contributed 3 % to total runoff during 1975–1998.

Kaser et al. (2010) performed the only global-scale study on the effects of glaciers on freshwater resources and provided a first-order estimate of the role of glaciers to water availability and their societal importance. For 18 large glacierized river basins, the fraction of runoff that is seasonally delayed by glaciers was computed based on gridded climatologies and theoretical considerations rather than glacier mass balance and runoff data. Monthly accumulation was computed as a function of elevation using gridded climatological data from the Climatic Research Unit (CRU). Assuming the glaciers to be in equilibrium with climate, an equal amount of annual ablation was distributed to each month based on monthly air temperatures. Any excess ablation beyond accumulation, for a given month, was considered seasonally delayed glacier runoff and weighted with population to assess the societal impact of delayed runoff. Results showed that seasonally delayed glacier runoff is most significant in seasonally arid regions and of moderate importance in midlatitude basin, but negligible in lowland basins affected by monsoon climates. This underlines the importance of climate regimes in determining the importance of glaciers on runoff.

6 Synthesis and discussion

Accelerated glacier wastage in many parts of the world and the resulting impacts on sea-level rise and water resources is a topic of global concern. Mass losses from the Earth's mountain glaciers and ice caps contribute to the freshwater influx to the ocean and make up one-third of recent sea-level rise (the remaining parts come in equal shares from ice sheet mass losses and thermal expansion of seawater). They also influence the runoff characteristics of glacierized basins with significant effects even at low levels of glacierization. The expected changes in glacier runoff may be larger than those generally projected for other components of the water cycle.

The main impacts of glacier wastage vary regionally. For sea-level rise, the most important regions are found in high-latitude regions where large ice volumes are typical, such as the Antarctic and Greenland peripheries, Canadian Arctic, Alaska, and the Russian Arctic (Gardner et al. 2013). In contrast, mid- and low-latitude regions (e.g., European Alps, Scandinavia, Tropical Andes, and Western Canada/USA) have relatively little ice cover and therefore (except for the High Asian Mountains) less potential impact on sea-level change. However, many of these regions have relatively large populations and the hydrological consequences of glacier wastage are of concern.

Assessing and projecting the effects of glaciers on sea level and terrestrial hydrology requires accurate assessments of the glacier mass balance and its components. In recent years, much progress has been made in measuring glacier mass changes on regional and global scales, mostly due to the launch of the ICESat and GRACE satellites in the beginning of the twenty-first century. For the first time, regional scale mass-balance observations were possible in regions with sparse local in situ observations. These results will be valuable for calibration and validation of global hydrology models. Although the traditional technique of extrapolating local observations is problematic in regions with sparse data, as it can bias global results (Gardner et al. 2013), in situ measurements are essential for calibration and validation of glacier mass-balance and runoff models. Unfortunately, the number of mass-balance monitoring glaciers has declined in recent years.

Until recently, the lack of basic inventory data was a major impediment in global mass-balance assessments and projections resulting in large uncertainties in the results due to necessary upscaling procedures or other workarounds (e.g., Raper and Braithwaite 2005; Radić and Hock 2010). The recently completed Randolph Glacier Inventory, the first globally complete glacier inventory (Arendt et al. 2012), is a major step forward toward reducing uncertainties in global-scale studies. Also, for the first time, it has become possible to model global mass balances for each glacier in the world individually (Radić and Hock 2011; Marzeion et al. 2012; Radić et al. 2013). However, there is a large range in the twenty-first century projections from the three independent studies (Marzeion et al. 2012; Radić et al. 2013 and Hirabayashi et al. 2013) that use the new inventory despite using the same climate forcings (RCP scenarios) and largely overlapping selection of GCMs. Hirabayashi et al. (2013) projections are at the low end. Results also indicate that previously found large uncertainty due to the choice of the GCMs (e.g., Radić and Hock 2011) has not been reduced. Glacier models also still suffer from the omission of frontal ablation (calving and submarine melt) due to the inherent difficulty in modeling these processes and the lack of data to develop parameterizations suitable for regional scales.

For sea-level change calculations, rates of regional or global glacier net mass loss are generally converted into sea-level equivalent simply by dividing the volume of water lost by the ocean area (362.5×10^{12} m^2, Cogley et al. 2011), thus neglecting the effects of altering ocean area and terrestrial hydrology. The effect of flow of meltwater into groundwater aquifers or enclosed basins rather than the oceans is virtually unknown and should be addressed by coupling glacier mass-balance models with global hydrology models. For future scenarios, it is important that hydrology models have the capacity to model glacier retreat. Few studies on local scales have incorporated simple parameterizations (Stahl et al. 2008; Huss et al. 2010) into their glacier runoff models; however, while there are examples of macroscale models using glacier models for local applications (Zhao et al. 2013), we are not aware of any current global-scale watershed models (e.g., Hanasaki et al. 2008; Wisser et al. 2010) incorporating glacier modeling in macroscale applications. The Randolph Glacier Inventory will further facilitate the inclusion of glacier mass changes into global hydrology models.

Many studies on various spatial scales have investigated the effects of glaciers on hydrology under a warming climate. Generally, annual glacier runoff is found to increase initially due to increased meltwater, followed by reduced flows as glaciers recede, and their ability to augment flows diminishes. However, contradictory results are reported with regard to the importance of glacier runoff relative to total runoff in glacierized catchments (e.g., Weber et al. 2010; Huss 2011). While this can at least partially be attributed to differences in physical factors such as climate regimes, catchment size, degree of glacierization, or glacier mass change rates, these differences also depend on the way the glacier runoff is quantified. In fact, studies on the relative importance of glaciers for runoff are difficult to compare, because authors use different concepts to compute the contribution of glaciers to runoff (Table 3). Definitions of glacier runoff fall into two principal categories (Comeau et al. 2009): (1) those that only consider the net mass loss component of a glacier due to glacier wastage, i.e., runoff is zero (Concept 4) or equal to P_l-E (Concept 5, Table 3) if the glacier is in balance or gains mass, and (2) those that consider all meltwater originating from a glacier no matter the magnitude or sign of the mass budget (Concepts 1–3, Table 3). It is obvious that for the concepts in (2), glacier runoff generally is much larger than for the concepts in (1), and consequently the relative importance of glacier runoff to total runoff will differ between these two categories.

Concepts based on net glacier mass loss are most useful over annual timescales as they provide a measure for how much water is added to (or withdrawn from) the hydrological cycle through glacier volume storage changes. In contrast, concepts considering all meltwater are useful on seasonal timescales in order to assess the effects of glaciers on seasonal hydrographs. Precipitation that has fallen as snow is released later during the melt season and hence modulates the seasonality of flow even if the glacier's annual mass budget is zero. Such concepts are also useful on longer timescales, for example when physical properties of the meltwater, such as temperature or conductivity, are of relevance.

Considering only ice or firn melt (Concept 3, Table 3) aims to isolate the effects of glaciers on seasonal or annual flows compared to non-glacierized catchments. Thus, melt of snow on the glacier surface is excluded from the glacier contribution because this component also occurs in unglacierized catchments. However, this approach is not unproblematic since typically some winter snow remains on the glacier by the end of each melt season, a necessity for a glacier to survive. Hence, in contrast to unglacierized regions, snowmelt from the glacier surface occurs over the entire length of the summer (Fig. 2) and therefore is a characteristic feature of a glacier that is eliminated in Concept 3. The surviving winter snow is also directly linked to the glacier system through subsequent transformation of snow to ice.

Overall, all concepts found in the literature are legitimate, and the choice of concept will depend on the purpose of the investigation. It is paramount that glacier runoff is clearly defined to avoid confusion and allow fair comparison between studies.

7 Conclusions

In light of strongly accelerated glacier wastage, there is an urgent need for further investigations quantifying and projecting the changes in glacier mass and runoff, and their importance for the Earth's hydrological cycle. We identify the following issues that need special attention:

- The current decline of the in situ glacier monitoring programs is a matter of concern. Although the remote sensing techniques have overcome many obstacles encountered by the traditional in situ observations, the latter are essential for calibration and validation of glacier mass-balance and runoff models.
- Despite the recent progress in the development of the global-scale glacier models, they still suffer from the omission of physics-based simulation of glacier dynamics and frontal ablation (calving and submarine melt).
- The effect of flow of meltwater into groundwater aquifers or enclosed basins is virtually unknown and should be addressed by coupling glacier mass-balance models with global hydrological models.
- For future scenarios, it is important that these hydrological models have the capacity to model glacier retreat.
- It is essential that glacier runoff is clearly defined in studies aiming to quantify the contribution of glacier runoff to streamflow to avoid confusion and facilitate fair comparison between studies.

Acknowledgments This study was supported by grants from NSF (EAR 0943742, EAR 1039008) and NASA (NNX11AO23G, NNX11AF41G). H. Feilhauer assisted with Fig. 2.

References

Abdalati W, Krabill W, Frederick E, Manizade S, Martin C, Sonntag J, Swift R, Thomas R, Yungel J, Koerner R (2004) Elevation changes of ice caps in the Canadian Arctic Archipelago. J Geophys Res 109 (F04007). doi:10.1029/2003JF000045

Adalgeirsdottir G, Johannesson T, Bjornsson H, Palsson F, Sigurdsson O (2006) Response of Hofsjokull and southern Vatnajokull, Iceland, to climate change. J Geophys Res 111(F03001). doi:10.1029/2005JF000388

Adhikari S, Marshall SJ (2012) Glacier volume-area relation for high-order mechanics and transient glacier states. Geophys Res Lett 39(L16505). doi:10.1029/2012GL052712

Anderson B, MacKintosh A, Stumm D, George L, Kerr T, Winter-Billington A, Fitzsimons S (2010) Climate sensitivity of a high-precipitation glacier in New Zealand. J Glaciol 56(195):114–128

Arendt A, Echelmeyer K, Harrison W, Lingle C, Valentine VB (2002) Rapid wastage of Alaska glaciers and their contribution to rising sea level. Science 297:382–386

Arendt A et al (2012) Randolph glacier inventory: a dataset of global glacier outlines version: 2.0. GLIMS Technical Report

Bahr DB, Meier MF, Peckham SD (1997) The physical basis of glacier volume-area scaling. J Geophys Res 102:20355–20362

Bahr DB, Dyurgerov M, Meier MF (2009) Sea-level rise from glaciers and ice caps: a lower bound. Geophys Res Lett 36:L03501. doi:10.1029/2008GL036309

Berthier E, Schiefer E, Clarke GKC, Menounos B, Remy F (2010) Contribution of Alaskan glaciers to sea-level rise derived from satellite imagery. Nature Geosci 3:92–95

Bhatia MP, Kujawinski EB, Das SB, Breier CF, Henderson PB, Charette MA (2013) Greenland meltwater as a significant and potentially bioavailable source of iron to the ocean. Nature Geo 6:274–278. doi:10.1038/ngeo1746

Bjornsson H (2002) Subglacial lakes and jokulhlaups in Iceland. Global Planet Change 35:255–271

Bjornsson H, Palsson F, Gudmundsson S, Magnusson E, Adalgeirsdottir G, Johannesson T, Berthier E, Sigurdsson O, Thorsteinsson T (2013) Contribution of Icelandic ice caps to sea level rise: trends and variability since the Little Ice Age. Geophys Res Lett 40:1–5. doi:10.1002/grl.50278

Bolch T, Sandberg Sørensen L, Simonsen SB, Molg N, Machguth H, Rastner P, Paul F (2013) Mass loss of Greenland's glaciers and ice caps 2003–2008 revealed from ICESat laser altimetry data. Geophys Res Lett 40:875–881. doi:10.1002/grl.50270

Braithwaite RJ (2002) Glacier mass balance: the first 50 years of international monitoring. Progress in Phys Geogr 26(1):76–95

Braithwaite RJ, Zhang Y (1999) Modelling changes in glacier mass balance that may occur as a result of climate changes. Geogr Ann 81A(4):489–496

Braithwaite RJ, Zhang Y (2000) Sensitivity of mass balance of five Swiss glaciers to temperature changes assessed by tuning a degree-day model. J Glaciol 46(152):7–14

Braun LN, Weber M, Schulz M (2000) Consequences of climate change for runoff from Alpine regions. Ann Glaciol 31(1):19–25

Bring A, Destouni G (2011) Relevance of hydro-climatic change projection and monitoring for assessment of water cycle changes in the Arctic. Ambio 40:361–369

Burgess EW, Forster RR, Larsen CF (2013) Flow velocities of Alaskan glaciers. Nat Commun 4:2146. doi:10.1038/ncomms3146

Carenzo M, Pellicciotti F, Rimkus S, Burlando P (2009) Assessing the transferability and robustness of an enhanced temperature-index glacier melt model. J Glaciol 55(190):258–274

Casassa G, Rivera A, Schwikowski M (2006) Glacier mass balance data for southern South America (30°S - 56°S)". KNIGHT, P.G., ed., Glacier Science and Environmental Change, Blackwell, Oxford, UK, In, pp 239–241

Chen J, Ohmura A (1990) On the influence of Alpine glaciers on runoff. In: Lang H, Musy A (Eds) Hydrology in Mountainous Regions I, IAHS Publ 193: 117-125

Chen JL, Tapley BD, Wilson CR (2006) Alaskan mountain glacial melting observed by satellite gravimetry. Earth Planet Sci Lett 248(1–2):368–378

Chen JL, Wilson CR, Tapley BD, Blankenship DD, Ivins ER (2007) Patagonia Icefield melting observed by Gravity Recovery and Climate Experiment (GRACE). Geophys Res Lett 34:L22501. doi:10.1029/2007GL031871

Clarke GKC, Anslow FS, Jarosch AH, Radić V, Menounos B, Bolch T, Berthier E (2012) Ice volume and subglacial topography for western Canadian glaciers from mass balance fields, thinning rates, and a bed stress model. J Clim, e-View. doi:10.1175/JCLI-D-12-00513.1

Cogley JG (2003) GGHYDRO—global hydrographic data, release 2.3. Trent Technical Note 2003-1, Department of Geography, Trent University, Peterborough, Ont. [http://www.trentu.ca/geography/glaciology.]

Cogley JG (2005) Mass and energy balances of glaciers and ice sheets, in M. G. Anderson, ed., Encyclopedia of Hydrological Sciences, p 2555–2573

Cogley JG (2009a) A more complete version of the World Glacier Inventory. Ann Glaciol 50(53):32–38

Cogley JG (2009b) Geodetic and direct mass-balance measurements: comparison and joint analysis. Ann Glaciol 50(50):96–100

Cogley JG (2011) The future of the world's climate (2011) Chapter 8

Cogley JG, Hock R, Rasmussen LA, Arendt AA, Bauder A, Braithwaite RJ, Jansson P, Kaser G, Möller M, Nicholson L, Zemp M (2011) Glossary of glacier mass balance and related terms, technical documents in hydrology No. 86, UNESCO-IHP, Paris

Colgan W, Pfeffer WT, Rajaram H, Abdalati W, Balog J (2012) Monte Carlo ice flow modeling projects a new stable configuration for Columbia Glacier, Alaska, c. 2020. The Cryosphere 6:1395–1409. doi:10.5194/tc-6-1395-2012

Collier E, Mölg T, Maussion F, Scherer D, Mayer C, Bush ABG (2013) High-resolution interactive modelling of the mountain glacier–atmosphere interface: an application over the Karakoram. The Cryosphere Discuss 7:103–144. doi:10.5194/tcd-7-103-2013

Comeau LEL, Pietroniro A, Demuth MN (2009) Glacier contribution to the North and South Saskatchewan Rivers. Hydrol Process 23:2640–2653. doi:10.1002/hyp.7409

de Woul M, Hock R (2005) Static mass balance sensitivity of Arctic glaciers and ice caps using a degree-day approach. Ann Glaciol 42:217–224

Deponti A, Pennati V, de Biase L, Maggi V, Berta F (2006) A new fully three-dimensional numerical model for ice dynamics. J Glaciol 52(178):365–377

Dowdeswell JA, Bassford RP, Gorman MR, Williams M, Glazovsky AF, Macheret YY, Shepherd AP, Vasilenko YV, Savatyuguin LM, Hubberten H-W, Miller H (2002) Form and flow of the Academy of

Sciences Ice Cap, Severnaya Zemlya. Russian High Arctic. J Geophys Res 107:B4. doi:10.1029/2000/JB000129

Dowdeswell J, Benham J, Strozzi T, Hagen JO (2008) Iceberg calving flux and mass balance of the Austfonna ice cap on Nordaustlandet. Svalbard. J Geophys Res 113:F03022. doi:10.1029/2007JF000905

Dyurgerov, MB (2002) Glacier mass balance and regime: data of measurements and analysis. In: Meier, M. F. and Armstrong, R., Institute of Arctic and Alpine Research 55, University of Colorado, Boulder

Dyurgerov MB (2003) Observational evidence of accelerated glacier wastage: Uncertainty in prediction. Workshop on Assessing Global Glacier Recession, 16–17 March 2003, Boulder: World Data Center for Glaciology

Dyurgerov MB (2010) Reanalysis of Glacier Changes: from the IGY to the IPY, 1960-2008. Data of Glaciological Studies 108:1–116

Dyurgerov MB, Carter CL (2004) Observational Evidence of Increases in Freshwater Inflow to the Arctic Ocean Arctic. Arct Antarct Alp Res 36(1):117–122

Dyurgerov MB, Meier MF (1997a) Mass balance of mountain and subpolar glaciers: a new global assessment for 1961–1990. Arct Antarct Alp Res 29:379–391

Dyurgerov MB, Meier MF (1997b) Year-to-year fluctuation of global mass balance of small glaciers and their contribution to sea level changes. Arct Antarct Alp Res 29:392–402

Dyurgerov MB, Meier MF (2005) Glaciers and the Changing Earth System: a 2004 Snapshot, Occasional Paper 58 Institute of Arctic and Alpine Research. University of Colorado, Boulder 118p

Dyurgerov MB, Meier MF, Bahr DB (2009) A new index of glacier area change: a tool for glacier monitoring. J Glaciol 55(192):710–716

Fleming SJ, Clarke GKC (2003) Glacial control of water resource and related environmental responses to climate warming: empirical analysis using historical streamflow data from northwestern Canada. Canadian Water Resources Journal 28(1):69–86

Fountain AG, Tangborn WV (1985) The effect of glaciers on streamflow variations. Water Resour Res 21(4):579–586

Gardner AS, Moholdt G, Wouters B, Wolken G, Burgess DO, Sharp MJ, Cogley JG, Braun C, Labine C (2011) Sharply increased mass loss from glaciers and ice caps in the Canadian Arctic Archipelago. Nature 437:357–360

Gardner A, Moholdt G, Cogley JG, Wouters B, Arendt A, Wahr J, Berthier E, Hock R, Pfeffer T, Kaser G, Ligtenberg S, Bolch T, Sharp M, Hagen JO, van den Broeke M, Paul F (2013) A Reconciled Estimate of Glacier Contributions to Sea Level Rise: 2003 to 2009. Science 340:852–857. doi:10.1126/science.1234532

Giesen RH, Oerlemans J (2013) Climate-model induced differences in the 21st century global and regional glacier contributions to sea-level rise. Clim Dyn doi. doi:10.1007/s00382-013-1743-7

Gleckler PJK, Taylor E, Doutriaux C (2008) Performance metrics for climate models. J Geophys Res 113:L06711. doi:10.1029/2007JD008972

Gregory JM, Oerlemans J (1998) Simulated future sea level sea level rise due to glacier melt based on regionally and seasonally resolved temperature changes. Nature 391:474–476

Hagg W, Braun LN, Weber M, Becht M (2006) Runoff modelling in glacierized Central Asian catchments for present-day and future climate. Nord Hydrol 37:93–105

Hanasaki N, Kanae S, Oki T, Masuda K, Motoya K, Shirakawa N, Shen Y, Tanaka K (2008) An integrated model for the assessment of global water resources - Part 1: model description and input meteorological forcing. Hydrol Earth Syst Sci 12:1007–1025

Hirabayashi Y, Kanae S, Struthers I, Oki T (2005) A 100-year (1901-2000) global retrospective estimation of the terrestrial water cycle. J Geophys Res 110:D19101. doi:10.1029/2004JD005492

Hirabayashi Y, Kanae S, Masude K, Motoya K, Döll P (2008) A 59-year (1948-2006) global near-surface meteorological data set for land surface models. Part I: development of daily forcing and assessment of precipitation intensity. Hydrological Research Letters 2:36–40. doi:10.3178/HRL.2.36

Hirabayashi Y, Doll P, Kanea S (2010) Global-scale modeling of glacier mass balances for water resources assessments: glacier mass changes between 1948 and 2006. J Hydrol 390(3–4):12

Hirabayashi Y, Zhang Y, Watanabe S, Koirala S, Kanae S (2013) Projection of glacier mass changes under a high-emission climate scenario using the global glacier model HYOGA2. Hydrological Research Letters 7(1):6–11. doi:10.3178/HRL.7.6

Hock R (2003) Temperature index melt modelling in mountain regions. J Hydrol 282(1–4):104–115. doi:10.1016/S0022-1694(03)00257-9

Hock R (2005) Glacier melt: a review on processes and their modelling. Prog in Phys Geogr 29(3):362–391

Hock R, Jansson P (2005) Modelling glacier hydrology. In: Anderson, M. G. and J. McDonnell (Eds.). Enzyclopedia of Hydrological Sciences, John Wiley & Sons, Ltd, hichester. 4: 2647–2655

Hock R, Jansson P, Braun L (2005) Modelling the response of mountain glacier discharge to climate warming. In: Huber UM, Reasoner MA, Bugmann H (eds) Global Change and Mountain Regions - A State of Knowledge Overview. Springer, Dordrecht, pp 243–252

Hock R, de Woul M, Radić V, Dyurgerov M (2009) Mountain glaciers and ice caps around Antarctica make a large sea-level rise contribution. Geophys Res Lett 36:L07501. doi:10.1029/2008GL037020

Hodson A, Anesio AM, Tranter M, Fountain A, Osborn M, Priscu J, Laybourn-Parry J, Sattler B (2008) Glacial Ecosystems. Ecol Monogr 78(1):41–67

Hood E, Berner L (2009) The effect of changing glacial coverage on the physical and biogeochemical properties of coastal streams in southeastern Alaska. J Geophys Res 114:G03001. doi:10.1029/2009JG000971

Hood E, Scott D (2008) Riverine organic matter and nutrients in southeast Alaska affected by glacial coverage. Nature Geosci 1:583–587

Hood E, Fellman J, Spencer RGM, Hernes PJ, Edwards R, D'Amore D, Scott D (2009) Glaciers as a source of ancient and labile organic matter to the marine environment. Nature 462:1044–1047. doi:10.1038/nature08580

Hopkinson C, Young GJ (1998) The effect of glacier wastage on the flow of the Bow River at Banff, Alberta, 1951-1993. Hydrol Process 12:1745–1762

Horton P, Schaefli B, Mezghani A, Hingray B, Musy A (2006) Assessment of climate-change impacts on alpine discharge regimes with climate model uncertainty. Hydrol Process 20:2091–2109. doi:10.1002/hyp.6197

Huss M (2011) Present and future contribution of glacier storage change to runoff from macroscale drainage basins in Europe. Water Resour Res 47:W07511. doi:10.1029/2010WR010299

Huss M, Farinotti D (2012) Distributed ice thickness and volume of all glaciers around the globe. J Geophys Res 117:F04010. doi:10.1029/2012JF002523

Huss M, Farinotti D, Bauder A, Funk M (2008) Modelling runoff from highly glacierized alpine drainage basins in a changing climate. Hydrol Processes 22:3888–3902. doi:10.1002/hyp.7055

Huss M, Funk M, Ohmura A (2009) Strong Alpine glacier melt in the 1940 s due to enhanced solar radiation. Geophys Res Lett 36:L23501. doi:10.1029/2009GL040789

Huss M, Jouvet G, Farinotti D, Bauder A (2010) Future high-mountain hydrology: a new parameterization of glacier retreat. Hydrol Earth Syst Sc 14:815–829

Immerzeel WW, van Beek LPH, Bierkens MFP (2010) Climate Change Will Affect the Asian Water Towers. Science 328:1382–1385. doi:10.1126/science.1183188

Immerzeel WW, Pelliciotti F, Bierkens MFP (2013) Rising river flows throughout the twenty-first century in two Himalayan glacierized watersheds. Nature Geosci 6:742–745. doi:10.1038/ngeo1896

Ivins ER, Watkins MM, Yuan D-N, Dietrich R, Casassa G, Rlke A (2011) On-land ice loss and glacial isostatic adjustment at the Drake Passage: 2003–2009. J Geophys Res 116:B02403

Jacob T, Wahr J, Pfeffer WT, Swenson S (2012) Recent contributions of glaciers and ice caps to sea level rise. Nature 482(7386):514–518. doi:10.1038/nature10847

Jansson P, Hock R, Schneider T (2003) The concept of glacier water storage - a review. J Hydrol 282(1–4):116–129. doi:10.1016/S0022-1694(03)00258-0

Jarosch AH, Schoof CG, Anslow FS (2012) Restoring mass conservation to shallow ice flow models over complex terrain. The Cryosphere 7:229–240. doi:10.5194/tc-7-229-2013

Jouvet G, Huss M, Blatter H, Picasso M, Rappaz J (2009) Numerical simulation of Rhonegletscher from 1874 to 2100. J Comput Phys 228(17):6426–6439. doi:10.1016/j.jcp.2009.05.033

Kaser G, Fountain A, Jansson P (2002) A manual for monitoring the mass balance of mountain glaciers. UNESCO, International Hydrological Programme, Technical Documents in Hydrology, No. 59.107 pp

Kaser G, Cogley JG, Dyurgerov MB, Meier MF, Ohmura A (2006) Mass balance of glaciers and ice caps: consensus estimates for 1961-2004. Geophys Res Lett 33:L19501

Kaser G, Grosshauser M, Marzeion B (2010) Contribution potential of glaciers to water availability in different climate regimes. Proc Natl Acad Sci U S 107:20223–20227. doi:10.1073/pnas.1008162107

Klok EJ, Oerlemans J (2002) Model study of the spatial distribution of the energy and mass balance of Morteratschgletscher. Switzerland. J Glaciology 48(163):505–518

Kobierska F, Jonas T, Zappa M, Bavay M, Magnusson J, Bernasconi SM (2013) Future runoff from a partly glacierized watershed in Central Switzerland: a two-model approach. Adv Water Resour 55:204–214

Koboltschnig GR, Schoner Wolfgang, Zappa M, Kroisleitner C, Holzmann H (2008) Runoff modelling of the glacierized Alpine Upper Salzach basin (Austria): multi-criteria result validation. Hydrol Process 22:3950–3964. doi:10.1002/hyp.7112

Koboltschnik GR, Schoner WS, Zappa M, Holzmann H (2007) Contribution of glacier melt to stream runoff: if the climatically extreme summer of 2003 had happened in 1979. Ann Glaciol 46:303–308

Konz M, Seibert J (2010) On the value of glacier mass balances for hydrological model calibration. J Hydrol 385:238–246. doi:10.1016/j.jhydrol.2010.02.025

Kyle RE, Brabets TB (2001) Water temperature of streams in the Cook Inlet basin, Alaska, and implications of climate change. US Geological Survey Water-Resources Investigation Report 01-4109

Lambrecht A, Kuhn M (2007) Glacier changes in the Austrian Alps during the last three decades, derived from the new Austrian glacier inventory. Ann Glaciol 46:177–184

Lambrecht A, Mayer C (2009) Temporal variability of the non-steady contribution from glaciers to water discharge in western Austria. J Hydrol 376:353–361

Lang H (1986) Forecasting meltwater runoff from snow-covered areas and from glacier basins. In Kraijenoff DA, Moll JR (Eds) River Flow Modelling and Forecasting. Reidel Publishing. Dordrecht pp 99–127

Lemke P, Ren J, Alley RB, Allison I, Carrasco J, Flato G, Fujii Y, Kaser G, Mote P, Thomas RH, Zhang T (2007) Observations: changes in snow, ice and frozen ground, in Solomon, S., et al., eds., Climate Change 2007: The Physical Science Basis. Contribution of Working Group I to the Fourth Assessment Report of the Intergovernmental Panel on Climate Change, 337–383. Cambridge University Press, Cambridge

Linsbauer A, Paul F, Haeberli W (2012) Modeling glacier thickness distribution and bed topography over entire mountain ranges with GlabTop: a new, fast and robust approach. J Geophys Res 117:F03007. doi:10.1029/2011JF002313

Lliboutry L, Morales Arnao B, Pautre A, Schneider B (1977) Glaciological problems set by the control of dangerous lakes in Cordillera Blance, Peru. I Historical failures of morainic dams, their causes and prevention. J Glaciol 18:239–254

Lüthi MP (2009) Transient response of idealized glaciers to climate change. J Glaciol 55(193):918–930

Luthcke SB, Arendt AA, Rowlands DD, McCarthy JJ, Larsen CF (2008) Recent glacier mass changes in the Gulf of Alaska region from GRACE mascon solutions. J Glaciol 54(188):767–777

MacDougall AH, Flowers GE (2011) Spatial and temporal transferability of a distributed energy-balance glacier melt-model. J Clim 24(5):1480–1498

Marzeion B, Jarosch AH, Hofer M (2012) Past and future sea-level change from the surface mass balance of glaciers. The Cryosphere 6:1295–1322. doi:10.5194/tc-6-1295-2012

Matsuo K, Heki K (2010) Time-variable ice loss in Asian high mountains from satellite gravimetry. Earth Planet Sci Lett 290:30–36

McNabb B, Hock R, O'Neel S, Rasmussen LA, Ahn Y, Conway H, Herreid S, Joughin I, Pfeffer T, Smith B, Truffer M (2012) Using surface velocities to infer ice thickness and bed topography: a case study at Columbia Glacier. Alaska. J Glaciol 58(212):1151–1164. doi:10.3189/2012JoG11J249

Meier M (1984) Contribution of Small Glaciers to Global Sea Level. Science 226(4681):1418–1421. doi:10.1126/science.226 4681.1418

Meier MF, Tangborn WV (1961) Distinctive characteristics of glacier runoff. US Geol Surv Prof Pap 424(B):14–16

Meier MF, Dyurgerov MB, Rick U, O'Neel S, Pfeffer WT, Anderson AS, Glazovsky AF (2007) Glaciers dominate eustatic sea-level rise in the 21st century. Science 317(5841):1064–1067. doi:10.1126/science.1143906

Mercanton PL ed. (1916) Vermessungen am Rhonegletscher/Mensuration au glacier du Rhone: 1874–1915. Neue Denkschr Schweiz Naturforsch Ges 52

Mernild S, Lipscomp W, Bahr D, Radić V, Zemp M (2013) Global glacier retreat: A revised assessment of committed mass losses and sampling uncertainties. The Cryosphere (in press)

Milner AM, Knudsen EE, Soiseth C, Robertson AL, Schell D, Phillips IT, Magnusson K (2000) Colonization and development of stream communities across a 200-year gradient in Glacier Bay National Park, Alaska, USA. Can J of Fish Aquat Sci 57(11):2319–2335

Moholdt G, Nuth C, Hagen JO, Kohler J (2010) Recent elevation changes of Svalbard glaciers derived from ICESat laser altimetry. Remote Sens Environ 114(11):2756–2767. doi:10.1016/j.rse.2010.06.008

Moholdt G, Wouters B, Gardner AS (2012) Recent mass changes of glaciers in the Russian High Arctic. Geophys Res Lett. doi:10.1029/2012GL051466, in press

Mölg T, Kaser G (2011) A new approach to resolving climate-cryosphere relations: downscaling climate dynamics to glacier-scale mass and energy balance without statistical scale linking. J Geophys Res 116:D16101. doi:10.1029/2011JD015669

Mölg T, Cullen NJ, Hardy DR, Winkler M, Kaser G (2009) Quantifying climate change in the tropical mid-troposphere over East Africa from glacier shrinkage on Kilimanjaro. J Clim 22:4162–4181

Müller-Lemans VH, Funk M, Aellen M, Kappenberger G (1994) Langjährige massenbilanzreihen von gletschern in der Schweiz, Z. Gletscherkd Glazialgeol 30:141–160

Neal EG, Walter MT, Coffeen C (2002) Linking the Pacific Decadal Oscillation to seasonal stream discharge patterns in southeast Alaska. J Hydrol 263:188–197

Neal EG, Hood E, Smikrud K (2010) Contribution of glacier runoff to freshwater discharge into the Gulf of Alaska. Geophys Res Lett 37:L06404. doi:10.1029/2010GL042385
Nuth C, Moholdt G, Kohler J, Hagen JO, Kääb A (2010) Svalbard glacier elevation changes and contribution to sea level rise. J Geophys Res 115:F01008. doi:10.1029/2008JF001223
Oerlemans J, Fortuin JPF (1992) Sensitivity of glaciers and small ice caps to greenhouse warming. Science 258(5079):115–117
Ohmura A (2001) Physical basis for the temperature-based melt-index method. J Appl Meteorol 40(4):753–761
Ohmura A (2004) Cryosphere during the twentieth century. In: Sparks, R.S.J. and Hawkesworth, C.J. (eds.) The state of the Planet: Frontiers and Challenges in Geophysics. Geophys Monogr Ser 150, AGU, Washington DC: 239–257
Ohmura A, Bauder A, Muller H, Kappenberger G (2007) Long-term change of mass balance and the role of radiation. Ann Glaciol 46(1):367–374
Østrem G, Brugman M (1991) Glacier mass-balance measurements: a manual for field and office work. NHRI Science Report pp 224
Paul F, Haeberli W (2008) Spatial variability of glacier elevation changes in the Swiss Alps obtained from two digital elevation models. Geophys Res Lett 35:L21502. doi:10.1029/2008GL034718
Picasso M, Rappaz J, Reist A, Funk M, Blatter H (2004) Numerical simulation of the motion of a two-dimensional glacier. Int J Numer Meth Engng 60:995–1009. doi:10.1002/nme.997
Pierce DW, Barnett TP, Santer BD, Gleckler PJ (2009) Selecting global climate models for regional climate change studies. Proc Natl Acad Sci U S 106:8441–8446. doi:10.1073/pnas.0900094106
Radić V, Hock R (2010) Regional and global volumes of glaciers derived from statistical upscaling of glacier inventory data. J Geophys Res 115:F01010. doi:10.1029/2009JF001373
Radić V, Hock R (2011) Regional differentiated contribution of mountain glaciers and ice caps to future sea-level rise. Nature Geo 4:91–94. doi:10.1038/NGEO1052
Radić V, Hock R, Oerlemans J (2007) Volume-area scaling vs flowline modelling in glacier volume projections. Ann Glaciol 46:234–240
Radić V, Hock R, Oerlemans J (2008) Analysis of scaling methods in deriving future volume evolutions of valley glaciers. J Glaciol 54(187):601–612
Radić V, Bliss A, Beedlow AC, Hock R, Miles E, Cogley JG (2013) Regional and global projections of 21st century glacier mass changes in response to climate scenarios from global climate models. Clim Dyn. doi:10.1007/s00382-013-1719-7
Raper SCB, Braithwaite RJ (2005) The potential for sea level rise: new estimates from glacier and ice cap area and volume distributions. Geophys Res Letters 32:L05502. doi:10.1029/2004GL021981
Raper SCB, Braithwaite RJ (2006) Low sea level rise projections from mountain glaciers and icecaps under global warming. Nature 439:311–313. doi:10.1038/nature04448
Rees HG, Collins DN (2006) Regional differences in response of flow in glacier-fed Himalayan rivers to climatic warming. Hydrol Process 20:2157–2169
Reijmer CH, Hock R (2008) A distributed energy balance model including a multi-layer sub-surface snow model. J Glaciol 54(184):61–72
Rignot E, Rivera A, Casassa G (2003) Contribution of the Patagonia Icefields of South America to sea level rise. Science 302(5644):434–437. doi:10.1126/science.1087393
Robinson CT, Uehlinger U, Hieber M (2001) Spatio-temporal variation in macroinvertebrate assemblages of glacial streams in the Swiss Alps. Freshwater Biol 46:1663–1672. doi:10.1046/j.1365-2427.2001.00851.x
Schiefer E, Menounos B, Wheate R (2007) Recent volume loss of British Columbia glaciers, Canada. Geophys Res Lett 34:L16503. doi:10.1029/2007GL030780
Shepard et al (2012) A reconciled estimate of ice-sheet mass balance. Science 338(6111):1183–1189. doi:10.1126/science.1228102
Slangen ABA, Katsman CA, van de Wal RSW, Vermeersen LLA, Riva REM (2012) Towards regional projections of twenty-first century sea-level change based on IPCC SRES scenarios. Clim Dyn 38(5–6):1191–1209. doi:10.1007/s00382-011-1057-6
Stahl K, Moore RD (2006) Influence of watershed glacier coverage on summer streamflow in British Columbia. Canada. Water Resour Res 42:W06201. doi:10.1029/2006WR005022
Stahl K, Moore RD, Shea JM, Hutchinson D, Cannon AJ (2008) Coupled modelling of glacier and streamflow response to future climate scenarios. Water Resour Res 44:W02422. doi:10.1029/2007WR005956
Tamisiea ME, Leuliette EW, Davis JL, Mitrovica JX (2005) Constraining hydrological and cryospheric mass flux in southeastern Alaska using space-based gravity measurements. Geophys Res Lett 32:L20501. doi:10.1029/2005GL023961

Tapley BD, Bettadpur S, Ries JC, Thompson PF, Watkins MM (2004) GRACE measurements of mass variability in the earth system. Science 305(5683):503–505. doi:10.1126/science.1099192

van der Wal RSW, Wild M (2001) Modelling the response of glaciers to climate change by applying volume-area scaling in combination with a high-resolution GCM. Clim Dyn 18:359–366

Weber M, Braun L, Mauser W, Prasch M (2010) Contribution of rain, snow and icemelt in the upper Danube today and in the future. Geogr Fis Din Quat 33:221–230

Willis MG, Melkonian AK, Pritchard ME, Rivera A (2012) Ice loss from the Southern Patagonian Ice Field, South America, between 2000 and 2012. Geophys Res Lett 39:L17501

Wisser D, Fekete BM, Vörösmarty CJ, Schumann AH (2010) Reconstructing 20th century global hydrography: a contribution to the Global Terrestrial Network- Hydrology (GTN-H). Hydrol Earth Syst Sc 14:1–24

Wouters B, Chambers D, Schrama EJO (2008) GRACE observes small-scale mass loss in Greenland. Geophys Res Lett 35:L20501. doi:10.1029/2008GL034816

Wu X, Heflin MB, Schotman H, Vermeersen BLA, Dong D, Gross RS, Ivins ER, Moore AW, Owen SE (2010) Simultaneous estimation of global present-day water transport and glacial isostatic adjustment. Nature Geosci 3:642–646. doi:10.1038/ngeo938

Yao TD, Pu JC, Lu AX, Wang YQ, Wu WS (2007) Recent glacial retreat and its impact on hydrological processes on the Tibetan Plateau, China and surrounding regions. Arc Antarc Alp Res 39:642–650

Zemp M, Hoelzle M, Haeberli W (2009) Six decades of glacier mass-balance observations: a review of the worldwide monitoring network. Ann Glaciol 50:101–111

Zemp M, Jansson P, Holmlund P, Gärtner-Roer I, Kobelt T, Thee P, Haeberli W (2010) Reanalysis of multi-temporal aerial images of Storglaciären, Sweden (1959-1999) – Part 2: comparison of glaciological and volumetric mass balances. The Cryosphere 4:345–357

Zemp M et al (2013) Uncertainties and re-analysis of glacier mass balance measurements. The Cryosphere Discuss 7: 789–839 (accepted)

Zhao Q, Ye B, Ding Y, Zhang S, Yi S, Wang J, Shangguan D, Zhao C, Han H (2013) Coupling a glacier melt model to the Variable Infiltration Capacity (VIC) model for hydrological modeling in north-western China. Environ Earth Sci 68(1):87–101. doi:10.1007/s12665-012-1718-8

DOI 10.1007/978-94-017-8789-5_16
Reprinted from *Surveys in Geophysics* Journal, DOI 10.1007/s10712-013-9269-4

Observing Global Surface Water Flood Dynamics

Paul D. Bates · Jefferey C. Neal · Douglas Alsdorf · Guy J.-P. Schumann

Received: 14 May 2013 / Accepted: 20 November 2013 / Published online: 17 December 2013

Abstract Flood waves moving along river systems are both a key determinant of globally important biogeochemical and ecological processes and, at particular times and particular places, a major environmental hazard. In developed countries, sophisticated observing networks and ancillary data, such as channel bathymetry and floodplain terrain, exist with which to understand and model floods. However, at global scales, satellite data currently provide the only means of undertaking such studies. At present, there is no satellite mission dedicated to observing surface water dynamics and, therefore, surface water scientists make use of a range of sensors developed for other purposes that are distinctly sub-optimal for the task in hand. Nevertheless, by careful combination of the data available from topographic mapping, oceanographic, cryospheric and geodetic satellites, progress in understanding some of the world's major river, floodplain and wetland systems can be made. This paper reviews the surface water data sets available to hydrologists on a global scale and the recent progress made in the field. Further, the paper looks forward to the proposed NASA/CNES Surface Water Ocean Topography satellite mission that may for the first time provide an instrument that meets the needs of the hydrology community.

Keywords Floods · Surface water · Floodplains · Rivers · Wetlands · Remote sensing · Surface water ocean topography (SWOT) mission

P. D. Bates (✉)
School of Geographical Sciences, University of Bristol, University Road, Bristol BS8 1SS, UK
e-mail: paul.bates@bristol.ac.uk

J. C. Neal
School of Geographical Sciences, University of Bristol, University Road, Bristol BS8 1SS, UK

D. Alsdorf
Byrd Polar Research Center, School of Earth Sciences, The Ohio State University, 123 Scott Hall, 1090 Carmack Road, Columbus, OH 43210-1308, USA

G. J.-P. Schumann
Jet Propulsion Laboratory, California Institute of Technology, Pasadena, CA 91109, USA

1 Introduction: Surface Water Floods in the Earth System

Flood waves moving along river systems are both a key determinant of globally important biogeochemical and ecological processes and, at particular times and particular places, a major environmental hazard. The time and length scales of such waves vary depending on river basin area, basin shape, basin slope, geology, vegetation and land use. In the very smallest urban catchments, rivers respond near instantaneously to rainfall, whilst in the world's largest river systems, there may a single annual flood pulse and river flood waves may span whole continents. Such waves therefore vary in length from perhaps 1 to 1,000 s of km and in duration from a few minutes to a whole year. Compared to their length, river flood waves are extremely low amplitude: for example, the Amazon flood wave in the middle reach of the river has a maximum amplitude of ~ 12 m for a wave thousands of kilometres in length. In most other basins, even catastrophic flash floods have amplitudes much less than this, and typically, flood waves are just a few metres in height. Flood waves are therefore shallow water phenomenon where typical horizontal length scales far exceed those in the vertical. Hydraulically, most flood waves are gradually varying sub-critical flows (Froude number <1) where the influence of downstream water level controls can propagate upstream (the so-called backwater effect). Sub-critical hydrodynamics occur because most river longitudinal slopes are low (typical river slopes are in the range 1–100 cm km^{-1}) and change only gradually. Flood waves are translated with speed or celerity, c, and attenuated by frictional losses such that in downstream sections, the hydrograph is flattened out. Wave speeds vary with discharge (see NERC 1975) such that maximum wave speed occurs at approximately two-thirds bankfull capacity (Knight and Shiono 1996). Typical observed values for c reported by NERC (1975) and Bates et al. (1998) for UK rivers are in the ranges 0.5–1.8 and 0.3–0.67 ms^{-1}, respectively.

Shallow water waves are described, in one dimension, by the Saint–Venant equation:

$$\underbrace{\frac{\partial u}{\partial t}}_{(i)} + \underbrace{u\frac{\partial u}{\partial x}}_{(ii)} u + g\left(\underbrace{\frac{\partial h}{\partial x}}_{(iii)} + \underbrace{S_{\mathrm{f}}}_{(iv)} - \underbrace{S_{\mathrm{o}}}_{(v)}\right) = 0 \tag{1}$$

where wave propagation is controlled by the balance of the various forces in Eq. 1. Here, (i) represents the local inertia (or acceleration), (ii) represents the advective inertia, (iii) represents the pressure differential, and (iv) and (v) account for the friction and bed slope, respectively. The relative magnitude of these terms for different types of shallow water flow is discussed in detail by Hunter et al. (2007), but in general, for sub-critical flow, the advective inertia term (ii) can be disregarded, and the pressure differential, friction slope and bed slope terms (iii, iv and v, respectively) are significantly more important than local inertia. For super-critical flow however, where shocks, hydraulic jumps and bores may exist, term (i) assumes much greater importance, and term (ii) cannot be disregarded.

This one-dimensional description is reasonable when flood waves are contained within defined river channels; however, when bankfull height is exceeded and water is transferred to floodplains and wetlands adjacent to the main channel, this description is insufficient. Here, water flow paths cannot be predicted a priori and such flows are clearly two-dimensional phenomena where flow spreads according to the hydraulic gradient and floodplain topography, which may be exceedingly complex (see, for example, Nicholas and Mitchell 2003). Floodplains and wetlands act as additional routes for flow conveyance or areas of water storage. Even when floodplains convey flow, the typically higher friction and shallower depth means that flow velocities are usually significantly smaller than those in the main channel. Typical river flows have velocities in the range 1–3 ms^{-1}, whilst floodplain flows in all but the most extreme events have velocity of <1 ms^{-1}. Floodplain storage therefore alters wave propagation and has important consequences for many physical processes. Hence, whilst floodplains and wetlands cover only approximately 4 % of the Earth's land surface, they exert a critical influence on global biogeochemical cycles (Richey et al. 2002; Frey and Smith 2005; Zhuang et al. 2009), terrestrial run-off to the world's oceans (Richey et al. 1989), sediment and nutrient transport (Beighley et al. 2008), basinwide flood response (Turner-Gillespie et al. 2003) and global biodiversity (Tockner and Stanford 2002) as a result of the multitude of landscapes generated by floodplain geomorphologic complexity (Mertes et al. 1996). Moreover, over longer timescales, floodplains and wetlands form sedimentary basins where significant oil and natural gas reserves are found. It follows from this that surface water processes occur within, and are mediated by, the wider catchment hydrological system (Destouni et al. 2010; Cvetkovic et al. 2012) which also includes significant human activity (Destouni et al. 2013), and these interactions become especially important when considering the role of water fluxes as drivers of biogeochemical and ecological processes (e.g., Lyon et al. 2010).

Extreme floods can also be a significant natural hazard. According to the World Health Organization EM-DAT natural hazards database (www.em-dat.be), in 2011 floods and related hydrological hazards (e.g., wet mass landslides) accounted for over half of all reported disasters and affected ~140 million people (Guha-Sapir et al. 2012). In the UK alone, 5 million people (i.e. 1 in 12 of the population) in 2 million properties live on coastal and fluvial floodplains, and over 200,000 of these properties have less than the standard of protection mandated by the UK government (1 in 75-year recurrence interval). The proportion of at-risk population is likely to be similar in many other developed countries and perhaps worse in developing nations, where risk is often poorly understood due to a lack of numerical modelling supported by suitable hydrometric and topographic data sets. Moreover, when they do inevitably occur, floods cause major social disruption, civil unrest, economic loss and insurance sector bankruptcies (e.g., the floods in Mozambique, 2000; New Orleans, 2005; Thailand, 2007; UK, 2007).

Surface water floods therefore play an important role in the Earth system, yet despite a number of groundbreaking studies (e.g., Alsdorf et al. 2000; Hamilton et al. 2002; Mertes et al. 1995), their dynamics at global scales remain poorly quantified through either ground observations, satellite observations or modelling. For example, current estimates of global inundated area from ground and satellite instruments vary from 1 to 12 million square kilometres (Zhuang et al. 2009) and do not capture seasonal variation adequately. As a consequence, estimates of the magnitude of other processes driven by such dynamics, such as methane emissions from flooded wetlands, which are a significant contributor to global atmospheric methane, also cannot be well estimated.

Given the importance of surface water floods in the Earth system over the last decade, an increasing volume of research has been undertaken to better observe and understand the above phenomena. The aim of this paper is to review this progress and look forward to future satellite missions, which may further add to our knowledge.

The paper is organized as follows. In Sect. 2, we review the data sets currently available to describe flood dynamics globally and the recent progress in combining these data to further our understanding. At present, there is no satellite mission dedicated to making these observations, and therefore, surface water scientists make use of a range of sensors developed for other purposes that are distinctly sub-optimal for the task in hand. Nevertheless, by careful combination of the data available from topographic mapping, oceanographic, cryospheric and geodetic satellites, progress in understanding some of the world's major river, floodplain and wetland systems can be made. In Sect. 3, we describe the proposed NASA/CNES Surface Water Ocean Topography (SWOT) satellite mission, which would provide the first dedicated observing system for surface water by measuring water height (h), water slope ($\partial h/\partial x$) and water height change over time ($\partial h/\partial t$) at ~100-m spatial resolution between 78°N and 78°S every 11 days. SWOT would not measure river discharge directly, and instead, this would need to be estimated using a hydraulic model driven by the SWOT water elevation observations. Constructing such a model requires knowledge of the channel bathymetry, which may be poorly known for many rivers, and estimation of the unknown channel friction. Section 4 therefore describes recent studies that have explored data assimilation techniques that could use the anticipated SWOT and existing observations to infer the unknown bathymetry and friction and hence estimate discharge. Section 5 summarizes progress to date and future prospects for research in this area.

2 Observing Global Flood Dynamics

An ideal set of measurements for observing global flood dynamics would comprise data describing channel bathymetry, floodplain topography, river discharge, inundation extent, water level and water storage at appropriate spatial and temporal resolutions as determined by our understanding of flood wave physics outlined in Sect. 1. Determining these resolutions in some situations, however, may not be straightforward. For example, we know that floodplains consist of features such as former channels, levees, pans and crevasse splays which give a complex microtopography that can affect both local-scale patterns and larger-scale flow routing (see, for example, Neal et al. 2012). Similarly, on the basis of a small number of unique and opportunistically acquired data sets, we know that water levels in inundated floodplains and wetlands show significant variability in time over periods of 24 h and in space over length scales down to 10–100 m (see, for example, Nicholas and Mitchell 2003; Bates et al. 2006; Alsdorf et al. 2007b), yet currently available observations are incapable of capturing this. Appropriate sampling density will therefore vary with event dynamics, which will be controlled to first order by basin size and climatology (Biancamaria et al. 2010) and complicated by such factors as basin shape, geology and land use. Measurements can be taken either through ground observations or using remote sensing platforms, and these are discussed in more detail below.

Ground observations of surface waters are made through discharge gauging stations; however, these are located on main rivers only where flow is confined to a single channel and can be fully sampled by a single measurement. Developed countries may have extensive ground gauging networks with long records but, worldwide, the number of gauges is declining (Vörösmarty 2002) and there can be significant barriers to data access. Moreover, floodplains and wetlands, which may convey a significant quantity of the total flow (e.g., Richey et al. 1989), are almost entirely ungauged. We therefore do not currently possess a comprehensive and globally consistent observing system for surface water. Nevertheless, at ground gauging sites, frequent water depth measurements can be taken to centimetric precision and made available in near real time with appropriate telemetry systems. If the gauge site is geodetically levelled, then absolute water elevation measurements referenced to a local ellipsoid or global geoid are possible. Flow rating curves constructed by fitting a relationship between repeated simultaneous measurements of flow cross-sectional area, velocity and depth can then be used to determine discharge through time. Where rating curves have been carefully constructed and flows remain in channel, then discharge can be estimated to an accuracy of perhaps 5–10 % (Fekete et al. 2012). However, where flow is out of bank, such that small increases in water height lead to large increases in discharge, where fewer observations are available to constrain the shape of the rating curve, or where flow is so high that the rating curve needs to be extrapolated, then errors may increase significantly. For example, Di Baldassarre and Montanari (2009) conducted a quantitative assessment of the effect of rating curve uncertainty on river discharge estimation for a reach of the River Po in Italy and found errors in the range 6.2–42.8 % at the 95 % significance level, with an average of 25.6 %. In an extensive previous study, Pelletier (1987) reviewed 140 publications that quantified uncertainty in river discharge and found errors in the range 8–20 %. Ground gauging stations typically record data at intervals between 15 min and 1 day and are located between tens and hundreds of kilometres apart, depending on the flashiness of the flow regime and the purpose for which the network is being used. Ground observations of inundation extent can be made, although the possible coverage is very limited and typically remote sensing platforms offer a much better solution for this variable. No global ground-based topography and channel bathymetry data sets currently exist, and this situation appears unlikely to change in the future.

Global coverage is clearly much easier to attain using remote sensing platforms; however, this may come at the expense of accuracy, and the orbit and instrument characteristics of existing systems may provide only a partial view of river, floodplain and wetland surface water dynamics. Indeed, satellite systems may often miss flood events entirely due to their particular orbital period/revisit times. This is largely because the satellite data used by surface water scientists come from either generic systems (e.g., the optical Landsat sensors) or more bespoke systems designed for applications in different geophysical fields such as oceanography, glaciology or geodesy. These systems are less than ideal for observing surface water floods, but can, if carefully employed, yield important insights at certain scales. Below, we discuss the available systems for measuring floodplain topography, water elevation, inundation extent and water storage. No current or planned future satellite system is capable of measuring either river bathymetry or discharge directly.

2.1 Remote Measurements of Floodplain Topography

For local-, regional- and national-scale studies, a number of high accuracy and fine spatial resolution systems are available for collecting remotely sensed terrain data. These include aerial stereo-photogrammetry (Baltsavias 1999; Lane 2000; Westaway et al. 2003), airborne laser altimetry or LiDAR (Krabill et al. 1984; Gomes-Pereira and Wicherson 1999) and airborne synthetic aperture radar (SAR) interferometry (Hodgson et al. 2003). LiDAR instruments in particular are now capable of generating data at sub-metre spatial resolution with vertical accuracy of ~5 cm root mean square error (RMSE) over wide areas and are ideal for flood modelling. For example, over 70 % of England and Wales is now mapped using LiDAR. Such data are capable of capturing the complexity of floodplain microtopography and have vertical errors much lower than typical flood wave amplitudes. Globally, however, comparable data do not exist, and the terrain data available to surface water scientists are of much lower resolution and accuracy. A number of near-global terrain models are available, but amongst the most useful for surface water scientists are the measurements from the NASA Shuttle Radar Topography Mission (SRTM, Farr et al. 2007). SRTM was captured using an interferometric synthetic aperture radar flown on board the space shuttle in February 2000. SRTM was used to produce a digital elevation model (DEM) from 56°S to 60°N at 3-arc-second (~90 m) spatial resolution. Average global height accuracies vary between 5 and 9 m (Farr et al. 2007) with pixel-to-pixel noise of ~6 m (Rodriguez et al. 2006), and this is problematic given typical flood amplitudes. The vertical error has been shown to be correlated with topographic relief with large errors and data voids over high-relief terrain, whilst in the low-relief sites, such as river valleys, floodplains and wetlands, errors are smaller (Falorni et al. 2005). However, despite better accuracy over low-relief terrain, pixel-to-pixel noise is not reduced and the X- and C-band radars used for the SRTM mission only partially penetrate vegetation canopies such that for forested floodplains, the DEM is corrupted by vegetation artefacts. Accordingly, the SRTM spatial resolution cannot capture the floodplain and wetland microtopography that can be critical to an understanding of flow dynamics (see Trigg et al. 2012) and at their native resolution have noise that can be larger than the flood "signal". Attempts at solving these problems by post-processing to remove the vegetation signal and spatial averaging to reduce uncorrelated noise (Paz et al. 2010; Paiva et al. 2011, 2013) have been attempted with limited success, and with careful handling, SRTM data have been shown to be useful for some flood modelling problems (Sanders 2007; Wilson et al. 2007; Di Baldassarre et al. 2009; Neal et al. 2012).

Other global terrain data sets include the ASTER GDEM (global digital elevation model), the SPOT 5 DEM and the forthcoming TanDEM-X products. ASTER GDEM is a 30-m spatial resolution DEM developed using stereo-photogrammetry and available from 83°S to 83°N. However, its accuracy of 17 m at the 95 % confidence level (Tachikawa et al. 2011) means that SRTM has significant advantages for most flood modelling studies. More promising perhaps is the TanDEM-X global DEM available from 2014 which will use X-band synthetic aperture radar interferometry to create a global DEM with ~12-m spatial resolution and target accuracy of better than 2 m. Whilst potentially of greater accuracy and resolution than SRTM, the use of X-band radars will mean that TanDEM-X may still be corrupted by vegetation artefacts that may be difficult to fully remove even with sophisticated processing techniques.

2.2 Remote Measurements of Inundation Extent

Globally available remote measurements of inundation extent are reviewed in detail by Marcus and Fonstad (2008) and Schumann et al. (2012), and these are made principally using (a) optical sensors; (b) passive microwave instruments; or (c) synthetic aperture radars. Visible-band satellite imagery (e.g., 30-m resolution Landsat or coarser 250-m resolution MODIS data) can detect floods (e.g., Bates et al. 1997); however, cloud cover and restriction to daytime only operation may limit the utility of these data. Passive microwave instruments, such as the scanning multichannel microwave radiometer (SMMR), have good temporal but limited spatial resolution (6-day revisit time and 0.25° pixels in the case of SMMR) that limits their use to particular types of study (see, for example, Hamilton et al. 2002). For these reasons, SAR data are often preferred for flood remote sensing.

SARs are active systems that emit microwave pulses at an oblique angle towards the target. Open water acts as a specular reflector, and the microwave energy is reflected away from the sensor so such areas appear as smooth areas of low backscatter in the resulting imagery. Terrestrial land surfaces, by contrast, reflect the energy in many directions, including back towards the sensor, and therefore appear as noisy high-backscatter zones. These differences allow flood extent to be mapped using a variety of techniques to an accuracy of ~1 pixel. Pixel sizes range from ~3 to ~100 m in space-borne imagery (e.g., Horritt 2000; Di Baldassarre et al. 2009), depending on the orbit revisit time, and can potentially be excellent for flood extent determination. Misclassification errors do occur however, with flattened and wet vegetation behaving, in certain situations, in the same way as open water, and emergent vegetation disrupting the specular reflections in shallow open water to appear more like dry land. Moreover, orbit repeat times may be low (3 days for ASAR wide swath mode, 7–10 days for RADARSAT and 35 days for ERS-1 and ERS-2) compared to the flood dynamics in many basins. From records of flood events all around the world since 1985 collected by the Dartmouth Flood Observatory (http://www.dartmouth.edu/~floods/archiveatlas/index.htm), it appears that the mean duration of floods is around 9.5 days and the median duration is 5 days. Accordingly, there may only be a low probability of a SAR overpass occurring simultaneously with a flood in all but the largest river systems. Moreover, SAR sensors are designed to be all-purpose instruments and may not be optimal for flood mapping (see, for example, Bates et al. 2004). Constellations of satellites are likely to be the only way to achieve a suitable combination of resolution and revisit frequency (García-Pintado 2013). For example, the COSMO–Sky-Med constellation can offer a revisit time as short as 12 h. The few studies to have obtained simultaneous aerial photograph and satellite SAR data have shown that the accuracy of satellite radars in classifying flood extent to be only of the order 80–85 % (Biggin and Blyth 1996). As a consequence, significant research effort has been expended in developing sophisticated techniques to classify SAR imagery into wet and dry areas (see, for example, Matgen et al. 2007; Mason et al. 2007; Giustarini et al. 2013).

Combining the observations of inundation extent available from optical, passive microwave and active microwave systems, a number of researchers have developed global floodplain and wetland inundation extent data sets. For example, Prigent et al. (2007) used passive microwave land surface emissivities calculated from SSM/I and ISCCP observations, ERS scatterometer responses, and AVHRR visible and near-infrared reflectances from 1993 to 2000 to calculate average monthly inundated fractions of equal-area grid cells (0.25° × 0.25° at the equator). Similarly, the Dartmouth Flood Observatory (http://floodobservatory.colorado.edu/, see Adhikari et al. 2010) uses 250 m resolution MODIS and other data, such as the SRTM Water Body Data set (SWBD), to map flooding in near real time and from this compile an archive of large floods. Such data sets provide a first comprehensive global view of surface water dynamics and flooding.

2.3 Remote Measurements of Water Elevation

Remote measurements of water surface elevation can be obtained from (a) profiling altimeters such as the JASON and Topex–Poseidon radar altimeters or the Geoscience Laser Altimeter System (GLAS) on board the ICESat satellite; (b) interferometric measurements of water surface elevation change using pairs of synthetic aperture radar images; and (c) the intersection of shorelines derived from inundation extent data (e.g., a satellite SAR scene) with a suitable digital elevation model.

Satellite radar altimeters were primarily designed from oceanic studies and have a footprint of ~2 km and vertical accuracy of decimetres to metres (Birkett et al. 2011). Such instruments also have wide (hundreds of kilometres) spacing between tracks which miss many of the world's rivers and most of the world's lakes. Over the continental land surface, such instruments therefore only record elevations over the very largest rivers; however, sophisticated retracking algorithms have recently been developed (e.g., Berry et al. 2005), which allow separation of water and other signals in mixed pixels. In this way, the elevation of smaller water bodies (~hundreds of metres across) can be obtained and used for flood model validation (e.g., Wilson et al. 2007). The GLAS laser aboard the ICESat satellite, although primarily designed to measure ice sheet topography, produced data with a footprint of ~70 m, which makes it more suitable for observing river water levels than radar altimeters. However, GLAS only operated between 2003 and 2009 and the laser instruments on board suffered from a number of technical issues such that only limited data with track spacing similar to radar altimeters are available. Nevertheless, the data have proved useful in particular areas of surface water science, such as geodetically levelling river gauges in remote basins (Hall et al. 2012) and determining water surface slopes in large unmonitored rivers (e.g., O'Loughlin et al. 2013).

As an alternative to profiling instruments, images of relative water height change over time ($\partial h/\partial t$) can be obtained from interferometric analysis of pairs of coherent SAR scenes taken from slightly different viewing geometries. Coregistration of the images to sub-pixel accuracy and subtraction of the complex phase and amplitude for each image allows surface displacement to be measured to centimetric accuracy. Such techniques were originally developed for ground deformation and glaciological studies (see, for example, Massonnet et al. 1993; Goldstein et al. 1993), but have subsequently been employed to map surface waters in particular circumstances (see Alsdorf et al. 2000, 2001a, b). For open water, specular reflection of the radar signal usually results in complete loss of temporal coherence, but for inundated floodplains, where there is emergent vegetation, Alsdorf et al. (2000) show that it is possible to obtain reliable repeat-pass interferometric measurements because of the so-called double bounce effect whereby the radar path includes both water and tree trunk surfaces. This allows relative water elevation change between images to be mapped to ~100-m resolution with centimetric accuracy and has been used to map complex water height change patterns in the Amazon floodplain (Alsdorf et al. 2007b) and to undertake rigorous testing of the ability of two-dimensional floodplain models to simulate the spatial and temporal dynamics of inundation (Jung et al. 2012).

Finally, from maps of inundation extent determined using the techniques outlined in Sect. 2.2, estimates of water elevation can be obtained by intersecting the shoreline vector with a suitable DEM. Such techniques are reviewed in detail by Schumann et al. (2009) who note their utility for constraining hydraulic models. The accuracy of water elevation data derived in this way clearly depends on both the quality of the image processing and the resolution and accuracy of the DEM, but Schumann et al. (2010) show that useful information for flood wave analysis can be obtained even when using low-resolution (75 m pixel) ASAR wide swath mode images and the SRTM DEM. Moreover, Mason et al. (2009) show that water elevations obtained by intersecting SAR imagery with DEM data are better at discriminating between competing model formulations than inundation extent data.

2.4 Remote Measurements of Water Storage

Change in water storage on the land surface can be measured either indirectly by calculating the implied volume difference between two flood extent measurements when intersected with a suitable DEM or directly using observations of the Earth's changing gravity field. Data on the latter are available from the GRACE and GOCE satellite missions, although with limited spatial (~hundreds of kilometres) and temporal (~monthly) resolution. In their raw state, such data may therefore not be terribly useful for surface flood studies; however, Alsdorf et al. (2010) show that by carefully combining GRACE data with information on precipitation, evaporation and inundation extent, it was possible to estimate floodplain inundation rate, water storage and drainage rate for six sections of the Amazon main stem in Brazil. Whilst satellite gravimetry is of lower resolution than many of the measurements discussed so far, water storage change is relevant to a range of hydraulic, biogeochemical and ecological processes, and therefore, such data add usefully to our knowledge of surface water processes.

3 The Proposed SWOT Satellite Mission

Section 2 demonstrates convincingly that no current satellite system can capture the detail of surface flows in rivers, floodplains and wetlands (e.g., Alsdorf et al. 2000, 2007a; Bates et al. 2006) and that we lack a comprehensive and consistent view of global surface water dynamics at a scale commensurate with known process variability. In the absence of reliable observations, it is also impossible to build, calibrate and validate models that can be applied with confidence to river, floodplain and wetland systems. Whilst progress can be made by carefully employing the data derived from topographic mapping, oceanographic, cryospheric and geodetic satellites, the lack of a dedicated surface water observing mission fundamentally limits our ability to map, model and understand surface water dynamics. Against this background, NASA and CNES are currently developing a new satellite mission to address this gap in the global observing system: the proposed Surface Water Ocean Topography mission or SWOT (see http://swot.jpl.nasa.gov/).

SWOT is being designed as a small version of SRTM. Both are "interferometers" that construct radar interferometric phase using one pass of two SAR antennae that are permanently connected by a fixed baseline. The interferometric phase is a measurement of surface elevations, i.e. topography of land and elevations of water. Because SWOT would use a Ka-band wavelength, which is shorter than the C-band and X-band wavelengths used by SRTM, the SWOT boom separating the two SAR antennae would be 10 m compared to the 60-m boom used by SRTM. SWOT would use a near-nadir viewing geometry with look angles of well less than 10°. In contrast, SRTM used $\sim$30°–58° look angles. This near-vertical geometry results in height accuracies that would be at least an order of magnitude better than those of SRTM. The proposed SWOT mission is expected to produce $\pm$50 cm height accuracies per sampling element (e.g., pixel). However, this viewing geometry would also result in a greater amount of layover for SWOT compared to SRTM (layover results when higher elevations are mapped by the SAR geometry into spatial locations closer to the radar). Also because of this viewing geometry, the spatial samples would vary in size from potentially as small as 2.5 m $\times$ 10 m to as large as 10 m $\times$ 70 m. The height error is normally distributed, so averaging samples improves the height accuracy by $1/\sqrt{(m)}$, where m is the number of samples. For example, a 250 m $\times$ 250 m lake sampled entirely by the finest spatial resolution would have a $\pm$1 cm height accuracy after averaging, whereas when sampled entirely by the coarsest spatial resolution, the height accuracy is reduced to about $\pm$5 cm.

The proposed SWOT mission is presently in "Phase A" of the NASA and CNES mission development life cycle. An international science definition team is working with SWOT project engineers and planners to define the required spatial, temporal and height accuracies. These requirements are expected to allow sampling of rivers at least as small as 100 m in channel width and perhaps smaller. Lakes and other water bodies 250 m $\times$ 250 m in size and perhaps even smaller are also under consideration for the mission design. To further help in defining the mission, an airborne version of SWOT has been created. AirSWOT has initiated test flights and is expected to sample rivers, lake and wetlands during 2013 and thereafter. Amongst the AirSWOT goals is to demonstrate the capability of the radar system to penetrate vegetation to the underlying water surface. Given the radar design, both SWOT and AirSWOT are expected to penetrate vegetation through canopy openings.

4 Inferring Remaining Unknown Variables Using DA

As noted above, no current or planned future satellite system is capable of measuring either river bathymetry or discharge directly, and to determine discharge from space requires that the river bathymetry and friction are known (see, for example, Smith 1997; Bjerklie et al. 2003). Discharge is a key variable for surface water science, for which we currently have no globally consistent and comprehensive data. However, by combining dynamical information on changing water level and flood extent derived from remote sensing with a suitable hydraulic model, it may be possible to infer the unknown bathymetry and friction and hence estimate discharge from space. Data assimilation provides the mathematical framework for this analysis as it allows for optimal estimation of the unknown variables given the observed data and the constraints provided by the physical laws encoded by the model. To first order the problem of estimating, discharge from space can be illustrated by the well-known Manning equation:

$$Q = \frac{AR^{2/3}S_f^{1/2}}{n} \tag{2}$$

where Q is the discharge; A is the channel cross-sectional area; R is the hydraulic radius; S_f is the water surface slope; and n is the Manning resistance coefficient which describes all the frictional losses. Clearly, only S_f is observable from space, yet to estimate discharge, we also need to know A, R and n. Early research in this area showed that if either friction or bathymetry was assumed to be known, it was relatively easy to estimate the remaining unknown variable (see, for example, Andreadis et al. 2007; Durand et al. 2008; Neal et al. 2009; Biancamaria et al. 2011; Yoon et al. 2012). However, Eq. 2 clearly shows that A, R and n trade-off against each other, which complicates the joint estimation problem. Joint estimation therefore requires considerably more dynamical information to isolate the differing effects of bathymetry and friction on water level dynamics. However, recent research (e.g., Lai and Monnier 2009; Hostache et al. 2010; Durand et al. 2010, submitted) is beginning to show that such joint estimation may indeed be possible because friction and bathymetry vary in distinctive and different ways in space and affect the various terms in the Saint–Venant equation (Eq. 1) in different ways. Changing friction or bathymetry has different "signature" effects on water surface height and slope change in time and space, and only a few combinations of both can fully explain observations of floods with different wave speeds or water surface slopes. Physically, water surface slopes respond only gradually to changing friction, whereas a sudden change in channel capacity or bed slope can have a much more immediate effect on the flow and wave propagation. Data assimilation methods can be developed to exploit these differences and hence estimate unknown bathymetry and friction simultaneously based only on repeated observations of water level and slope and an appropriate dynamical model to obtain discharge. Key research questions are therefore exactly how much water elevation and slope data are required and how much dynamical variation in the observations is necessary to obtain a (near) unique solution.

5 Conclusions

This paper has reviewed our understanding of global surface water flood dynamics and the role such waves play in the Earth system. Flood waves are both a key determinant of globally important biogeochemical and ecological processes and, at particular times and particular places, a major environmental hazard. Despite this, the current global observing system cannot capture the detail of surface flows in rivers, floodplains and wetlands, and we lack a comprehensive and consistent view of global surface water dynamics at a scale commensurate with known process variability. The paper demonstrates that by careful use of the data obtained from remote sensing instruments designed for different geophysical applications, progress can be made in our understanding of the surface water dynamics of a number of major floodplain and wetland systems. Ultimately, however, a detailed understanding would only be possible with the launch of a dedicated satellite mission for surface water carrying an instrument capable of capturing data of the right resolution and accuracy. The proposed SWOT satellite mission would have the potential to address this need and help answer new and exciting science questions that would be likely to revolutionize our view of hydrology.

Acknowledgments Part of this research was carried out at the Jet Propulsion Laboratory, California Institute of Technology, under a contract with the National Aeronautics and Space Administration.

References

Adhikari P, Hong Y, Douglas KR, Kirschbaum D, Gourley JJ, Adler RF, Brakenridge GR (2010) A digitized global flood inventory (1998–2008): compilation and preliminary results. Nat Hazards 55:405–422

Alsdorf DE, Melack JM, Dunne T, Mertes LAK, Hess LL, Smith LC (2000) Interferometric radar measurements of water level changes on the Amazon floodplain. Nature 404:174–177

Alsdorf D, Birkett C, Dunne T, Melack J, Hess L (2001a) Water level changes in a large Amazon lake measured with spaceborne radar interferometry and altimetry. Geophys Res Lett 28:2671–2674

Alsdorf DE, Smith LC, Melack JM (2001b) Amazon water level changes measured with interferometric SIR-C radar. IEEE Trans Geosci Remote Sens 39:423–431

Alsdorf DE, Rodriguez E, Lettenmaier D (2007a) Measuring surface water from space. Rev Geophys 45(2):RG2002

Alsdorf DE, Bates PD, Melack JM, Wilson MD, Dunne T (2007b) The spatial and temporal complexity of the Amazon flood measured from space. Geophys Res Lett 34:L08402

Alsdorf D, Han S-C, Bates P, Melack J (2010) Seasonal water storage on the Amazon floodplain measured from satellites. Remote Sens Environ 114:2448–2456

Andreadis KM, Clark EA, Lettenmaier DP, Alsdorf DE (2007) Prospects for river discharge and depth estimation through assimilation of swath-altimetry into a raster-based hydrodynamics model. Geophys Res Lett 34. Paper no L10403

Baltsavias EP (1999) A comparison between photogrammetry and laser scanning. ISPRS J Photogramm Remote Sens 54(2–3):83–94

Bates PD, Horritt M, Smith C, Mason D (1997) Integrating remote sensing observations of flood hydrology and hydraulic modelling. Hydrol Process 11:1777–1795

Bates PD, Stewart MD, Siggers GB, Smith CN, Hervouet J-M, Sellin RHJ (1998) Internal and external validation of a two-dimensional finite element model for river flood simulation. Proc Inst Civ Eng Water Marit Energy 130:127–141

Bates PD, Horritt MS, Aronica G, Beven K (2004) Bayesian updating of flood inundation likelihoods conditioned on flood extent data. Hydrol Process 18:3347–3370

Bates PD, Wilson MD, Horritt MS, Mason D, Holden N, Currie A (2006) Reach scale floodplain inundation dynamics observed using airborne synthetic aperture radar imagery: data analysis and modelling. J Hydrol 328:306–318

Beighley RE, Dunne T, Melack JM (2008) Impacts of climate variability and land use alterations on frequency distributions of terrestrial runoff loading to coastal waters in southern California. J Am Water Resour Assoc 44(1):62–74

Berry PAM, Garlick JD, Freeman JA, Mathers EL (2005) Global inland water monitoring from multi-mission altimetry. Geophys Res Lett 32 (16), article no. L16401

Biancamaria S, Andreadis KM, Durand M, Clark EA, Rodriguez E, Mognard NM, Alsdorf DE, Lettenmaier DP, Oudin Y (2010) Preliminary characterization of SWOT hydrology error budget and global capabilities. IEEE JSTARS 3(1):6–19

Biancamaria S, Durand M, Andreadis K, Bates PD, Boone A, Mognard NM, Rodriguez E, Alsdorf DE, Lettenmaier D, Clark E (2011) Assimilation of virtual wide swath altimetry to improve Arctic river modelling. Remote Sens Environ 115(2):373–381

Biggin DS, Blyth K (1996) A comparison of ERS-1 satellite radar and aerial photography for river flood mapping. J Chart Inst Water Eng Manag 10:59–64

Birkett C, Reynolds C, Beckley B, Doorn B (2011) From research to operations the USDA global reservoir and lake monitor. In: Vignudelli S, Kostianoy AG, Cipollini P, Benveniste J (eds) Coastal altimetry. Springer, New York, pp 19–50

Bjerklie DM, Dingman SL, Vorosmarty CJ, Bolster CH, Congalton RG (2003) Evaluating the potential for measuring river discharge from space. J Hydrol 278:17–38

Cvetkovic V, Carstens C, Selroos J-O, Destouni G (2012) Water and solute transport along hydrological pathways. Water Resour Res 48:W06537

Destouni G, Persson K, Prieto C, Jarsjö J (2010) General quantification of catchment-scale nutrient and pollutant transport through the subsurface to surface and coastal waters. Environ Sci Technol 44:2048–2055

Destouni G, Jaramillo F, Prieto C (2013) Hydroclimatic shifts driven by human water use for food and energy production. Nat Clim Change 3:213–217

Di Baldassarre G, Montanari A (2009) Uncertainty in river discharge observations: a quantitative analysis. Hydrol Earth Syst Sci 13:913–921

Di Baldassarre G, Schumann G, Bates PD (2009) Near real time satellite imagery to support and verify timely flood modelling. Hydrol Process 23:799–803

Durand M, Andreadis KM, Alsdorf DE, Lettenmaier DP, Moller D, Wilson M (2008) Estimation of bathymetric depth and slope from data assimilation of swath altimetry into a hydrodynamic model. Geophys Res Lett 35:L20401

Durand M, Rodriguez E, Alsdorf DE, Trigg M (2010) Estimating river depth from remote sensing swath interferometry measurements of river height, slope, and width. IEEE J Sel Top Appl Earth Obs Remote Sens 3(1):20–31

Durand M, Neal J, Rodríguez E, Andreadis K, Smith L, Yoon Y (submitted) Estimating reach-averaged discharge for the River Severn from measurements of river water surface elevation and slope. J Hydrol

Falorni G, Teles V, Vivoni ER, Bras RL, Amaratunga KS (2005) Analysis and characterization of the vertical accuracy of digital elevation models from the shuttle radar topography mission. J Geophys Res 110:F02005

Farr TG, Caro E, Crippen R, Duren R, Hensley S, Kobrick M, Paller M, Rodriguez E, Rosen P, Roth L, Seal D, Shaffer S, Shimada J, Umland J, Werner M, Burbank D, Oskin M, Alsdorf D (2007) The shuttle radar topography mission. Rev Geophys 45(2):RG2004

Fekete BM, Looser U, Pietroniro A, Robarts RD (2012) Rationale for monitoring discharge on the ground. J Hydrometeorol 13(6):1977–1986

Frey KE, Smith LC (2005) Amplified carbon release from vast west Siberian peatlands by 2100. Geophys Res Lett 32:L09401

García-Pintado J, Neal JC, Mason DC, Dance S, Bates PD (2013) Scheduling satellite-based SAR acquisition for sequential assimilation of water level observations into flood modelling. J Hydrol 495:252–266

Giustarini L, Hostache R, Matgen P, Schumann G, Bates PD, Mason DC (2013) A change detection approach to flood mapping in urban areas using TerraSAR-X. IEEE Trans Geosci Remote Sens 51(4):2417–2430

Goldstein RM, Engelhardt H, Kamb B, Frolich RM (1993) Satellite radar interferometry for monitoring ice sheet motion: application to an Antarctic ice stream. Science 262:1525–1530

Gomes-Pereira LM, Wicherson RJ (1999) Suitability of laser data for deriving geographical data: a case study in the context of management of fluvial zones. Photogramm Remote Sens 54:105–114

Guha-Sapir D, Vos F, Below R, Ponserre S (2012) Annual disaster statistical review 2011: the numbers and trends. CRED, Brussels, 52 pp

Hall AC, Schumann GJ-P, Bamber JL, Bates PD, Trigg MA (2012) Geodetic corrections to Amazon River water level gauges using ICESat altimetry. Water Resour Res 48. Paper W06602
Hamilton SK, Sippel SJ, Melack JM (2002) Comparison of inundation patterns among major South American floodplains. J Geophys Res Atmos 107(D20). Article no 8308
Hodgson ME, Jensen JR, Schmidt L, Schill S, Davis B (2003) An evaluation of LIDAR- and IFSAR-derived digital elevation models in leaf-on conditions with USGS Level 1 and Level 2 DEMs. Remote Sens Environ 84(2):295–308
Horritt MS (2000) Calibration of a two-dimensional finite element flood flow model using satellite radar imagery. Water Resour Res 36(11):3279–3291
Hostache R, Lai X, Monnier J, Puech C (2010) Assimilation of spatially distributed water levels into a shallow-water flood model. Part II: use of a remote sensing image of Mosel River. J Hydrol 390(3–4):257–268. doi:10.1016/j.jhydrol.2010.07.003
Hunter NM, Bates PD, Horritt MS, Wilson MD (2007) Simple spatially-distributed models for predicting flood inundation: a review. Geomorphology 90:208–225
Jung HC, Jasinski M, Kim J-W, Shum CK, Bates P, Neal J, Lee H, Alsdorf D (2012) Calibration of two-dimensional floodplain modeling in the Atchafalaya River Basin using SAR interferometry. Water Resour Res 48. Paper W07511
Knight DW, Shiono K (1996) River channel and floodplain hydraulics. In: Anderson MG, Walling DE, Bates PD (eds) Floodplain processes. Wiley, Chichester, pp 139–182
Krabill WB, Collins JG, Link LE, Swift RN, Butler ML (1984) Airborne laser topographic mapping results. Photogramm Eng Remote Sens 50:685–694
Lai X, Monnier J (2009) Assimilation of spatially distributed water levels into a shallow-water flood model. Part I: Mathematical method and test case. J Hydrol 377(1–2):1-11. doi:10.1016/j.jhydrol.2009.07.058, ISSN:0022-1694
Lane SN (2000) The measurement of river channel morphology using digital photogrammetry. Photogramm Rec 16(96):937–957
Lyon SW, Mörth M, Humborg C, Giesler R, Destouni G (2010) The relationship between subsurface hydrology and dissolved carbon fluxes for a sub-Arctic catchment. Hydrol Earth Syst Sci 14:941–950
Marcus WA, Fonstad MA (2008) Optical remote mapping of rivers at sub-meter resolutions and watershed extents. Earth Surf Proc Land 33:4–24
Mason DM, Horritt MS, Dall'Amico JT, Scott TR, Bates PD (2007) Improving river flood extent delineation from synthetic aperture radar using airborne laser altimetry. IEEE Trans Geosci Remote Sens 45(12):3932–3943
Mason DC, Bates PD, Dall'Amico JT (2009) Calibration of uncertain flood inundation models using remotely sensed water levels. J Hydrol 368:224–236
Massonnet D, Rossi M, Carmona C, Adragna F, Peltzer G, Feigl K, Rabaute T (1993) The displacement field of the Landers earthquake mapped by radar interferometry. Nature 364:138–142
Matgen P, Schumann G, Henry J, Hoffmann L, Pfister L (2007) Integration of SAR-derived inundation areas, high precision topographic data and a river flow model toward real-time flood management. Int J Appl Earth Obs Geoinf 9(3):247–263
Mertes LAK et al (1995) Spatial patterns of hydrology, geomorphology and vegetation on the floodplain of the Amazon River in Brazil: a remote sensing perspective. Geomorphology 13:215–232
Mertes LAK, Dunne T, Martinelli LA (1996) Channel–floodplain geomorphology along the Solimoes-Amazon River, Brazil. Geol Soc Am Bull 108(9):1089–1107
Neal J, Schumann G, Bates P, Buytaert W, Matgen P, Pappenberger F (2009) A data assimilation approach to discharge estimation from space. Hydrol Process 23(25):3641–3649
Neal J, Schumann GJ-P, Bates PD (2012) A simple model for simulating river hydraulics and floodplain inundation over large and data sparse areas. Water Resour Res 48. Paper no W11506
NERC (1975) Flood studies report, 5 volumes. Natural Environment Research Council, London
Nicholas AP, Mitchell CA (2003) Numerical simulation of overbank processes in topographically complex floodplain environments. Hydrol Process 17(4):727–746
O'Loughlin F, Schumann GJ-P, Trigg M, Bates PD (2013) Hydraulic characterization of the middle reach of the Congo River. Water Resour Res 49(8):5059–5070
Paiva RCD, Collischonn W, Tucci CEM (2011) Large scale hydrologic and hydrodynamic modeling using limited data and a GIS based approach. J Hydrol 406(3–4):170–181
Paiva RCD, Collischonn W, Buarque DC (2013) Validation of a full hydrodynamic model for large-scale hydrologic modelling in the Amazon. Hydrol Process 27(3):333–346
Paz AR, Bravo JM, Allasia D, Collischonn W, Tucci CEM (2010) Large-scale hydrodynamic modeling of a complex river network and floodplains. J Hydrol Eng 15(2):152–165

Pelletier MP (1987) Uncertainties in the determination of river discharge: a literature review. Can J Civ Eng 15:834–850
Prigent C, Papa F, Aires F, Rossow WB, Matthews E (2007) Global inundation dynamics inferred from multiple satellite observations, 1993–2000. J Geophys Res 112:D12107
Richey JE, Mertes LAK, Dunne T, Victoria RL, Forsberg BR, Tancredi ACNS, Oliveira E (1989) Sources and routing of the Amazon River flood wave. Glob Biogeochem Cycles 3:191–204
Richey JE, Melack JM, Aufdenkampe AK, Ballester VM, Hess LL (2002) Outgassing from Amazonian rivers and wetlands as a large tropical source of atmospheric CO_2. Nature 416:617–620
Rodriguez E, Morris CS, Belz JE (2006) A global assessment of the SRTM performance. Photogramm Eng Remote Sens 72(3):249–260
Sanders BF (2007) Evaluation of on-line DEMs for flood inundation modelling. Adv Water Resour 30(8):1831–1843
Schumann G, Bates PD, Horritt M, Matgen P, Pappenberger F (2009) Progress in integration of remote sensing derived flood extent and stage data and hydraulic models. Rev Geophys 47:RG4001
Schumann G, Di Baldassarre G, Alsdorf DE, Bates PD (2010) Near real-time flood wave approximation on large rivers from space: application to the River Po, Northern Italy. Water Resour Res 46. Paper no W05601
Schumann GJ P, Mason DC, Di Baldassarre G, Bates PD (2012) The use of radar imagery in riverine flood inundation studies. In: Piegay H, Carbonneau P (eds) Fluvial remote sensing for science and management. Wiley, Chichester, pp 115–140
Smith LC (1997) Satellite remote sensing of river inundation area, stage, and discharge: a review. Hydrol Process 11:1427–1439
Tachikawa T, Kaku M, Iwasaki A, Gesch D, Oimoen M, Zhang Z, Danielson J, Krieger T, Curtis B, Haase J, Abrams M, Crippen R, Carabajal C (2011) ASTER global digital elevation model version 2—summary of validation results. NASA, 27 pp. Available from http://www.jspacesystems.or.jp/ersdac/GDEM/ver2Validation/Summary_GDEM2_validation_report_final.pdf
Tockner K, Stanford JA (2002) Riverine flood plains: present state and future trends. Environ Conserv 29(1):308–330
Trigg MA, Bates PD, Wilson MD, Schumann G (2012) Floodplain channel morphology and networks of the middle Amazon River. Water Resour Res 48. Paper no W10504
Turner-Gillespie DF, Smith JA, Bates PD (2003) Attenuating reaches and the regional flood response of an urbanising drainage basin. Adv Water Resour 26:673–684
Vörösmarty CJ (2002) Global water assessment and potential contributions from earth systems science. Aquat Sci 64(4):328–351
Westaway RM, Lane SN, Hicks DM (2003) Remote survey of large-scale braided, gravel-bed rivers using digital photogrammetry and image analysis. Int J Remote Sens 24(4):795–815
Wilson MD, Bates PD, Alsdorf D, Forsberg B, Horritt M, Melack J, Frappart F, Famiglietti J (2007) Modeling large-scale inundation of Amazonian seasonally flooded wetlands. Geophys Res Lett 34. Paper no L15404
Yoon Y, Durand M, Merry CJ, Clark EA, Andreadis KM, Alsdorf DE (2012) Estimating river bathymetry from data assimilation of synthetic SWOT measurements. J Hydrol 464–465:363–375
Zhuang Q, Melack JM, Zimov S, Walter KM, Butenhoff CL, Khalil MAK (2009) Global methane emissions from wetlands, rice paddies, and lakes. EOS Trans Am Geophys Union 90(5):37–44

DOI 10.1007/978-94-017-8789-5_17
Reprinted from *Surveys in Geophysics* Journal, DOI 10.1007/s10712-013-9267-6

Arctic Climate and Water Change: Model and Observation Relevance for Assessment and Adaptation

Arvid Bring · Georgia Destouni

Received: 15 February 2013 / Accepted: 14 November 2013 / Published online: 24 December 2013

Abstract The Arctic is subject to growing economic and political interest. Meanwhile, its climate and water systems are in rapid transformation. In this paper, we review and extend a set of studies on climate model results, hydro-climatic change, and hydrological monitoring systems. Results indicate that general circulation model (GCM) projections of drainage basin temperature and precipitation have improved between two model generations. However, some inaccuracies remain for precipitation projections. When considering geographical priorities for monitoring or adaptation efforts, our results indicate that future projections by GCMs and recent observations diverge regarding the basins where temperature and precipitation changes currently are the most pronounced and where they will be so in the future. Regarding late twentieth-century discharge changes in major Arctic rivers, data generally show excess of water relative to precipitation changes. This indicates a possible contribution to sea-level rise of river water that was previously stored in permafrost or groundwater. The river contribution to the increasing Arctic Ocean freshwater inflow is similar in magnitude to the separate contribution from glaciers, which underlines the importance of considering all possible sources of freshwater when assessing sea-level change. We further investigate monitoring systems and find a lack of harmonized water chemistry data, which limits the ability to understand the origin and transport of nutrients, carbon and sediment to the sea. To provide adequate information for research and policy, Arctic hydrological and hydrochemical monitoring needs to be extended, better integrated and made more accessible. Further water-focused data and modeling efforts are required to resolve the source of excess discharge in Arctic rivers. Finally, improvements in climate model parameterizations are needed, in particular for precipitation projections.

Keywords Hydrology · Monitoring · Arctic · Climate change · Adaptation

A. Bring (✉) · G. Destouni
Department of Physical Geography and Quaternary Geology, Stockholm University, 106 91 Stockholm, Sweden
e-mail: arvid.bring@natgeo.su.se

A. Bring · G. Destouni
Bert Bolin Centre for Climate Research, Stockholm University, Stockholm, Sweden

1 Introduction

A multitude of global changes, including climate change, are currently transforming the Earth system. This is clearly evident in the Arctic, where surface temperatures over the last half-century have increased at a rate of 50 % higher than the Northern Hemisphere average (McBean et al. 2005), and where future climate change is expected to be the most pronounced (Kattsov et al. 2005). Coupled to the changes in the physical environment, the Arctic is also increasingly becoming a focal point of economic and geopolitical interest.

These transformations present a considerable challenge. The critical role of the Arctic in the global climate system implies that Arctic changes will have far-reaching consequences for, and feedbacks to, the entire Earth system (McGuire et al. 2006). For the Arctic region, in turn, global-scale changes, regional and local environmental changes, and geopolitical and economic changes all contribute to the societal need for adaptation, and to the need for information and monitoring to guide that adaptation (Azcárate et al. 2013).

As an integrating, propagating and regulating factor, water plays a central role in the changing Arctic and the wider global climate system. It is a shared component in the most recognized Arctic indications of global change. These indications include rapidly diminishing extent of sea ice (Comiso et al. 2008; Stroeve et al. 2012a), increased mass loss from glaciers (Kaser et al. 2006; Gardner et al. 2011), increasing river flows (Peterson et al. 2002, 2006; McClelland et al. 2006; Shiklomanov and Lammers 2009; Overeem and Syvitski 2010) and increasing groundwater contribution to those flows (Smith et al. 2007), permafrost degradation (Hinzman et al. 2005; White et al. 2007; Lyon and Destouni 2010; Brutsaert and Hiyama 2012), ecosystem regime shifts (Smol et al. 2005; Karlsson et al. 2011), and shorter extent of snow cover season (Brown et al. 2010; Callaghan et al. 2011). All these constitute water changes, some of which are fundamentally caused by the overarching driving force of climate change. Some water changes, however, may also arise due to direct human interference, for example, through freshwater abstractions for or losses by food and energy production (Destouni et al. 2013), or clear-cutting of forests (Seitz et al. 2013). Whether the driving forces are local or global, water changes require local adaptation, for example, of infrastructure for energy, transport and buildings, of agricultural and forestry practices, and of measures for food and water security (Nilsson et al. 2013).

A primary information basis for projections of large-scale climate change is the ensemble of general circulation models (GCMs) that underlie the assessment reports of the Intergovernmental Panel on Climate Change (IPCC). The two most recent full reports are the Third Assessment Report (TAR; 2001) and the Fourth Assessment Report (AR4; 2007). The fifth IPCC report (AR5) is currently being released, with the first working group contribution published in September 2013.

The performance of GCMs in the Arctic has been the subject of extensive discussion. Assessments of TAR and AR4 model performance have shown improvements between successive generations of models, but also indicated that significant shortcomings remain in simulating observed climate parameters (Christensen et al. 2007). There have been several assessments of simulations of sea ice processes (Zhang and Walsh 2006; Overland and Wang 2007; Stroeve et al. 2007; Eisenman et al. 2007; Holland et al. 2010; Stroeve et al. 2012b), the surface radiation budget (Sorteberg et al. 2007; Boé et al. 2009) and surface temperature (Lui et al. 2008) over the Arctic Ocean, but fewer studies of GCM performance related to the continental part of the Arctic hydrological cycle on drainage basin scales. Kattsov et al. (2007) analyzed the output of the AR4 model ensemble for four major basins in the Pan-Arctic Drainage Region (PADB), and Roesch (2006) evaluated

AR4 simulations of snow cover. Holland et al. (2007) investigated ten AR4 GCMs to estimate change in the freshwater budget of the Arctic Ocean, including pan-Arctic scale runoff, and Rawlins et al. (2010) used the same GCMs combined with reanalysis and observational data in a pan-Arctic analysis of Arctic hydrological cycle intensification.

However, no basin-wise investigation and comparison of GCM results between the TAR and AR4 has been performed for a larger set of basins within the PADB, and no benchmark of AR4 GCM performance over such a set of basins exists for comparison with the AR5 set of models. An evaluation of GCM projections in a hydrological context is therefore motivated and would inform both the parameterization of land surface schemes in GCMs, and the developers and end users of regional climate model and their results. In the end, it would benefit anyone whose decisions are influenced by the reliability of water change projections. Previous investigations into GCM agreement with hydro-climatic observations, for other regions than the Arctic, have for instance revealed that GCM projections of evapotranspiration may be more uncertain than originally thought (Mueller et al. 2011).

Even if projections of the atmospheric components of the Arctic hydrological cycle (AHC) were satisfactory, it is the translation of changes in these components to water system changes in the landscape that is central to adaptation. Spatial planning, infrastructure dimensioning and water resources planning all depend on reliable understanding and projection of changes to water availability, river flows, flood and drought frequencies. Therefore, considerable efforts have been directed toward understanding the complex changes in Arctic surface and groundwater systems in the recent decades. Several integrative system assessments (e.g., Vörösmarty et al. 2001; Serreze et al. 2006; Slater et al. 2007; Rawlins et al. 2009, 2010), together with numerous site-specific studies, have greatly improved knowledge of the AHC. Nevertheless, a number of inconsistencies, gaps in understanding and open questions still remain (Arctic-HYDRA consortium 2010). Definite understanding of several AHC components, and how they are linked, is still lacking.

To remediate these shortcomings, and advance the development of GCMs and our understanding of hydro-climatic change, relevant and accessible observations have a central role. Advances in theories, models, scenarios and projections fundamentally rely on observational data. The importance of data and observation systems has also recently been emphasized at intergovernmental summits (GEO 2010) and recognized as one of five "grand challenges" for Earth system science and science policy over the next decade (Reid et al. 2010; ICSU 2010). Constraints in the availability of data limit the ability to evaluate projections, climate model parameterizations, and hypothesized changes and functions of environmental systems. There is therefore a strong link between the output of GCMs, observation systems and, in the end, the ability of society to plan for and adapt to hydrological changes that affect industry, agriculture and water resource availability.

A primary class of environmental data required to assess AHC change is river discharge. In addition, sediment and water chemistry data are needed to estimate the waterborne mass fluxes of constituents in global biogeochemical cycles, such as carbon, nitrogen and phosphorus. Besides their role in the global cycles, these elements are also directly related to societal impacts through their links to eutrophication (Darracq et al. 2008), aquatic habitat and ecosystem changes (Palmer et al. 2009), and feedbacks to climate change (Lyon et al. 2010). The collection of such hydrological and hydrochemical data is normally conducted through continuous monitoring programs by various government agencies. However, in many countries, hydrological observation systems have been in decline during recent decades. The extent of monitoring generally peaked around 1980, after the significant increases in monitoring efforts during the International Hydrological Decade

1965–1974. Since then, budget constraints and failure to maintain existing systems have resulted in a general decline in monitoring, something that has been reported in several studies (Brown 2002; Fekete and Vörösmarty 2002; Maurer 2003; Hannerz 2008; FAO 2009).

The global trend of declining discharge monitoring is also evident in the Arctic (Lammers et al. 2001; Shiklomanov et al. 2002; Hinzman et al. 2005; Walsh et al. 2005, Arctic-HYDRA consortium 2010). The particular situation in the Arctic has received relatively much attention in the scientific community, partly due to the rapid changes and the general scientific interest in the region. Also, the accessibility to discharge data has been improved more in the Arctic than in many other parts of the world. These improvements are mostly due to a few concentrated international collaborations, several of which were coordinated from the University of New Hampshire. Nevertheless, some of these efforts are now several years in the past, and the accessibility to recent discharge data that they initially provided has not always been sustained.

In contrast to the situation for discharge monitoring, a clear picture of the status of water chemistry data has been lacking. Previous efforts have tried to assess the state of affairs for certain parameters (e.g., Holmes et al. 2000, 2002; Raymond et al. 2007) and estimated the quality of existing data (Zhulidov et al. 2000, 2003; Holmes et al. 2001). Although specific data sets have been made accessible for parts of the PADB (e.g., Holmes et al. 2000; Holmes and Peterson 2002), and in at least one case for the wider PADB (McClelland et al. 2008; data at http://www.arcticgreatrivers.org), no international repository and data host exists for all accessible Arctic water chemistry data. Neither has any initiative yet been launched to develop a common set of indicators, such as the Millennium Development Goals-related UN Federated Water Monitoring System (FWMS) and its Key Water Indicators Portal (KWIP).

The decline in station numbers and the lack of integrated hydrological and hydrochemical information, together with the grand challenge of improving Earth observation systems, constitute an imperative to develop Arctic hydrological monitoring networks and to ensure their relevance under conditions of climate change. This would enable improvements in both GCMs and hydro-climatic change understanding.

The importance of the continental water system in Arctic and global change means that the hydrological drainage basin is a fundamental and relevant spatial scale unit, both for water management and adaptation, and for basic research (Pahl-Wostl 2007; UNECE 2009). In this paper, we therefore survey and review several components required to provide reliable water information, and to do this consistently at drainage basin scales. Basin-scale water information also has ensuing applications in understanding coupled changes across other terrestrial, atmospheric and marine systems (e.g., Karlsson et al. 2011).

This review paper addresses the following three overarching topics:

- The reliability of GCM projections on the scale of main Arctic river basins, as base information for understanding and for societal adaptation to Arctic climate and water change;
- The recently observed changes to water flow and water budgets in the Arctic hydrological cycle, and their potential consequences for both societal adaptation and freshwater input to the Arctic Ocean; and
- The representativeness, accessibility and relevance of hydrological and hydrochemical observation systems for assessing changes to water, sediment and hydrochemical fluxes in the Arctic hydrological cycle.

Furthermore, an ensuing question that relates to all of the above is: What critical gaps and key limitations exist in each of these topics, and is there a basis to rationally prioritize how to address those limitations by hydrological monitoring development?

We here synthesize information on these topics by integrating a set of previous studies of Arctic hydro-climatic change (Bring and Destouni 2009, 2011, 2013; Dyurgerov et al. 2010). We also extend the results from these reports with an additional analysis of GCM projection performance for major Arctic hydrological basins.

2 Methods Summary

In the following, we summarize the data and methods used in the synthesized previous studies (Bring and Destouni 2009, 2011, 2013; Dyurgerov et al. 2010) and the novel extensions in this paper. For a more detailed account of the methods used in the previous studies, we refer to the aforementioned publications. We have in this review re-evaluated some of the previous results by correcting observed precipitation data sets for gauge undercatch and orographic effects. To this end, we used two global data sets by Adam and Lettenmaier (2003) and Adam et al. (2006). However, as precipitation corrections in other cases have been known to result in overestimated precipitation (Mächel et al. 2012), we here treat the originally reported observations and the fully corrected values as two ends of a range. As we have no knowledge of the probability distribution of values within this range, we treat the middle of the range (i.e., the average of the original and the corrected values) as the best estimate of observed precipitation.

In general, the hydrological drainage basin has been the fundamental basis for the investigations in this study. The drainage basin constitutes a physically consistent boundary for closing the flow balance of water and the mass balances of constituents transported by water, which makes it a relevant scale for addressing both science and management problems.

In the paper, we focus primarily on the 14 largest watersheds in the PADB. Each of them is at least 200,000 km^2 in size, and their combined drainage covers 13.8 million km^2 (green and blue basins in the map of the study area in Fig. 1). These 14 basins are sufficiently large to allow an analysis of GCM data. Figure 2 shows the characteristic temperature and precipitation conditions, and recent observed changes for those parameters, in the 14 major basins.

The analyses in the surveyed and synthesized publications concern slightly varying time periods and, in general, involve comparisons of time periods of different lengths. Although it would be desirable to always compare the same time periods, the differences between the surveyed publications are relatively small in this regard. Several publications also study the longest possible period since 1990 for which there are data available, and compare that period with the reference of 1961–1990. Since the recent period from 1991 to (near-) present time is considerably shorter than the 30 years of the climatological reference period 1961–1990, we here regard the changes from the latter to the recent period as deviations from the reference-time climate and not necessarily as climate change. With inter-annual variability of various magnitudes for different parameters and basins, the future 30-year climate starting from 1991 may differ from these shorter-term deviations.

2.1 GCM Projections Across Arctic Basins

We first examine two successive generations of GCMs, which form the basis for the two latest available full IPCC assessment reports, the TAR and AR4. We here compare the

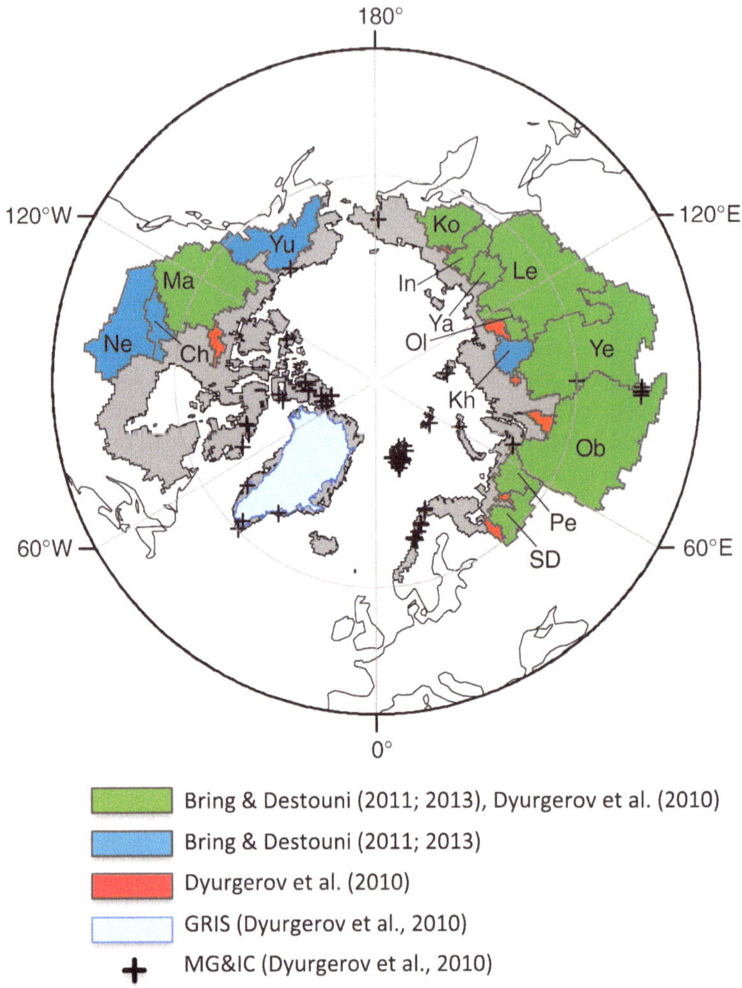

Fig. 1 Map indicating studied areas of the PADB. Bring and Destouni (2009) includes all areas within the PADB, demarcated by the combined area of all colored basins and the *dark gray area*. MG&IC: Mountain glaciers and ice caps; GRIS: Greenland ice sheet. Basin names are abbreviated as *Ne* Nelson, *Ch* Churchill, *Ma* Mackenzie, *Ko* Kolyma, *Kh* Khatanga, *Ol* Olenek, *Le* Lena, *Ya* Yana, *In* Indigirka, *Ye* Yenisey, *SD* Severnaya Dvina, *Pe* Pechora and *Yu* Yukon. Lambert azimuthal equal-area projection; approximate scale 1:150,000,000. Spatial resolution of basin outlines is 0.5°

14-basin average values of TAR and AR4 temperature and precipitation (downloaded from http://www.ipcc-data.ch; included models summarized in Table 1) with observations (CRU TS 2.1, Mitchell and Jones 2005) for the periods 1961–1990, 1991–2002 (observations) and projections for 2010–2039 (GCMs).

We then investigate in more detail the AR4 model results for the historical 20C3M scenario and the period 1961–1990, by calculating three separate measures of model performance: mean absolute error (MAE), mean bias error (MBE) and an index of model performance (d_r). For a given drainage basin, the MAE is defined as

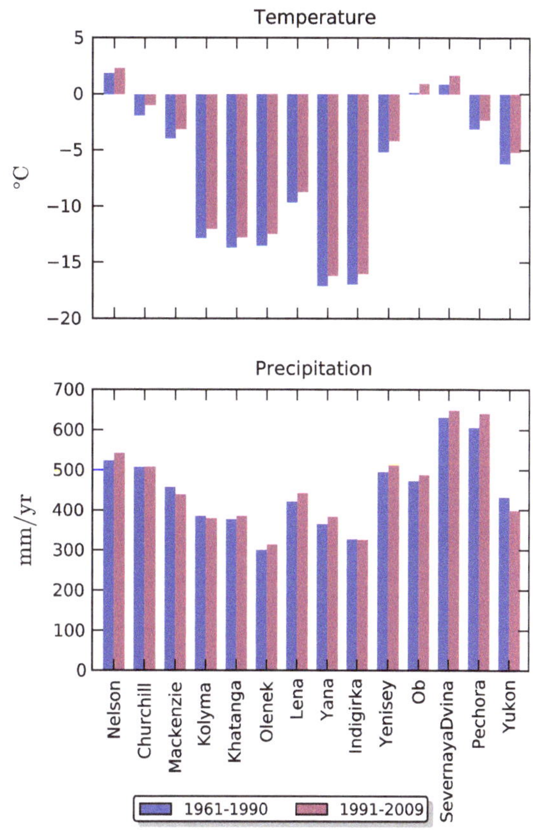

Fig. 2 Climate characteristics of the 14 major Arctic basins. The *top panel* shows average annual temperature; the *bottom* one shows annual average precipitation, for the periods 1961–1990 and 1991–2009. Data from CRU (Harris et al. 2013) and Willmott & Matsuura (http://climate.geog.udel.edu/~climate). Precipitation values are averages of original data and data with orographic and undercatch corrections (Adam and Lettenmaier 2003; Adam et al. 2006)

$$\mathrm{MAE} = \frac{1}{W}\sum_{i=1}^{n} w_i|e_i|$$

where $e_i = P_i - O_i$ is the difference between the model-projected value P_i and the observed value O_i for all cells $i = 1, 2, \ldots, n$ within the basin. The term w_i denotes the area-relative weight of each cell and W the sum of these weights. Similarly, the MBE describes the area-weighted sum of the deviations of model projections from observations, but retains the signs of the differences:

$$\mathrm{MBE} = \frac{1}{W}\sum_{i=1}^{n} w_i e_i$$

In comparing models across several drainage basins or for several parameters, a dimensionless index of model performance can complement the MAE and MBE, which are both defined in the units of the studied model parameter. We therefore also calculate such a dimensionless measure: the refined index of model performance d_r, with values on a unitless scale from −1 to 1 (worst to best). The d_r index is comprehensively defined in Willmott et al. (2012), and in their words, "[i]t indicates the sum of the magnitudes of the differences between the model-predicted and observed deviations about the observed mean relative to the sum of the magnitudes of the perfect model ($P_i = O_i$, for all i) and observed deviations about the observed mean."

Table 1 GCMs included in the analysis

TAR models	AR4 models
CCCma	BCM2
CSIRO	CGHR
ECHAM4	CNCM3
GFDL99	CSMK3
HADCM3	ECHOG
NIES99	FGOALS
	GFCM20
	GFCM21
	GIAOM
	GIER
	HADCM3
	HADGEM
	INCM3
	IPCM4
	MIHR
	MIMR
	MPEH5
	MRCGCM
	NCCCSM
	NCPCM

Note that not all models have data for all scenarios and parameters investigated

For the analysis of MAE, MBE and d_r, we use the mean of the CRU TS 3.1 (Harris et al. 2013) and Willmott & Matsuura 3.1 data sets (http://climate.geog.udel.edu/~climate) as a benchmark against which to evaluate GCM output. By averaging two separate data sets, our confidence in the reliability of the precipitation estimates increases, as different data sets use slightly different approaches to calculate the same estimate.

Climate projections are critical to adaptation planning, for example, for long-term planning of infrastructure, power production and agriculture. We therefore test whether these projections agree with the observations of recent deviations with regard to which Arctic drainage basins will be most affected by climate change. The aim is to investigate whether a set of highly divergent future climate projections, together with recent observations of climate deviations, can form a consistent basis for prioritizing monitoring. Under limited resources, one either has to rely on some degree of certainty in the distribution of future changes and prioritize monitoring based on this distribution, or try to harmonize the different possible bases for rational monitoring prioritization that may arise from uncertain change projections.

In order to include as wide a spectrum of future climate change as possible over the next half-century, we formulate for this analysis two alternate final stages of possible climates. We select the five warmest or wettest models from the most severe IPCC Special Report on Emission Scenarios (SRES) scenario (A2), and the five coldest or driest models from the least severe scenario (B1). For details on the procedure and the selected models, we refer to Bring and Destouni (2013). For these two cases, we analyze the relative distribution of projected climate change severity across the 14 largest Arctic drainage basins and compare this distribution with recent observed deviations from the 1961 to 1990 climate. We emphasize here that, in absolute terms, we do expect divergence between the scenarios, and between each scenario and observations. However, in terms of the geographical distribution of relative change or deviation intensity, divergence is not necessarily expected; if such divergence in relative intensity prevails, it presents a main challenge for geographical monitoring prioritization, as discussed further in Sect. 6.3.

2.2 Arctic Hydro-Climatic Change

Precipitation output from GCMs is used to further model other hydro-climatic changes, most importantly runoff changes, at regional and finer scales. The usefulness of precipitation projections for adaptation planning, such as dimensioning of infrastructure and drainage systems, depends on this modeling and understanding of how the precipitation changes are (and have previously) transferred to runoff and other hydro-climatic changes in the landscape. We therefore here investigate how recently observed precipitation deviation has related to runoff deviation for 13 major basins where runoff data are accessible for 1961–1990 and 1991–2002, and compare these data on runoff deviation with corresponding observations of precipitation deviation from the CRU TS 2.1 database.

To gain a more comprehensive understanding of total freshwater flux changes in the Arctic system, we also perform an integrated assessment of freshwater inflow from both rivers and glaciers to the Arctic Ocean. Separation of the freshwater flux contributions from rivers and glaciers is here possible because the glaciated area of the major Arctic river basins is very small. Instead, glaciers and ice caps mostly contribute their melt water directly to the coast or through smaller watersheds. For the river component in this analysis, we therefore consider only major basins with negligible glacier area that drain to the proper Arctic Ocean with a discharge of at least 10 km^3/year. This drainage is a subset of the whole PADB and includes 17 basins, of which 11 (green basins in Fig. 1) are in common with the 14 major basins (green and blue basins in Fig. 1) in the GCM comparison discussed above, and the remainder are additional smaller river basins (red basins in Fig. 1). We specifically investigate the periods 1961–1992 and 1993–2006, with the period break coinciding with a marked increase in glacier mass loss. We combined discharge data from the R-ArcticNET (Lammers et al. 2001), ArcticRIMS (http://rims.unh.edu) and Water Survey of Canada HYDAT (Environment Canada 2004) databases. Glacier data consist of data on annual direct mass balance observations carried out for mountain glaciers and ice caps (MG&IC; marked with crosses in Fig. 1; Dyurgerov and Meier 2005; Glazovsky and Macheret 2006; Fluctuations of Glaciers (FoG) 2008), and of several recent modeling studies for the Greenland ice sheet (GRIS; light blue in Fig. 1; Rignot et al. 2008; Hanna et al. 2008, 2009; Box et al. 2006; Mernild et al. 2009).

2.3 Pan-Arctic Drainage Basin Monitoring

Observations of discharge on various scales allow testing of water budgets, both for different landscape types and for the land surface area in general, which is useful in evaluating model assumptions and parameterizations. Water chemistry observations give information on upstream sources, sinks and hydrological transport pathways of biogeochemical constituents, and their changes. We therefore also synthesize and evaluate all accessible discharge and water chemistry data for the PADB. The extent of accessible data is presented in map form, illustrating the maximum length of time series and latest data year for the PADB. We further summarize the characteristics of monitored and unmonitored areas in North America, Europe and Asia and compare the differences between them.

This synthesis shows whether the properties of areas that are monitored are representative of the unmonitored areas. This question is important, considering that extrapolation from monitored to unmonitored areas will always be needed to some degree.

3 Results for GCM Projections Across Arctic Basins

Figure 3 illustrates GCM projections and Climatic Research Unit (CRU) observations for the 14 major Arctic drainage basins (green and blue basins in Fig. 1; adapted from Bring and Destouni 2011). Temperature projections are compatible with observed deviations during the late twentieth and early twenty-first century, both for the TAR and AR4, while precipitation projections indicate an increase that is hitherto not evident in observations. The absolute GCM results compare well with the observations for temperature, but the models generally overestimate precipitation. However, while the difference between temperature projections and observations has increased between the TAR and AR4 model ensembles, precipitation projections have come closer to observations. Generally, GCM projections for temperature and precipitation have become more precise in the AR4 ensemble, as the models in AR4 converge more closely on the mean than do the TAR ensemble models.

Results from the error and performance analysis of the AR4 GCMs over the 14 largest Arctic watersheds indicate that there is a large spread in error and relative performance between the models used in the AR4 (Fig. 4). Although the inter-model variation in MAE is similar for temperature and precipitation, the difference between the systematic overestimation of precipitation, evident in the above-zero MBE for all models but one, and the relatively smaller systematic underestimation of temperature, evident in MBEs closer to zero, mean that the model performance index is less variable and, for most models, better for temperature than for precipitation.

For some basins, and for the pan-Arctic in general, an above-average model performance in temperature simulation does not correlate particularly strongly with performance in simulating precipitation (negative rank correlations for MAE and d_r; Table 2). This implies that choosing a "best" climate model for the Arctic or any of its major drainage

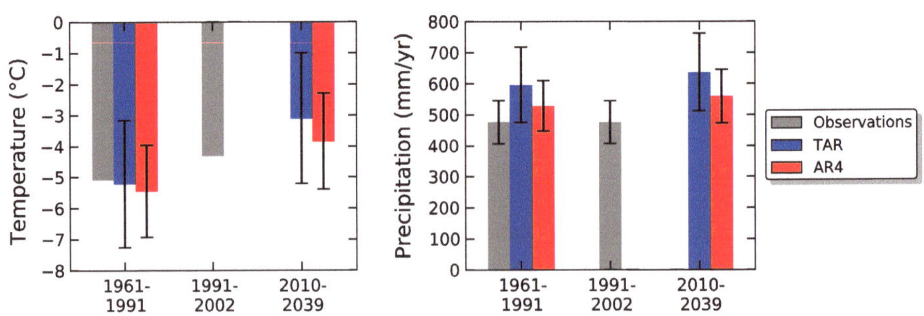

Fig. 3 Temperature (*left*) and precipitation (*right*) values for 1961–1990 and 1991–2002 for observations and for 1961–1990 and 2010–2039 for GCM projections across 14 major Arctic basins. *Error bars* for GCM projections indicate one standard deviation of different GCM results from the model ensemble mean. *Error bars* for precipitation observations indicate upper and lower estimates, corresponding to uncorrected and bias-corrected values, respectively

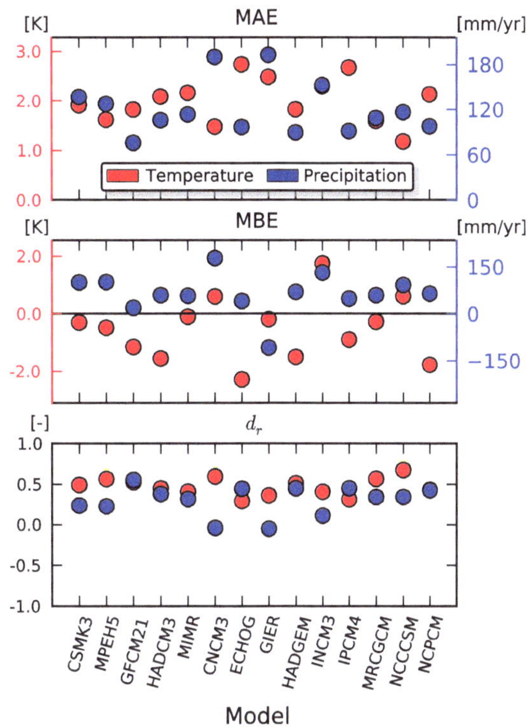

Fig. 4 Mean absolute error (MAE; *top panel*), mean bias error (MBE; *middle panel*) and model performance index (d_r; *bottom panel*) for 14 GCMs across the PADB for 1961–1990

Table 2 Spearman's rank correlation coefficient ρ between the pan-Arctic temperature and precipitation values of MAE, MBE and d_r

	MAE	MBE	d_r
ρ	−0.13	0.41	−0.13

basins is difficult if one wants to have consistently good model performance for both of these parameters.

Model performance also varies considerably between the studied basins. The large Mackenzie and Yenisey basins exhibit consistently good d_r values for temperature, while the d_r of some models is considerably smaller for the Ob and Lena basins (not shown). In contrast, temperature d_r values for the relatively smaller and neighboring basins of Pechora and Severnaya Dvina are poor for most models, indicating difficulties with temperature simulations west of the Ural Mountains. This may be due to the relatively smaller size of these basins, or the presence of natural low-frequency variation in the region (Hurrell and van Loon 1997). Precipitation d_r values are overall lower, with the lowest values for the smaller Olenek and Severnaya Dvina basins, followed by the medium-sized Yukon basin, and the highest for the large Ob, medium-sized Nelson and small Khatanga basins.

Furthermore, results from an analysis of bases for prioritization monitoring under conditions of climate change (Bring and Destouni 2013) indicate that the basins with the highest recently observed deviations of temperature and precipitation are not the same basins that have the highest projected changes in future climate (rank correlations of basin orders are close to zero, or negative; Table 3). Therefore, prioritizing monitoring to basins

Table 3 Spearman's correlation of GCM-projected climate change ranking with observed climate change ranking

Scenario	ρ
Temperature	
Cool B1	0.01
Hot A2	−0.03
Precipitation	
Dry B1	−0.13
Wet A2	−0.19

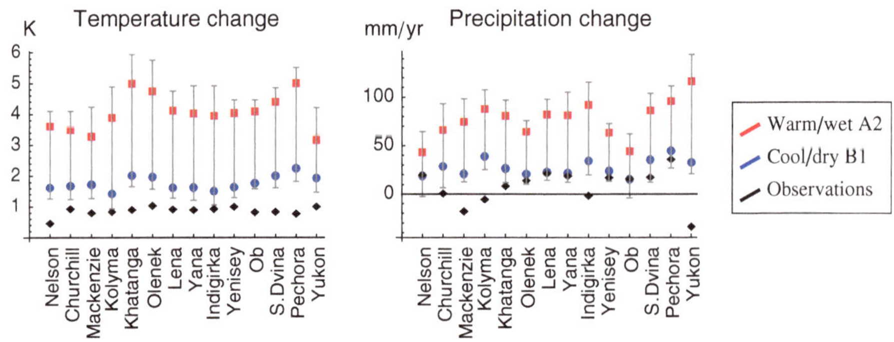

Fig. 5 Changes to temperature (*left*) and precipitation (*right*) from 1961–1990 to 1991–2009 for observations and to 2040–2069 for GCM projections across 14 major Arctic basins. *Error bars* indicate the range of different GCM results for both scenarios for each basin. *Top ends* of error bars extend one standard deviation of model means from the A2 scenario ensemble mean, and *bottom ends* extend one standard deviation of model means from the B1 scenario ensemble mean

with the strongest observed climate deviations is not reconcilable with prioritizing basins with greatest anticipated future climate change. In this analysis, the question is not whether models are able to reproduce observations, but whether the most (least) severe observed deviations occur in the same places as the projected future most (least) severe changes. To bring observed deviations in line with projections, continued increases are generally needed but, for precipitation, the direction of deviations must in several cases be reversed (Fig. 5). With ongoing climate change, continued increases in the same direction are naturally expected, as observations and projections concern different time periods in this analysis. Developing monitoring to capture such monotonous changes in magnitude is then one possible strategy. However, the disagreement in relative severity across basins between observations and GCM projections implies that alternative monitoring prioritization considerations are also possible and rational, as discussed further in Sect. 6.3.

4 Results for Arctic Hydro-Climatic Change

In Bring and Destouni (2011), we have compared precipitation and discharge deviations from the 1961 to 1990 climate average for the 14 major Arctic basins (green and blue basins in Fig. 1). A key result from that analysis is that the discharge deviation relations to the corresponding deviations in precipitation vary widely for the major basins in the PADB (Table 4). The majority of basins exhibit excess flow deviation in relation to the

Table 4 Changes to precipitation (P) and discharge (Q) from 1961–1991 to 1991–2002 for 13 major Arctic basins (mm/year)

Basin	ΔP	ΔQ	ΔQ − ΔP
Churchill	−49.8	−69.7	−19.9
Indigirka	−38.3	−3.0	35.3
Kolyma	−13.9	−8.8	5.2
Lena	−7.6	0.9	8.5
Mackenzie	−32.1	−3.4	28.7
Nelson	−1.0	0.1	1.1
Ob	18.3	9.6	−8.7
Olenek	−2.7	80.7	83.4
Pechora	23.8	47.3	23.5
S Dvina	5.9	26.9	21.0
Yana	−34.2	0.0	34.2
Yenisey	9.7	19.7	9.9
Yukon	−26.7	16.6	43.3
Average basin value	−11.4	9.0	20.4
Area-weighted pan-Arctic average	−3.7	7.5	11.1

Positive values in the last column indicate excess discharge, in relation to precipitation

precipitation deviation; that is, after the 1961–1990 climate period, discharge has increased more (or decreased less) than precipitation within each basin. These disagreements between precipitation and discharge deviations are in some cases large. At the same time, temperature has increased in the basins, which should lead to increased evapotranspiration and therefore less runoff, not more. Previous studies have indeed indicated that Arctic evapotranspiration has increased (Serreze et al. 2002; Park et al. 2008; Rawlins et al. 2010). Therefore, an essential question for forthcoming investigation is: where does the extra discharge water come from?

In the present synthesis, we note that the same question also arises from an integrated assessment of freshwater inflow to the Arctic Ocean (Dyurgerov et al. 2010). This assessment is in line with other studies, which also indicate an increase in freshwater inflow to the Arctic Ocean. However, the Dyurgerov et al. (2010) analysis further highlights that the magnitude of the increase in total flows from the 1961–1992 to the 1993–2006 period is similar for the Arctic river contributions (87 km^3/year) as for the Arctic glacier contributions (56 km^3/year). Figure 6 shows a main implication of that result for sea-level rise, underlining the importance of also accounting for river discharge changes that are not related to glacial mass balance changes as a possible contributing source for sea-level rise and freshening of the Arctic Ocean. However, determining whether also the eustatic components of these increases are of similar magnitude requires further and more refined analysis and modeling.

5 Results for Pan-Arctic Drainage Basin Monitoring

The earlier comprehensive analysis of pan-Arctic hydrological and hydrochemical monitoring by Bring and Destouni (2009) has shown a considerable difference between accessible water chemistry monitoring and discharge monitoring. This difference concerns both the total extent and the characteristics of the data. In general, discharge data are

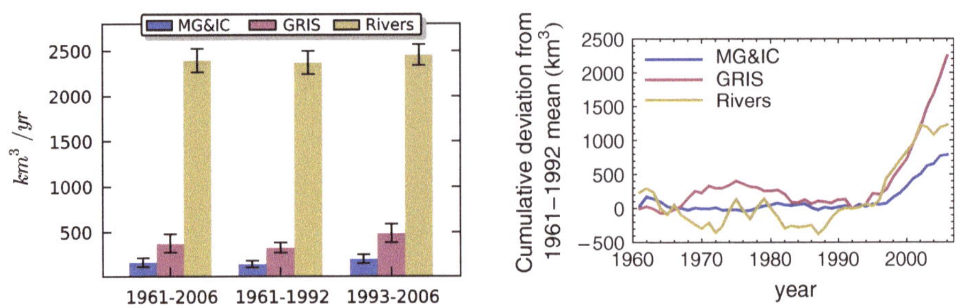

Fig. 6 *Left* Total meltwater runoff and total river runoff into the Arctic Ocean. *Right* Cumulative deviations in annual freshwater flows from mountain glaciers and ice caps (MG&IC), Greenland ice sheet (GRIS), and rivers relative to average values for 1961–1992

available for a wide range of different hydrological basins, from small catchments of a few square kilometers in size to the major river basins in the PADB. In contrast, accessible data from water chemistry monitoring are limited to a much smaller set of stations, which cover a significantly smaller area and also a much less complete range of basin sizes.

Figure 7 summarizes the accessible length of time series for discharge, and waterborne sediment and carbon data in hydrologically monitored Arctic areas. The average length of the time series for the various parameters, and the corresponding share of the PADB that is monitored, is further summarized in Table 5.

Results from this analysis further show a marked difference in the characteristics of monitored and unmonitored areas (Fig. 8). For example, monitored areas are distinctly dominated by the taiga eco-region, while unmonitored areas are generally strongly defined by tundra-type vegetation. This tendency is particularly evident for discharge monitoring in all regions and for all monitoring parameters in Asia. The most balanced monitoring, in terms of eco-region proportions, is for carbon monitoring in Europe, while the most unbalanced monitoring is found in North America and Asia. Such regional monitoring differences complicate the interpretation of observation data differences between different parts of the PADB.

6 Discussion

Robust scientific understanding of climate and water systems requires access to relevant information on changes to flows of water and waterborne constituents. Gaps in understanding and unreliability in GCM projections imply, for example, that costly dimensioning decisions for infrastructure and buildings may be more or less risk-prone than expected from the model results, or that changes in agricultural, and food and water security conditions are not properly accounted for.

In this paper, we have aimed to contribute to a more complete picture of several of the central components required for AHC change assessment and adaptation planning at drainage basin scales. Specifically, we have investigated three overarching topics pertaining to the relevance of GCM projections for Arctic drainage basins, the understanding of hydro-climatic change in these basins beyond just precipitation and the spatiotemporal basin coverage of Arctic hydrological and hydrochemical monitoring systems.

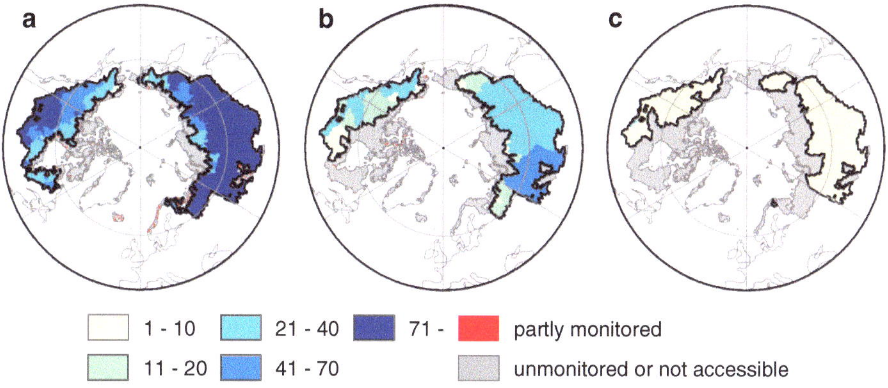

Fig. 7 Overview of the maximum length of accessible data series (years) for the pan-Arctic monitoring of **a** water discharge, **b** sediment and **c** carbon. Cells containing stations with drainage areas smaller than five cells are indicated as partly monitored

Table 5 Average length of time series and monitored share of total area for discharge, carbon and sediment data in the PADB

Parameters	Average time series length (years)	Area monitored (%)
Discharge	29	73
Sediment	7	63
Carbon	5	51

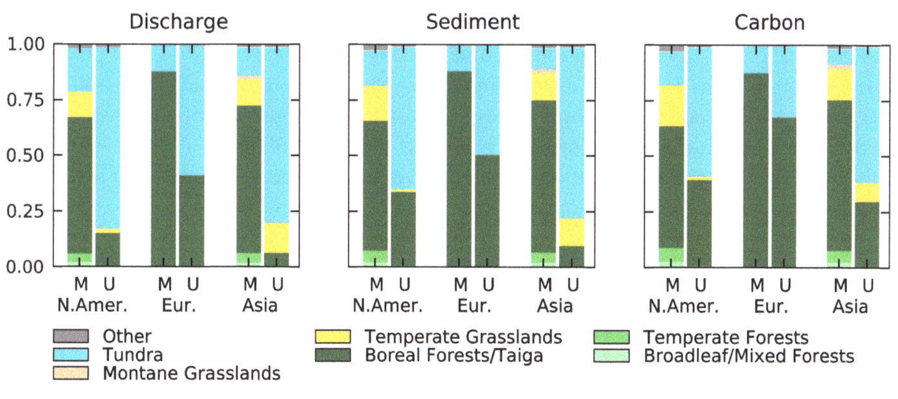

Fig. 8 Distribution of terrestrial eco-regions in monitored (M) and unmonitored (U) areas, by monitoring parameter and continent

6.1 GCM Projections Across Arctic Basins

Regarding the first investigation topic, on the relevance of GCM projections, the results highlight the improvements in basin-scale climate model simulation of temperature and precipitation between the TAR and AR4 generation of GCMs. However, the analysis also identifies some remaining limitations in precipitation simulations.

Considerable advances have been made between the two successive generations of model ensemble runs that underlie the latest available full IPCC Assessment Reports, the TAR and the AR4. Examples include improved parameterization of land surface schemes and snowpack, as well as the inclusion of canopy processes in most models (Randall et al. 2007). However, remaining uncertainties include the representation of cryospheric feedbacks, which explain part of the range of model responses at mid- and high latitudes (Randall et al. 2007).

These improvements are suggested by the fact that the results indicate a lower spread among the ensemble members for the AR4 models and, in the case of precipitation, a marked improvement in the agreement with absolute observed values. However, despite the improvement in precision, accuracy remains insufficient for precipitation projections, which, even if they were correct, are still difficult to translate to subsequent changes in the water cycle without further and finer resolved hydrological modeling. The uncertainty in observations, particularly gauge undercatch of solid precipitation due to wind (e.g., Yang et al. 2005; Tian et al. 2007), may explain part of the remaining difference between observations and the improved AR4 results.

The large span in model performance, evident from the explicit error and performance analysis of AR4 models, also indicates that large uncertainties and shortcomings remain for reliable simulations of hydro-climatic parameters on basin scales. The analysis furthermore underlines that models may yield good output results on these scales for the wrong reasons. For example, the pan-Arctic bias error (MBE) of temperature for the GIER model is close to zero. This would place it at the top of a basin-scale simulation performance ranking based on this measure. At the same time, the absolute error (MAE) and the performance index (d_r) for the same model are among the worst of the ensemble, implying large deviations from observations at individual grid points, even though the deviations happen to almost cancel out across the PADB.

The results for climate deviation and change severity across different basins further indicate that the relative distribution of climate change simulated by AR4 GCM projections does not agree with the relative distribution of currently observed climate deviations across the major Arctic basins. This complicates the choice of a robust prioritization strategy for hydrological monitoring development based on the reconciliation of observations and projections for the assessment of which Arctic river basins that are/will be most affected by climate change; this complication is further discussed in Sect. 6.3.

6.2 Arctic Hydro-Climatic Change

In the Arctic, certain counterintuitive hydro-climatic changes have caused extensive investigation and discussion in the scientific community. The fact that most of the 13 major rivers we study can be termed so-called excess rivers (i.e., the increase in discharge has been greater than the increase in precipitation) is consistent with findings by Milliman et al. (2008) for a related set of Arctic rivers. This and other studies (e.g., Dyurgerov and Carter 2004; McClelland et al. 2006; Smith et al. 2007; Lyon et al. 2009; Brutsaert and Hiyama 2012) indicate that both specific climatic and subsurface processes pertaining to the Arctic, such as permafrost degradation, as well as other anthropogenic changes and general atmospheric patterns, including increased moisture transport from outside the PADB, are responsible for the range in discharge patterns. Changes in glacier mass balance cannot explain the results of increasing river flows as the vast majority of Arctic glacier area and volume is located outside the major river basins (Dyurgerov et al. 2010). However, the relative importance of the different contributing factors is not definitely established.

Undercatch of snow, in combination with increases in winter precipitation (Bulygina et al. 2009; Rawlins et al. 2009), is also a potential contributing factor.

The wide spread in discharge responses to precipitation changes indicates a need for further concerted modeling and field investigation efforts aimed at increasing the process knowledge of hydrological regimes particular to the Arctic domain, including the simulations of evapotranspiration and permafrost processes, considering also the effects of scale (Rennermalm et al. 2012). Improved parameterization, in particular with precipitation focus, is also critical to the development of GCMs, and continued water-balance assessments and closure experiments in Arctic catchments are motivated.

The extended water budget change over the entire PADB and its drainage into the Arctic Ocean, including glacier contributions, also has implications for the global climate system. The fact that the increase in the river freshwater contribution is of the same order of magnitude as the increase in meltwater from MG&IC or from GRIS underlines the importance of better understanding the reasons for the river flow changes, and the potentially contributing components of frozen and liquid water storage changes in major Arctic basins (e.g., Muskett and Romanovsky 2009). Such changes could have large implications for the pan-Arctic hydro-climatic system, for example, through ground subsidence from permafrost degradation and/or altered soil moisture conditions, but also beyond the Arctic through contribution to sea-level rise and thermohaline circulation.

6.3 Pan-Arctic Drainage Basin Monitoring

The analysis of hydrological and hydrochemical data accessibility points to some major shortcomings, but also to opportunities and results of relevance for future research and monitoring improvements.

The synthesis of monitoring data shows a particular lack of water chemistry data, whereas discharge data are more extensively accessible. The range of spatial monitoring coverage, sampling frequency and length of time series for water chemistry data overall compare negatively with the corresponding attributes for discharge data. Together, these shortcomings imply that the full potential of translating existing discharge data to also calculate mass fluxes of biogeochemically important water constituents is hindered.

Furthermore, the difference in characteristics of hydrologically monitored areas evident from this analysis shows that existing monitoring data are not representative of the PADB as a whole. This constitutes a limitation to the input/validation data, and thereby also to reliability of the modeling that must be used to interpret and project AHC changes in unmonitored areas. It also limits the ability to improve GCM parameterizations and land surface schemes for the region, due to the lack of data to establish a ground truth to compare with.

From the perspective of an integrative pan-Arctic analysis, the limited accessibility to water chemistry data is remarkable, given the high profile of and international commitment to research into Arctic environmental changes. While discharge data have been compiled into pan-Arctic data sets [most importantly, R-ArcticNET (Lammers et al. 2001) and the ARDB (http://ardb.bafg.de)] and also made accessible in near real time through the ArcticRIMS project (http://rims.unh.edu), water chemistry data remain fragmented, although the recent continuation of the PARTNERS project as the Arctic Great Rivers Observatory (http://www.arcticgreatrivers.org) constitutes a substantial improvement.

Several factors may explain these results. Firstly, although the research community emphasizes a system perspective on the pan-Arctic domain, no international body has formal responsibility for an integrated water monitoring system. Instead, monitoring efforts must build upon undertakings by separate government agencies in at least the eight different nations of the Arctic Council, and for hydrology also Mongolia and Kazakhstan, which constitute parts of the PADB. Even if all national hydrometeorological agencies were committed to promoting a coordinated Arctic observation effort, domestic budget limitations and conflicting information goals may still interfere with ambitions being met.

Secondly, public agencies that apply cost-recovery principles to their environmental data may be reluctant to undermine this rule by freely sharing their data with international repositories. Thirdly, in a previously isolated Arctic that is now rapidly becoming more accessible to activities such as natural resource exploration, shipping and tourism, water data may be viewed as increasingly sensitive information from both political and economic perspectives. Occasionally, pressure from international organizations on member states to disclose water chemistry information viewed as sensitive has rebounded and instead caused delays in the progress of sharing other water data (Vladimir Ryabinin, personal communication).

Despite these potential obstacles, member states of the Arctic Council have in the Tromsø Declaration recently committed to facilitate data access. The organizational form for this commitment is the Sustaining Arctic Observation Networks (SAON) process, which is currently in its implementation phase. The European Commission has also expressed support for SAON (European Commission 2008, 2012). It remains to be seen to which extent the SAON process can contribute to increased accessibility to monitored water chemistry data in the PADB, but it is now established as a platform for policy dialog on Arctic monitoring issues, for example, through recurring Arctic Observation Summits.

To improve monitoring systems while considering climate change, and with limited resources, one must develop a strategy to decide which areas that should be prioritized. If GCM projections and observations were in agreement on which basins that are the most affected by climate change, a natural prioritization basis would be the rank of these basins by their relative intensity of change. However, the results in this study show that projections and observations diverge in this regard.

An alternative strategy may be to instead prioritize basins where the disagreement between observations and projections is particularly large, as such a strategy would yield important information on the hydro-climatic system functioning and changes regardless of whether there is actual convergence of observations and projections in the end or not. Alternatively, one could argue for prioritizing monitoring of basins with greater observed deviations so far, as these are based on the actual measurements, and increased efforts at understanding and adapting to them can be intrinsically motivated. With respect to precipitation deviations, such a prioritization is to some degree evident in the present distribution of monitoring (greater observed deviations correlate with greater monitoring effort), although this situation is most likely by coincidence rather than by design.

Based on the results in this study, one can thus argue for different rationales and prioritization bases when planning for increased hydro-climatic monitoring efforts under climate change conditions. These different rationales and prioritization bases point in diverging directions, which underlines the importance of attempting to formulate win–win or no-regret solutions (UNECE 2009) that also incorporate other parameters, in addition to temperature and precipitation outputs of climate models, and to explicitly formulate water information goals to be achieved. The results presented here can inform observation assessments connected to strategic Arctic initiatives and programs, such as SAON and the upcoming Third International Conference on Arctic Research Planning (ICARP III) in 2015, where continued evaluation of monitoring efforts will be a priority.

6.4 General Discussion

The review of literature and analysis in this paper confirms a picture of advances in our knowledge about the Arctic water environment and its functioning during recent decades. Climate model advances have improved in precision and alignment with observations in the Arctic, and numerous studies have contributed to a fuller understanding of hydro-climatic changes. International efforts at making discharge information accessible, e.g., through the ArcticRIMS project, have contributed to a more timely access to river flow data for a number of the major Arctic basins and sub-basins. Similarly, the PARTNERS monitoring campaign (http://ecosystems.mbl.edu/partners), now semi-permanently established as the Arctic Great Rivers Observatory (http://www.arcticgreatrivers.org), has provided a multi-season data set of concentrations of a range of biogeochemical water chemistry constituents for six major Arctic rivers. These developments, which coincide in time with a growing focus on the Arctic as a place of large-scale geophysical and geo-political changes, imply that our ability to understand the AHC has increased.

Nevertheless, considerable challenges for understanding and managing the rapidly changing Arctic water system still remain. A critical priority must be to continue improving the accessibility to water data, in particular for water chemistry, for the PADB. Observations must be made at well-chosen places, with the choices and improvements being based on clearly stated information goals, striving in their achievement for win–win and no-regret approaches that do not rely on some single, most likely model scenario of future hydro-climatic conditions. Improved coverage of unmonitored areas in the northern rims of the PADB is also motivated, particularly as these areas are expected to become increasingly accessible and also subject to exploration with a warming climate (Andreeva 1998). In addition, improved monitoring here may provide better insight into the total flux of water constituents to the ocean from these areas, as near-ocean catchments have in other regions been shown to contribute a disproportionately large share of coastal pollution in relation to the drainage basin as a whole (Destouni et al. 2008).

7 Conclusions

In this paper, we have synthesized and investigated climate model projections, hydro-climatic change understanding and adequacy of water flow and water chemistry observations in the PADB. The purpose has been to establish a quantitative picture of the status of model results, hydro-climatic links and observations needed to understand and manage water cycle changes in the Arctic. The main findings of the paper can be summarized in the following conclusions:

- The precision in climate model projection of precipitation and temperature change on drainage basin scales has improved between successive generations of the IPCC model ensemble. Individual model performance varies greatly, and models can be right for the wrong reasons when relatively large errors cancel out on large basin scales for some models. Further investigation and benchmarking of model performance in the GCMs that underlie the IPCC AR5 will show whether certain hydro-climatological model shortcomings in the Arctic have been addressed.
- From the investigation into geographical consistency in the relative distribution of climate deviations and changes, it follows that establishing regional priorities for hydrological monitoring systems, with regard to the specific issue of climate changes in

the Arctic, can currently not be achieved based solely on a reconciliation of observations and projections. When taking different data and system/change perspectives as starting points, different conclusions about what constitutes rational monitoring priorities, and related strategies, arise.

- Hitherto observed deviations in precipitation do not translate into similar deviations in discharge even over the large scales of the 13 studied basins. The discrepancy between precipitation and discharge deviations, and the fact that this discrepancy principally is expressed as excess in discharge compared to available precipitation, indicates that a component of the discharge variation may be due to changes in permafrost or groundwater storage. Potentially, these storage changes may also be a factor in the general deviations between observations and GCM simulations noted above.
- Even though the increase in river inflow to the Arctic Ocean during 1993–2006 is small in relative terms compared to the average flow for 1961–1992, in absolute terms it is of the same order of magnitude as the meltwater increase from glaciers. As the river flow changes are independent of the glacier contributions studied here, this underlines the importance of also accounting for river discharge changes as a potentially contributing source for sea-level rise and Arctic Ocean freshening.
- There is a lack of long-term and accessible water chemistry data for large parts of the PADB, and discharge data are also limited for considerable areas. The data that are accessible do not constitute a representative sample of the whole PADB environment.

Acknowledgments The research in this paper has been funded through grants from the Swedish research council Formas (Project Number 2007-1263) and the Swedish Research Council (VR; Project Number 2007-8393), and has been linked to the Bolin Centre for Climate Research (in turn supported by VR and Formas through a Linnaeus grant) and the strategic research project EkoKlim at Stockholm University. The study was inspired by and written for the documentation of the ISSI conference on *The Earth's hydrological cycle*, Bern, Switzerland, February 6–10, 2012.

References

Adam JC, Lettenmaier DP (2003) Adjustment of global gridded precipitation for systematic bias. J Geophys Res 108:1–14

Adam JC, Clark EA, Lettenmaier DP, Wood EF (2006) Correction of global precipitation products for orographic effects. J Clim 19:15–38

Andreeva EN (1998) The Russian Arctic coastal zone management problems: past lessons and new realities. Ocean Coast Manag 41:237–256

Arctic-HYDRA consortium (2010) The Arctic hydrological cycle monitoring, modelling and assessment programme: Science and implementation plan. ISBN 978-9979-9975-0-4

Azcárate J, Balfors B, Bring A, Destouni G (2013) Strategic environmental assessment and monitoring: Arctic key gaps and bridging pathways. Environ Res Lett 8:044033

Boé J, Hall A, Qu X (2009) Current GCMs' unrealistic negative feedback in the Arctic. J Clim 22:4682–4695

Box JE, Bromwich DH, Veenhuis BA, Bai L-S, Stroeve JC, Rogers JC, Steffen K, Haran T, Wang S-H (2006) Greenland ice sheet surface mass balance variability (1988–2004) from calibrated Polar MM5 output. J Clim 19:2783–2800

Bring A, Destouni G (2009) Hydrological and hydrochemical observation status in the pan-Arctic drainage basin. Polar Res 28:327–338

Bring A, Destouni G (2011) Relevance of hydro-climatic change projection and monitoring for assessment of water cycle changes in the Arctic. Ambio 40:361–369
Bring A, Destouni G (2013) Hydro-climatic changes and their monitoring in the Arctic: observation-model comparisons and prioritization options for monitoring development. J Hydrol 492:273–280
Brown K (2002) Water scarcity: forecasting the future with spotty data. Science 297:926–927
Brown R, Derksen C, Wang L (2010) A multi-data set analysis of variability and change in Arctic spring snow cover extent, 1967–2008. J Geophys Res 115:D16111
Brutsaert W, Hiyama T (2012) The determination of permafrost thawing trends from long-term streamflow measurements with an application in eastern Siberia. J Geophys Res 117:D22110
Bulygina ON, Razuvaev VN, Korshunova NN (2009) Changes in snow cover over Northern Eurasia in the last few decades. Environ Res Lett 4:045026
Callaghan TV, Johansson M, Brown RD, Groisman PY, Labba N, Radionov V, Barry RG, Bulygina ON, Essery RLH, Frolov DM, Golubev VN, Grenfell TC, Petrushina MN, Razuvaev VN, Robinson DA, Romanov P, Shindell D, Shmakin AB, Sokratov SA, Warren S, Yang D (2011) The changing face of Arctic snow cover: a synthesis of observed and projected changes. Ambio 40:17–31
Christensen JH, Hewitson B, Busuioc A, Chen A, Gao X, Held I, Jones R, Kolli RK, Kwon W-T, Laprise R, Magaña Rueda V, Mearns L, Menéndez CG, Räisänen J, Rinke A, Sarr A, Whetton P (2007) Regional climate projections. In: Solomon S, Qin D, Manning M, Chen Z, Marquis M, Averyt KB, Tignor M, Miller HL (eds) Climate change 2007: the physical science basis. Contribution of working group I to the fourth assessment report of the intergovernmental panel on climate change. Cambridge University Press, Cambridge
Comiso JC, Parkinson CL, Gersten R, Stock L (2008) Accelerated decline in the Arctic sea ice cover. Geophys Res Lett 35:L01703
Darracq A, Lindgren G, Destouni G (2008) Long-term development of phosphorus and nitrogen loads through the subsurface and surface water systems of drainage basins. Global Biogeochem Cycles 22:GB3022
Destouni G, Hannerz F, Prieto C, Jarsjö J, Shibuo Y (2008) Small unmonitored near-coastal catchment areas yielding large mass loading to the sea. Global Biogeochem Cycles 22:GB4003
Destouni G, Jaramillo F, Prieto C (2013) Hydroclimatic shifts driven by human water use for food and energy production. Nat Clim Chang 3:213–217
Dyurgerov MB, Carter CL (2004) Observational evidence of increases in freshwater inflow to the Arctic Ocean. Arct Antarct Alp Res 36:117–122
Dyurgerov MB, Meier MF (2005) Glaciers and the changing Earth system: a 2004 snapshot. Occasional Paper 58. Institute of Arctic and Alpine Research, Boulder, Colorado, pp 117
Dyurgerov MB, Bring A, Destouni G (2010) Integrated assessment of changes in freshwater inflow to the Arctic Ocean. J Geophys Res 115:D12116
Eisenman I, Untersteiner N, Wettlaufer JS (2007) On the reliability of simulated Arctic sea ice in global climate models. Geophys Res Lett 34:L10501
Environment Canada (2004) HYDAT version 2004-20.04. Water Survey of Canada, Ottawa
European Commission (2008) Communication from the commission to the European parliament and the council—the European union and the Arctic region. COM 763
European Commission (2012) The inventory of activities in the framework of developing a European Union Arctic Policy. Joint Staff Working Document 182
FAO (2009) UN-Water task force on indicators, monitoring and reporting final report. Monitoring progress in the water sector: a selected set of indicators. Food and Agriculture Organization of the United Nations, Rome
Fekete BM, Vörösmarty CJ (2002) The current status of global river discharge monitoring and potential new technologies complementing traditional discharge measurements. In: Predictions in Ungauged Basins: PUB Kick-off (Proceedings of the PUB Kick-off meeting held in Brasilia, 20–22 November 2002). IAHS Publication 349. International Association of Hydrological Sciences, Paris
Fluctuations of Glaciers (FoG) 2000–2005 (2008) ICSU(CCS)-UNEP-UNESCO 2008, vol IX. World Glacier Monitoring Service, Zürich
Gardner AS, Moholdt G, Wouters B, Wolken GJ, Burgess DO, Sharp MJ, Cogley JG, Braun C, Labine C (2011) Sharply increased mass loss from glaciers and ice caps in the Canadian Arctic Archipelago. Nature 473:357–360
GEO (2010) The GEO Beijing declaration: Observe, share, inform. Beijing ministerial summit document. Group on Earth Observations, Geneva
Glazovsky AF, Macheret YY (2006) Evraziyskaya Arktika. In: Kotlyakov VM (ed) Oledenenie Evrazii v proshlom, nastoyashchem i blizhayshem budushchem, vol 1. Nauka, Moscow, pp 97–114

Hanna E, Huybrechts P, Steffen K, Cappelen J, Huff R, Shuman C, Irvine-Fynn T, Wise S, Griffits M (2008) Increased runoff from melt from the Greenland ice sheet: a response to global warming. J Clim 21:331–341. doi:10.1175/2007JCLI1964.1

Hanna E, Cappelen J, Fettweis X, Huybrechts P, Luckman A, Ribergaard MH (2009) Hydrologic response of Greenland ice sheet: The role of oceanographic warming. Hydrol Process 23. doi:10.1002/hyp.7090

Hannerz F (2008) Making water information relevant on local to global scale-the role of information systems for integrated water management. PhD thesis. Stockholm University, Sweden

Harris I, Jones PD, Osborn TJ, Lister DH (2013) Updated high-resolution grids of monthly climatic observations—the CRU TS3.10 dataset. Int J Climatol. doi:10.1002/joc.3711

Hinzman LD, Bettez ND, Bolton WR, Chapin FS, Dyurgerov MB, Fastie CL, Griffith B, Hollister RD, Hope A, Huntington HP, Jensen AM, Jia GJ, Jorgenson T, Kane DL, Klein DR, Kofinas G, Lynch AH, Lloyd AH, McGuire AD, Nelson FE, Oechel WC, Osterkamp TE, Racine CH, Romanovsky VE, Stone RS, Stow DA, Sturm M, Tweedie CE, Vourlitis GL, Walker MD, Walker DA, Webber PJ, Welker JM, Winker KS, Yoshikawa K (2005) Evidence and implications of recent climate change in northern Alaska and other Arctic regions. Clim Chang 72:251–298

Holland MM, Finnis J, Barrett AP, Serreze MC (2007) Projected changes in Arctic Ocean freshwater budgets. J Geophys Res 112. doi:10.1029/2006JG000354

Holland MM, Serreze MC, Stroeve J (2010) The sea ice mass budget of the Arctic and its future change as simulated by coupled climate models. Clim Dyn 34:185–200

Holmes RM, Peterson BJ (2002) Eurasian river historical nutrient and sediment flux data. Digital media. National Snow and Ice Data Center, Boulder, Colorado

Holmes RM, Peterson BJ, Gordeev VV, Zhulidov AV, Meybeck M, Lammers RB, Vörösmarty CJ (2000) Flux of nutrients from Russian rivers to the Arctic Ocean: can we establish a baseline against which to judge future changes? Water Resour Res 36:2309–2320

Holmes RM, Makkaveev PN, Stunzhas PA, Kosmenko LS, Köhler GH, Shiklomanov AI (2001) Nutrient chemistry of the Ob and Yenisey rivers, Siberia: results from June 2000 expedition and evaluation of long-term data sets. Mar Chem 75:219–227

Holmes RM, McClelland JW, Peterson BJ, Shiklomanov IA, Shiklomanov AI, Zhulidov AV, Gordeev VV, Bobrovitskaya NN (2002) A circumpolar perspective on fluvial sediment flux to the Arctic Ocean. Global Biogeochem Cycles 16:1098

Hurrell JW, van Loon H (1997) Decadal variations in climate associated with the North Atlantic Oscillation. In: Climatic change at high elevation sites. Springer, Berlin, pp 69–94

ICSU (2010) Earth system science for global sustainability: The grand challenges. International Council for Science, Paris

Karlsson JM, Bring A, Peterson GD, Gordon LJ, Destouni G (2011) Opportunities and limitations to detect climate-related regime shifts in inland Arctic ecosystems through eco-hydrological monitoring. Environ Res Lett 6:014015

Kaser G, Cogley JG, Dyurgerov MB, Meier MF, Ohmura A (2006) Mass balance of glaciers and ice caps: consensus estimates for 1961–2004. Geophys Res Lett 33:L19501

Kattsov VM, Källén E, Cattle H, Christensen J, Drange H, Hanssen-Bauer I, Jóhannesen T, Karol I, Räisänen J, Svensson G, Vavulin S, Chen D, Polyakov I, Rinke A (2005) Future climate change: modeling and scenarios for the Arctic. In: Arctic climate impact assessment. Cambridge University Press, Cambridge, pp 99–150

Kattsov VM, Walsh JE, Chapman WL, Govorkova VA, Pavlova TV, Zhang X (2007) Simulation and projection of Arctic freshwater budget components by the IPCC AR4 global climate models. J Hydrometeorol 8:571–589

Lammers RB, Shiklomanov AI, Vörösmarty CJ, Fekete BM, Peterson BJ (2001) Assessment of contemporary Arctic river runoff based on observational discharge records. J Geophys Res 106:3321–3334

Lui J, Zhang Z, Hu Y, Chen L, Dai Y, Ren X (2008) Assessment of surface air temperature over the Arctic Ocean in reanalysis and IPCC AR4 model simulations with IABP/POLES observations. J Geophys Res 113:D10105

Lyon SW, Destouni G (2010) Changes in catchment-scale recession flow properties in response to permafrost thawing in the Yukon river basin. Int J Climatol 30. doi:10.1002/joc.1993

Lyon SW, Destouni G, Giesler R, Humborg C, Mörth M, Seibert J, Karlsson J, Troch PA (2009) Estimation of permafrost thawing rates in a sub-Arctic catchment using recession flow analysis. Hydrol Earth Syst Sci 13:595–604

Lyon S, Morth M, Humborg C, Giesler R, Destouni G (2010) The relationship between subsurface hydrology and dissolved carbon fluxes for a sub-arctic catchment. Hydrol Earth Syst Sci 14:941–950

Mächel H, Rudolf B, Maurer T, Hagemann S, Hagenbrock R, Kitaev L, Førland EJ, Rasuvaev V, Tveito OE (2012) Observed hydrological cycle. In: Lemke P, Jacobi H-W (eds) Arctic climate change: the ACSYS decade and beyond, Springer, Berlin, pp 199–246

Maurer T (2003) Development of an operational internet-based near real time monitoring tool for global river discharge data. GRDC Report 30. Global Runoff Data Centre, Koblenz

McBean G, Alekseev G, Chen D, Foerland E, Fyfe J, Groisman PY, King R, Melling H, Vose R, Whitfield PH (2005) Arctic climate: past and present. In: Arctic climate impact assessment. Cambridge University Press, Cambridge, pp 21–60

McClelland JW, Déry SJ, Peterson BJ, Holmes RM, Wood EF (2006) A pan-Arctic evaluation of changes in river discharge during the latter half of the 20th century. Geophys Res Lett 33:L06715

McClelland JW, Holmes RM, Peterson BJ, Amon R, Brabets T, Cooper L, Gibson J, Gordeev VV, Guay C, Milburn D, Staples R, Raymond PA, Shiklomanov I, Striegl R, Zhulidov A, Gurtovaya T, Zimov S (2008) Development of a pan-Arctic database for river chemistry. EOS Trans Am Geophys Union 89. doi:10.1029/2008EO240001

McGuire AD, Chapin FS, Walsh JE, Wirth C (2006) Integrated regional changes in Arctic climate feedbacks: implications for the global climate system. Annu Rev Environ Resour 31:61–91

Mernild SH, Liston GE, Hiemestra CA, Steffen K, Hanna E, Christensen JH (2009) Greenland ice sheet surface mass-balance modeling and freshwater flux for 2007, and in a 1995–2007 perspective. Hydrol Process 23. doi:10.1002/hyp.7354

Milliman J, Farnsworth K, Jones P, Xu K, Smith L (2008) Climatic and anthropogenic factors affecting river discharge to the global ocean, 1951–2000. Global Planet Change 62:187–194

Mitchell TD, Jones PD (2005) An improved method of constructing a database of monthly climate observations and associated high-resolution grids. Int J Climatol 25:693–712

Mueller B, Seneviratne SI, Jimenez C, Corti T, Hirschi M, Balsamo G, Ciais P, Dirmeyer P, Fisher JB, Guo Z, Jung M, Maignan F, McCabe MF, Reichle R, Reichstein M, Rodell M, Sheffield J, Teuling AJ, Wang K, Wood EF, Zhang Y (2011) Evaluation of global observations-based evapotranspiration datasets and IPCC AR4 simulations. Geophys Res Lett 38:L06402

Muskett RR, Romanovsky VE (2009) Groundwater storage changes in arctic permafrost watersheds from GRACE and in situ measurements. Environ Res Lett 4:045009

Nilsson LM, Destouni G, Berner J, Dudarev AA, Mulvad G, Odland JO, Parkinson A, Tikhonov C, Rautio A, Evengård B (2013) A call for urgent monitoring of food and water security based on relevant indicators for the Arctic. Ambio 42. doi:10.1007/s13280-013-0427-1

Overeem I, Syvitski JPM (2010) Shifting discharge peaks in Arctic rivers, 1977–2007. Geografiska Ann Ser A Phys Geogr 92:285–296

Overland JE, Wang M (2007) Future regional Arctic sea ice declines. Geophys Res Lett 34:L17705

Pahl-Wostl C (2007) Transitions towards adaptive management of water facing climate and global change. Water Resour Manage 21:49–62

Palmer MA, Lettenmaier DP, Poff NL, Postel SL, Richter B, Warner R (2009) Climate change and river ecosystems: protection and adaptation options. Environ Manage 44:1053–1068

Park H, Yamazaki T, Yamamoto K, Ohta T (2008) Tempo-spatial characteristics of energy budget and evapotranspiration in the eastern Siberia. Agric For Meteorol 148:1990–2005

Peterson BJ, Holmes RM, McClelland JW, Vörösmarty CJ, Lammers RB, Shiklomanov AI, Shiklomanov IA, Rahmstorf S (2002) Increasing river discharge to the Arctic Ocean. Science 298:2171–2173

Peterson BJ, McClelland J, Curry R, Holmes RM, Walsh JE, Aagaard K (2006) Trajectory shifts in the Arctic and subarctic freshwater cycle. Science 313:1061–1066

Randall DA, Wood RA, Bony S, Colman R, Fichefet T, Fyfe J, Kattsov V, Pitman A, Shukla J, Srinivasan J, Stouffer RJ, Sumi A, Taylor KE (2007) Climate models and their evaluation. In: Solomon S, Qin D, Manning M, Chen Z, Marquis M, Averyt KB, Tignor M, Miller HL (eds) Climate change 2007: The physical science basis. Contribution of working group I to the fourth assessment report of the intergovernmental panel on climate change. Cambridge University Press, Cambridge

Rawlins MA, Ye H, Yang D, Shiklomanov A, McDonald KC (2009) Divergence in seasonal hydrology across northern Eurasia: emerging trends and water cycle linkages. J Geophys Res 114:D18119

Rawlins MA, Steele M, Holland MM, Adam JC, Cherry JE, Francis JA, Groisman PY, Hinzman LD, Huntington TG, Kane DL, Kimball JS, Kwok R, Lammers RB, Lee CM, Lettenmaier DP, McDonald KC, Podest E, Pundsack JW, Rudels B, Serreze MC, Shiklomanov A, Skagseth O, Troy TJ, Vörösmarty CJ, Wensnahan M, Wood EF, Woodgate R, Yang D, Zhang K, Zhang T (2010) Analysis of the Arctic system for freshwater cycle intensification: observations and expectations. J Clim 23:5715–5737

Raymond PA, McClelland JW, Holmes RM, Zhulidov AV, Mull K, Peterson BJ, Striegl RG, Aiken GR, Gurtovaya TY (2007) Flux and age of dissolved organic carbon exported to the Arctic Ocean: a carbon isotopic study of the five largest Arctic rivers. Global Biogeochem Cycles 21:GB4011

Reid W, Chen D, Goldfarb L, Hackmann H, Lee Y, Mokhele K, Ostrom E, Raivio K, Rockström J, Schellnhuber H, Whyte A (2010) Earth system science for global sustainability: grand challenges. Science 330:916–917

Rennermalm AK, Bring A, Mote TL (2012) Spatial and scale-dependent controls on North American pan-Arctic minimum river discharge. Geogr Anal 44:202–218

Rignot E, Box JE, Burgess E, Hanna E (2008) Mass balance of the Greenland ice sheet from 1958 to 2007. Geophys Res Lett 35:L20502. doi:10.1029/2008GL035417

Roesch A (2006) Evaluation of surface albedo and snow cover in AR4 coupled climate models. J Geophys Res 111:D15111

Seitz NE, Westbrook CJ, Dubé MG, Squires AJ (2013) Assessing large spatial scale landscape change effects on water quality and quantity response in the lower Athabasca River Basin. Integr Environ Assess Manag 9. doi:10.1002/ieam.1336

Serreze MC, Clark MP, Etringer AJ, Zhang T, Bromwich DH, Lammers R (2002) The large-scale hydro-climatology of the terrestrial Arctic drainage system. J Geophys Res 107. doi:10.1029/2001JD000919

Serreze MC, Barrett AP, Slater AG, Woodgate RA, Aagaard K, Lammers RB, Steele M, Moritz R, Meredith M, Lee CM (2006) The large-scale freshwater cycle of the Arctic. J Geophys Res 111. doi:10.1029/2005JC003424

Shiklomanov AI, Lammers RB (2009) Record Russian river discharge in 2007 and the limits of analysis. Environ Res Lett 4:045015

Shiklomanov AI, Lammers RB, Vörösmarty CJ (2002) Widespread decline in hydrological monitoring threatens pan-Arctic research. EOS Trans Am Geophys Union 83:13–17

Slater AG, Bohn TJ, McCreight JL, Serreze MC, Lettenmaier DP (2007) A multimodel simulation of pan-Arctic hydrology. J Geophys Res 112:G04S45

Smith LC, Pavelsky TM, MacDonald GM, Shiklomanov AI, Lammers RB (2007) Rising minimum daily flows in northern Eurasian rivers: a growing influence of groundwater in the high-latitude hydrologic cycle. J Geophys Res 112:G04S47

Smol JP, Wolfe AP, Birks HJB, Douglas MSV, Jones VJ, Korhola A, Pienitz R, Rühland K, Sorvari S, Antoniades D, Brooks SJ, Fallu M-A, Hughes M, Keatley BE, Laing TE, Michelutti N, Nazarova L, Nyman M, Paterson AM, Perren B, Quinlan R, Rautio M, Saulnier-Talbot É, Siitonen S, Solovieva N, Weckström J (2005) Climate-driven regime shifts in the biological communities of Arctic lakes. Proc Natl Acad Sci 102:4397–4402

Sorteberg A, Kattsov V, Walsh J, Pavlova T (2007) The Arctic surface energy budget as simulated with the IPCC AR4 AOGCMs. Clim Dyn 29:131–156

Stroeve J, Holland MM, Meier W, Scambos T, Serreze M (2007) Arctic sea ice decline: faster than forecast. Geophys Res Lett 34:9501

Stroeve JC, Serreze MC, Holland MM, Kay JE, Malanik J, Barrett AP (2012a) The Arctic's rapidly shrinking sea ice cover: a research synthesis. Clim Chang 110:1005–1027

Stroeve JC, Kattsov V, Barrett A, Serreze M, Pavlova T, Holland M, Meier WN (2012b) Trends in Arctic sea ice extent from CMIP5, CMIP3 and observations. Geophys Res Lett 39:L16502

Tian X, Dai A, Yang D, Xie Z (2007) Effects of precipitation-bias corrections on surface hydrology over northern latitudes. J Geophys Res 112:D14101

UNECE (2009) Guidance on water and adaptation to climate change. United Nations Economic Commission for Europe, Geneva, p 144

Vörösmarty CJ, Hinzman LD, Peterson BJ, Bromwich DH, Hamilton LC, Morison J, Romanovsky VE, Sturm M, Webb RS (2001) The hydrologic cycle and its role in Arctic and global environmental change: A rationale and strategy for synthesis study. Arctic Research Consortium of the U.S., Fairbanks, Alaska, p 84

Walsh J, Anisimov O, Hagen JO, Jakobsson T, Oerlemans J, Prowse TD, Romanovsky V, Savelieva N, Serreze M, Shiklomanov A, Shiklomanov I, Solomon S (2005) Crysophere and hydrology. In: Arctic climate impact assessment. Cambridge University Press, Cambridge, pp 183–242

White D, Hinzman L, Alessa L, Cassano J, Chambers M, Falkner K, Francis J, Gutowski WJ, Holland M, Holmes RM, Huntington H, Kane D, Kliskey A, Lee C, McClelland J, Peterson B, Scott Rupp T, Straneo F, Steele M, Woodgate R, Yang D, Yoshikawa K, Zhang T (2007) The Arctic freshwater system: changes and impacts. J Geophys Res 112:G04S54

Willmott CJ, Robeson SM, Matsuura K (2012) A refined index of model performance. Int J Climatol 32. doi:10.1002/joc.2419

Yang D, Kane D, Zhang Z, Legates D, Goodison B (2005) Bias corrections of long-term (1973–2004) daily precipitation data over the northern regions. Geophys Res Lett 32:19

Zhang X, Walsh JE (2006) Toward a seasonally ice-covered Arctic Ocean: scenarios from the IPCC AR4 model simulations. J Clim 19:1730–1747

Zhulidov AV, Khlobystov VV, Robarts RD, Pavlov DF (2000) Critical analysis of water quality monitoring in the Russian Federation and former Soviet Union. Can J Fish Aquat Sci 57:1932–1939
Zhulidov AV, Robarts RD, Holmes RM, Peterson BJ, Kämäri J, Meriläinen JJ, Headley JV (2003) Water quality monitoring in the former Soviet Union and the Russian Federation: Assessment of analytical methods. The Finnish Environment Report No. 620. Finnish Environment Institute, Helsinki

DOI 10.1007/978-94-017-8789-5_18
Reprinted from *Surveys in Geophysics* Journal, DOI 10.1007/s10712-013-9223-5

Irrigation Effects on Hydro-Climatic Change: Basin-Wise Water Balance-Constrained Quantification and Cross-Regional Comparison

Shilpa M. Asokan · Georgia Destouni

Received: 17 September 2012 / Accepted: 13 February 2013 / Published online: 13 March 2013

Abstract Hydro-climatic changes driven by human land and water use, including water use for irrigation, may be difficult to distinguish from the effects of global, natural and anthropogenic climate change. This paper quantifies and compares the hydro-climatic change effects of irrigation using a data-driven, basin-wise quantification approach in two different irrigated world regions: the Aral Sea drainage basin in Central Asia and the Indian Mahanadi River Basin draining into the Bay of Bengal. Results show that irrigation-driven changes in evapotranspiration and latent heat fluxes and associated temperature changes at the land surface may be greater in regions with small relative irrigation impacts on water availability in the landscape (here represented by the Mahanadi River Basin) than in regions with severe such impacts (here represented by the Aral region). Different perspectives on the continental part of Earth's hydrological cycle may thus imply different importance assessments of various drivers and impacts of hydro-climatic change. Regardless of perspective, however, actual basin-wise water balance constraints should be accounted to realistically understand and accurately quantify continental water change.

Keywords Hydro-climatic change · Irrigation · Evapotranspiration · Surface temperature · Hydrological cycle · Hydrological catchment · Aral Sea · India

1 Introduction

Regional changes in land use and water use impact the Earth's hydrological cycle in different ways, which need to be distinguished and understood (Destouni et al. 2013), not least in view of increasing food production demands for Earth's growing human population (Gordon et al. 2003). The hydrological impacts of land–water use changes, however, may

S. M. Asokan (✉) · G. Destouni
Department of Physical Geography and Quaternary Geology, Bert Bolin Centre for Climate Research, Stockholm University, 106 91 Stockholm, Sweden
e-mail: shilpa.asokan@natgeo.su.se

G. Destouni
e-mail: georgia.destouni@natgeo.su.se

be difficult to distinguish among different change drivers (e.g., agriculture, forest, irrigation developments, various land-cover changes and/or land productivity changes) and from the impacts of global, natural and anthropogenic climate change. Nevertheless, among different regional land–water use changes, irrigation in particular has been found to greatly enhance evapotranspiration (ET) (Gordon et al. 2005; Shibuo et al. 2007; Asokan et al. 2010), and thereby affect regional climate (Boucher et al. 2004; Kueppers et al. 2007; Bonfils and Lobell 2007; Lobell et al. 2009; Destouni et al. 2010; Lee et al. 2011), as well as future hydrological responses to forthcoming climate change (Jarsjö et al. 2012).

With regard to regional irrigation effects on hydro-climatic change, Destouni et al. (2010) developed a hydrological approach to estimating irrigation-driven changes in ET, latent heat flux (*F*) and surface temperature (*T*) and distinguishing these change contributions from the regional effects of global climate change. The approach is data-driven and relatively simple, utilizing seasonal differences in actually observed historic *T* changes within a hydrological drainage basin, along with fundamental water balance constraints that corresponding, historic water flux observations imply for the basin. Destouni et al. (2010) specially developed and applied this basin-wise data-driven approach to the Central Asian system of the Aral Sea and its drainage basin.

Irrigation effects on ET and other hydrological changes, however, can differ greatly among hydrological basins in different world regions with different hydro-climatic conditions (Destouni et al. 2013). As noted by Destouni et al. (2010), irrigation effects may also differ between hydrological basins of different types (Vörösmarty et al. 2000), such as an endorheic[1] basin, draining into terminal surface waters like the Aral Sea, which depends greatly on that specific regional runoff, and an exorheic[2] type of basin, draining into the ocean, which does not depend so much on the runoff from just one specific basin.

In view of such cross-regional differences, the present paper aims at comparatively testing the applicability of the basin-wise data-driven approach proposed by Destouni et al. (2010) and comparing its hydro-climatic change results across two different world regions: the Aral region in Central Asia that includes the terminal Aral Sea and the endorheic Aral Sea drainage basin draining into it, which has already been studied using this approach, and the Mahanadi River Basin (MRB) in India, which is a novel application of the approach. The MRB is an exorheic basin, draining into the Bay of Bengal, with monsoon-driven hydro-climatic conditions and seasonality (Asokan et al. 2010) that differ greatly from those in the Aral Sea region (Shibuo et al. 2007; Destouni et al. 2010). In addition to extending and testing the applicability of the approach across different world regions, another main aim of this comparative study is to investigate and identify which regional conditions may lead to large and important hydro-climatic changes and impacts driven by irrigation. In the following sections, we first provide a literature review (Sect. 2), followed by a description of sites and methodology in Sect. 3, cross-regional results and their comparison in Sect. 4 and main conclusions in Sect. 5.

2 Literature Review

In general, vegetation interacts with climate to regulate the partitioning of precipitation (*P*) into ET and runoff (*R*) at the land surface (Gordon et al. 2005; Douglas et al. 2006;

[1] An endorheic basin is a closed drainage basin that retains water and allows no outflow to other external bodies of water such as oceans, but converges instead into lakes. Here the water loss is through evaporation, evapotranspiration and seepage.

[2] An exorheic basin is a drainage basin that discharges into ocean.

Donohue et al. 2007; Shibuo et al. 2007; Asokan et al. 2010; Destouni et al. 2013). In addition to impacts from global climate change, human changes in land uses and water uses that affect vegetation will also affect the Earth's water cycle (Foley et al. 2005; Shibuo et al. 2007; Piao et al. 2007; Weiskel et al. 2007; Wisser et al. 2010; Destouni et al. 2013). In particular, all agricultural developments that increase the cultivated area or change the biomass production in a given land area, for instance by irrigation, are associated with vegetation changes that will then also affect regional ET rates (Kvalevag et al. 2010; Destouni et al. 2013).

Previous studies have shown different types of land-use and water-use changes that may lead to considerable hydro-climatic change. For instance, deforestation may decrease ET and increase *R*, while opposite impacts may result from new forest establishment on previously sparsely vegetated land (Vanlill et al. 1980; Gordon et al. 2005; Loarie et al. 2011). Furthermore, the conversion of natural unplowed land to cultivated land may often increase ET (Loarie et al. 2011; Destouni et al. 2013), but such conversions may under some conditions also decrease it (Schilling et al. 2008). A change from agriculture to forest may further initially decrease ET (Qiu et al. 2011) and later increase it (Donohue et al. 2007). Regarding irrigation of agricultural areas, the direct withdrawal of freshwater for the irrigation, in addition to the actual land irrigation itself, has been found to affect water and vapor fluxes at the land surface, both globally (Foley et al. 2005; Gordon et al. 2005) and regionally (Shibuo et al. 2007; Lobell et al. 2009; Asokan et al. 2010; Destouni et al. 2010; Lee et al. 2011; Törnqvist and Jarsjö 2012; Jarsjö et al. 2012). With close to one billion people living in regions where irrigation is already used and may be used increasingly in the future to further enhance agricultural yields and ensure food safety for growing populations, it is important to understand and distinguish different aspects and magnitudes of hydro-climatic changes and their water resource impacts driven by irrigation under different regional conditions (Keiser et al. 2005; Lobell and Field 2007).

In order to realistically understand, project and efficiently mitigate or adapt to the adverse hydrological effects of both global climate change and regional changes in land use and water use, including irrigation, the dominant drivers, processes and effects need to be understood and quantified across different scales, including the relevant water management scales, which are commonly those of regional drainage basins. Regionally, the hydrological impacts from changes in global atmospheric circulation and climate overlap with the impacts from regional irrigation and other land-use and water-use changes (Lobell and Field 2007). This overlap makes it difficult to distinguish different hydrological cause and effect relations (Milly et al. 2002; Piao et al. 2007; Destouni et al. 2008). However, the topographical water divides that define regional drainage basins are also physical boundaries, which constrain reasonably well the flows of water and waterborne substances through the landscape, and associated environmental impacts of man-made changes to these flows (Jarsjö and Destouni 2004; Darracq et al. 2005; Shibuo et al. 2007; Destouni and Darracq 2009; Törnqvist et al. 2011; Jarsjö et al. 2012; Visser et al. 2012; Destouni et al. 2013). The understanding of and distinction between different hydrological change components and their drivers and effects can therefore be greatly aided and improved by honoring and accounting for the water flux constraints implied by the fundamental water balance quantification $\mathrm{ET} = P - R - \mathrm{DS}$, which applies to all hydrological drainage basins. Here, DS is water storage change within the basin, the effects of which are often small so that this term may be neglected for long-term changes over temporal scales that are much larger than the annual scale (Destouni et al. 2013).

The basin-wise water balance constraints imply that the ET, which is difficult to measure and quantify on large scales, can be derived from directly measured *P* data across the basin surface and R data at the basin outlet, with storage change DS neglected for long-term average ET. The focus on the quantification of large-scale ET is relevant both because this is the main unmonitored and unknown water flux term in a basin-wise water balance equation and because it is essential to quantify from different change perspectives on the Earth's hydrological cycle. One perspective may focus on the ET change effects on vapor and latent heat fluxes, and associated temperature changes at the land surface, based on a primary climatic change interest. Another perspective may focus on the ET change effects on water availability and runoff in the landscape, based on a primary interest in water resource changes and considering that ET represents the main consumptive loss of water from the landscape. At any rate, without the basin-wise water balance constraints, ET models can, for instance, yield errors of 30–50 % (Kite and Droogers 2000), compared with errors of 10–15 % involved in the ET estimation from basin-scale water balance closure (Asokan et al. 2010).

Recent technological possibilities, such as MODIS (King et al. 1992) and ET algorithm techniques based on satellite remote sensing (Zhang et al. 2010), provide more tools for ET change distinction and quantification. However, they can only be used for studying changes in relatively recent time periods that overlap with the accessibility to these technologies (Douglas et al. 2006; Cheng et al. 2011; Loarie et al. 2011). Furthermore, expected increasing hydrological impacts of forthcoming global and regional changes constitute another great quantification challenge (Milly et al. 2005; Groves et al. 2008; Bengtsson 2010; Jarsjö et al. 2012). Several scientific questions are open and require further investigation in the context of future hydro-climatic change projection, including the large spatial scale discrepancy between typical hydrological drainage basins and the coarse resolution of general circulation models (GCMs) (Milly et al. 2005; Groves et al. 2008). However, also in the context of future change, Jarsjö et al. (2012) have showed that historic hydro-climatic change understanding, based on basin-scale, water balance-constrained hydrological modeling, can be used for extending and constraining projections of future hydro-climatic change. They compiled climate (T and P) change projections from an ensemble of different GCMs, on a basin large enough to be sufficiently well resolved by the GCMs, and linked these projections with basin-scale hydrological modeling, conditioned by data and water balance-constrained resolution of historic hydro-climatic change drivers and effects in the basin, in order to constraint future ET and R change quantifications.

With specific regard to irrigation as a driver of hydro-climatic change, the approach developed by Destouni et al. (2010) could distinguish and quantify its effects among different change drivers of changes in ET, latent heat flux (F) and T when applied to the whole Central Asian system of the Aral Sea and its drainage basin. They used changes in ET from the drainage basin area (ΔET), as previously calculated through distributed hydrological modeling (Shibuo et al. 2007), and additional calculated changes of evaporation (E) from the Aral Sea itself (ΔE), as well as of the shift from Aral Sea E to dried seabed ET as the Aral Sea shrank from the beginning to the end of the twentieth century. Based on these results, they further calculated corresponding changes in latent heat flux (ΔF) and surface temperature (ΔT) for the whole regional system of the Aral Sea and its drainage basin. For the calculation of different ΔT components and their association to different change drivers, Destouni et al. (2010) used a difference in observed change of seasonal temperature between the growing and the non-growing season in the region. This difference reflected an irrigation cooling effect during the growing season, which was not present in the non-growing season without irrigation.

In the Aral Sea region, this cooling effect due to increased water vapor flux by evapotranspiration from the irrigated areas was quantified to be −0.6 °C, a considerable effect compared to the warming from the regional manifestation of global climate change, which was also quantified to be 1.2 °C (Destouni et al. 2010). However, in that region, an additional warming effect of 0.5 °C also occurred, because the water withdrawal for irrigation reduced the river runoff from the drainage basin into the terminal Aral Sea by 80 %. This led to dramatic Aral Sea shrinkage and associated warming due to smaller *E* from the smaller Aral Sea surface, which to a large degree counteracted the regional cooling effect of irrigation (Destouni et al. 2010). In the following, we will particularly survey the cross-regional applicability and implications of this relatively simple, yet potentially widely useful, basin-scale quantification approach to hydro-climatic irrigation effects, by extending and applying it also to the exorheic and hydro-climatologically monsoon-driven MRB in India and comparing the results with the previously investigated endorheic Aral Sea drainage basin and region in Central Asia.

3 Methodology

3.1 Comparative Site Data

Figure 1 shows location maps, and Table 1 summarizes basin-scale data and water fluxes for the MRB (with an area of 135,084 km^2) and the Aral region (with a total area of 1,888,810 km^2 for the Aral Sea and its drainage basin). The analysis focuses on the twentieth-century historic time span of available hydro-climatic data and compares changes from the beginning to the end of the century (see specific periods in Table 1). The compared time periods are chosen to represent the hydro-climatic situation and associated major land-use and water-use changes from the time before to the time after the construction of Hirakud dam (shown in Fig. 1) for the MRB, and before and after the irrigation development in the Aral Sea region. For the present new application of the approach to the MRB, we use the same climate database (CRU TS 2.1 by Mitchell and Jones 2005) as in the previous hydrological investigation and distributed modeling of this basin by Asokan et al. (2010) and extend here the data analysis to also consider and utilize the seasonality of surface temperature (*T*), precipitation (*P*) and their changes, as previously done and reported for the Aral Sea region by Destouni et al. (2010).

With regard to irrigation, agricultural areas in the MRB are cultivated throughout the whole year. The cropping seasons are broadly divided into *Kharif*, the Wet Season, and

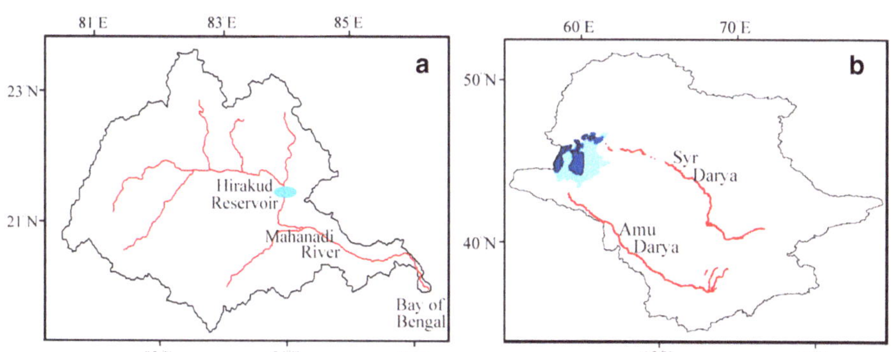

Fig. 1 Maps of **a** Mahanadi River Basin (MRB) in India and **b** Aral region in Central Asia. The *black line* shows the water divides of the basins. The *red lines* show the major rivers. The *light* and *dark shades* of *blue* in **b** illustrate the outline of Aral Sea before and after (in 2005) its major shrinkage

Table 1 Summary of basin-scale data and flux results for the Mahanadi river basin (MRB) (from Asokan et al. (2010) for 1901–1955, and new simulations with the same distributed hydrological model of the basin for 1956–2000) and the Aral region (from Shibuo et al. (2007)—average results from two different evapotranspiration models in that study; and Destouni et al. (2010), who extended the analysis and results from the previous study to also include the Aral Sea itself and the evaporation from it)

	Pre-reservoir	Climate–irrigation	Climate: hypothetical scenario of only climate change
Mahanadi River Basin	1901–1955	1956–2000	1956–2000
Data-given average temperature (°C)	25.19	25.45	25.45
Data-given annual average precipitation (mm yr^{-1})	1,334	1,274	1,274
Total modeled ET (mm yr^{-1})	668	706	656
Data-given irrigation water-use within the basin (mm yr^{-1})	–	81	–
Modeled runoff at basin outlet (mm yr^{-1})	666	568	618
Observed runoff at basin outlet (mm yr^{-1})	–	515[a]	–
Aral region	1901–1950	1983–2002	1983–2002
Data-given average temperature (°C)	7.5	8.6	8.6
Data-given annual average precipitation (mm yr^{-1})	249	260	260
Total modeled ET from whole region—Aral Sea and its drainage basin (mm yr^{-1})	250	265	260
Modeled irrigation water-use within the basin (mm yr^{-1})	–	23	–
Modeled runoff from drainage basin into Aral Sea (mm yr^{-1})	35	7	36
Observed runoff from drainage basin into the Aral Sea (mm yr^{-1})	38	6	–

[a] Average runoff from available observations for the period 1990–2000 (Asokan 2005)

Rabi, the Dry Season. The Wet Season extends from June to November, and the Dry Season extends from December to May. Out of the total annual irrigation water of 11 km^3 (81 mm if normalized with drainage basin area) that is used for agriculture in the MRB, 7 km^3 (52 mm) is available and utilized in the Wet Season and 4 km^3 (29 mm) in the Dry Season (Asokan 2005). For the Aral region, the growing season is from March to October, while the remaining months of the year constitute the non-growing season. Figure 2 illustrates basin-average time series of annual average T (Fig. 2a) and P (Fig. 2b), and Table 2 further shows the average seasonal T and P at the beginning and the end of twentieth century for both regions.

The change (increase) in annual average T from the beginning to the end of the century is 0.26 °C in the MRB and 1.1 °C in the Aral region. With regard to annual average P, its change is a decrease of 60 mm per year in the MRB and an increase of 11 mm per year in the Aral region. In the MRB, about 93 % of the total P occurs during the Wet Season. The Dry Season crops hence depend mainly on irrigated agriculture even though less irrigation water is available and used in this season (4 km^3 or 29 mm) than in the Wet Season (7 km^3 or 52 mm) (Asokan 2005). In the Aral region, P is much more evenly distributed between the growing and the non-growing season, the temporal extents of which are primarily determined by T rather than by P (Table 2).

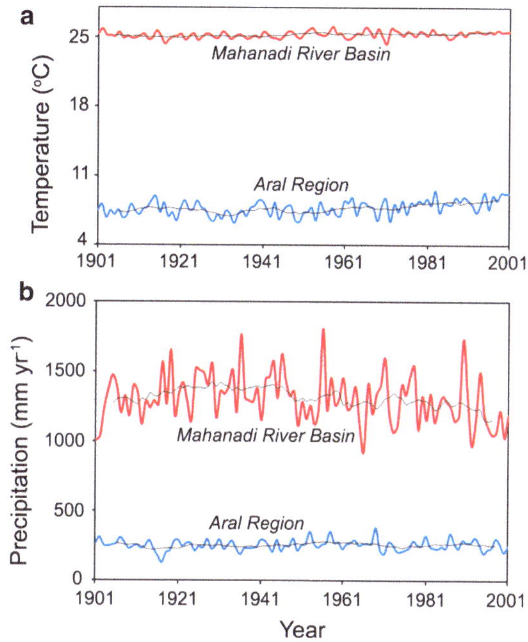

Fig. 2 Annual average temperature in degrees Celsius (**a**) and precipitation in mm per year (**b**) within the Mahanadi River Basin (*color red*) and the Aral Sea Region (*color blue*), based on data from Mitchell and Jones (2005). Running average (10 years) is shown by the *black thin lines*

Table 2 Average seasonal temperature and precipitation at the beginning and the end of twentieth century for the Mahanadi River Basin (MRB) and the Aral region

	Temperature (°C)		Precipitation (mm yr^{-1})	
Mahanadi River Basin	1901–1955	1956–2000	1901–1955	1956–2000
Dry season	24.44	24.74	198	190
Wet season	25.93	26.16	2,470	2,358
Aral region	1901–1950	1983–2002	1901–1950	1983–2002
Growing season	14.7	15.5	192	193
Non-growing season	−0.1	1.54	354	381

Figure 3 shows the seasonal T changes and their spatial distribution in both regions. Destouni et al. (2010) have already reported how they used the difference in seasonal T change for the Aral region to distinguish and quantify different regional T change components (ΔT) and their different drivers, and associated changes in ET (ΔET) and latent heat flux (ΔF) in that region. In the following, we will follow the same approach to also distinguish, quantify and directly compare with the previous Aral region results corresponding change components and drivers within the MRB.

3.2 Estimating Irrigation Effects from Temperature Seasonality

In the MRB, the basin-average T increase during the Wet Season is 0.23 °C, which is smaller than the T change of 0.30 °C during the Dry Season. From these seasonal change differences, which are consistent with more irrigation water being used and thereby leading to more cooling in the Wet Season (7 km^3) than in the Dry Season (4 km^3), along with the

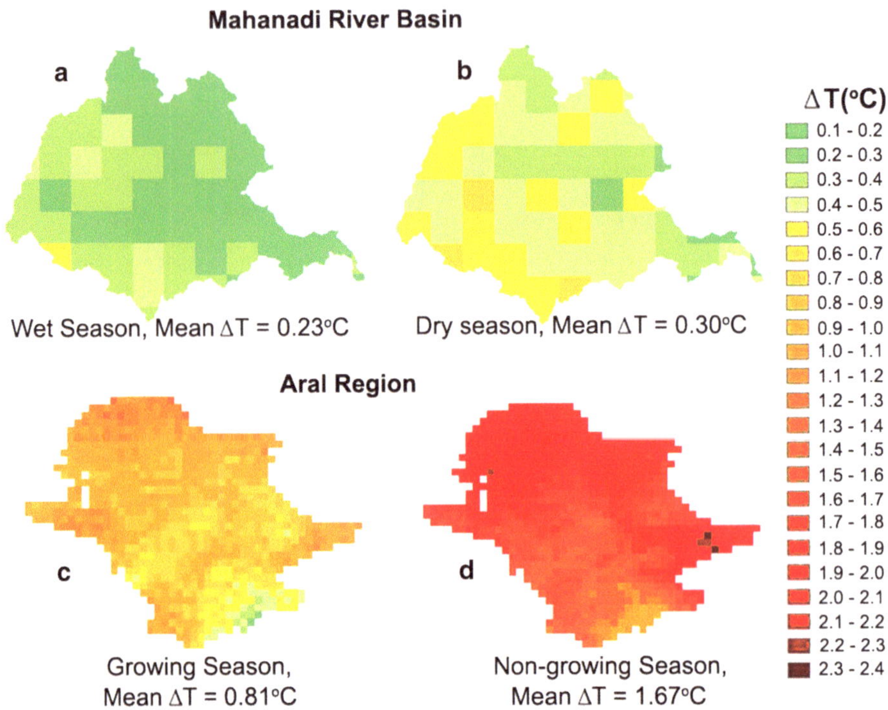

Fig. 3 Changes in average seasonal surface temperature (ΔT) in degrees Celsius for the Mahanadi River Basin (MRB) and the Aral region including the Aral Sea Drainage Basin and the Aral Sea. **a** The change from 1901–1955 to 1956–2000 for the Wet Season in MRB. **b** The change from 1901–1955 to 1956–2000 for the Dry Season in MRB. **c** The change from 1901–1950 to 1983–2002 for the growing season in the Aral region. **d** The change from 1901–1950 to 1983–2002 for non-growing season in the Aral region. All these seasonal temperature changes (ΔT) are significant at least at the confidence level $p = 0.999$ (see Appendix, Table 3). *Source* Mitchell and Jones (2005)

annual average T change (of 0.26 °C), we can quantify different ΔT components by expressing the seasonal T changes as follows:

$$\Delta T_{\mathrm{WS}} = \Delta T_{\mathrm{cl}} + \Delta T_{\mathrm{WS-irr}} \quad (1)$$

$$\Delta T_{\mathrm{DS}} = \Delta T_{\mathrm{cl}} + \Delta T_{\mathrm{DS-irr}} \quad (2)$$

where $\Delta T_{\mathrm{WS\text{-}irr}}$ and $\Delta T_{\mathrm{DS\text{-}irr}}$ are the T change components due to the irrigation in the Wet Season and the Dry Season, respectively; these are and should be different to reflect different cooling due to the difference in the amount of water used for irrigation between the two seasons. Furthermore, ΔT_{cl} is the T change component due to the regional manifestation of global climate change. As assumed by Destouni et al. (2010) for the Aral region, the seasonality of the latter, climate change component ΔT_{cl} can also in the MRB be assumed to be relatively small (i.e., the change of T assumed to be more or less similar even if T itself differs between different seasons), compared to the seasonality of T change due to irrigation, which differs between the Wet Season and the Dry Season. With a ΔT_{cl} component that is relatively similar between the Wet Season and Dry Season, subtraction of Eq. (2) from Eq. (1) yields:

$$\Delta T_{\mathrm{WS}} - \Delta T_{\mathrm{DS}} = \Delta T_{\mathrm{WS-irr}} - \Delta T_{\mathrm{DS-irr}} \tag{3}$$

Inserting the data-given values of $\Delta T_{\mathrm{WS}} = 0.23$ °C and $\Delta T_{\mathrm{DS}} = 0.30$ °C in Eq. (3) yields:

$$\Delta T_{\mathrm{WS-irr}} - \Delta T_{\mathrm{DS-irr}} = -0.07^{\circ}\mathrm{C} \tag{4}$$

The T change components due to the Wet Season and Dry Season irrigation are further related to corresponding changes in latent heat flux $\Delta F_{\mathrm{WS\text{-}irr}}$ and $\Delta F_{\mathrm{DS\text{-}irr}}$ as follows:

$$\Delta F_{\mathrm{WS-irr}} = \frac{M_{\mathrm{a}} C_{\mathrm{p}} \Delta T_{\mathrm{WS-irr}}}{\tau_{\mathrm{a}}} \tag{5}$$

$$\Delta F_{\mathrm{DS-irr}} = \frac{M_{\mathrm{a}} C_{\mathrm{p}} \Delta T_{\mathrm{DS-irr}}}{\tau_{\mathrm{a}}} \tag{6}$$

where M_{a} is the regional air mass that is cooled by the extra ET of the applied irrigation water, C_{p} is the specific heat capacity of the air and τ_{a} is the average regional residence time of that air. Assuming, as done in Destouni et al. (2010), that the seasonal F and T changes affect more or less the same regional air mass implies that

$$\Delta T_{\mathrm{DS-irr}} = \Delta T_{\mathrm{WS-irr}} * \frac{\Delta F_{\mathrm{DS-irr}}}{\Delta F_{\mathrm{WS-irr}}} \tag{7}$$

The latent heat flux change due to irrigation in either season can in turn be calculated as follows:

$$\Delta F_{\mathrm{irr}} = \Delta \mathrm{ET}_{\mathrm{irr}} \rho_{\mathrm{W}} L \tag{8}$$

where $\Delta \mathrm{ET}_{\mathrm{irr}}$ is the corresponding evapotranspiration change due to irrigation, ρ_{w} is the water density assumed in present calculations to be 1,000 kg/m^3 and L is the latent heat of vaporization of 2,260 kJ/kg. Inserting Eq. (8) in Eq. (7) yields the seasonal ΔT components as functions of the seasonal ΔET components:

$$\Delta T_{\mathrm{DS-irr}} = \Delta T_{\mathrm{WS-irr}} * \frac{\Delta \mathrm{ET}_{\mathrm{DS-irr}}}{\Delta \mathrm{ET}_{\mathrm{WS-irr}}} \tag{9}$$

The ratio of the seasonal ET changes required to evaluate equation (9) can be estimated from the quantities of irrigation water applied during the respective seasons, that is, 4 km^3 for Dry Season and 7 km^3 for Wet Season (Asokan 2005). To investigate different possibilities of how much of the applied irrigation water leads to actual ET increase in each season, we consider two different scenarios for the seasonal ET change ratio. (1) The base scenario, assuming zero water storage change over each season so that all applied irrigation water feeds into the ET of each season and $\frac{\Delta \mathrm{ET}_{\mathrm{DS-irr}}}{\Delta \mathrm{ET}_{\mathrm{WS-irr}}} = \frac{4}{7}$. (2) The test scenario, which accounts for the possibility that some part of the total applied irrigation water of 7 km^3 during the Wet Season increases groundwater storage and/or adds to the runoff from the basin rather than feeding into the ET and then assumes for comparative purposes the non-ET-contributing part to be 2 km^3; this yields the ET change ratio as $\frac{\Delta \mathrm{ET}_{\mathrm{DS-irr}}}{\Delta \mathrm{ET}_{\mathrm{WS-irr}}} = \frac{4}{5}$ in this scenario.

The assumption of 2 km^3 in the test scenario (2) represents a large (near maximum possible) contribution of the used irrigation water during the Wet Season to storage change in the basin and/or runoff from the basin, instead of to ET. This contribution is assumed to be zero in the base scenario (1; assuming that all irrigation water feeds into ET), and if it were assumed ≥3 km^3 in the comparative test scenario, the corresponding seasonal ET change ratio would be $\frac{\Delta \mathrm{ET}_{\mathrm{DS-irr}}}{\Delta \mathrm{ET}_{\mathrm{WS-irr}}} \geq 1$, implying equal or greater loss of irrigation water by ET during the Dry Season, when less irrigation water is used, than during the Wet Season when more irrigation water is used. This would in turn mean a greater cooling effect by an irrigation-driven increase in ET and associated latent heat flux during the Dry Season than during the Wet Season, which is inconsistent with available temperature observation data (Tables 1, 2; Fig. 3). For the Dry Season, it is physically reasonable to assume in both scenarios that all of the used irrigation water feeds into ET because runoff from the basin is negligible and water storage is decreasing rather than increasing in the basin during this season.

For the base scenario (1), Eq. (9) yields

$$\Delta T_{\mathrm{DS-irr}} = \frac{4}{7} \Delta T_{\mathrm{WS-irr}} \tag{10}$$

In addition, inserting Eq. (10) into Eq. (4) yields $\Delta T_{\text{WS-irr}} = -0.16$ °C and $\Delta T_{\text{DS-irr}} = -0.09$ °C for the base scenario. Similarly, for the test scenario, Eq. (9) yields

$$\Delta T_{\mathrm{DS-irr}} = \frac{4}{5} \Delta T_{\mathrm{WS-irr}} \tag{11}$$

which, inserted in Eq. (4), yields $\Delta T_{\text{WS-irr}} = -0.35$ °C and $\Delta T_{\text{DS-irr}} = -0.28$ °C for the test scenario.

For both scenarios, the annual average T change due to irrigation, $\Delta T_{\text{ann-irr}}$, is obtained as the temporal average of $\Delta T_{\text{DS-irr}}$ and $\Delta T_{\text{WS-irr}}$. The regional manifestation of global climate change, ΔT_{cl}, can be calculated from either one of the Eqs. (1) or (2), given the obtained $\Delta T_{\text{WS-irr}}$ and $\Delta T_{\text{DS-irr}}$ values, as well as from the relation $\Delta T_{\text{cl}} = \Delta T_{\text{a}} - \Delta T_{\text{ann-irr}}$, where ΔT_{a} is the data-given value of annual average T change. There are thus three different possibilities for calculating ΔT_{cl}, and all of these possibilities must and do provide the same resulting ΔT_{cl} value for each scenario. Furthermore, with seasonal (and thereby also corresponding average annual) ΔET components given directly from the assumptions of the different scenarios [see the seasonal temperature change ratios that define the different scenarios in Eqs. (10)–(11)], associated seasonal (and average annual) latent heat flux changes ΔF can be calculated from Eq. (8).

With regard to the ΔET and ΔF changes that are driven by the regional manifestation of global climate change, however, one cannot calculate them in the same way as the irrigation-driven flux changes, from the climate-driven temperature change ΔT_{cl}. The reason is that ΔT_{cl} quantifies the total climate-driven T change and not only the latent heat-related ΔT contribution, whereas the ΔT components driven by irrigation are entirely due to the latent heat flux changes implied by the regional irrigation. In order to also compare the irrigation-driven ΔET and ΔF changes with those driven purely by the observed climate change, we must use results from more complex, distributed hydrological modeling of a hypothetical scenario of only climate (i.e., actually observed T and P) change, the results from which are summarized in Table 1. The ET change result from this hypothetical climate scenario is then used to quantify the climate-driven ΔET, and the associated latent heat flux change ΔF can further be calculated from Eq. (8).

The total annual average changes of ET and F can finally be calculated as the sums of the different (irrigation and climate change) components of ΔET and ΔF, respectively. This ΔET result for the simple base and test scenarios can be compared with the corresponding ΔET result from the more complex, distributed hydrological modeling for the realistic climate–irrigation simulation results (accounting for both climate and irrigation changes) in Table 1. Furthermore, with the actual change in precipitation (ΔP) given directly from observed data (Table 1), corresponding changes in runoff (ΔR) can also be calculated from water balance consideration as $\Delta R = \Delta P - \Delta \mathrm{ET}$ for the base and test scenarios and compared with the corresponding ΔR result from the climate–irrigation simulation listed in Table 1. A comparison between the different ΔET and ΔR results for the MRB quantifies the uncertainty associated with making different physically possible ΔET assumptions: those in the base and test scenario of the presently applied, relatively simple quantification approach, and that associated with the more complex hydrological modeling of the MRB (results listed in Table 1, based on Asokan et al. 2010). Table 1 shows that also the latter, more complex model results are uncertain, as indicated by the difference between the model result for R in the period 1956–2000 (568 mm/year) and available R observation data for part of that time period (515 mm/year).

4 Cross-Regional Result Comparison and Discussion

For the MRB base scenario, the average annual irrigation effect on ΔT is obtained as $\Delta T_{\text{ann-irr}} = -0.13$ °C. From this result and the data-given total average annual T change, $\Delta T_{\text{a}} = 0.26$ °C [significant ($p = 0.999999$), Appendix, Table 3], the ΔT contribution from the regional manifestation of global climate change is obtained as $\Delta T_{\text{cl}} = 0.39$ °C. Similarly, for the test scenario, $\Delta T_{\text{ann-irr}} = -0.32$ °C and $\Delta T_{\text{cl}} = 0.58$ °C. Figure 4 summarizes and illustrates the MRB results and compares them with those previously obtained for the Aral region (with data-given total average annual T change here being $\Delta T_{\text{a}} = 1.1$ °C [significant ($p = 0.999999999$), Appendix, Table 3].

The MRB scenario comparison in Fig. 4 shows greater T changes, both irrigation driven and climate driven, in the test than in the base scenario. Both scenarios imply a considerable regional cooling effect from irrigation in comparison with the total and the purely climate-driven T changes in MRB and the irrigation-driven T change in the Aral region. Furthermore, the total and purely climate-driven T changes are larger in the Aral region, even though the net total cooling effect, when accounting for both the irrigation itself and

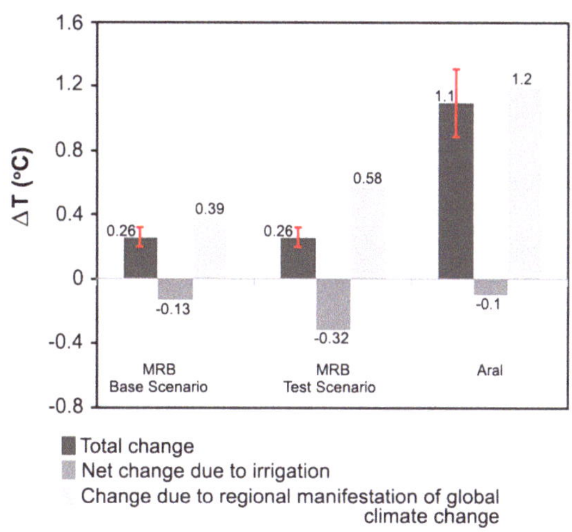

Fig. 4 Changes in surface temperature (ΔT) in degrees Celsius between same periods as in Fig. 3 for the Mahanadi River Basin (MRB) and the Aral region. *Error bars* show 80 % confidence intervals for the data-given total average annual T changes in both regions (Appendix, Table 3)

the associated Aral Sea desiccation, is much smaller in that region than in the MRB, both in absolute terms and in relation to the climate-driven ΔT contribution. This cross-regional comparison result is a direct effect of the endorheic nature of the Aral Sea drainage basin; the pure irrigation-driven ΔT contribution is here a cooling of −0.6 °C (Destouni et al. 2010), and thus larger than in the MRB if one neglects the major surface water (Aral Sea) shrinkage, which does not occur in an exorheic basin like the MRB.

Figure 5 further shows the cross-regional comparison of ΔET (significant at $p = 0.90$ confidence level in the Aral region and at $p = 0.68$ and $p = 0.63$ for the base and test scenario, respectively, in the MRB) along with the associated ΔF results. The irrigation-driven flux change contributions are much greater than the climate-driven ones in the MRB, and somewhat smaller in the test than in the base scenario. The climate-driven flux decreases in the MRB are due to the P decrease in that region (Table 1; significant at $p = 0.95$ confidence level, Appendix, Table 3), which dominates the climate-driven ΔET (and thereby also the corresponding ΔF) over the opposite ΔET effect of the regional T increase. The corresponding flux changes in the Aral region are overall much smaller, mainly because water fluxes, and thereby also the availability of annually renewable water, are generally much smaller in that region than in the MRB (see Table 1; see also in Appendix, Table 3 that the estimated R changes in the MRB are significant at $p = 0.95$ and

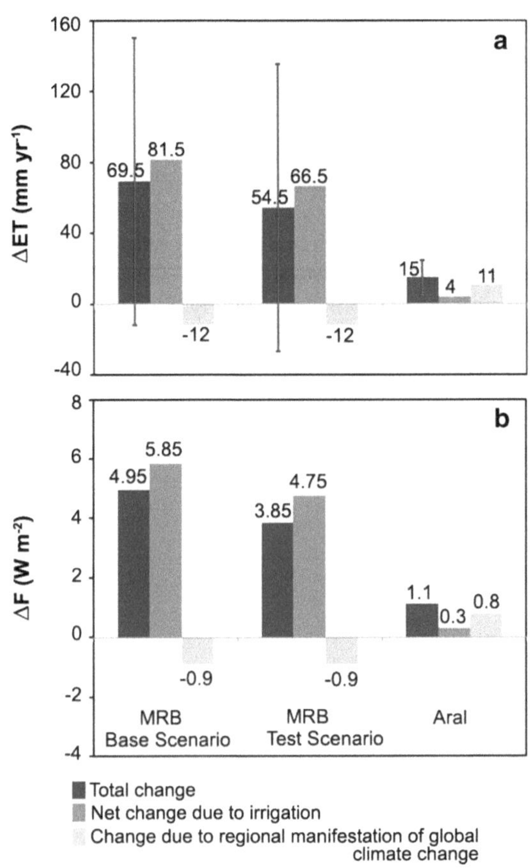

Fig. 5 Changes in **a** evapotranspiration (ΔET) in mm per year and **b** latent heat flux (ΔF) in watts per meter squared between same periods as in Fig. 3 for the Mahanadi River Basin (MRB) and the Aral region. *Error bars* show 80 % confidence intervals for ΔET in both regions (Appendix, Table 3)

$p = 0.90$ for the base and test scenario, respectively, and in the Aral region at $p \approx 1$, while the P change in the latter is significant at $p = 0.80$).

Figure 6 further emphasizes the major differences in water availability between the two regions, and the importance of this difference for the regional change magnitudes and their implications from different perspectives on hydro-climatic change. In absolute terms, all water flux changes are much smaller in the Aral region than in the MRB. Yet, the small R and associated other water flux changes in the Aral region have led to one of the world's worst environmental disasters in modern times: the Aral Sea desiccation and associated ecosystem collapse (Gaybullaev et al. 2012). Note then that the terminal Aral Sea is entirely dependent on the runoff R into it from its own endorheic drainage basin; furthermore, R constitutes an internal flux in that total land region, including both the Aral Sea and its drainage basin, with net total water flux balance P – ET still being negative and thus imbalanced at the end of the twentieth century (Table 1), so that the Aral Sea continued to shrink also after that time. In contrast, R from the MRB into the ocean, which is not particularly dependent on just that basin discharge into it, is an external flux from the MRB land system, with a net total water flux balance being P – ET – R, which was in this region essentially zero and thus balanced at the end as in the beginning of the twentieth century (Table 1).

In Fig. 6, the comparison of results for the MRB from the present, relatively simple scenario approach and the more complex hydrological modeling (accounting also for uncertainty in the latter implied by available regional R data listed in Table 1) indicates the test scenario as the least extreme of the two simple scenarios with regard to R and water availability changes in the landscape. However, with regard to T changes, the test scenario implies a greater cooling effect of irrigation that masks an also greater climate-driven T change (Fig. 4).

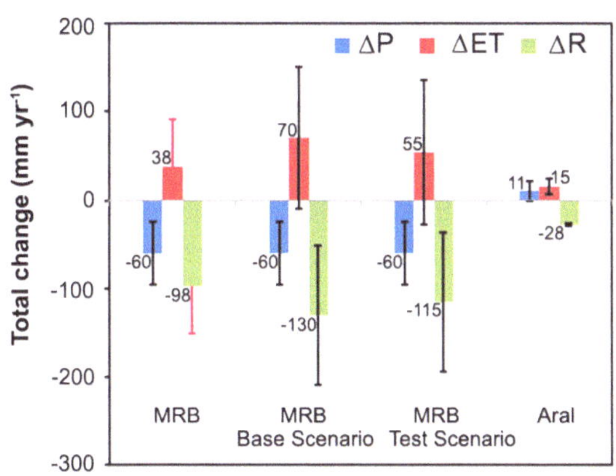

Fig. 6 Water flux changes (for precipitation, ΔP, evapotranspiration, ΔET, and runoff, ΔR) in mm per year between the same periods as in Fig. 3 for the Mahanadi River Basin (MRB) and the Aral region. The results for MRB that do not refer to any scenario, and those for the Aral region are from previous distributed hydrological modeling as listed in Table 1. The magenta error bars for ΔR and ΔET from the previous distributed hydrological modeling of MRB (*left*) show the possible error range associated with these results if considering the available runoff observation data for part of the 1956–2000 instead of the modeled runoff result for that period (Table 1); to maintain basin-scale water balance, the ΔR error range must also be mirrored as a corresponding ΔET error range. *Black error bars* further show 80 % confidence intervals (Appendix, Table 3) for the other ΔET and ΔR results, as well as for ΔP, for both regions

5 Conclusion

The present quantification and cross-regional comparison of hydro-climatic change has shown that significant differences in seasonal temperature changes, indicating the cooling effects of irrigation, are exhibited in different land regions. The relatively simple approach proposed by Destouni et al. (2010) may therefore be widely useful in distinguishing and quantifying such irrigation effects. An important insight gained from the present comparative study is that irrigation-driven changes to the climatically important vapor and latent heat fluxes and associated temperature change effects at the land surface may be greater in land regions with small relative changes to water resource availability in the landscape (like the MRB) than in regions with large such changes (like the Aral region).

The role and importance of various natural and anthropogenic change drivers and their impacts may thus differ greatly both among different regions and between different perspectives on continental water change. A focus on vapor and latent heat fluxes and temperature changes at the land–atmosphere interface represents one perspective on the continental part of Earth's hydrological cycle. Focus on the availability and flow of continental water in the landscape represents another such perspective. Regardless of perspective, however, the change constraints that are implied by basin-wise water balance should be accounted for in order to accurately understand and quantify continental hydro-climatic changes and their variability across different land regions.

Acknowledgments This work has been carried out within the framework of the strategic environmental research project EkoKlim at Stockholm University. The study was inspired by and written for the documentation of the ISSI conference *The Earth's Hydrological Cycle*, Bern, Switzerland, February 6–10, 2012.

Appendix: Significance Testing

The null hypothesis tested is that there has been no change in the long-term average values of investigated variables from the time period in the beginning to that in the end of the twentieth century, with the periods being those listed in Table 1 for each region. The null hypothesis is expressed as $\mu_b = \mu_e = \mu$, where μ_b and μ_e are the average values of each investigated variable in the beginning and the end of the twentieth century, respectively. The hypothesized same average value μ for the two periods is estimated from available data for the end-of-century period, that is, as $\mu = \mu_e$. The alternative hypothesis that there is significant change in the long-term average values between the two averaging time periods is expressed as $\mu_b \neq \mu_e$.

The standard normal test variable is as follows:

$$z = \frac{|\bar{x} - \mu|}{\sigma/\sqrt{n}}$$

which is normally distributed with mean 0 (when the null hypothesis is true) and variance σ^2/n, where σ is the standard deviation of each investigated variable, estimated consistently with μ from available data for the end-of-century period with n being the sample size (number of years with data in that period), $\bar{x}$ is the sample mean value in the beginning-of-century time period and $|\bar{x} - \mu|$ is the absolute value of the difference between $\bar{x}$ and μ. The value of the standard normal test variable z is computed and listed for different investigated variables in Table 3, along with the confidence level (p) at which the null hypothesis of no change is rejected, and hence, the change in variable average value indicated by $|\bar{x} - \mu|$ is significant.

Table 3 Parameters for testing the null hypothesis of no change, with notation as explained in the Appendix text, and with $|x^* - \mu|$ being the absolute value of the (positive and negative) limits of the 80 % confidence interval ($\frac{|x^*-\mu|}{\sigma/\sqrt{n}} = 1.282$) around μ

	$\lvert\bar{x} - \mu\rvert$	σ	n	z (confidence level p)	$\lvert x^* - \mu\rvert$ ($p = 0.80$)
Mahanadi River Basin					
Seasonal average temperature in dry season (°C)	0.3	0.47	45	4.4 ($p = 0.99999$)	0.09
Seasonal average temperature in wet season (°C)	0.23	0.48	45	3.2 ($p = 0.999$)	0.09
Annual average temperature (°C)	0.26	0.34	45	5.1 ($p = 0.999999$)	0.06
Annual precipitation (mm yr^{-1})	60	187	45	2.2 ($p = 0.95$)	36
Annual runoff (mm yr^{-1}) *(upper: base scenario; lower: test scenario)*	130	204	11	2.1 ($p = 0.95$)	79
	115			1.9 ($p = 0.90$)	
Annual evapotranspiration (mm yr^{-1}) *(upper: base scenario; lower: test scenario)*	70	209	11	1.1 ($p = 0.68$)	81
	55			0.9 ($p = 0.63$)	
Aral region					
Seasonal average temperature in growing season (°C)	0.81	0.85	20	4.3 ($p = 0.99999$)	0.24
Seasonal average temperature in non-growing season (°C)	1.67	1.06	20	7.0 ($p = 0.999999999$)	0.30
Annual average temperature (°C)	1.1	0.75	20	6.6 ($p = 0.999999999$)	0.21
Annual precipitation (mm yr^{-1})	11	38	20	1.3 ($p = 0.80$)	11
Annual runoff (mm yr^{-1})	28	4	20	31 ($p \approx 1$)	1.1
Annual evapotranspiration (mm yr^{-1})	15	36	20	1.9 ($p = 0.90$)	10

References

Asokan SM (2005) Water resources analysis under projected climate conditions in the Mahanadi River Basin, India, Master thesis No. WM-04-13, Asian Institute of Technology, Thailand

Asokan SM, Jarsjö J, Destouni G (2010) Vapor flux by evapotranspiration: effects of changes in climate, land use and water use. J Geophys Res 115:D24102. doi:10.1029/2010JD04417

Bengtsson L (2010) The global atmospheric water cycle. Environ Res Lett 5:025001. doi:10.1088/1748-9326/5/2/025001

Bonfils C, Lobell D (2007) Empirical evidence for a recent slow-down in irrigation-induced cooling. Proc Natl Acad Sci USA 104:13,582–13,587. doi:10.1073/pnas.0700144104

Boucher O, Myhre G, Myhre A (2004) Direct human influence of irrigation on atmospheric water vapor and climate. Clim Dyn 22:597–603. doi:10.1007/s00382-004-0402-4

Cheng L, Xu Z, Wang D, Cai X (2011) Assessing interannual variability of evapotranspiration at the catchment scale using satellite-based evapotranspiration data sets. Water Resour Res 47:W09509, p 11
Darracq A, Greffe F, Hannerz F, Destouni G, Cvetkovic V (2005) Nutrient transport scenarios in a changing Stockholm and Malaren valley region. Sweden Water Sci Technol 51(3–4):31–38
Destouni G, Darracq A (2009) Nutrient cycling and N2O emissions in a changing climate: the subsurface water system role. Environ Res Lett 4:035008, p 7
Destouni G, Hannerz F, Prieto C, Jarsjö J, Shibuo Y (2008) Small unmonitored near coastal catchment areas yielding large mass loading to the sea. Global Biogeochem Cycles 22:GB4003, p 10. doi: 10.1029/2008GB003287
Destouni G, Asokan SM, Jarsjö J (2010) Inland hydroclimatic interaction: effects of human water use on regional climate. Geophys Res Lett 37:L18402. doi:10.1029/2010GL044153
Destouni G, Jaramillo F, Prieto C (2013) Hydroclimatic shifts driven by human water use for food and energy production. Nat Clim Change. doi:10.1038/NCLIMATE1719 (in press)
Donohue RJ, Roderick ML, McVicar TR (2007) On the importance of including vegetation dynamics in Budyko's hydrological model. Hydrol Earth Syst Sci 11:983–995
Douglas EM, Niyogi D, Frolking S, Yeluripati JB, Pielke RA, Niyogi N, Vörösmarty CJ, Mohanty UC (2006) Changes in moisture and energy fluxes due to agricultural land use and irrigation in the Indian Monsoon Belt RID A-5015-2009. Geophys Res Lett 33, p 5
Foley JA, DeFries R, Asner GP, Barford C, Bonan G, Carpenter SR, Chapin FS, Coe MT, Daily GC, Gibbs HK, Helkowski JH, Holloway T, Howard EA, Kucharik CJ, Monfreda C, Patz JA, Prentice IC, Ramankutty N, Snyder PK (2005) Global consequences of land use. Science 309(570–574):2005
Gaybullaev B, Chen SC, Kuo YM (2012) Large-scale desiccation of the Aral Sea due to over-exploitation after 1960. J Mt Sci 9:538–546. doi:10.1007/s11629-012-2273-1
Gordon L, Dunlop M, Foran B (2003) Land cover change and water vapour flows: learning from Australia. Philos Trans Biol Sci 358:1973–1984
Gordon L, Steffen W, Jonsson B, Folke C, Falkenmark M, Johannessen A (2005) Human modification of global water vapor flows from the land surface RID A-4614-2010 RID C-7651-2011. Proc Natl Acad Sci USA 102:7612–7617
Groves DG, Yates D, Tebaldi C (2008) Developing and applying uncertain global climate change projections for regional water management planning. Water Resour Res 44:W12413. doi:10.1029/2008 WR006964
Jarsjö J, Destouni G (2004) Groundwater discharge into the Aral Sea after 1960. J Mar Syst 47:109–120
Jarsjö J, Asokan SM, Prieto C, Bring A, Destouni G (2012) Hydrological responses to climate change conditioned by historic alterations of land-use and water-use. Hydrol Earth Syst Sci Discuss 8:7595–7620. doi:10.5194/hessd-8-7595-2011
Keiser J, De Castro MC, Maltese MF, Bos R, Tanner M, Singer BH, Utzinger J (2005) Effect of irrigation and large dams on the burden of malaria on a global and regional scale. Am J Trop Med Hyg 72:392–406
King M, Kaufman Y, Menzel W, Tanre D (1992) Remote-sensing of cloud, aerosol, and water-vapor properties from the moderate resolution imaging spectrometer (MODIS) RID C-7153-2011 RID B-8306-2011. IEEE Trans Geosci Remote Sens 30:2–27
Kite GW, Droogers P (2000) Comparing evapotranspiration estimates from satellites, hydrological models and field data. J Hydrol 229:3–18
Kueppers LM, Snyder MA, Sloan LC (2007) Irrigation cooling effect: regional climate forcing by land-use change. Geophys Res Lett 34:L03703. doi:10.1029/2006GL028679
Kvalevag MM, Myhre G, Bonan G, Levis S (2010) Anthropogenic land cover changes in a GCM with surface albedo changes based on MODIS data RID A-3598-2008. Int J Climatol 30:2105–2117
Lee E, Sacks WJ, Chase TN, Foley JA (2011) Simulated impacts of irrigation on the atmospheric circulation over Asia. J Geophys Res-Atmos 116:D08114, 30. doi:10.1029/2010JD014740
Loarie SR, Lobell DB, Asner GP, Mu Q, Field CB (2011) Direct impacts on local climate of sugar-cane expansion in Brazil RID G-5695-2010. Nat Clim Change 1:105–109
Lobell D, Field CB (2007) Global scale climate-crop yield relationships and the impacts of recent warming. Environ Res Lett 2:014002. doi:10.1088/1748-9326/2/1/014002
Lobell D, Bala G, Mirin A, Phillips T, Maxwell R, Rotman D (2009) Regional differences in the influence of irrigation on climate. J Clim 22(8):2248–2255. doi:10.1175/2008JCL12703.1
Milly PCD, Wetherald RT, Dunne KA, Delworth TL (2002) Increasing risk of great floods in a changing climate. Nature 415:514–517
Milly PCD, Dunne KA, Vecchia AV (2005) Global pattern of trends in streamflow and water availability in a changing climate. Nature 438:347–350

Mitchell TD, Jones PD (2005) An improved method of constructing a database of monthly climate observations and associated high-resolution grids. Int J Climatol 25:693–712. doi:10.1002/joc.1181

Piao S, Friedlingstein P, Ciais P, De Noblet-Ducoudre N, Labat D, Zaehle S (2007) Changes in climate and land use have a larger direct impact than rising CO2 on global river runoff trends. Proc Natl Acad Sci USA 104:15242–15247

Qiu GY, Yin J, Tian F, Geng S (2011) Effects of the "Conversion of Cropland to Forest and Grassland Program" on the Water Budget of the Jinghe River Catchment in China. J Environ Qual 40:1745–1755

Schilling KE, Jha MK, Zhang YK, Gassman PW, Wolter CF (2008) Impact of land use and land cover change on the water balance of a large agricultural watershed: historical effects and future directions. Water Resour Res 44, p 12

Shibuo Y, Jarsjö J, Destouni G (2007) Hydrological responses to climate change and irrigation in the Aral Sea drainage basin. Geophys Res Lett 34:L21406. doi:10.1029/2007GL031465

Törnqvist R, Jarsjö J (2012) Water savings through improved irrigation techniques: basin-scale quantification in semi-arid environments. Water Resour Manage 26:949–962. doi:10.1007/s11269-011-9819-9

Törnqvist R, Jarsjö J, Karimov B (2011) Health risks from large-scale water pollution: trends in Central Asia. Environ Int 37:435–442

Vanlill W, Kruger F, Vanwyk D (1980) The effect of afforestation with Eucalyptus-Grandis Hill Ex Maiden and Pinus-Patula Schlecht Et Cham on streamflow from experimental catchments. J Hydrol 48:107–118

Visser A, Kroes J, Van Vliet MTH, Blenkinsop S, Fowler HJ, Broers HP (2012) Climate change impacts on the leaching of a heavy metal contamination in a small lowland catchment. J Contam Hydrol 127:47–64. doi:10.1016/j.jconhyd.2011.04.007

Vörösmarty CJ, Fekete BM, Meybeck M, Lammers RB (2000) Global system of rivers: its role in organizing continental land mass and defining land-to-ocean linkages. Global Biogeochem Cycles 14:599–621. doi:10.1029/1999GB900092

Weiskel PK, Vogel RM, Steeves PA, Zarriello PJ, DeSimone LA, Ries KG III (2007) Water use regimes: characterizing direct human interaction with hydrologic systems. Water Resour Res 43:W04402. doi:10.1029/2006WR005062

Wisser D, Fekete BM, Vörösmarty CJ, Schumann AH (2010) Reconstructing 20th century 20 global hydrography: a contribution to the Global Terrestrial Network-Hydrology (GTN-H). Hydrol Earth Syst Sci 14:1–24. doi:10.5194/hess-14-1-2010

Zhang K, Kimball JS, Nemani RR, Running SW (2010) A continuous satellite-derived global record of land surface evapotranspiration from 1983 to 2006 RID B-3227-2012. Water Resour Res 46, p 21

MIX
Papier aus verantwortungsvollen Quellen
Paper from responsible sources
FSC® C105338

If you have any concerns about our products,
you can contact us on
ProductSafety@springernature.com

In case Publisher is established outside the EU,
the EU authorized representative is:
Springer Nature Customer Service Center GmbH
Europaplatz 3, 69115 Heidelberg, Germany

Printed by Libri Plureos GmbH
in Hamburg, Germany